Seeds

The Ecology of Regeneration in Plant Communities

3rd Edition

Seeds

The Ecology of Regeneration in Plant Communities
3rd Edition

_____

Edited by

Robert S. Gallagher

Independent Agricultural Consultant

www.cabi.org

CABI is a trading name of CAB International

CABI
Nosworthy Way
Wallingford
Oxfordshire OX10 8DE
UK

CABI
38 Chauncey Street
Suite 1002
Boston, MA 02111
USA

Tel: +44 (0)1491 832111
Fax: +44 (0)1491 833508
E-mail: info@cabi.org
Website: www.cabi.org

Tel: +1 800 552 3083 (toll free)
Tel: +1 (0)617 395 4051
E-mail: cabi-nao@cabi.org

A catalogue record for this book is available from the British Library, London, UK.

Library of Congress Cataloging-in-Publication Data

Seeds: the ecology of regeneration in plant communities / edited by Robert S. Gallagher. -- 3rd ed.
    p. cm.
  Ecology of regeneration in plant communities
  Includes bibliographical references and index.
  ISBN 978-1-78064-183-6 (hbk)
1. Seeds--Ecology. 2. Regeneration (Botany) I. Gallagher, Robert S. II. Title: Ecology of regeneration in plant communities.

  QK661.S428 2013
  581.4'67--dc23

                      2013024762

ISBN-13: 978 1 78064 183 6

Commissioning editor: Vicki Bonham
Editorial assistant: Emma McCann
Production editor: Lauren Povey

Typeset by SPi, Pondicherry, India.
Printed and bound in the UK by CPI Group (UK) Ltd, Croydon, CR0 4YY.

Contents

Contributors

J. Derek Bewley, Department of Molecular and Cellular Biology, University of Guelph, Guelph, Ontario, Canada. E-mail: dbewley@ uoguelph.ca

Mark B. Burnham, Department of Biology, West Virginia University, Morgantown, WV 26506, USA. E-mail: mburnha1@mix.wvu.edu

Nathalie Colbach, INRA, UMR1347 Agroécologie, EcolDur, BP 86510, 21000 Dijon, France. E-mail: nathalie.colbach@dijon.inra.fr

Michael J. Crawley, Department of Biology, Imperial College London, Silwood Park, Ascot, Berkshire SL5 7PY, UK. E-mail: m.crawley@imperial.ac.uk

E. Patrick Fuerst, Department of Crop and Soil Science, Washington State University, Pullman, WA 99164, USA. E-mail: pfuerst@wsu.edu

Robert S. Gallagher, Independent Agricultural Consultant, 202 Maple Street, Clinton, SC 29324, USA. E-mail: rsgallagher61@gmail.com

Kristen L. Granger, Department of Crop and Soil Sciences, The Pennsylvania State University, University Park, PA 16801, USA. E-mail: kgranger1@gmail.com

Ruben Heleno, Centre for Functional Ecology (CEF-UC), Department of Life Sciences, University of Coimbra, PO Box 3046, 3001-401 Coimbra, Portugal. E-mail: rheleno@uc.pt

Henk W.M. Hilhorst, Wageningen Seed Lab, Laboratory of Plant Physiology, Wageningen University, Wageningen, the Netherlands. E-mail: henk.hilhorst@wur.nl

Pedro Jordano, Integrative Ecology Group, Estación Biológica de Doñana, CSIC-EBD Avda. Americo Vespucio S/N, Isla de La Cartuja, E-41092 Sevilla, Spain. E-mail: jordano@ebd.csic.es

Alistair J. Murdoch, School of Agriculture, Policy and Development, University of Reading, Earley Gate, PO Box 237, Reading, Berkshire RG6 6AR, UK. E-mail: a.j.murdoch@reading.ac.uk

Manuel Nogales, Island Ecology and Evolution Research Group (CSIC-IPNA), 38206 La Laguna, Tenerife, Canary Islands, Spain. E-mail: mnogales@ipna.csic.es

Thijs L. Pons, Department of Plant Ecophysiology, Institute of Environmental Biology, Utrecht University, Padualaan 8, 3508 CH Utrecht, the Netherlands. E-mail: t.l.pons@uu.nl

Peter Poschlod, LS Biologie VIII, Universität Regensburg, Universitätsstraße 31, 93040 Regensburg, Germany. E-mail: peter.poschlod@biologie.uni-regensburg.de

Arne Saatkamp, Institut Méditerranéen de Biodiversité et d'Ecologie (IMBE UMR CNRS 7263), Université d'Aix-Marseille, Faculté St Jérôme case 421, 13397 Marseille cedex 20, France. E-mail: arne.saatkamp@imbe.fr

Elwira Sliwinska, Department of Plant Genetics, Physiology and Biotechnology, University of Technology and Life Sciences, Bydgoszcz, Poland. E-mail: elwira@utp.edu.pl

Anna Traveset, Mediterranean Institute of Advanced Studies (CSIC-UIB), Terrestrial Ecology Group, 07190 Esporles, Mallorca, Balearic Islands, Spain. E-mail: atraveset@imedea.uib-csic.es

D. Lawrence Venable, Department of Ecology and Evolutionary Biology, University of Arizona, Tucson, AZ 85721, USA. E-mail: venable@u.arizona.edu

Rui Zhang, Harvard Forest, Harvard University, 324 N Main St., Petersham, MA 01366, USA. E-mail: ruizhang0410@gmail.com

Preface

CABI and I are pleased to offer the third edition of *Seeds: The Ecology of Regeneration in Plant Communities,* building on the first two outstanding editions by Michael Fenner in 1992 and 2000. It was an honour to serve as the editor for this edition, as the first two editions played an important role in shaping my thinking and study of seed banks and seed ecology/physiology during my graduate studies and the early stages of my academic career. As in the previous editions, the goal of this edition is to provide a detailed overview of seed ecophysiology and its role in shaping plant communities.

In this edition, I encouraged many of the contributors of previous editions to also contribute here. As such, I am pleased to see updated return contributions on frugivory (Chapter 2), seed dispersal (Chapter 3), seed predation (Chapter 4), light-mediated germination responses (Chapter 5), chemical regulation of germination (Chapter 6) and seed dormancy (Chapter 7). New chapter contributions with a mechanistic emphasis include an overview of seed development, anatomy and morphology (Chapter 1), and the chemical ecology of seed persistence (Chapter 8). New chapter contributions that have more of a synthesis emphasis cover the implications of climate change on the regeneration by seeds (Chapter 9) and the functional role of seed banks in agricultural (Chapter 10) and natural (Chapter 11) ecosystems. In all, 19 authors have contributed to this edition, with a mix of well-established and emerging young scientists.

As editor, I tried to give the contributing authors considerable creative and intellectual licence. As such, the chapters are not stylistically identical. I have also tried to provide a good level of continuity between the chapters, adding notes to link overlapping concepts and subject matter among the chapters. In the end, however, each chapter represents the thinking of the contributing authors and not this editor.

I was very pleased with how thoroughly the contributing authors integrated the more classic foundation literature with that more recently published. As such, I hope this book will serve as a highly comprehensive review resource for new students of seed ecology and plant community regeneration, as well as an update for the more seasoned and experienced scientist.

Robert S. Gallagher

1 Overview of Seed Development, Anatomy and Morphology

Elwira Sliwinska[1]* and J. Derek Bewley[2]

[1]*Department of Plant Genetics, Physiology and Biotechnology, University of Technology and Life Sciences, Bydgoszcz, Poland;* [2]*Department of Molecular and Cellular Biology, University of Guelph, Guelph, Ontario, Canada*

Introduction

The spermatophytes, comprised of the gymnosperms and angiosperms, are plants that produce seeds that contain the next generation as the embryo. Seeds can be produced sexually or asexually; the former mode guarantees genetic diversity of a population, whereas the latter (apomictic or vegetative reproduction) results in clones of genetic uniformity. Sexually produced seeds are the result of fertilization, and the embryo develops containing, or is surrounded by, a food store and a protective cover (Black et al., 2006). Asexual reproduction is probably important for the establishment of colonizing plants in new regions.

Seeds of different species have evolved to vary enormously in their structural and anatomical complexity and size (the weight of a seed varies from 0.003 mg for orchids to over 20 kg for the double coconut palm (*Lodoicea maldvica*)). Nevertheless, their development can be divided conveniently into three stages: Phase I – formation of the different tissues within the embryo and surrounding structures (histodifferentiation, characterized by extensive cell divisions); Phase II – cell enlargement and expansion predominates (little cell division, dry weight increase due to reserve deposition, water content decline); and Phase III – dry matter accumulation slows and ceases at physiological maturity (Black et al., 2006). In many species, physiological maturity is followed by maturation drying as the seed loses water to about 7–15% moisture content and becomes desiccation tolerant (orthodox seeds) and thus able to withstand adverse environmental conditions. At this stage a seed is quiescent (expressing little metabolic activity); in some cases it may also be dormant. In contrast, some species (about 7% of the world's flora, e.g. water lily (*Nymphaea* spp.), chestnut (*Aesculus* spp.), sycamore (*Acer* spp.), coconut (*Cocos nucifera*), Brazilian pine (*Araucaria angustifolia*)) produce recalcitrant seeds that do not undergo maturation drying and are desiccation sensitive at shedding. These seeds rapidly lose viability when water is removed by drying and therefore are difficult to store.

Pollination and Fertilization

In seed plants, the female gametophyte (embryo sac, megagametophyte) is located

---

* E-mail: elwira@utp.edu.pl

within an ovule, inside the pistil. It is surrounded by one or two integuments and the nucellus (megasporangium; Black *et al.*, 2006). In the *Polygonum* type, the developmental pattern exhibited by most angiosperm species, the embryo sac is formed when a single diploid (2*n*) cell (megasporocyte) in the nucellus undergoes meiosis (reduction division) which results in the production of four haploid (*n*) megaspores. Three of them undergo programmed cell death (PCD) and the remaining one undergoes mitosis three times, without cytokinesis, to produce a megaspore with eight haploid nuclei (Yadegari and Drews, 2004). Subsequently, cell walls form around these, resulting in a cellularized female gametophyte. During cellularization, two nuclei (one from each pole) migrate toward the centre of the embryo sac and fuse, resulting in a diploid central cell. Three cells that migrate to the micropylar end of the embryo sac become the egg apparatus with an egg cell (the female gamete) in the centre flanked by two synergids, and the remaining three are located to the opposite (chalazal) end forming antipodal cells (Fig. 1.1).

Male gametes (sperm cells) are produced as pollen grains inside the pollen sacs in the anthers (microsporangia). A mature pollen grain contains a large haploid vegetative (tube) cell and a smaller haploid generative cell that divides mitotically and produces two sperm cells.

Pollination involves the transfer of pollen grains to the stigma, the receptive surface of the pistil, which is followed by growth of a pollen tube through the style to the egg cell. Pollination can take place within the same plant (self-pollination, autogamy) or the pollen can be delivered from a different plant (cross-pollination, allogamy). Autogamy leads to the production of true-to-type offspring; this is disadvantageous when selfed offspring harbour recessive traits, but it may be advantageous for reproduction under unfavourable environmental conditions (Darwin, 1876). On the other hand, allogamy can introduce traits that increase resistance to diseases and predation, as well as seed and fruit yield. To prevent self-pollination and to promote the generation of genetic diversity, plants often exhibit self-incompatibility (SI), which causes the rejection of pollen from the same flower or plant. In many angiosperms, SI is controlled by a single multi-allelic locus S; incompatibility occurs when there is a match between the S alleles present in the pollen and style (Golz *et al.*, 1995). SI is determined by a protein secreted over the stigma surface, or in the style, which is a barrier that prevents pollen germination or pollen tube penetration through the style.

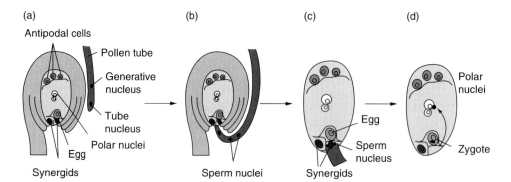

Fig. 1.1. Double fertilization in an angiosperm ovule. (a) Arrangement of cells in the mature embryo sac and in the pollen tube prior to fertilization. (b) Pollen tube enters the micropyle, with two sperm cells to be released to effect double fertilization. (c) Sperm cells are released from the pollen tube into a degenerated synergid cell flanking the egg cell and (d) they migrate to the egg cell and central (polar) cell resulting in: sperm cell (*n*) + egg cell (*n*) → zygote (2*n*) → embryo (2*n*); and sperm cell (*n*) + central cell (2*n*) → endosperm (3*n*). (From Purves *et al.*,1998.)

While pollination can be effected by wind, water or insects (sometimes by animals such as bats or birds), the majority (over 70% of angiosperms) depend upon insects for cross-pollination (Faheem *et al.*, 2004). The effectiveness of pollinators depends upon flower characteristics such as colour, scent, shape, size and nectar and pollen production. Wind pollination probably evolved from insect pollination in response to pollinator limitation and changes in the abiotic environment, especially in families with small, simple flowers and dry pollen (Culley *et al.*, 2002).

When pollen comes in contact with the stigma, it adheres, hydrates and germinates by developing a pollen tube. The two immotile sperm cells are carried by tube tip growth to the embryo sac (Lord and Russell, 2002). Fertilization involves two male gametes and therefore is called double fertilization. This occurs by discharge of the sperm nuclei from the pollen tube and their delivery into the embryo sac, whereafter one fuses with the egg cell nucleus and the second with the two polar nuclei of the central cell (Fig. 1.1; Hamamura *et al.*, 2012). As a result, a diploid embryo cell (zygote) and a triploid endosperm cell are created.

The embryo (Greek: *embryon* – unborn fetus, germ) is the future plant, and the endosperm, which may or may not be persistent, supplies nutrition to it. A single embryo usually develops per seed, although polyembryony occurs in some species (see next section).

Apomixis

Apomixis is a type of asexual reproduction, in which there is embryogenesis without the prior involvement of gametes; it occurs in about 35 families of angiosperms, including the Asteraceae (e.g. dandelion (*Taraxacum officinale*)), Rosaceae (e.g. blackberry (*Rubus fruticosus*)), Poaceae (e.g. Kentucky bluegrass (*Poa pratensis*)), Orchidaceae (e.g. *Nigritella* spp.) and Liliaceae (e.g. *Lilium* spp.). Its role in evolution is incompletely understood (Carneiro *et al.*, 2006). This type of reproduction leads to the formation of offspring

that are genetically identical to the maternal plant, although they tend to maintain high heterozygosity. There is no meiosis prior to the embryo sac formation (apomeiosis), and double fertilization does not occur, even though there is often normal pollen production. There are two types of apomictic development, sporophytic and gametophytic.

Sporophytic apomixis (adventitious embryony) is initiated late in ovule development and usually occurs in mature ovules (Bhat *et al.*, 2005). The embryo is formed directly from individual diploid cells of the nucellus or inner integument (cells external to the megagametophyte) and is not surrounded by the embryo sac. The formation of the endosperm occurs without fertilization of the central cell (autonomously). Most genera with adventitious embryony are diploids ($2n = 2x$). This type of apomixis is mainly found in tropical and subtropical woody species with multiseeded fruits (Carneiro *et al.*, 2006).

In gametophytic apomixis the megagametophyte develops from an unreduced (usually diploid) megaspore (diplospory) or from a cell in the nucellus (apospory; Bhat *et al.*, 2005; Carneiro *et al.*, 2006). In contrast to adventitious embryony, the cell that later develops into the embryo, after three mitotic divisions, forms a megasporophyte-like structure, similarly to during sexual reproduction. However, there is no fusion of female and male gametes, and the diploid egg cell develops by parthenogenesis (autonomously) to form an embryo. Endosperm development can be autonomous or it requires fertilization of the central cell (pseudogamous). Most plants with gametophytic apomixis are (allo) polyploids ($2n > 2x$), while the members of the same or closely related species that reproduce sexually usually are diploids. Diplospory is dominant in the Asteraceae and apospory in the Poaceae and Roseaceae.

Most apomicts are facultative because they retain their capacity for sexual reproduction. Depending on the mode of apomixis, the events of sexual reproduction might occur in the same ovule or in different ovules of the same apomictic plant (Koltunow and Grossniklaus, 2003). In many species apomixis is associated with polyspory or polyembryony (Carneiro *et al.*, 2006). In polysporic (bisporic or tetrasporic) species, ovule development is disrupted, but a reduced egg cell is formed and its fertilization occurs. In polyembryony multiple embryos are formed in one ovule, either from the synergids as a result of there being multiple embryo sacs in an ovule, or by cleavage of the zygote cells. Thus seeds with more than one embryo are produced; this polyembryony can be indicative of apospory or adventitious embryony. It is characteristic of *Poa, Citrus* and *Opuntia* spp., and of conifers such as fir (*Abies* spp.), pine (*Pinus* spp.) and spruce (*Picea* spp.).

One of the advantages of apomixis is that there is assured reproduction in the absence of pollinators, and it is a phenomenon often restricted to narrow ecological niches or challenging environments. Disadvantages include the inability to control the accumulation of deleterious genetic mutations, and lack of ability to adapt to changing environments.

Embryogenesis

During embryogenesis the central portion of the megagametophyte breaks down to form a cavity into which the embryo expands (Bewley and Black, 1994). The time it takes for the embryo to develop varies among different species and can be from several days to many months, even years. It also is influenced by the prevailing environmental conditions. Early embryo development includes the acquisition of an apico-basal polarity, differentiation of the epidermis and the formation of a shoot and root meristem (Dumas and Rogowsky, 2008). Polarity is established after the first division of the zygote. Polarity is usually perpendicular to the embryo axis and is asymmetric; a basal and apical cell are created.

In *Arabidopsis* (*Arabidopsis thaliana*), a model dicot, the two-celled embryo undergoes a number of mitotic divisions leading to the formation of a suspensor, root precursor and a proembryo (Black *et al.*, 2006). The suspensor facilitates the transport of nutrients from the endosperm to the embryo and undergoes PCD as the embryo matures. Its uppermost cell produces the root meristem. During development the embryo progresses through the globular, heart and torpedo stages to reach maturity (Fig. 1.2). During the transition from the globular to the heart stage the cotyledon structure and number are established. The mature embryo

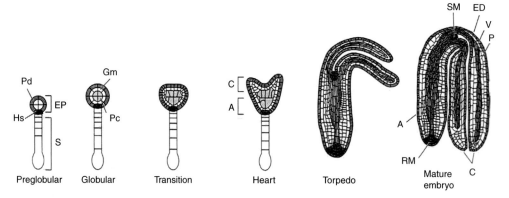

Fig. 1.2. Stages of development of a dicot embryo from the fertilized egg cell (zygote), as typified by *Arabidopsis thaliana*. Initially this cell divides to form an apical and a basal cell (not shown); the former divides to become the embryo (EP) and the latter the suspensor (S), as is evident at the preglobular stage. The protoderm (Pd) develops into the epidermal layer (Ed) of the mature embryo, the ground meristem (Gm) into the storage parenchyma (P), the procambium (Pc) into the vascular tissue (V) and the hypophysis (Hs) into the root and shoot meristems (RM, SM). A, axis; C, cotyledons. (From Bewley *et al.*, 2000.) Reproduced with permission of the American Society of Plant Biologists.

consists of two cotyledons borne on a hypocotyl, the collet (hypocotyl–radicle transition zone) and the radicle; seed desiccation then occurs.

In the Poaceae (monocot), after the first division of the zygote, the basal cell does not divide but forms the terminal cell of the suspensor (Black *et al.*, 2006). The other few suspensor cells and the embryo are produced from the axial cell. After further divisions a globular-shaped embryo is formed (Fig. 1.3). It elongates to a club-like shape with radial symmetry and then becomes bilaterally symmetrical through differentiation of the absorptive scutellum and coleoptile, which covers the first foliage leaf. During the coleoptile stage of development the embryo axis and suspensor are established. Later, during the leaf stage, the shoot and root meristems are defined and leaf primordia differentiate. The embryonic cells cease to divide during maturation, the embryo desiccates and becomes quiescent. Mature graminaceous embryos also contain a specialized thin tissue, the coleorhiza, which covers the radicle.

Species which undergo maturation drying at the termination of seed development and remain viable in the dry state are termed orthodox, but in some their seeds are unable to survive low water content and are known as recalcitrant (see Introduction). An extreme of this is exhibited by certain mangroves, e.g. *Avicennia* spp., in which there

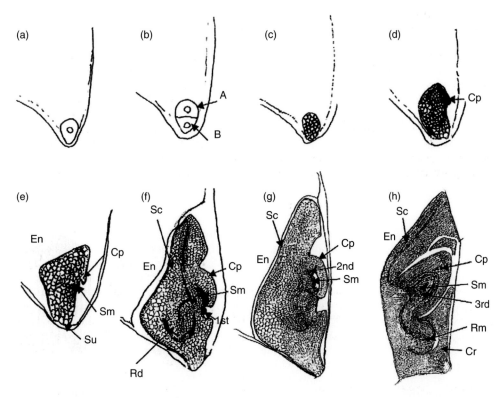

Fig. 1.3. Stages of development of a monocot embryo, such as in rice. (a) The fertilized egg cell (zygote) undergoes (b) mitosis to form an axial (A) and a basal (B) cell. (c) A multicellular proembryo is produced from the former, and then (d) an early coleoptile (Cp) stage. Further cell divisions result in (e) a late coleoptile stage when the shoot meristem (Sm) is visible. (f) By the first leaf (1st) stage other tissues are differentiated including the scutellum and the radicle, and by (g) the second leaf (2nd) stage differentiation is more advanced. In the mature embryo (h) all of the tissues are evident, with the addition of a third leaf (3rd), root meristem (Rm) and the structure covering this, the coleorhiza (Cr). En, endosperm; Su, suspensor. (From Black *et al.*, 2006.)

is no maturation drying and growth of the embryo continues from development into germination and early seedling growth while still on the mother plant (Black *et al.*, 2006). This is an example of vivipary and is a mechanism that helps in the dispersal of the seedling in an aquatic environment. For instance, in the grey mangrove (*A. marina*), seedlings may grow up to 30 cm long before they drop off the tree into the water or underlying mud. Vivipary is different from pre-harvest sprouting, which is the germination of a mature dry seed on the parent plant due to wetting by rain or when the environment is very humid.

Endosperm Development

Endosperm development is essential for proper development of the embryo and a viable seed. It usually precedes embryo development; a gap between the first division of the nucleus of the endosperm and the embryo takes usually several hours but it can be prolonged even to a few months (e.g. in *Colchicum*). Sometimes divisions in both the embryo and endosperm occur simultaneously (e.g. some aquatic species) or a zygotic division precedes the division of the endosperm nucleus. Endosperm development consists of four stages: syncytial, cellularization, differentiation, and PCD, although the last two stages have been jointly termed maturation (Li and Berger, 2012). After fertilization of the central cell the endosperm undergoes a cellular, nuclear or mixed (helobial, found for instance in the Alismatales) type of development (Dumas and Rogowsky, 2008). In cellular development, cell wall formation follows nuclear divisions; it is typical of the Acoraceae. In nuclear development, which is the most common mode in angiosperms and typical of the Brassicaceae (*Arabidopsis*) and Poaceae (grasses), the endosperm nucleus divides repeatedly without cytokinesis or cell wall formation (Fig. 1.4). As a result, a giant single cell with a central vacuole surrounded by nuclei is formed (coenocytic or syncytial endosperm). For instance, in *Arabidopsis*, after eight rounds of division

an endosperm with over 200 nuclei is created. Cellularization starts in the micropylar endosperm at the final mitosis, progressing as a wave through the peripheral and chalazal endosperm (the region close to the chalazal endosperm may not become completely cellular, e.g. in *Arabidopsis*). At this time the embryo is usually at the heart stage. After cellularization is completed the *Arabidopsis* endosperm directly enters endoreduplication (a process during which nuclei undergo repeated rounds of DNA synthesis without mitosis, resulting in endopolyploid cells), while in grasses numerous additional mitoses occur prior to endoreduplication. In heliobal development the embryo sac is divided into two unequal chambers after the first mitosis, with the smaller one (chalazal) developing into non-cellular endosperm, and the larger (micropylar) first becomes coenocytic and then multicellular. A deviation from these patterns occurs in coconut, which produces a liquid endosperm, a clear fluid with suspended free nuclei (Black *et al.*, 2006). Later the nuclei become associated with cytoplasm as free spherical cells and migrate to the periphery and cellular endosperm is formed ('coconut meat'); however, the central cavity remains filled with liquid ('coconut water').

The fully developed endosperm of the grasses consists of four major cell types: (i) those containing starch (starchy endosperm); (ii) aleurone layer cells; (iii) transfer cells; and (iv) those of the embryo-surrounding region (ESR; Black *et al.*, 2006). During seed filling the starchy endosperm expands rapidly and is filled with starch granules, interspersed with protein storage vacuoles. As the reserves are synthesized water is displaced from this region of the endosperm and the cells undergo PCD. In contrast, the aleurone layer cells (arranged in one to several layers located immediately below the pericarp) are not starch storing and retain their viability after maturation drying. These cells may contain anthocyanins that are responsible for seed colour. The transfer cells, which develop in the basal endosperm over the main vascular tissue of the mother plant, are responsible during seed filling

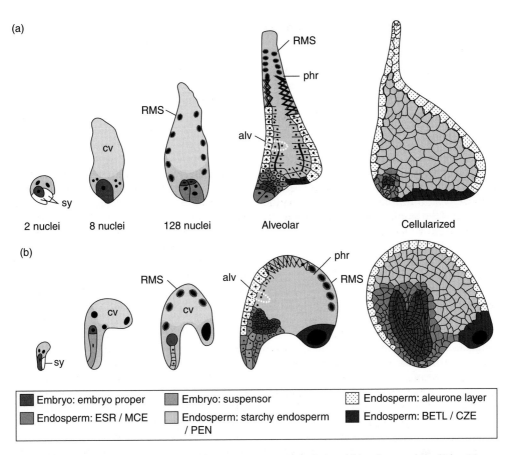

(a)

RMS

phr

RMS

alv

CV

RMS

sy

| 2 nuclei | 8 nuclei | 128 nuclei | Alveolar | Cellularized |

(b)

phr

RMS

alv

RMS

CV

CV

sy

| Embryo: embryo proper | Embryo: suspensor | Endosperm: aleurone layer |
| Endosperm: ESR / MCE | Endosperm: starchy endosperm / PEN | Endosperm: BETL / CZE |

Fig. 1.4. Development of the endosperm of (a) a monocot grain (maize) and (b) a dicot seed (*Arabidopsis*) concomitant with the development of the embryo. The endosperm cell prior to fertilization contains two nuclei, which after fusion with a sperm nucleus undergoes mitoses to form many nuclei (e.g. 128 nuclei stage, although later there may be over 200 nuclei) before cellularization begins at the alveolar stage. The central vacuole (cv) of the single-celled coenocytic (free-nuclear) endosperm initiates cell wall formation that spreads from the periphery, after the completion of mitoses, to form what becomes the cellular endosperm with a peripheral embryo (maize) or embedded embryo (*Arabidopsis*). At maturity the endosperm of maize is non-living except for a single layer of living cells at the periphery, the aleurone layer. In mature *Arabidopsis* seeds the endosperm is resorbed as the embryo undergoes development and only a single layer of living cells remains in the mature seed. sy, synergid; RMS, radial microtubules around the free nuclei as a prelude to cell wall initiation; alv, alveolus showing the start of cellularization; phr, phragmoplast – the site of cell wall formation. The embryo is shown in grey and the different regions of the developing endosperm are shown in shades as indicated. ESR, embryo-surrounding region; BETL, basal endosperm transfer layer in *Arabidopsis* and CZE, chalazal zone endosperm in maize, which are closest to the site of attachment to the mother plant; MCE, micropylar cellularized endosperm; PEN, peripheral resorbed endosperm in *Arabidopsis* or central (starchy) persistent endosperm in maize. (From Dumas and Rogowsky, 2008.)

for transferring amino acids and sugars to the endosperm and the embryo from the conducting tissue. The ESR cells at early endosperm development line the cavity in which the embryo develops; they probably play a role in the transfer of nutrients to the embryo and/or to establish a physical barrier between the embryo and the endosperm.

In contrast to the endosperm in grasses, the endosperm in many dicot species is ephemeral and is not a major component of the mature seed (non-endospermic) (Black et al., 2006). During development, the deposition of reserves within the endosperm is frequently uneven, e.g. castor bean (*Ricinus communis*) in which storage protein deposition is completed in the micropylar and peripheral regions several days before this occurs close to the cotyledons. The embryo grows into the endosperm, from which the reserves are remobilized to support its growth. In non-persistent endosperms mobilization of reserves is complete and the cells are degraded due to PCD, sometimes leaving several layers of crushed cell walls to the interior of the testa, or a single surrounding living layer of residual endosperm, sometimes called the aleurone layer (e.g. soybean (*Glycine max*)). In seeds with a persistent endosperm (endospermic) the cells are usually living at seed maturity. In some species, there is a persistent endosperm that can be very hard at maturity; this occurs in some legumes (e.g. carob (*Ceratonia siliqua*), Chinese senna (*Cassia tora*)), date (*Phoenix dactylifera*), coffee (*Coffea* spp.), ivory nut palm (*Phytelephas macrocarpa*) and asparagus (*Asparagus officinalis*). This is due to the synthesis of (galacto)mannans deposited as thickened secondary cell walls which occlude much of the cell interior, although some cytoplasm remains. In fenugreek (*Trigonella foenum-graecum*), however, the cytoplasm is completely occluded except in an outer single layer of aleurone cells in which there is no cell wall thickening.

Morphology and Anatomy of Mature Seeds

Embryo

The mature angiosperm embryo consists of an embryonic axis with a single cotyledon (monocots) or a pair of cotyledons (dicots) (Black *et al.*, 2006). The embryonic axis bears the radicle, hypocotyl, epicotyl and plumule (shoot apex); often there is a transition zone between the radicle and hypocotyl. The radicle contains the root meristem and after germination is completed gives rise to the embryonic root. It is usually adjacent to

the micropyle. The hypocotyl is a stem-like region of the axis delimited by the radicle at the basal end and by the cotyledon(s) at the proximal end; the epicotyl is the first shoot segment above the cotyledons. Cotyledons may be well developed and serve as storage organs (non-endospermic seeds) for reserves, or thin and flattened (endospermic seeds) (Bewley and Black, 1994). The embryos of many coniferous species contain several cotyledons, while these structures are absent from seeds of many parasitic plants. In the Poaceae, the single cotyledon is reduced and forms the scutellum, and the basal sheath of the cotyledon is elongated to form a coleoptile that covers the first leaves.

The embryo can be peripheral, axial or rudimentary (Martin, 1946; Fig. 1.5). Peripheral embryos are small (less than 25% of the seed volume) and restricted to the lower half of the seed, or are elongate and large (75% or more of the volume). Axial embryos range from small to occupying the entire seed interior, and may be central and straight, curved, coiled, bent or folded. Rudimentary embryos are small, globular to oval-oblong, usually with rudimentary or obscure cotyledons. Some mature embryos are green while in others there is a colour change late in maturation. For example, those of *Arabidopsis* turn pale or brownish yellow (Black *et al.*, 2006).

Endosperm

The endosperm can have a radial or antero-posterior (AP; from the micropyle to the chalaza) symmetry (Berger, 2003), as occurs in most angiosperms. Radial symmetry distinguishes the outer aleurone and subaleurone layers from the inner mass of starchy endosperm. Along the AP axis three regions are organized: (i) the micropylar endosperm (*Arabidopsis*) or embryo-surrounding region (maize (*Zea mays*)); (ii) the central, largest portion of the endosperm; and (iii) the posterior chalazal (*Arabidopsis*) or basal endosperm transfer layer (maize). The micropylar endosperm (endosperm cap) encloses the radicle tip, which in some species acts as a restraint to radicle emergence and

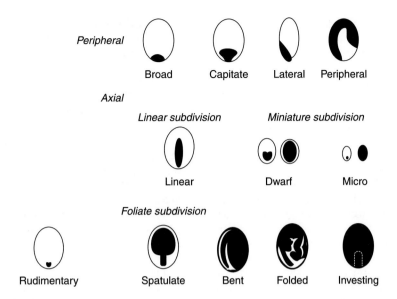

Fig. 1.5. Classification according to Martin (1946) of the internal morphology of seeds based on the position, size and shape of the embryo. (From Black *et al.*, 2006.)

imposes dormancy (e.g. fierce thornapple (*Datura ferox*)) (Black *et al.*, 2006). In some other species, such as Chinese senna or fenugreek, the micropylar endosperm has only a thin, non-restricting wall; in contrast, they have much thicker cells in the lateral endosperm, surrounding the cotyledons (Gong *et al.*, 2005). In date seeds the endosperm cap is an operculum (called also imbibition or germination lid), which is pushed off as the radicle emerges. In seeds of the Poaceae there is no micropylar endosperm.

The relative size of the endosperm varies considerably at maturity, depending on how much of the reserves have been transferred to the embryo during seed development (Black *et al.*, 2006). In seeds with a peripheral embryo the endosperm is abundant and starch laden, while starch is not the predominant reserve when the embryo is axial (except in some monocots) (Martin, 1946).

Perisperm

In some species the nucellar region of the ovule gives rise to a storage tissue called the perisperm. It is usually present in mature seeds together with the endosperm, in various proportions, locations and shapes (Black *et al.*, 2006). A perisperm is present in some monocots (e.g. Zingiberaceae) as well as certain dicots (e.g. Piperaceae, Nymphaeaceae, Cactaceae, Amaranthaceae, Chenopodiaceae). In a few species it is the major source of the storage reserves, the endosperm being absent (e.g. Quinoa and Yucca spp.) (Bewley and Black, 1994).

Seed coat

The seed coat (testa) is a maternal tissue that surrounds the embryo, endosperm and perisperm (if present). It protects the internal structures from biotic stresses and against desiccation and mechanical injury, assists in gas exchange and water uptake, provides a conduit for nutrients for the embryo and endosperm during their development, and may be a means for seed dispersal (Black *et al.*, 2006). Its structure influences seeds' viability, longevity and germinability/dormancy. Sometimes the seed coat can serve as a useful characteristic for taxonomy (e.g. in the Fabaceae).

During seed development the integuments, which are an initial seed coat tissue, undergo differentiation into layers containing specialized cells. Some cell layers in the seed coats may accumulate large quantities of compounds, such as mucilage, phenolics or pigments (Moise *et al.*, 2005). In the Brassicaceae and Fabaceae, some cell layers do not undergo any significant differentiation and remain parenchymatous (they are often crushed at maturity), while others undergo a slight thickening of the cell wall and become collenchymatous. Some cell layers undergo extensive secondary thickening of parts of the cell walls and become sclerified (palisade layers).

The seed coat of the Brassicaceae consists of four distinct layers: (i) the outermost, or epidermal layer of the outer integument, most frequently one cell thick, which may or may not contain mucilage; (ii) a subepidermal layer (in the centre of the outer integument; in some species absent), one or more cells thick, typically parenchymatous, although it may be collenchymatous or sclerified; (iii) the palisade layer (the innermost layer of the outer integument), often characterized by thickened inner tangential and radial walls; and (iv) the pigment layer (derived from the inner integument), formed of parenchyma cells compressed at maturity (Moise *et al.*, 2005).

In the Fabaceae the outermost layer of the seed coat is a waxy cuticle of variable thickness (Moise *et al.*, 2005). The next layer is the epidermis consisting of a single layer of palisade cells (macroscleriads) that are elongated perpendicular to the surface of the seed. Inside the palisade layer is an hourglass cell layer, which is composed of thick-walled osteosclereids. The innermost multicellular layer is composed of partially flattened parenchyma.

Besides its functional aspects, the seed coat can also develop characteristics or specialized structures that assist dispersal (Black *et al.*, 2006). In some angiosperm species the coat becomes fleshy and brightly coloured, which attracts animals, mainly birds. It is called the sarcotesta and is typical of 17 dicot and monocot families,

e.g. Annonaceae, Cucurbitaceae, Euphorbiaceae, Liliaceae, Magnoliaceae and Palmae. Coats of species such as flax (*Linum usitatissumum*) develop a mucilaginous epidermis, which facilitates the passage through a dispersing animal's intestines. The aril, a specialized outgrowth of the ovule or funiculus, can create a fruit-like structure attractive to fruit-eating birds (e.g. longan (*Dimocarpus longan*), akee (*Blighia sapida*), lleuque (*Prumopitys andina*)). Wind-dispersed seeds often possess wings (e.g. Asphodelaceae, Bignoniaceae, Cucurbitaceae, Hippocrateaceae). Also trichomes (hairs) produced by the seed coat, funiculus or placenta assist in wind dispersal (e.g. *Populus* and *Gossypium* spp.).

The seed coat may be undifferentiated or degraded, especially when mechanical protection of the seed is achieved by the fruit wall (pericarp, e.g. Poaceae) or inner region (endocarp, e.g. Apiaceae, Fagaceae, Urticaceae) or in small, wind-dispersed seeds of dehiscent fruits (e.g. in Orchidaceae) (Black *et al.*, 2006).

Seed shape, size and number

Developing seeds are surrounded by a pericarp, which is the wall of the ovary (Black *et al.*, 2006). In some fruits this develops relatively little and remains closely adhered to the seed (e.g. achene, nut, capsule, caryopsis). Consequently, an intact fruit is often a diaspore (the part of a plant that becomes separated and dispersed for reproduction) instead of a true seed; it is commonly referred to as a seed, however (Table 1.1).

Seeds vary greatly in shape between species. They can be round, round-oblate, oval, ellipsoid, oblong, flattened, reniform, spindly, triangular or curved; the longitudinal axis of the seed may be straight (anatropous or hemianatropous) or bent (campylotropous). The surface texture of a seed depends on the testa or pericarp structure and can be smooth, shiny, matt, grooved or concave.

Seed size varies over ten orders of magnitude. In general, larger plants produce larger seeds. However, certain habitats and

Table 1.1. Examples of dispersal units as seeds or fruits.

Seed	Fruit	
	Type	Examples

Acacia spp.
Black nightshade
  (*Solanum nigrum*)
	Achene	Buttercup (*Ranunculus asiaticus*), rose (*Rosa* spp.),
		  cottongrass (*Eriophorum vaginatum*)

Coffee (*Coffea arabica*) Cress (*Lepidium sativum*)	Berry	Barberry (*Berberis vulgaris*), crowberry (*Empetrum nigrum*), persimmon (*Diospyros* spp.)
Darwin's cotton (*Gossypium darwinii*)	Caryopsis	Grasses (Poaceae)
Lupin (*Lupinus* spp.) Pepper (*Capsicum* spp.)	Cypsela	Daisy (*Bellis perennis*), dandelion (*Taraxacum officinale*) and other Compositae

Quinoa (*Chenopodium quinoa*)
Scarlet flax (*Linum rubrum*)
	Drupe	Mulberry (*Morus* spp.), Arizona fan palm
		  (*Washingtonia filifera*), wild cashew (*Anacardium
		  excelsum*)

Swamp lily (*Crinum pedunculatum*) Wild onion (*Allium canadense*)	Nut	Oak (*Quercus* spp.), beech (*Fagus sylvatica*), alder (*Alnus glutinosa*), hornbeam (*Carpinus betulus*)
	Nutlet	Salvia (*Salvia* spp.), mint (*Mentha* spp.)
	Samara	Maple (*Acer* spp.), elm (*Ulnus americana*), ash (*Fraxinus excelsior*), hoptree (*Ptelea trifoliata*)

plant life history patterns are associated with the production of larger or smaller seeds (Lundgren, 2009). They tend to be larger when produced by plants growing in harsher habitats. Shady and/or dry habitats favour larger-, and high altitudes smaller-seeded species. When dispersed by wind, water or explosive pods, smaller seeds travel greater distances because they are lighter and fall more slowly. Seedlings growing from larger seeds, however, are able to emerge after burial at greater soil depths and better tolerate defoliation by herbivores. Seed size declines by two to three orders of magnitude between the equator and 60° (Moles and Westoby, 2003). Some species have a constant seed size (e.g. carob (*Ceratonia siliqua*) which is the historical basis of the carat unit of weight used for gold), but most species exhibit a wide variation around the average, typically four-fold (Black *et al.*, 2006).

Seed size is usually negatively related to seed number. Also important in determining seed number is the amount of resources that a plant allocates to reproduction, which is greater in annual plants than in perennials. Some wild species are especially fertile and can produce hundreds or even hundreds of thousands of seeds, e.g. field poppy (*Papaver rhoeas*) 80,000, Canadian horseweed (*Conyza canadensis*) 250,000, flixweed (*Sisymbrium sophia*) over 700,000. There is yearly variation in seed production, usually caused by such factors as pollination limitation, different degrees of attack by seed predators and pathogens or low nutrient or water availability. However, in some species, called masting species, an extreme yearly variation in seed production is observed, synchronized among individuals over large areas (e.g. beech (*Fagus* spp.), oak (*Quercus* spp.), tropical trees of the Dipterocarpaceae family). This is explained as a defensive mechanism against seed predators: keeping their numbers low in poor production years, and producing more than can be predated in favourable ones.

Dormancy

Dormancy is discussed in detail in Chapter 7 of this volume, but it is relevant to mention here the role of the embryonic coverings, collectively termed the 'coat' (although this can include the seed and/or fruit coat, endosperm, perisperm or megagametophyte) in imposing dormancy on the mature seed. Their contribution to dormancy can be physical, chemical or physiological, or a combination of these (Baskin and Baskin, 1998).

Physical dormancy occurs when the surrounding structures delay germination, usually by impeding water uptake or limiting oxygen availability to the embryo. This is attributable to the presence of one or more layers of sclerified palisade cells in the seed or fruit coat, and frequently also water-repellent compounds such as cutin, waxes and suberin. In addition to this overall impermeability, many seeds contain specialized coat structures that regulate water uptake, such as the micropyle, hilum or chalazal region. A chalazal cap is present in seeds of the Malvaceae and Cistaceae, a strophiole (located between the hilum and chalaza) in the Papilionoideae and Mimosoideae, and an operculum cap (a modified micropyle) in the Musaceae. Physical dormancy is usually broken in response to exposure of seeds to extremes of temperature, or temperature fluctuations. An extreme of this is the exposure of seeds to fire, causing cracking or softening of the coat, or disruption of the chalazal or stophiolar structure. Abrasion of the coat by sand under conditions of unusually high rain and run-off may contribute to germination of seeds of some arid zone species. Generally, dormancy is broken when the coat becomes permeable to water, leading to the weakening or unplugging of the natural opening between the exterior and the embryo.

Chemically induced dormancy is claimed to be achieved by compounds such as abscisic acid in the coat that inhibit the germination of the embryo. The structure of the coat may slow or impede their leaching upon imbibition until an adequate volume of water is available to wash them out, and be sufficient to support subsequent seedling growth. As discussed in Chapter 8 of this volume, phenolics are present in high quantities in the coats of some species, although their role in dormancy inhibition is questioned: they may be more important in dissuading pathogen attack.

Physiological dormancy includes cases in which the surrounding structures (frequently the endosperm) impose a constraint on the embryo after imbibition, preventing it from completing germination. Often seeds with thick-walled endosperms contain thin-walled non-restraining cells in the region juxtaposed to the radicle, through which it can extend. More commonly, mechanical restraint is observed by relatively thin and thin-walled endosperms, such as in lettuce (*Lactuca sativa*), tomato (*Solanum lycopersicum*) and cress (*Lepidium sativum*), or in the megagametophyte (*Picea glauca*). Its resistance is overcome by the synthesis of cell-wall hydrolysing enzymes in the endosperm itself, at least in some species, in response to a hormone signal from the germinating embryo or, in others, an increase in the growth potential of the axis (Nonogaki *et al.*, 2010).

Morphological dormancy relates to the shedding of seeds in which the embryo is still immature and incapable of germinating until it completes its development following imbibition of the dispersed seed. Conditions for completing maturity are also favourable for subsequent germination. This type of dormancy is common in many families, including the Apiaceae, although it is a more usual feature of those growing in tropical or sub-tropical climates. A variation on this is seeds with undifferentiated embryos, which occur in a few families that produce micro seeds. Members of the Orchidaceae represent an extreme in that at the time of shedding, the embryo of some species may consist of as few as two cells, and must grow to a critical size before being able to germinate. Requirements for other dormancy-breaking treatments may also be a feature of morphologically dormant seeds after their acquisition of maturity, which must then be satisfied before germination can occur.

Storage Compounds in Seeds

Following the completion of germination, the growing seedling is dependent upon stored reserves in the seed to provide nutrients to support its biosynthetic activities and energy requirements, until such time as it becomes autotrophic. These reserves are deposited during seed development following histodifferentiation to form the embryo and endosperm. In seeds with non-persistent endosperms the reserves are first deposited in this structure, to be remobilized and reutilized for reserve synthesis in the cotyledons. Usually, in seeds with a persistent endosperm (or perisperm) there is some deposition of similar reserves within the cotyledons but in much lower amounts.

Storage reserves fall into three major categories, carbohydrates, oils (triacylglycerols) and proteins, with phytin as a minor reserve. Other non-storage chemicals are present in lower amounts, including in the testa or pericarp, some of which may be important for dispersal of the seed or for their survival by discouraging predation. The approximate storage reserve content of seeds of some wild species is given in Table 1.2.

Carbohydrates

Starch is the most prevalent of the insoluble carbohydrates in seeds, although cell-wall-associated hemicelluloses are present in some species as the main carbohydrate reserve. Frequently present also are the raffinose-family oligosaccharides (RFOs) in the embryo, but in relatively much smaller amounts compared with the polymeric carbohydrates (Black *et al.*, 2006).

Starch is a polymer of glucose composed of long chains of amylose and the more abundant and larger multi-branched chains of amylopectin, the relative amounts of each being genetically determined. Both are synthesized in cellular bodies called amyloplasts where the starch is formed into discrete granules, with large and small ones becoming densely packed within the cell. In the endosperms of grasses they occupy much of the cell volume, to the exclusion of the living cytosol (Tetlow, 2011). Their importance is as a source of carbon for respiration and the synthesis of carbon-containing polymers (e.g. cellulose) during early growth of the seedling, until such time as this can be provided autotrophically by photosynthesis.

Usually starch is absent from seeds that store hemicelluloses; these are deposited to the inside of the primary wall of the endosperm to form a thickened secondary wall. The most common hemicellulose, galactomannan, is composed of long chains of mannose with variable numbers of unit galactose side chains. Those with few side chains are responsible for seed hardness, as

Table 1.2. Percent storage reserve content of seeds of some wild species.

Species	Carbohydrate (%)	Oil (%)	Protein (%)
Amaranthus spp.	62–69	5–8	14–19
Arabidopsis thaliana	–	45	20–40
Avena fatua	70	11	15
Echinochloa spp.	65–70	–	11–13
Ginkgo biloba	63	11–13	–
Quercus spp.	56–63	–	10–16
Pinus spp.	6	48	35
Trigonella foenum-graecum	45*	6–15	28
Silybum marianum	38	25–30	23

Carbohydrate reserve is mainly starch, except for * which indicates hemicellulose (galactomannan). – indicates not in measurable quantities, or not measured.
Comprehensive analyses of seed reserve content in non-agronomic species are rare. These values were obtained from web searches and many of the data are not from peer-reviewed scientific journals. Nor were all sources clear as to what the term % indicated (% dry weight, % total reserves, etc.). Hence the values should be regarded as approximations, at best.

in coffee (*Coffea* spp.) and ivory nut palm (*Phytelephas macrocarpa*), while those with many galactose units present become mucilaginous when imbibed. An example of the latter is the fenugreek seed in which the endosperm encloses the embryo. This species typically grows in dry climates and the ability of the hydrated seed galactomannans to form mucilage and retain water may be an adaptation to prevent the embryo from becoming stressed during germination if water availability declines. Thus these galactomannans may play a dual role, as a storage compound and as a retainer of water (Reid and Bewley, 1979). Other hemicelluloses, such as glucomannans, are present in the cell walls of seeds of some members of the Liliaceae and Iridaceae, xyloglucans can occur in those of the Caesalpinoideae, and some wild grasses (e.g. *Brachypodium distachyon*) contain endosperm walls thickened with glucans. They are all hydrolysed following germination to provide a source of carbon to support seedling growth.

Low molecular weight sugars occur in the cytosol of many species and account for between 1% in grasses and 16% in some legumes of dry mass of the embryos. The disaccharide sucrose and RFO members raffinose (galactosyl sucrose) and stachyose (digalactosyl sucrose) are the most common (Horbowicz and Obendorf, 1994). Their importance is as an early source of respirable substrate during germination and early seedling growth, before mobilization of the major starch or hemicellulose reserves commences, but they may also play a role in desiccation tolerance of the mature seed.

Oils (triacylglycerols)

Oils, which are insoluble in water, are esters of three fatty acids (acyl chains) attached to a glycerol backbone (Weiss, 2000). Fatty acids are variable in length, and their particular combination in an oil determines its properties; their distribution in oils is species specific, and while some fatty acids are common to many species, some are present in the oils of only one family of plants or just a few species. Oil-rich seeds do not store either

of the major carbohydrates in appreciable quantities, and vice versa; oils are alternative forms of stored carbon. Small wind-dispersed seeds tend to store oils rather than starch, for the former contain more calories per unit of weight than the latter.

Oils are deposited following their synthesis in discrete subcellular organelles termed oil bodies, into the membrane of which are embedded proteins, oleosin. These stabilize the membrane and during maturation drying prevent the oil bodies from coalescing.

There is a wide and diverse range of unusual fatty acids in the oils of seeds of wild species (Dyer *et al.*, 2008). Many of these fatty acids are present in low quantities and it is not possible to cultivate the source plants on an economic scale using modern agricultural practices. Several of the fatty acids have considerable industrial potential as, for example, high-temperature lubricants, pharmaceuticals, nutritional and health supplements, feedstocks and cosmetics, and as raw materials from which plastics and paints can be manufactured. Thus, attempts are being made to genetically engineer crop species to synthesize these on a scale that is commercially viable.

Proteins

Proteins are a source of stored carbon, nitrogen and sulfur and are a highly variable and complex group of polymers containing chains of approximately 20 different amino acids (Shewry and Casey, 1999). They may be composed of a single chain or a number of associated chains of similar or different sizes, linked by weak or strong bonds. Storage proteins are of three major types, classically determined by their solubility in solvents, although not all proteins adhere to these rigid definitions: albumins are soluble in water or dilute buffers, globulins require salt solutions to solubilize them, and prolamins are soluble in strong aqueous alcohols. Prolamins are often the major storage protein in grasses, and are absent from dicots and gymnosperms. Some grass seeds, however, contain globulins as the major storage protein (oat (*Avena sativa*), rice (*Oryza sativa*)), and these are the most common form of protein in seeds of non-monocot species. Albumins are invariably present in seeds, but usually in lower amounts than globulins and/or prolamins; an exception is rye (*Secale cereale*).

Storage proteins are sequestered in protein storage vacuoles (PSVs) during their synthesis; some may contain only one type of storage protein whereas in others there may be several different ones.

Some of the minor storage proteins have a role in addition to that as a source of nutrients for the growing seedling; coincidentally they have anti-nutritional properties for humans and domesticated animals (Akande *et al.*, 2010). Albumins, particularly of the lectin (sugar-binding) class are present in seeds that exhibit enzyme-inhibitory properties that reduce the effectiveness of food-hydrolysing enzymes in the digestive tract of animals and insects, thus dissuading them from predating the seed. Some lectins are highly toxic, notably ricin D from castor bean and abrin from the rosary pea (*Abrus precatorius*). Mandelonitrile lyase is a glycoprotein (one with sugar molecules attached to specific amino acids) present in seeds of black cherry (*Prunus serotina*) and is an enzyme that hydrolyses the cyanogenic disaccharide amygdalin. It may be important in this and seeds of other species of the Prunoideae and Maloideae in producing hydrogen cyanide (HCN) for protection when tissues are damaged by pathogens. Many seeds contain proteins that may be a part of their defence mechanism against pests and predators, e.g. in wild species of *Phaseolus* the glycoprotein arcelin is present, which confers resistance to brucid beetles. In domesticated cultivars of this species an inhibitor of the starch-degrading enzyme α-amylase is present, disrupting the digestive abilities of the predating insects. Chitinase, an enzyme that increases resistance to fungal attack because it degrades the mycelial cell walls, is present in mature seeds of several monocots and dicots (e.g. muskmelon (*Cucumis pepo*)) (Witmer *et al.*, 2003).

Phytin

Phytin is an insoluble mixed potassium, calcium and magnesium salt of *myo*-inositol hexaphosphoric acid (phytic acid), and although regarded as a minor storage component compared with the previous three classes of reserves, it is an important source of phosphate and mineral elements, which are released from the compound following germination. Other ions, manganese, iron and copper, can also be associated in phytin in lesser amounts (Lott, 1981). Phytin is often present as discrete dense aggregations in PSVs, but not necessarily within all of them in the storage cells. Nor is the mineral element content of phytin uniform in all regions of the seed; in dicot seeds calcium is bound into the molecule in globoids of the embryonic radicle and hypocotyls, but it is virtually absent from those in the cotyledons.

Other compounds

There is a plethora of minor constituents of seeds, most of which cannot be regarded strictly as storage compounds because they are not utilized during or following germination as a source of nutrients. Many of these are known because of their deleterious effects on the metabolism of humans and domestic animals, or because they can be used as pharmaceuticals for medical purposes. Their role in nature is often not known, or is surmised, but some potent toxic alkaloids such as strychnine and brucine in *Strychnos nux-vomica* may dissuade predation of the seeds, as may non-protein amino acids such as canavanine in *Dioclea megacarpa*, and *Griffonia* spp.; the latter may be used as a source of nitrogen after germination. Phenolic compounds (e.g. coumarin and chlorogenic acid) and their derivatives (e.g. caffeic and sinapic acids) occur in the testa of many seeds (also see Chapter 8 of this volume). They may prevent germination of the seed that contains them, until leached out into the soil where they may inhibit germination of surrounding seeds (a form of allelopathy).

References

Akande, K.E., Doma, U.D., Agu, H.O. and Adamu, H.M. (2010) Major antinutrients found in plant protein sources: Their effect on nutrition. *Pakistan Journal of Nutrition* 9, 827–832.

Baskin, C.C. and Baskin, J.M. (1998) *Seeds: Ecology, Biogeography, and the Evolution of Dormancy and Germination.* Academic Press, London.

Berger, F. (2003) Endosperm: the crossroad of seed development. *Current Opinion in Plant Biology* 6, 42–50.

Bewley, J.D. and Black, M. (1994) *Seeds: Physiology of Development and Germination.* Plenum Press, New York and London.

Bewley, J.D., Hempel, F.D., McCormick, S. and Zambryski, P. (2000) Reproductive development. In: Buchanan, B., Gruissem, W. and Jones, R.L. (eds) *Biochemistry and Molecular Biology of Plants.* American Society of Plant Biologists, Rockville, MD, pp. 988–1043.

Bhat, V., Dwivedi, K.K., Khurna, J.P. and Sopory, S.K. (2005) Apomixis: An enigma with potential applications. *Current Science* 89, 1879–1893.

Black, M., Bewley, J.D. and Halmer, P. (2006) *The Encyclopedia of Seeds. Science, Technology and Uses.* CAB International, Wallingford, UK.

Carneiro, V.T.C., Dusi, D.M.A. and Ortiz, J.P.A. (2006) Apomixis: Occurrence, applications and improvements. In: da Silva, J.A.T. (ed.) *Floriculture, Ornamental and Plant Biotechnology I.* Global Science Books Ltd., London, pp. 564–571.

Culley, T.M., Weller, S.G. and Sakai, A.K. (2002) The evolution of wind pollination in angiosperms. *Trends in Ecology and Evolution* 17, 361–369.

Darwin, C.R. (1876) *The Effects of Cross and Self Fertilisation in the Vegetable Kingdom.* John Murray, London.

Dumas, C. and Rogowsky, P. (2008) Fertilization and early seed formation. *Comptes Rendus Biologies* 331, 715–725.

Dyer, J.M., Styme, S., Greene, A.G. and Carlsson, A.S. (2008) High-value oils from plants. *Plant Journal* 54, 640–655.

Faheem, M., Aslam, M. and Razaq, M. (2004) Pollination ecology with special reference to insects – a review. *Journal of Research (Science)* 15, 395–409.

Golz, J.F., Clarke, A.E. and Newbigin, E. (1995) Self-incompatibility in flowering plants. *Current Opinion in Genetics and Development* 5, 640–645.

Gong, X., Bassel, G.W., Wang, A., Greenwood, J.S. and Bewley, J.D. (2005) The emergence of embryos from hard seeds is related to the structure of the cell walls of the micropylar endosperm, and not to endo-β-mannanase activity. *Annals of Botany* 96, 1165–1173.

Hamamura, Y., Nagahara, S. and Higashiyama, T. (2012) Double fertilization on the move. *Current Opinion in Plant Biology* 15, 70–77.

Horbowicz, M. and Obendorf, R.L. (1994) Seed desiccation tolerance and storability; Dependence on flatulence-producing oligosaccharides and cyclitols – review and survey. *Seed Science Research* 4, 385–405.

Koltunow, A.M. and Grossniklaus, U. (2003) Apomixis: A developmental perspective. *Annual Reviews of Plant Biology* 54, 547–574.

Li, J. and Berger, F. (2012) Endosperm: food for humankind and fodder for scientific discoveries. *New Phytologist.* DOI:10.1111/j.1469-8137.2012.04182.x

Lord, E.M. and Russell, S.D. (2002) The mechanisms of pollination and fertilization in plants. *Annual Reviews of Cell and Developmental Biology* 18, 81–105.

Lott, J.N.A. (1981) Protein bodies in seeds. *Nordic Journal of Botany* 1, 421–432.

Lundgren, J.G. (2009) Seed nutrition and defense. In: Lundgren, J.G. (ed.) *Relationships of Natural Enemies and Non-prey Foods*, Progress in Biological Control 7, Springer Science + Business Media B.V., New York, pp. 183–209.

Martin, A.C. (1946) The comparative internal morphology of seeds. *American Midland Naturalist* 36, 513–660.

Moise, J.A., Han, S., Gudynaite-Savitch, L., Johnson, D.A. and Miki, B.L.A. (2005) Seed coats: structure, development, composition, and biotechnology. *In Vitro Cellular and Developmental Biology – Plant* 41, 620–644.

Moles, A.T. and Westoby, M. (2003) Latitude, seed predation and seed mass. *Journal of Biogeography* 30, 105–128.

Nonogaki, H., Bassel, G.W. and Bewley, J.D. (2010) Germination – still a mystery. *Plant Science* 179, 574–581.

Purves, W.K., Orians, G.H., Heller, H.C. and Sadava, D. (1998) *Life: The Science of Biology*, 5th edn. Sinauer Associates, W. H. Freeman and Company, Sunderland, MA.

Reid, J.S.G. and Bewley, J.D. (1979) A dual role for the endosperm and its galactomannan reserves in the germinative physiology of fenugreek (*Trigonella foenum-graecum* L.), an endospermic leguminous seed. *Planta* 147, 145–150.

Shewry, P.R. and Casey, R. (1999) *Seed Proteins*. Kluwer Academic, Dordrecht.

Tetlow, I. (2011) Starch biosynthesis in developing seeds. *Seed Science Research* 21, 5–32.

Weiss, E.A. (2000) *Oil Crops*, 2nd edn. Blackwell Science, Oxford.

Witmer, X., Nonogaki, H., Beers, E.P., Bradford, K.J. and Welbaum, G.E. (2003) Characterization of chitinase activity and gene expression in muskmelon seeds. *Seed Science Research* 13, 167–178.

Yadegari, R. and Drews, G.N. (2004) Female gametophyte development. *Plant Cell* 16, S133–S141.

2 Fruits and Frugivory

Pedro Jordano*

*Integrative Ecology Group, Estación Biológica de Doñana,
CSIC-EBD, Sevilla, Spain*

Introduction

The pulp of fleshy fruits, with the soft, edible, nutritive tissues surrounding the seeds, is a primary food resource for many frugivorous animals, notably mammals and birds, but also reptiles and fish, which are able to obtain energy and nutrients from it (Howe, 1986). These animals either regurgitate, defecate, spit out or otherwise drop undamaged seeds away from the parent plants; they are the seed dispersers that establish a dynamic link between the fruiting plant and the seed/seedling bank in natural communities. Therefore, frugivory is a central process in plant populations where natural regeneration is strongly dependent upon animal-mediated seed dispersal.

Early conceptual contributions to the study of frugivory emphasized dichotomies in frugivory patterns and fruit characteristics that presumably had been originated by co-evolved interactions (Snow, 1971; McKey, 1975; Howe and Estabrook, 1977; van der Pijl, 1982). Fruits with pulps of a high energetic content and nutritive value surrounding a single large seed would be

one extreme of specialization by interacting with specialized frugivores providing high-quality dispersal; fruits with succulent, watery, carbohydrate-rich pulps occupy the other extreme by having their numerous small seeds dispersed by opportunist frugivores. Subsequent work during the last three decades has centred around these seminal ideas and there is a bulk of information about patterns of frugivory in particular taxa, variation in fruit characteristics, and detailed descriptions of plant–frugivore interactions for particular plant species or communities (see Estrada and Fleming, 1986; Fleming and Estrada, 1993; Corlett, 1998; Levey *et al.*, 2002; Dennis *et al.*, 2007, for reviews). However, studies of frugivory have rarely been linked conceptually with demographic patterns in the plant population. Also, the evolutionary consequences of frugivore choices, fruit processing and movement patterns have seldom been examined in an explicit evolutionary context, where fitness differentials in plant populations are measured and associated with individual variation in dispersal-related traits. Predictive frameworks that link frugivory patterns, associated

---

* E-mail: jordano@ebd.csic.es

differences in seed/seedling mortality and differential reproductive success with demographic patterns in natural plant populations are very scarce (Howe, 1989; Jordano and Herrera, 1995; Schupp and Fuentes, 1995; Wenny and Levey, 1998).

For most frugivores, fleshy fruits are a non-exclusive food resource that is supplemented with animal prey, vegetative plant parts, seeds, etc. (Hladik, 1981; Moermond and Denslow, 1985; Fleming, 1986; Howe, 1986; Willson, 1986; Corlett, 1998, 2011). Very few vertebrates rely totally on fruit food but many species are 'partial' frugivores that consume other prey together with various amounts of fruit; dietary habits among these species range between sporadic fruit consumption to almost totally frugivorous diets. For example, only 17 families of birds (15.6%) can be considered as strictly frugivorous, yet at least 21 families (19.3%) consume a mixed diet with a large proportion of fruits and a minor contribution of animal prey; and 23 families (21.1%) mix, in roughly equal proportions, fruits and other material in their diets (see Snow, 1981). Total frugivory among mammals is non-existent. Among bats, only pteropodids (Old World bats) and phyllostomids (New World fruit bats) can be considered largely frugivorous (Gardner, 1977; Marshall, 1983; Fleming, 1986; Muscarella and Fleming, 2007; Kunz et al., 2011), supplementing fruit food with insects (Courts, 1998) and/or leaves (Kunz and Diaz, 1995). Fruit is the most widely used type of food among primates, found in the diets of 91% of the species examined to date (Harding, 1981; Hladik, 1981; Lambert and Garber, 1998; Lambert, 2011) and certain frugivorous forest ungulates such as brocket deer (*Mazama* spp.) and African cephalophines (*Cephalophus* spp.) can include up to 85% of fruit material in their diet (Dubost, 1984; Bodmer, 1989a, 1990). However, partially frugivorous mammals include opossums, phalangers, kangaroos, lemurs, lorises, apes, foxes, bears, elephants, horses and other ungulates (Harding, 1981; Janzen, 1983; Howe, 1986). Finally, among reptiles,

turtles, lizards and iguanids can have an important role as seed dispersers even with infrequent and non-obligate frugivory (Barquín and Wildpret, 1975; Losos and Greene, 1988; Olesen and Valido, 2004). Fish are extremely important frugivores in some habitats, such as Amazonian várzea forest or the Pantanal, subject to periodic inundation (Horn et al., 2011).

Frugivorous animals, relying sporadically or obligately on fruits for food, have a central role in demography, population genetics and plant community evolution because: (i) their interaction with plants takes place at the final stage of each plant reproductive episode, having a potential to 'screen off' or nullify previous effects of the pollination and fruit growth phases (Jordano and Herrera, 1995; Nathan and Muller-Landau, 2000; Wang and Smith, 2002); (ii) by directing the early spatial distribution of the seeds, i.e., the 'seed shadow' (Janzen et al., 1976), they provide a template over which future spacing patterns of adult plants will build up; and (iii) seed deposition patterns by frugivores directly affect patterns of early seed survival and seedling establishment (Howe et al., 1985; Katusic-Malmborg and Willson, 1988; Schupp, 1988; Willson, 1988; Herrera et al., 1994).

The purpose of this chapter is to dissect this fleshy-fruit–frugivore interface, which brings up both characteristics of the fruits as 'prey items', that must be searched, handled and efficiently processed, and the ability of frugivores to perform these tasks with consequences for the plants themselves (Martin, 1985a). Throughout the chapter, any mention of fruits will be with reference to fleshy fruits, loosely defined to include any structure enclosing seeds surrounded by a fleshy, edible, pulp layer (Howe and Smallwood, 1982). Most references to frugivorous animals will be to birds, primates, ungulates and bats that behave as seed dispersers. The first section of the chapter describes fruits as prey items from the perspective of the foraging animal, and examines their characteristics, temporal and spatial patterns of availability, and intrinsic traits such as design and nutritive value. The second part reviews frugivore traits that influence fruit choice, fruit and seed processing, and foraging movements that have implications for seed deposition patterns.

Fruit Production and Availability

Fleshy fruits are, for the organisms consuming them, discrete food items available in an extremely diverse array of spatial and temporal configurations. The various characteristics (Table 2.1) include those that define their spatial distribution and the temporal patterns of availability, both seasonally and between years, and their food value as prey, which must be processed as discrete items. Availability characteristics influence overall abundance of frugivores in particular habitat patches, their foraging movements, and important aspects of the annual cycles. Intrinsic features determine fruit and seed processing and, consequently, how the seeds reach the ground. Both groups of traits ultimately influence seed deposition patterns, because they determine the movement patterns of frugivores foraging for fruits in relation to the mosaic of habitat patches.

Production and abundance of fruits

Variation between communities in the frequency of endozoochorous seed dispersal is broadly associated with variation in precipitation and moisture (Gentry, 1982)

and a latitudinal gradient is also evident. Vertebrate seed dispersal is very common among woody plants in neotropical (70–94% of woody species), Australian (82–88%) and African rainforests (approximately 80%) (Table 2.2). Mediterranean scrubland and some tropical dry and humid forests and woodlands usually range between 50 and 70%; temperate coniferous and broad-leaved forests vary within 30–40% of woody species animal dispersed. Frugivory and endozoochorous seed dispersal are virtually absent or unimportant in grasslands, extreme deserts, alpine vegetation and certain types of scrublands on nutrient-poor sites.

This range of variation is also exemplified when considering between-community variation in production of fleshy fruits, both in numbers and biomass. Overall levels of fruit production in particular habitats are strongly associated with the relative importance of zoochory as an adaptation for the dispersal of seeds (Fig. 2.1), but the rigorous estimation of absolute abundance is subject to numerous potential biases (Blake *et al.*, 1990; Chapman *et al.*, 1992b, 1994; Zhang and Wang, 1995). Fruit production in temperate forests of the northern hemisphere is always below 10^5 fruits ha^{-1}, representing less than 10 kg ha^{-1} (dry mass). Mediterranean scrublands have productions similar to some tropical forests, in general around 80 kg (dry mass) ha^{-1}, but fruit density might reach more than 1.4×10^6 fruits ha^{-1} in good crop years (Herrera, 1984b; Jordano, 1985); however, high-elevation Mediterranean scrublands have productions

Table 2.1. Summary of major characteristics of fleshy fruits as food resources for frugivorous vertebrates.

A. Availability characteristics
   a. Marked seasonal changes in abundance
   b. Non-renewable in the short term
   c. Strong between-years changes in availability for certain species
   d. Heterogeneous spatial distribution: highly clumped; local superabundance;
      few species available at the same particular location
B. 'Intrinsic' characteristics as prey items
   a. High water content
   b. Strong imbalance between energetic and protein components
   c. Presence of voluminous mass of indigestible material (seeds)
   d. Presence of secondary metabolites

Table 2.2. Percentages of woody species adapted for endozoochorous seed dispersal by vertebrates in different vegetation types.

Vegetation type	Mean (Range)	References[a]
Temperate coniferous forest	41.8 (33.3–56.5)	1–4
Temperate deciduous forest	35.4 (9.5–53.8)	1–5
Savannah woodland	41.2	6
Mediterranean scrubland (Spain)	43.9 (31.7–64.3)	7–8
Mediterranean scrubland (Chile)	41.9 (20.0–55.1)	9
Mediterranean scrubland (California)	34.4 (16.7–43.3)	9
Mediterranean scrubland (Australia)	22.5 (10.0–50.0)	9–11
Neotropical dry forest	46.2 (27.0–58.7)	12–14
New Zealand lowland forest	64.0	15
Subtropical humid forest	69.4 (65.2–73.5)	16–17
Neotropical and palaeotropical humid forest	74.7 (62.1–82.1)	5, 18–22
Tropical rainforest	89.5 (70.0–93.5)	5, 22–24

[a]References: 1, Johnson and Landers (1978); 2, Marks and Harcombe (1981); 3, Schlesinger (1978); 4, Franklin *et al.* (1979); 5, Howe and Smallwood (1982) and references therein; 6, Poupon and Bille (1974); 7, Herrera (1984b); 8, Jordano (1984); 9, Hoffmann and Armesto (1995); 10, Milewski (1982); 11, Milewski and Bond (1982); 12, Gentry (1982); 13, Frankie *et al.* (1974b); 14, Daubenmire (1972); 15, Burrows (1994); 16, Frost (1980); 17, Boojh and Ramakrishnan (1981); 18, Charles-Dominique *et al.* (1981); 19, Alexandre (1980); 20, Lieberman (1982); 21, Tanner (1982); 22, Willson *et al.* (1989) and references therein; 23, Putz (1979); 24, Janson (1983).

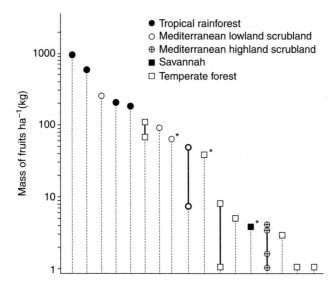

Fig. 2.1. Total production (per unit area) of fleshy fruits in different plant communities (placed in order of decreasing magnitude of production). Symbols with asterisks indicate biomass figures as wet mass, all others are dry mass. Thick lines join values for several localities. References: Leigh (1975); Johnson and Landers (1978); Baird (1980); Stransky and Halls (1980); Charles-Dominique *et al.* (1981); Hladik (1981); Sorensen (1981); Guitián (1984); Herrera (1984b,c); Jordano (1984, 1985, 1988, 1993).

more similar to those of temperate forests (Fig. 2.1). Tropical rainforests range widely in production, usually between 180 and approximately 1000 kg ha^{-1} (dry mass). For additonal data, see Blake *et al.* (1990).

Extreme between-year variations in production of fleshy fruits have been found (Davies, 1976; Foster, 1982; Jordano, 1985, 1993; Levey, 1988a; Herrera 1998, among others) but a direct, causal, relation between these fluctuations and frugivore numbers has been rarely documented. In most instances, studies with long-term data are lacking and inferences about causal associations due to the plant–frugivore interaction are unwarranted or are established without a proper evaluation of the influence of external variables (e.g., climate, food resource levels outside the study area, etc.). Between-year variations in availability of fruits, paralleled or not by variations in frugivore numbers, add an important stochastic component to plant–frugivore interactions and long-term data are needed to begin a realistic assessment of their demographic implications.

Seasonality

The overall production figures outlined above illustrate broad patterns of variation in fruit abundance but mask actual availability for frugivores, which frequently face seasonal and annual shortages of this food resource. Figure 2.2 summarizes variation in the phenology of ripe fruit availability in six major community types. In general, fruiting peaks occur during periods of low photosynthetic activity or after periods of high rates of reserve accumulation towards the end of the growing season (French, 1992; see review by Fenner, 1998). Fruiting peaks occur at the end of the dry seasons, matching generalized increases in precipitation and these trends are evident even without shifting the graphs to compensate for latitudinal differences. Unimodal fruiting peaks of the highly seasonal forests are not replicated in the very humid rainforests where several peaks of different importance occur as a result of both variations in rainfall intensity within the rainy season and delays in the phenological responses of different growth forms (Frankie et al., 1974a; Croat, 1978; Opler et al., 1980). Several authors point out the absence of significant flowering and fruiting seasonality in certain rainforests of South-East Asia (Koelmeyer, 1959; Putz, 1979), Colombia (Hilty, 1980), and South-East Brazil (Morellato et al.,

2000). Seasonality in the number of plant species bearing ripe fruits decreases from temperate to tropical forests, largely as a result of the increase in the average duration of the fruiting phenophase (although the seasonal pattern can be strikingly similar in some cases; see Fig. 2.2). Average duration of period of ripe fruit availability for a given species is always less than 1.5 months (mean = 0.6–1.3 months) in temperate forests and always more than 4 months (mean = 4.3–5.8 months) in tropical forests (Herrera, 1984c; see also references in Table 2.2). Lowland Mediterranean scrublands (Herrera, 1984c; Jordano, 1984) have intermediate averages of 2.2–4.0 months. It would be interesting to know if these consistent patterns of variation reflect similar environmental influences or if, as evidenced for the flowering seasons of temperate forest plants, they are largely attributable to phylogenetic affinities (Kochmer and Handel, 1986; Fenner, 1998; Staggemeier et al., 2010).

These differences in the seasonal patterns of fruit availability between the tropics and temperate zones define important differences in frugivory patterns. Temperate frugivory is a strongly seasonal phenomenon among migrant birds (Thompson and Willson, 1978; Stiles, 1980; Herrera, 1982; Jordano, 1985; Wheelwright, 1986, 1988; Willson, 1986; Snow and Snow, 1988; Noma and Yumoto, 1997; Parrish, 1997) and mammal species, such as carnivores (Debussche and Isenmann, 1989) or warm-temperate pteropodid bats (Funakoshi et al., 1993) which show marked seasonal shifts in diet composition. Tropical frugivores usually exploit fruit food during the whole year, but important seasonal dietary shifts also take place (Snow, 1962a,b,c; Decoux, 1976; Hilty, 1977; Worthington, 1982; Terborgh, 1983; Leighton and Leighton, 1984; Sourd and Gautier-Hion, 1986; Fleming, 1988; Erard et al., 1989; Rogers et al., 1990; Williamson et al., 1990; Conklin-Brittain et al., 1998; Wrangham et al., 1998).

Seasonality of fruit availability causes dietary shifts by frugivorous animals which 'track' the changes in the fruit supply (Loiselle and Blake, 1991). For whole-year resident frugivores, this type of

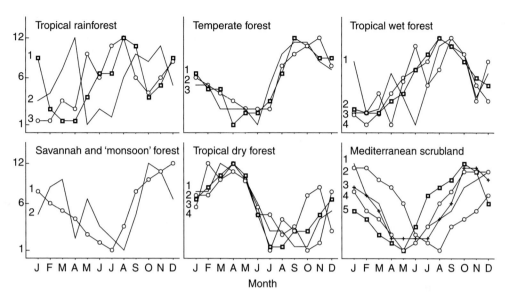

Fig. 2.2. Seasonal patterns in availability of ripe fleshy fruits in several habitat types. Months have been ranked (scores from 1 to 12 in vertical axis) according to proportion of woody species with ripe fruit available. References: Tropical rainforest, Davis (1945)[1,3]; Hilty (1980)[2]. Temperate forest, Halls (1973)[1]; Sorensen (1981)[2]; Guitián (1984)[3]. Tropical wet forest, Frankie *et al.* (1974)[1]; Crome (1975)[2]; Alexandre (1980)[3]; Medway (1972)[4]. Savannah and monsoon forest, Poupon and Bille (1974)[1]; Boojh and Ramakrishnan (1981)[2]. Tropical dry forest, Daubenmire (1972)[1]; Frankie *et al.* (1974a)[2]; Morel and Morel (1972)[3]; Lieberman (1982)[4]. Mediterranean scrubland, Herrera (1984c)[1,3]; Mooney *et al.* (1977), California[2], Chile[4]; Jordano (1984)[5].

resource tracking involves the sequential consumption of a great variety of fruit species, with a major effect on nutrient dietary balance and nutrient intake (Witmer and van Soest, 1998; Wrangham *et al.*, 1998). Important aspects of the annual cycles of frugivores, such as reproduction, breeding, migratory movements, etc., are associated with seasonal fruiting peaks. However, in most cases a direct causal link between both cyclic phenomena cannot be established. The long-term studies by Crome (1975) and Innis (1989) in the rainforests of Queensland (Australia) clearly show that seasonal patterns of abundance of certain fruit pigeons are strongly associated with the seasonal patterns of fruit ripening. Similarly, Leighton and Leighton (1984) found a good correlation between local densities of major frugivorous vertebrates (fruit pigeons, hornbills, primates and ungulates) and fruit abundance in a Bornean rainforest; regional migration,

nomadism, exploitation of aseasonal fruit types (e.g. *Ficus*) or alternative food resources were means of escaping seasonal fruit scarcity in time and space (see also Whitney and Smith, 1998, for African *Ceratogymna* hornbills). Wheelwright (1983) describes marked shifts in habitat selection by resplendent quetzals that track the seasonal sequence of ripe fruit availability among Lauraceae. Migratory or nomadic movements among Megachiroptera (Marshall, 1983) can be associated with changes in the fruit supply. Also, the annual cycle of frugivorous bird abundance in Mediterranean scrubland has been found to track closely the abundance and biomass cycle of ripe fruits (Jordano, 1985). On the other hand, Reid (1990) showed no clear relation between the seasonal abundance patterns of the mistletoe bird (*Dicaeum hirundinaceum*) and its preferred fruit, *Amyema quadang* (Loranthaceae) in Australia. The breeding seasons of certain tropical frugivorous birds (e.g. Snow, 1962a,b; Worthington, 1982), bats (Marshall, 1983; Fleming, 1988) and primates (e.g. Terborgh, 1983) all match local maxima of ripe fruit availability. Loiselle and Blake (1991) found that frugivorous birds bred when the fruit supply was low but, after the breeding season, moved to areas where fruit was more abundant. Seasonal use of fruits as an alternative food resource for temperate passerines is probably the major impelling influence in the evolution of long-distance migratory movements in the Nearctic and Palaearctic (Levey and Stiles, 1992).

The evidence outlined by these studies suggests that seasonal fruiting patterns can have a great effect on the annual cycles of most frugivores (van Schaik *et al.*, 1993). Frugivorous animals, on the other hand, probably have a negligible effect in shaping the abundance patterns of fleshy fruits in time. Thus, for Western European bird-dispersed plants, Fuentes (1992) found parallel seasonal trends in bird abundance and the number and biomass of fruits, but not in the proportion of species with ripe fruit; frugivores might favour the seasonal displacement of fruit availability by positive demographic effects on particular plant species fruiting when birds are most abundant. Major patterns of convergence in community-level fruiting patterns strongly support the findings of previous studies showing: (i) a complex role of climate (alternation of drought/rainfall seasons) in shaping the fruiting curves at a community level in relation to flowering and leafing activity (Janzen, 1967; Borchert, 1983; Gautier-Hion *et al.*, 1985a; Hopkins and Graham, 1989); (ii) a prominent role of germination requirements at the start of the rainy season (Garwood, 1983); (iii) phylogenetic constraints in the timing and duration of the fruiting phenophase (Kochmer and Handel, 1986; Gorchov, 1990; Staggemeier *et al.*, 2010); (iv) the effect of physiological constraints derived from the integration of flowering, fruit growth, ripening and seed dispersal phases of the reproductive cycle (Primack, 1987; Fenner, 1998); and (v) potential effects of frugivores in shaping

fruit availability patterns but not the fruiting phenophase itself (Debussche and Isenmann, 1992; Fuentes, 1992).

Spatial distribution

Relative to other food resources like animal prey (e.g. insects), fruits are extremely aggregated in space, usually in relatively isolated patches with high local abundance. In addition to the intrinsic spacing patterns of the adult trees that determine the spacing patterns of the fruits themselves, the spatial distribution of fruits as food resources for foraging animals is constrained by two major factors: (i) successional characteristics of the patch; and (ii) relative frequency of fruit-bearing trees in the patch. Fruit abundance increases in gaps and secondary growth of temperate forests (Thompson and Willson, 1978; Willson *et al.*, 1982; Martin, 1985b), and fruiting individuals of a given species usually bear larger crops when growing in open sites rather than the forest interior (Piper, 1986a; Denslow, 1987). Work in tropical rainforest (De Foresta *et al.*, 1984; Levey, 1988a,b; Murray, 1988; Restrepo and Gómez, 1998) showed that patchiness in fruit availability is predictably associated with treefall gaps and other disturbances. Individual plants growing in Costa Rican treefall gaps produced more fruit over a longer period of time than conspecifics growing in intact forest understory; the diversity of fruiting plants also increased in gaps (Levey, 1988b, 1990).

The same pattern exists in temperate forests where mature stands are dominated by *Quercus* spp., *Fagus* spp., *Acer* spp., among others, and fleshy-fruited shrubs and treelets are characteristics of early successional stages and forest gaps (Marks, 1974; Smith, 1975; Kollmann and Poschlod, 1997). Forest gaps of temperate forest are sites of increased local concentration of fruits (Sherburne, 1972; Sorensen, 1981; Blake and Hoppes, 1986; Martin and Karr, 1986). For example, Blake and Hoppes (1986) found average fruit abundance at the start of the fruiting season (September) of approximately 50 fruits 80 m^{-2} in Illinois forest gaps versus approximately 5 fruits 80 m^{-2}

in forest interior plots. Among the reasons for these trends in both tropical and temperate forests are: (i) increased abundance of individual plants in gaps; (ii) increased diversity of fleshy-fruit producing species; and (iii) increased crop sizes among individuals growing in gaps.

In Mediterranean shrubland however, pioneer, successional species with dry fruits and capsules are progressively substituted by endozoochorous species, which eventually dominate the late-successional stands (Bullock, 1978; Houssard *et al.*, 1980; Debussche *et al.*, 1982; Herrera, 1984d). For example, average cover of fleshy-fruited species in southern Spanish Mediterranean lowland shrubland mature stands (Jordano, 1984) is 96.88%; and it is 62.00% in open, successional stands.

Two additional sources of local patchiness in fruit availability have seldom been considered. First, abundance will be influenced by the frequent association of dioecism with production of fleshy fruits (Givnish, 1980; Donoghue, 1989). In Mediterranean shrubland, the relative cover of female individuals can vary on local patches between 20 and 95%, and increasing local abundance of male, non-fruiting plants is associated with decreased fruit availability (Jordano, 1984). This factor is probably irrelevant as a source of patchiness in fruit abundance in temperate forests, but might prove to be important in tropical habitats where dioecism is relatively frequent. Secondly, fleshy-fruiting plants are frequently associated with particular patches below the closed canopy of taller trees, probably because of increased recruitment in these foci as a result of increased seed rain beneath trees (McDonnell and Stiles, 1983; Tester *et al.*, 1987; Hoppes, 1988; Izhaki *et al.*, 1991; Holl, 1998; Clark *et al.*, 2004). Bat roosts, nests of frugivorous birds, fruiting plants where frugivores defend feeding territories, traditional perches for sexual displays and latrines of certain 'carnivore' mammals, are among the many types of sites that create recruitment foci with seed densities orders of magnitude greater than sites elsewhere in the forest (Lieberman and Lieberman, 1980;

Stiles and White, 1986; Dinerstein and Wemmer, 1988; Théry and Larpin, 1993; Fragoso, 1997; Kinnaird, 1998). In addition, seed rain of fleshy-fruited species is significantly higher beneath female, fruit-bearing plants compared with male plants of dioecious species (Herrera *et al.*, 1994), a result of preferential foraging by fruit-seeking frugivores. All these processes generate predictable spatial patterns of fruit availability which, in turn, influence the pattern of patch use by foraging frugivores.

Using a spatially explicit approach García *et al.* (2011) dissected the predictable spatial patterns of bird abundance and seed predation rate at three hierarchical spatial scales (broad, intermediate and fine). Scale-specific spatial distributions were explained by the response of animals to plant resource availability and habitat structure, with a hierarchically nested response of frugivores to the scales of fruit availability. Birds tracked fruits at large spatial scales and, within some systems, even across consecutive scales. Seed predation distribution was more responsive to habitat features than to resource availability. This suggests that consistent responses of frugivory patterns within and across spatial scales (García and Ortiz-Pulido, 2004) may condition the redundancy of seed dispersal as an ecosystem function.

Fruit Characteristics

Fruits are particulate foods that frugivorous animals usually harvest, handle and swallow as individual items. Relevant traits of fleshy fruits, from the perspective of the foraging animal, include design (e.g. size, number and size of seeds, mass of pulp relative to fruit mass), nutrient content (relative amounts of lipids, protein, carbohydrates and minerals per unit mass of fruit processed) and secondary metabolites (Table 2.1B; van der Pijl, 1982). These traits influence the overall, intrinsic profitability of fruits, by determining both the total amount of pulp ingested per fruit handled and the nutrient concentration of the ingesta (Herrera, 1981a), but the profitability of a given fruit should be examined in the context of an interaction with a particular frugivore species (Martin, 1985a; Martínez del Rio and Restrepo, 1993).

Fruit size and design

The ability to handle, swallow and process a given fruit efficiently depends on fruit size relative to body size of the frugivorous animal, particularly the gape width and mouth size. These types of constraints are similar to those found among gape-limited predators seeking particulate food and, from the plant perspective, they restrict the potential range and diversity of frugivores and dispersers (Pratt and Stiles, 1985; Wheelwright, 1985). Consumption of extremely large-seeded fruits (e.g. family Lauraceae, Palmae, etc.) by frugivorous birds is largely confined to large-bodied (toucans, trogons, bellbirds; Wheelwright, 1985; also see Pratt, 1984) or terrestrial species (trumpeter (*Psophia crepitans*): Erard and Sabatier, 1988; cassowary (*Casuarius casuarius*): Pratt, 1983; Stocker and Irvine, 1983). Bonaccorso (1979) reported a significant positive relationship between body mass variation among individual phyllostomid bats of three species and the mass of individual fruits taken (also see Kalko et al., 1996). Extremely large seeds (>3 cm length) have been reported to be dispersed exclusively by large mammals (apes, rhinos and elephants: Tutin et al., 1991; Chapman et al., 1992a). Most oversized fruit species in the paleotropics and neotropics, however, are most likely characteristic of the 'megafauna' syndrome, i.e. adaptations to animal frugivores extinct during the late Pleistocene (Janzen and Martin, 1982; Guimarães et al., 2008).

The maximum and mean diameter of fruit species included in the diets of Costa Rican birds is positively correlated with gape width, and the number of bird species feeding on the fruits of a particular species of Lauraceae was inversely correlated with fruit diameter (Wheelwright, 1985). Reduced species richness of avian frugivores visiting large-fruited species was also reported by Green (1993) in subtropical Australian rainforest. Lambert (1989a,b) found that seven species of frugivorous pigeons in Malaysia fed on at least 22 *Ficus* species, and a positive relation exists between body size and mean fig diameter of the species consumed. Fig size choice by different bird species was influenced by body size, in spite of the fact that the structure of the syconium enables exploitation by birds of all sizes (Jordano, 1983; Lambert, 1989a). Snow and Snow (1988) reported a decrease in fruit handling success (percent fruits dropped or rejected) with fruit diameter/bill width ratios greater than 1.0 (Fig. 2.3a). In turn, gape width strongly limited the size and variety of fruits included in the diet of six warbler species (genus *Sylvia*) in southern Spain (Jordano, 1987b) (Fig. 2.3b,c). The average fruit size consumed (calculated by weighting the fruit diameter of each fruit species by the relative consumption) was positively correlated with gape width (however see Johnson et al., 1985 for North American migrant birds). In addition, the average percentage of fruits dropped during short feeding bouts decreased in the larger species (Fig. 2.3c), indicating increasingly larger handling costs for smaller species (Snow and Snow, 1988). Rey and Gutiérrez (1996) reported that blackcaps switch from swallowing whole wild olive fruits to fruit pecking in the olive orchards, where seeds are twice as large; as a result, only 4.9% of faecal samples from orchards contained seeds, but 58.1% of those from the wild contained wild olive seeds. In a more exhaustive set of experiments with several Mediterranean passerine species, Rey et al. (1997) showed that fruit size determined a shift from swallowing to pecking, as pecking frequency increased with the enlargement of the fruit size; all the species showed increased fruit-handling failure rate when trying to swallow increasingly large fruits. These trends reflect the increase in handling cost associated with picking, seizing and positioning in the bill of increasingly larger fruits, but the main effect of fruit size on handling success, especially in drupes and other single-seeded fruits, is due to seed size and not to fruit size.

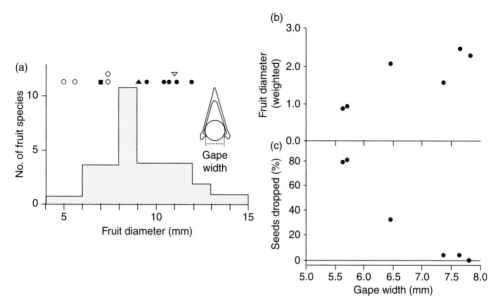

Fig. 2.3. (a) Frequency distribution of fruit diameter for bird-dispersed plants in southern England, and gape widths (width of the bill measured at the commissures) of the main seed dispersers (Snow and Snow, 1988). Filled circles, *Turdus* spp.; open circles, warblers; square, *Erithacus rubecula*; filled triangle, *Sturnus unicolor*; open triangle, *Bombycilla garrulus*. (b) Relationship between mean gape width of six species of *Sylvia* warblers and mean fruit size in the diet; fruit size of each plant species weighted by the frequency of consumption. (c) The mean percentage of fruits that are dropped during feeding sequences at *Prunus mahaleb*, a tree species with average fruit diameter of 8.4 mm. Data from Jordano (1987b), Jordano and Schupp (2000). Circles, in order of increasing gape width, indicate *S. conspicillata*, *S. cantillans*, *S. melanocephala*, *S. atricapilla*, *S. communis* and *S. borin*.

Few studies have concentrated, however, on intraspecific comparisons of fruit removal as related to fruit size variation among individual plants. Bonaccorso (1979) reported strong selectivity by individual bats of figs of *Ficus insipida* differing in size, which suggest strong fruit size selection limited by aerodynamic constraints on fruit transport on the wing. These results have been validated more recently for the whole bat community, where small bats preferentially ate small-fruited and strangler figs while large bats consumed mostly large-fruited and free-standing figs (Kalko *et al.*, 1996; Wendeln *et al.*, 2000). Howe (1983) reported that an average of 62% of variation in seed removal of *Virola surinamensis* by birds was accounted for by the aril/seed ratio of individual trees; 78% of the variation in seed size of this species is among individual crops (Howe and Richter,

1982). Intraspecific variation in fruit and seed size is thus sufficient for selective animal frugivores to exert strong phenotypic selection on fruit and seed size (Wheelwright, 1993; Jordano, 1995a). Significant correlations are frequently obtained between seed dispersal efficiency (the percentage of the seed crop dispersed) and both fruit and seed size, although the sign most probably varies as a result of the degree of gape limitation of the particular set of frugivores interacting with a plant species (Herrera, 1988; White and Stiles, 1991; Sallabanks, 1992; Herrera *et al.*, 1994; Jordano, 1995b).

The potential selective pattern on fruit seediness differs with seed size and seed packaging, and complex allocation patterns to flesh, seed endocarp and seed content exist in fleshy fruits (Lee *et al.*, 1991). For multiseeded fruits, the fraction of total fruit mass allocated to seeds increases with seed number, and frugivores are expected to select few-seeded fruits (Herrera, 1981b). In drupes and other single-seeded fruits, seed burden per unit pulp mass increases with increasing fruit size, and frugivores are expected to select small fruits, especially if gape limited (Jordano, 1987b, 1995a; Snow and Snow 1988; Rey et al., 1997). However, small-bodied frugivores might cause strong negative selection pressures on fruit and seed size by differentially dispersing the more accessible smaller seeds (Jordano, 1995a). Future studies should bridge the gap in our knowledge of the demographic effect of these types of selective pressures on the plant populations by considering simultaneously the effect of fruit size and seed size on germination and early seedling vigour and survival and the potential for evolutionary shifts mediated by frugivores.

Allocating many small seeds within a given fruit increases the potential diversity of dispersers by allowing small frugivores to ingest pulp pieces and seeds. Levey (1987) found that the percentage of seeds dropped during feeding trials with several tanager (Thraupidae) species in captivity increased as a function of seed size; birds consistently dropped more than 60% of seeds which were greater than 2.0 mm in length. These birds are 'mashers', which crush all fruits in their bills; the largest seeds are worked to the edge of the bill and dropped and the smallest seeds are swallowed along with pulp pieces. In contrast, manakins (Pipridae) are 'gulpers', which swallow the whole fruits and defecate all seeds up to the 10 mm threshold imposed by their gape width; however, the percentage of fruits taken by manakins decreased as seed size increased. See Rey and Gutiérrez (1996) for a similar example of switching between 'gulper' and 'masher' behaviour.

The same trend is also exhibited by other taxonomic groups. The smallest species of African forest frugivorous ungulates of genus Cephalophus (C. monticola, 4.9 kg) take no fruit above 3 cm diameter and the largest, C. sylvicultor, consumes fruit up to 6 cm in diameter (Dubost, 1984). Similar size-related constraints have been found in bats (Fleming, 1986) and primates (Hylander, 1975; Terborgh, 1983; Corlett and Lucas, 1990; Tutin et al., 1996; Kaplin and Moermond, 1998). For example, seed size strongly influences whether seeds are swallowed or spat out or dropped in situ by long-tailed macaques (Macaca fascicularis); seeds of most species with individual seeds less than 4.0 mm width are swallowed (Corlett and Lucas, 1990; see also Gautier-Hion, 1984). Kaplin and Moermond (1998) report that most seeds >10 mm are dropped by Cercopithecus monkeys, but variability in behaviour as seed predators or legitimate dispersers was observed. In summary, all this evidence indicates that small frugivores are limited in the largest fruit they can efficiently handle and process and, on the other hand, increase in fruit size generally limits the range of potential seed dispersers to the largest frugivores. Both assertions are especially true for drupes or other single-seeded fruits, and have important implications for the resulting seed dispersal pattern, the evolution of fruit and seed shape and their biogeographical patterns (Mack, 1993). Thus, evidence of negative allometry in the development of large-fruited species (e.g. Lauraceae) has been interpreted as an adaptation to gape-limited avian frugivores (Mazer and Wheelwright, 1993; however see Herrera, 1992).

As stated by Wheelwright (1985), fruit size alone does not explain the wide variability in the number of frugivore species feeding at different plant species that have fruits of the same size. Studies examining interspecific trends in fruit structural characteristics have also found that overall size provides the main source of functional variation in fruits relative to the types of frugivores consuming them, but additional important traits were the number of seeds per fruit, the mass of each seed, and the mass of pulp per seed (Janson, 1983; Wheelwright et al., 1984; Gautier-Hion et al., 1985b; O'Dowd and Gill, 1986; Debussche et al., 1987; Herrera, 1987; Debussche, 1988). However, only fruit size among another 15 fruit traits examined by Jordano (1995a; see Appendix to this chapter) was associated with a major type of seed disperser when accounting for phylogenetic affinities in a comparative analysis of a large data set of angiosperms.

Nutrient content of the pulp

Comparative studies of nutrient content of fleshy fruits have revealed that most variation in components can be explained by a few major patterns of covariation that have a major correlate with phylogeny, especially at the family and genus level (Jordano, 1995a). Herrera (1987) found by means of factor analysis that 46.5% of the variance in nutrient content among 111 species of the Iberian Peninsula was accounted for by the strong negative correlation between lipid and non-structural carbohydrate (NSC) content; three additional factors accounted for 51.1% of variance. Therefore, rather than the succulence continuum suggested by some authors, pulp composition patterns included: high lipid–low NSC–low fibre; low lipid–high NSC–low fibre; and medium lipid–medium NSC–high fibre. Variation in protein and water content was independent of these pulp types. Similar patterns have been described by other authors (Wheelwright

et al., 1984; Gautier-Hion *et al.*, 1985b; Johnson *et al.*, 1985; O'Dowd and Gill, 1986; Debussche *et al.*, 1987; Kitamura *et al.*, 2002; Traveset *et al.*, 2004; Galetti *et al.*, 2011) and are probably caused by the great variation in lipid content among angiosperm fruit pulps relative to other constituents and its strong inverse correlation with carbohydrate content.

The pulp of fruits has been considered repeatedly as deficient in certain nutrients, especially nitrogen and protein (Snow, 1971; Morton, 1973; White, 1974; Berthold, 1977; Thomas, 1984). Relative to other dietary items usually consumed by vertebrate frugivores (Table 2.3), the fruit pulp shows the highest concentration of soluble carbohydrates and the lowest relative amount of protein. Lipid content is relatively high but shows extreme interspecific variation. The importance of the mineral fraction is relatively constant among food types, but the content of particular cations is very variable (Nagy and Milton, 1979; Piper, 1986b; Herrera, 1987; Pannell and Koziol, 1987). Fruits are extremely poor in protein in comparison with leaves and insects. However, their energetic value in terms of soluble carbohydrates and lipids exceeds any other food type (Table 2.3). Therefore, the combination of traits that best characterizes the fruit

Table 2.3. Summary of nutrient contents of different food types consumed by vertebrate frugivores. Figures are mean and range of % of each component relative to dry mass. Data for seeds refer to wet mass.

Food type	Water	Protein	Lipids	Non-structural carbohydrates	Minerals
Insects[1]	63.7	68.3	16.8	14.9	8.9
	(56.8–70.4)	(59.9–75.9)	(9.4–21.2)	(0.5–20.0)	(3.1–19.0)
Seeds[2]	11	11	4	69	2.2
	(4–12)	(6–14)	(0.3–9)	(61–73)	(1.1–5.3)
Neotropical fruits[3]	71.3	7.8	18.5	67.8	5.6
	(38.0–95.2)	(1.2–24.5)	(0.7–63.9)	(5.6–98.3)	(1.3–19.4)
Mediterranean fruits[4]	69.9	6.4	9.0	80.1	4.6
	(36.9–90.1)	(2.5–27.7)	(3.7–58.8)	(33.2–93.7)	(1.1–13.1)
Mature leaves[5]	59.4	12.6	3.3	6.9	4.9
	(46.2–76.2)	(7.1–26.1)	(0.7–10.7)	(1.9–14.7)	(1.5–11.3)
Young leaves[5]	71.9	18.2	3.2	15.4	5.0
	(54.0–82.3)	(7.8–36.3)	(0.7–6.3)	(1.8–32.7)	(3.4–7.5)

References: [1], White (1974); [2], Jenkins (1969) cited in Moermond and Denslow (1985); [3], see references in Appendix; [4], Herrera (1987); [5], Hladik (1978), Oates (1978), Oates *et al.* (1980), Waterman *et al.* (1980).

pulp nutritive content is the excess of digestible energy relative to protein, the high water content, and the extreme deficiency in some compounds relative to others (i.e. imbalance between components).

The Appendix to this chapter summarizes most of the information available at present on nutrient content of the pulp of the main angiosperm families dispersed by vertebrate frugivores (Jordano, 1995b). Detailed reports for local or regional floras include: Hladik *et al.* (1971); Sherburne (1972); White (1974); Crome (1975); Frost (1980); Stiles (1980); Viljoen (1983); Wheelwright *et al.* (1984); Johnson *et al.* (1985); O'Dowd and Gill (1986); Piper (1986b); Debussche *et al.* (1987); Herrera (1987); Fleming (1988); Snow and Snow (1988), Eriksson and Ehrlén (1991); Hughes *et al.* (1993); Corlett (1996); Witmer (1996); Heiduck (1997); Ko *et al.* (1998); Kitamura *et al.* (2002); Traveset *et al.* (2004); and Galetti *et al.* (2011), among others.

In the case of frugivorous birds virtually nothing is known about the protein demand in natural conditions, although recent efforts have been made to understand the nutritional limitations of fruits (Sorensen, 1984; Karasov and Levey, 1990; Martínez del Rio and Karasov, 1990; Levey and Grajal, 1991; Levey and Duke, 1992; Witmer, 1996, 1998a; Witmer and van Soest, 1998). Information available, mostly from domestic, granivorous species, indicates that a diet with 4–8% protein (wet mass) is necessary for maintenance (several authors cited in Moermond and Denslow, 1985), by providing a daily consumption of 0.43 g N kg$^{-0.75}$ day^{-1} (Robbins, 1983). Considering that the high amount of water in the pulp of fleshy fruits acts as a 'solvent' of the included nutrients, most fruits contain amounts of protein, relative to dry mass of pulp, within the limits adequate for maintenance. Thus, average protein content for a sample of angiosperm fleshy fruits (Appendix to this chapter) is 6.12 ± 4.47% (mean ± SD, *n* = 477 species), ranging between 0.1 and 27.7%.

These nutrient levels are adequate if the fruit supply in nature is not limiting, but this is an infrequent situation (Foster, 1977; Witmer, 1996, 1998a). Dinerstein (1986)

found that protein, content of the fruits consumed by frugivorous bats (*Artibeus, Sturnira*) in Costa Rican cloud forest (mean = 6.7% protein, dry mass) was apparently sufficient to sustain the protein demands of lactating females; otherwise females could be depending on previously accumulated protein reserves. The data available regarding *Carollia perspicillata* (Herbst, 1986; Fleming, 1988) indicate that dietary mixing of a protein-rich fruit, such as *Piper* spp. (Piperaceae) and an energy-rich fruit, such as *Cecropia peltata* (Cecropiaceae), adequately balanced the daily net energy and nitrogen requirements. In contrast to these phyllostomid bats, totally frugivorous pteropodid bats relying on low-quality *Ficus* fruit food (less than 4.0% protein, dry mass) obtain sufficient protein by overingesting energy from fruits, but are unable to supplement this diet with animal prey (Thomas, 1984). In other pteropodids (*Rousettus*), Korine *et al.* (1996) reported a positive nitrogen balance on a totally fruit diet due to exceptionally low nitrogen demands (55% lower than expected from allometry), apparently as an adaptation to periods of low fruit availability. Overingestion of energy to meet the protein needs has been reported for the totally frugivorous oilbird *Steatornis caripensis* (Steatornithidae) (White, 1974). Early findings by Berthold (1976) that lipids and protein in fruits were insufficient for maintenance and migratory fat deposition by warblers (*Sylvia* spp.) have been challenged by the experiments of Simons and Bairlein (1990) demonstrating significant body mass gain by *Sylvia borin* when fed on a totally frugivorous diet, although additional work has confirmed loss of body mass and nitrogen on diets of sugary fruits for some species (Izhaki and Safriel, 1989; Witmer, 1996, 1998a; Witmer and van Soest, 1998). Several studies reveal positive nitrogen balance of specialized frugivorous birds, such as phainopeplas or waxwings, when feeding on fruits with protein content greater than 7.0% dry mass (Walsberg, 1975; Berthold and Moggingen, 1976; Studier *et al.*, 1988; Witmer, 1998a).

Therefore, the poor value of fruits as a unique food largely results from the internal imbalance of major nutritive components relative to others – basically the extreme protein and nitrogen deficiency relative to energy content. Thus, it is paradoxical that certain neotropical fruits qualified as highly nutritious had calorie/protein ratios greater than 1500 (Moermond and Denslow, 1985), when others, considered as poor (Rubiaceae, Melastomataceae), had ratios more similar to those of insects. The main effect of these types of relative deficiencies for frugivorous animals is that the assimilation of a particular nutrient can be limited by the impossibility of processing enough food material to obtain it, and not by the scarcity of the nutrient itself. That is, the effect is due to a digestive bottleneck (Kenward and Sibly, 1977; Sibly, 1981). Consumption of minor amounts of animal prey provides the necessary nitrogen input to escape the constraint imposed by the overingestion of energy, as demonstrated by field studies of phyllostomid bats and frugivorous warblers (Fleming, 1988; Jordano, 1988; see also Bowen *et al.*, 1995).

Direct interaction among different components present in the pulp, such as secondary metabolites, can limit nutrient digestibility and assimilation (Herrera, 1981a; Izhaki and Safriel, 1989; Mack, 1990; Cipollini and Levey, 1992, 1997; Izhaki, 2002). The presence of tannins, together with alkaloids and saponins, is particularly frequent among Mediterranean species (Jordano, 1988, and references therein). The presence of tannins in the pulp may cause lower assimilation of proteins and damage the digestive epithelium (Hudson *et al.*, 1971; Swain, 1979). Experiments by Sherburne (1972) demonstrate that other types of secondary compounds, such as glycosides or alkaloids, have a direct effect on frugivore foraging by preventing feeding or drastically reducing the palatability of unripe fruits. However, little is known about the effects of metabolites that act like tannins and phenols, reducing the assimilation efficiency (Izhaki and Safriel, 1989; Mack, 1990; Cipollini and Levey, 1997).

Finally, the content in the fruit pulp of cations and microelements, such as calcium, phosphorus, iron, manganese, and zinc, is frequently below the requirements of frugivorous birds, and situations of negative balance in wild birds have been reported (Studier *et al.*, 1988). These types of effects should be controlled in experiments assessing the nutritional limitation of fruit food for frugivores.

Frugivory

Frugivory appears to be a feeding mode that is open to many types of organism. No special adaptations, such as deep beaks or special digestive processing of the ingesta, are necessary to consume fruit, but certain morphological, anatomical and physiological characteristics determine an animal's ability to rely extensively on fruit food. The purpose of this section is to review patterns of anatomical and physiological variation associated with exclusive or extensive frugivory.

At least three basic types of frugivory can be defined, relative to their potential consequences for seed dispersal. First, legitimate dispersers swallow whole fruits and defecate or regurgitate seeds intact. Secondly, pulp consumers tear off pulp pieces while the fruit is attached to its peduncle or mandibulate fruits and ingest only pulp by working the seed(s) out. Finally, seed predators may extract seeds from fruits, discard the pulp, crack the seed, and ingest its contents or can swallow whole fruits and digest both pulp and seeds. From the plant's perspective, these categories define a wide gradient of seed dispersal 'quality' (Snow, 1971; McKey, 1975; Howe, 1993; Schupp, 1993; Schupp *et al.*, 2010), from frugivores that deliver seeds unharmed (dispersers) to those that destroy seeds (granivores), with no clear-cut limits between them (Jordano and Schupp, 2000). Single traits such as body size, wing form or bill width are not satisfactory predictors of frugivory intensity or the type of frugivorous behaviour shown by a species, and simultaneous consideration of a number of traits is needed. Herrera (1984a) found that a multiple discriminant analysis of body mass and six ratios describing bill shape accurately predicted the assignment of Mediterranean scrubland birds to three frugivory types. Seed dispersers showed larger body size and flatter and wider bills than non-frugivores and pulp-seed predators. Consumers of pulp that discarded the seeds beneath plants (finches, emberizids and parids) were characterized by smaller size, deeper beaks and narrower gapes. Non-frugivores showed more slender bills than the other two groups. Actually, species of seed dispersers, pulp-seed predators and non-frugivores occupy a continuum along the discriminant function, emphasizing the absence of clear limits between categories.

Whether a given frugivore behaves as a seed disperser, pulp predator or seed predator in a particular interaction with plants is not only dependent on frugivore ecomorphology and behaviour, but also on fruit characteristics (especially seed size) of the plants in the specific situation. Detailed descriptions of these categories and associated behavioural patterns are given by, among others: Hladik and Hladik (1967); Hladik (1981); Janzen (1981a,b,c, 1982); Fleming (1982); Herrera (1984c); Moermond and Denslow (1985); Levey (1986, 1987); Bonaccorso and Gush (1987); Snow and Snow (1988); Bodmer (1989a), Corlett and Lucas (1990); Green (1993); Corlett (1998, 2011); Jordano and Schupp (2000). It is apparent from these studies that the different types of frugivory are present in all groups of vertebrate frugivores, but in markedly different proportions.

Anatomical Characteristics of Frugivores

Frugivore size and form

Body mass is a major determinant of intensity of frugivory. The relative importance of fruit in the diet of Mediterranean passerines is strongly correlated with body mass (Herrera, 1984a; Jordano, 1984, 1987c). Smaller birds, such as those in genera *Phylloscopus, Saxicola, Hippolais* and *Acrocephalus*, only sporadically consume fruits. Fruit makes up 30–70% of diet volume among medium-sized *Phoenicurus,*

Luscinia, the smaller *Sylvia* warblers and *Erithacus* and always more than 80% in the larger species (*Sylvia atricapilla, S. borin, Turdus* spp., *Cyanopica cyanus* and *Sturnus* spp.). Katusic-Malmborg and Willson (1988) found a similar relationship for eastern North American frugivorous birds, but Willson (1986) found no consistent differences in body size between frugivores and non-frugivores in a number of habitats in this region.

Body size affects frugivory intensity by limiting the maximum number of fruits that can be swallowed or otherwise processed in feeding bouts (e.g. during short visits to plants) and the maximum amount of pulp mass that can be maintained within the gut, since gut capacity is strongly correlated with body mass. Thus, average number of fruits ingested per feeding visit to *Prunus mahaleb* plants is 1.5 for *Phoenicurus ochruros* (16.0 g), 9.0 for *Turdus viscivorus* (107.5 g), and 21.0 for *Columba palumbus* (460.0 g) (Jordano and Schupp, 2000). The number of fruits consumed per visit by frugivorous birds has been found to be strongly correlated with body mass in a number of studies (Fig. 2.4). Therefore, body size alone sets an upper limit to the potential maximum number of seeds that a given frugivore can disperse after a feeding bout. Note that sporadic visits by large frugivores can have a far greater effect on crop removal than consistent visitation by small frugivores, but the net result on seed dispersal also depends on differences in postforaging movements between small and large frugivores (Schupp, 1993).

Body size differs markedly among species showing different types of frugivory and influences fruit and seed handling prior to ingestion or immediately after it. Usually, small species tend to be pulp consumers rather than legitimate dispersers, mostly by their inability to handle fruits efficiently and swallow them intact. Thus, fruit and seed swallowing among frugivorous primates is restricted to large hominoids and cebids (Corlett and Lucas, 1990); smaller species either spit out seeds (some cercopithecines) or consume only pulp and discard seeds (Terborgh, 1983), although some

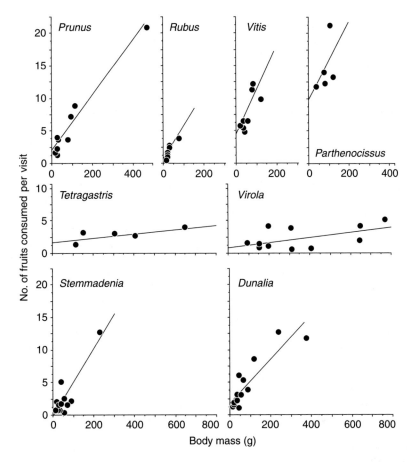

Fig. 2.4. Relationship between number of fruits consumed per visit and body mass of frugivorous birds in different plant species. Data from Jordano (1982) (*Rubus ulmifolius*); Howe and Vande Kerckhove (1981) (*Virola surinamensis*); Howe (1980) (*Tetragastris panamensis*); McDiarmid *et al.* (1977) (*Stemmadenia donnell-smithii*); Cruz (1981) (*Dunalia arborescens*); Jordano and Schupp (2000) (*Prunus mahaleb*) and Katusic-Malmborg and Willson (1988) (*Vitis vulpina* and *Parthenocissus quinquefolia*).

small species such as *Saguinus* can swallow very large seeds (Garber, 1986). Among Mediterranean mammal species the range of frugivory types generates a broad gradient between the extremes of antagonism (most seeds ingested are destroyed) and mutualism as outcomes of the interactions (Fig. 2.5; Perea *et al.*, 2013).

The use by frugivores of different foraging manoeuvres to reach fruits on plants is constrained by external morphology and body proportions, which can be considered in most cases as preadaptations to other forms of prey use. Fitzpatrick (1980) showed that fruit use among tyrannid flycatchers is restricted to three groups of genera with generalist foraging modes and fruit-feeding techniques that reflect the typical insect-foraging manoeuvres. Among Mediterranean frugivorous birds, the relative importance of fruits in the diet is significantly larger for foliage-gleaning species than for those with more specialized or stereotyped means of prey capture, such as sallyers, flycatchers and trunk foragers (Jordano, 1981). Therefore, it is reasonable to conclude that the ecomorphological configuration of a species is a preadaptation limiting feeding on fruit food, especially for those partial frugivores that consume other

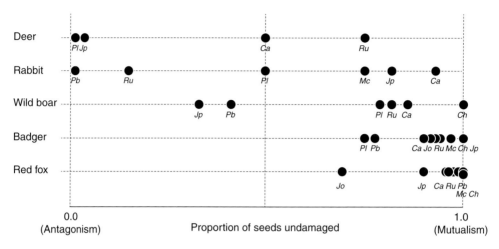

Fig. 2.5. Variation in the proportion of seeds dispersed undamaged by different species of Mediterranean mammal frugivores. Each dot represents the fraction of seeds undamaged recovered from scats of each animal species, corresponding to several plant species in their diets: Pl, *Pistacia lentiscus*; Jp, *Juniperus phoenicea*; Ca, *Corema album*; Ru, *Rubus ulmifolius*; Pb, *Pyrus bourgaeana*; Mc, *Myrtus communis*; Ch, *Chamaerops humilis*; and Jo, *J. oxycedrus*. (Modified from Perea *et al.*, 2013.)

prey types; functional and behavioural predisposition, rather than specific adaptations, are expected (Herrera, 1984a; however see Moermond and Denslow, 1985).

Differences in fruit capture modes among frugivores show strong ecomorphological correlations, especially with wing morphology, bill form or dental characteristics, and locomotory morphology (Hylander, 1975; Karr and James, 1975; Moermond and Denslow, 1985; Moermond *et al.*, 1986; Bonaccorso and Gush, 1987; Levey, 1987; Snow and Snow, 1988; Corlett and Lucas, 1990). Fleming (1988) reported relatively more elongated wings and higher wing loadings (g cm^{-2} of wing surface) among plant-visiting phyllostomid bats, which are more able to perform rapid, straight flights and hovering than insectivorous or carnivorous species. Frugivorous bats are quite conservative in the way they reach fruits, major differences being found in fruit handling and postforaging movements. The ecomorphological patterns that define the patterns of habitat selection among groups of these species (canopy-dwelling stenodermines and ground-storey carollines and glossophagines) strongly influence frugivory patterns, fruit selectivity and fruit

foraging behaviour (Bonaccorso and Gush, 1987; Fleming, 1988; see also Marshall and McWilliam, 1982 and Marshall, 1983 for information on Old World pteropodids).

Among frugivorous birds, fruits may be taken from a perch or on the wing (Herrera and Jordano, 1981; Moermond and Denslow, 1985; Foster, 1987; Snow and Snow, 1988; Jordano and Schupp, 2000). Ground-foraging frugivorous birds are larger and rarely use branches (Erard and Sabatier, 1988), but some perching species also forage for fruits on the ground (e.g. *Turdus* spp., Snow and Snow, 1988). The description that follows relies heavily on detailed accounts and experiments reported by Denslow and Moermond (1982); Levey *et al.* (1984); Santana and Milligan (1984); Moermond and Denslow (1985); Levey (1986, 1987); Moermond *et al.* (1986); Foster (1987); Snow and Snow (1988); Green (1993); and Jordano and Schupp (2000). In addition to reaching from a perch, Moermond and Denslow (1985) describe four distinct flight manoeuvres by which birds pluck fruits: hovering, the method used by manakins, flycatchers and small tanagers; stalling, used by trogons and similar to hovering; swooping and stalling, involving a continuous movement from perch to perch plucking the fruit on the way, which is the method used by most cotingids; and taking fruit from perches by picking, reaching and hanging. The first two manoeuvres are the two most commonly used, but those species that take most fruit on the wing are unable to reach well from a perch.

From the plant's perspective, the patterns described above have important implications for seed dispersal. These studies demonstrated that consistent choices between fruit species are made by foraging birds, based on accessibility restrictions that set different foraging costs, depending on anatomical characteristics. Consequences for seed dispersal are important because small changes in accessibility override preferences for particular fruits; hence non-preferred fruits are consumed when accessibility to preferred fruits decreases. Other things being equal, decreasing fruit accessibility to legitimate dispersers would increase fruit retention time on branches and the probability of resulting damage or consumption by non-disperser frugivores (Denslow and Moermond, 1982; Jordano, 1987a). The ability to access and pick fruits of a given species by different frugivores varies, depending on the positions of the fruits within the infructescence or their locations relative to the nearest perch (and the thickness of that perch). In turn, differences in feeding techniques may influence dietary diversity by affecting which specific types of fruit displays are accessible. For example, frugivorous birds that take fruit on the wing show lower diet diversity and are more selective than species that pick fruits from perches (Wheelwright, 1983; Levey *et al.*, 1984; Wheelwright *et al.*, 1984; Moermond *et al.*, 1986). An ecomorphologically diverse array of visitors might result in a more thorough removal of the crop if different species predominantly take fruits from different positions in the canopy differing in accessibility to their foraging mode (Kantak, 1979; Herrera and Jordano, 1981; Santana and Milligan, 1984; Jordano and Schupp, 2000). In addition, if microhabitat selection is related to ecomorphological

variation, individual trees differing in their relative position within a given habitat can differ markedly in the particular frugivore assemblage visiting the tree (see, for example, Manasse and Howe, 1983; Traveset, 1994; Carlo and Morales, 2008).

Once the fruit is plucked, differences in dental characteristics, mouth size and bill shape among frugivores have important consequences for external seed treatment and seed dispersal. Two basic handling modes, gulping and mashing, originally described for frugivorous birds (Levey, 1987) can probably be expanded to accommodate fruit handling behaviour by most vertebrate frugivores. For example, phyllostomid bats (*Artibeus* spp.) take single bites out of fruits (*Ficus* spp.), slowly masticating the pulp and then pressing the food bolus against the palate with the tongue; thus, they squeeze the juice and expectorate the pulp along with seeds (Morrison, 1980; Bonaccorso and Gush, 1987). In contrast, *Carollia* species masticate the pulp and swallow it along with the seeds and discard the fruit skin (Bonaccorso and Gush, 1987; Fleming, 1988). Both behaviours are functionally similar to mashing, but the consequences for the plant depend on frugivore movement after fruit plucking. Many ungulates swallow whole fruits and defecate seeds (Alexandre, 1978; Merz, 1981; Short, 1981; Lieberman *et al.*, 1987; Dinerstein and Wemmer, 1988; Bodmer, 1989b; Sukumar, 1990; Chapman *et al.*, 1992a; Fragoso, 1997) and others spit out seeds (Janzen, 1981c, 1982). Seed spitting is a common behaviour among primates, especially cercopithecines that use cheek pouches to store food and later spit out the seeds, but whether a particular seed is defecated, spat out or destroyed is strongly dependent upon seed size and fruit structure (Corlett and Lucas, 1990; Tutin *et al.*, 1996; Kaplin and Moermond, 1998; Perea *et al.*, 2013). New World apes (ceboids) and Old World hominoids apparently swallow and defecate intact most seeds (Hladik and Hladik, 1967; Hladik *et al.*, 1971; Hladik, 1981; Garber, 1986; Idani, 1986; Janson *et al.*, 1986; Rogers *et al.*, 1990; Tutin *et al.*, 1991, 1996; Wrangham *et al.*, 1994; Corlett, 1998;

Lambert and Garber, 1998; Lambert, 2011) but some species mash fruits or tear off pulp pieces and can spit out or destroy seeds (Howe, 1980; Terborgh, 1983). Colobines and some cercopithecines destroy most seeds they consume (McKey *et al.*, 1981; Davies *et al.*, 1988), but at least some *Cercopithecus* can disperse relatively large seeds by dropping or defecating them unharmed (Kaplin and Moermond, 1998).

In summary, frugivore ecomorphology *per se* determines, from the plant perspective, the position of each frugivore species along a gradient ranging between zero and 1.0 survival probability for the seeds after interaction (Fig. 2.5); and the main result of the studies discussed above is that vertebrate frugivore ecomorphologies are not distributed at random over this gradient. Our task is to search for these patterns and measure their consequences in plant–frugivore interactions.

Digestion of fruits

The bizarre digestive structures of some specialized frugivorous birds have been documented long ago by ornithologists (Forbes, 1880; Wetmore, 1914; Wood, 1924; Desselberger, 1931; Cadow, 1933; Docters van Leeuwen, 1954; Walsberg, 1975; Decoux, 1976). Typically in birds, an oesophagus, which may or may not be dilated into a crop, is continued in a stomach with a glandular proventriculus and a muscular ventriculus or gizzard. Common traits of modified digestive systems of frugivorous birds (Fig. 2.6, also including *Ducula* and *Ptilinopus* pigeons, Cadow, 1933) are: (i) absence or extreme reduction and simplification of the crop and/or proventriculus; (ii) presence of a thin-walled, non-muscular gizzard; (iii) lateral position of the simplified gizzard as a 'diverticulum' and an almost direct continuation of the oesophagus into the duodenum; and (iv) short intestines relative to body size. Despite the absence of a distinct crop, some specialized frugivorous birds, such as waxwings, can store fruits in the distensible oesophagus (Levey and Duke, 1992). This ability to store fruits oral to the gizzard somewhat offsets the problem of process-rate limitation, by allowing ingestion of two meals of fruit in a single foraging bout.

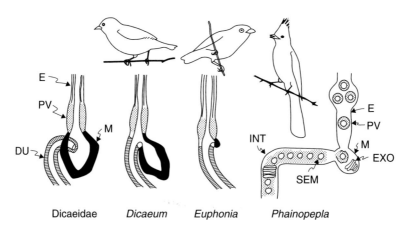

| Dicaeidae | *Dicaeum* | *Euphonia* | *Phainopepla* |

Fig. 2.6. Schematic representation of several types of proventriculus and gizzard configurations in specialized frugivorous birds. Left, arrangement of a relatively differentiated 'normal' muscular gizzard (M), stomach and associated oesophagus (E), proventriculus (PV) and duodenum in insectivorous Dicaeidae (after Desselberger, 1931). Note the normal approximation of the cardiac and pyloric ends of the stomach similar to most birds. Extreme simplification of the gizzard, with thinner walls and lack of hard epithelium, and location of the gizzard as a lateral diverticulum along the oesophagus–duodenum axis is characteristic of frugivorous dicaeids (*Dicaeum*) and *Euphonia* tanagers (Forbes, 1880). Right, arrangement in phainopeplas *Phainopepla nitens*, with schematic view of ingested fruits, exocarps (EXO) being accumulated in the simplified gizzard and seeds (SEM) passing to the small intestine (INT). (From Walsberg, 1975.)

Frugivorous bats also show a typical stomachal structure where the oesophagus leads into a cardiac vestibule and the rest of the stomach is an elongated tube with a conspicuous and large fundic caecum (Bhide, 1980 and references therein; see also Fleming, 1988).

Extreme diversification is also found in the anatomy of the digestive tract among non-volant, mammalian frugivores (Langer, 1986). Aside from ruminant artiodactyls, which consume fleshy fruits only sporadically (Bodmer, 1990), the digestive processing by non-ruminant frugivores differs chiefly between foregut and hindgut fermenters. To my knowledge, no comparative assessment has been made of the differential consequences for seed survival within the gut between these two types of digestive strategies (however see Bodmer, 1989a,b) and what fruit or seed traits, if any, are consistently associated with safe seed delivery by these frugivorous mammals. However, it is well known that fore-stomach fermenters usually crack seeds before ingestion (e.g. some colobine monkeys, and peccaries) and some hindgut fermenters also destroy most seeds they ingest (e.g. tapirs and suids, Janzen, 1981a; Corlett, 1998).

These digestive patterns are perhaps extreme examples of specialization not found in partial frugivores. Pulliainen *et al.* (1981) examined the digestive systems of three European granivorous birds and three seed dispersers and found no difference except for *Bombycilla garrulus*, which is a specialized frugivore (Berthold and Moggingen, 1976; Voronov and Voronov, 1978) and showed the largest liver mass. Eriksson and Nummi (1982) reported higher liver activity and detoxification ability in *B. garrulus* relative to granivorous and omnivorous species. However, Herrera (1984a) showed no significant differences in relative mass of gizzard and liver and relative intestine length among avian seed dispersers and pulp/seed consumers and non-frugivores (for additional data, see Magnan, 1912; Cvitanic, 1970). The largest livers were found among muscicapid warblers and would have preadapted them to frugivory by enabling efficient detoxification of the secondary metabolites present in the pulp. In addition, a closer examination of variation in frugivory among six *Sylvia* warblers (Jordano, 1987b) revealed that most variation in fruit consumption across species was accountable by considering only external morphology. Functional modulation of gut morphology allowing constant digesta retention and extraction efficiency usually requires prolonged time periods and does not seem an alternative open to frugivores, which frequently face local and short-term changes in fruit supply (Karasov, 1996; McWilliams and Karasov, 1998). Therefore, rather than elaborate morphological transformations one finds more functional adaptations to digest a soft, dilute food with low nutrient density that has a great caloric content relative to protein (Herrera, 1984a; Moermond and Denslow, 1985; Karasov and Levey, 1990; Afik and Karasov, 1995; Karasov, 1996).

There are marked functional differences among different diet types from the perspective of the digestion process (Table 2.4). Ruminant diets are characteristically high in structural hexose and pentose polymers requiring special pregastric microbial digestion, which, in addition, detoxifies many secondary plant substances (Morris and Rogers, 1983). In contrast with this slow digestion process, the digestive processing of the fruit pulp is much more rapid and more similar to digestion of vegetative plant parts by non-ruminant herbivores. In general, both forage and fruit diets show much lower digestibilities than diets based on animal prey. In addition, a sizeable fraction of the fruit food mass ingested by frugivores (the seeds) is actually indigestible and causes gut displacement (Levey and Grajal, 1991; Witmer, 1998b). Herbivore diets, and fruits are no exception, pose a frequent problem by creating digestive bottlenecks (Kenward and Sibly, 1977), which prevent frugivores increasing fruit intake to compensate for low fruit quality. The energy requirements can be adequately met but the food processing rate is too slow to meet the demand for micronutrients or nitrogen, which are deficient in the fruit pulp, and an alternative source is needed (Foster, 1978; Moermond and Denslow, 1985).

Table 2.4. Some characteristics of ruminant, carnivore and frugivore diets from the perspective of digestive physiology (modified after Morris and Rogers, 1983).

Characteristics	Ruminant diets	Animal prey	Frugivore diets
Nature of diet	Structural and photosynthetic parts of plants	Animal tissue	Fruit pulp
Digestibility	Cell wall components are refractory to mammalian enzymes	Readily digested by mammalian and avian enzymes	Readily digested, but presence of indigestible seeds
Food passage through the gut	Very slow	Slow	Very rapid
Organic matter digestibility (%)	Most forages <65	>85	~60–80
Presence of natural toxins	Generalized	None in species normally eaten	Generalized
Proximate constituents of the diet:			
Lipids	Low	High	Variable–low
Protein	Low (generally)	Very high	Very low
Non-structural carbohydrates	Low	Very low	Very high
Structural carbohydrates	Very high	–	Variable–low

Frugivores, as monogastric herbivores, base their feeding on rapid processing of their poor-quality food and maximization of the ingestion rate. They thus appear to be process-rate limited, because ingestion rate is limited by the processing of the previous meal (Sorensen, 1984; Worthington, 1989; Levey and Grajal, 1991; Levey and Duke, 1992). Throughput rate (i.e. the rate of flow of digesta past a specified point in the gut) is a function of both gut capacity (intestine length) and food retention time (Sibly, 1981; Hume, 1989; Levey and Grajal, 1991). Rapid processing of separate pulp and seed fractions, rapid passage of seeds, partial emptying of the rectal contents, rectal antiperistalsis and nutrient uptake in the rectum are all characteristics of the digestive process of frugivores to cope with nutrient-poor fruit pulp (Levey and Duke, 1992). For frugivores that defecate seeds, high throughput rates of this indigestible material must be achieved, with minimum costs for pulp digestion and assimilation. Karasov and Levey (1990) have demonstrated that this cost exists as a lower digestive efficiency, due to the absence of compensatory high

rates of digestive nutrient transport, among frugivores (however see Witmer, 1998b). In consequence, an important functional adaptation among strong frugivores would be a relatively large gut (e.g. long intestine) and extremely short throughput times; therefore, nutrient assimilation is maximized with high throughput rates. Holding constant the throughput rate, a larger gut allows processing of a greater volume of digesta at the same processing speed.

Among strongly frugivorous vertebrate species, high throughput rates are achieved by extreme shortening of throughput times (e.g. Turcek, 1961; Milton, 1981; Sorensen, 1983; Herrera, 1984a; Levey, 1986, 1987; Jordano, 1987b; Worthington, 1989; Karasov and Levey, 1990; Levey and Grajal, 1991). Seeds are processed much more quickly than pulp, either by rapid regurgitation or by 'selective' processing and defecation (however see Levey and Duke, 1992), indicating that they limit fruit processing by gut displacement and that frugivores void them selectively in order to maximize gut capacity for digestible pulp. Time to regurgitate seeds by frugivorous birds is very rapid, frequently 5–20 min, while throughput times for seed defecation are much longer, usually in the range 0.3–1.5 h (Levey, 1986; Snow and Snow, 1988; Worthington, 1989; Levey and Grajal, 1991). Levey (1986) also showed that pulp throughput times are longer than seed retention times. In some species, such as the phainopeplas (Fig. 2.6), an active mechanism for selective pulp retention is used; but in most instances differences in throughput times might be caused by the differences in specific gravity between pulp and seeds.

Relative intestine length is greater among Mediterranean frugivorous *Sylvia* warblers than among non-frugivorous muscicapid warblers (Jordano, 1987b), although gut passage time is shorter in the former. For a sample of Mediterranean scrubland frugivorous passerines, variation across species in the relative importance of fruit in the diet was positively correlated with food throughput rate ($r^2 = 0.465$; $F = 8.69$; $P = 0.015$; $n = 38$ species including *Turdus* spp., *Sylvia* spp., *Erithacus rubecula*, and several other muscicapids; Jordano, pers. obs.). This indicates that the ability to modulate retention time of digesta to achieve a high throughput rate might be important for sustained frugivory. Similarly, McWilliams and Karasov (1998) reported that compensatory modulation of retention time or digesta mixing (and not rate of hydrolysis and absorption) explained the remarkably constant digestive efficiency in waxwings exposed to varied fruit-feeding costs. Rapid fruit handling and processing thus appear to be very limiting for sustained frugivory.

Evidence that the size of indigestible seed material limits feeding rates by causing gut displacement and represents an important foraging cost for frugivores mostly comes from observations in captivity (Bonaccorso and Gush, 1987; Levey, 1987; Fleming, 1988; Snow and Snow, 1988; Corlett and Lucas, 1990; Levey and Duke, 1992; however see Witmer, 1998b) which revealed: (i) negative correlations between seed size and the number of seeds ingested per feeding bout; (ii) continuous feeding rates of birds and bats, resulting in at least one ingested seed retained in the gut;

(iii) selective throughput times for seeds and pulp; and (iv) immediate consumption of new fruits after defecation or regurgitation, implying that ingested seeds in the crop limited ingestion of additional fruits. Apparently, however, frugivores might compensate for these costs to achieve adequate intake of basic nutrients (Levey and Duke, 1992; Witmer, 1998b; Witmer and van Soest, 1998). These costs of internal handling of seed ballast are obviously overcome by frugivorous mashers and spitters, as well as by pulp consumers, which manage seeds externally; however, these frugivores have increased handling costs and lower rates of pulp ingestion per fruit handled.

Foraging for fruits and seed transport

Most seed movement away from the parent trees of fleshy-fruited species is a direct consequence of movement patterns by frugivores. The interaction of frugivore movement patterns and complex landscapes creates the template on which plant regeneration occurs, an ecological process that links movement ecology with plant dispersal patterns (Nathan *et al.*, 2008). When adequately integrated, these aspects of movement and landscape can be useful to model animal-mediated seed dispersal and predict long-distance dispersal events (Levey *et al.*, 2008). Frugivore movements take place on a habitat template with numerous microhabitats, patches, safe sites or other potential 'targets' for seed delivery. These patches differ in potential 'quality' for plant recruitment, measured as the probabilities for early survival of seeds, germination and seedling establishment (Schupp, 1993; Schupp *et al.*, 2010). From the plant perspective, the potential evolutionary and demographic relevance of the interaction with a particular disperser depends on the number of seeds it moves and how they are delivered over this habitat template, which includes a non-random distribution of patches of variable probability for establishment and survival of the plant propagules. Therefore, the two main aspects of frugivory that influence the resulting seed dispersal are the seed processing behaviour (both external and digestive) and the ranging behaviour of the frugivore (Nathan _et al._, 2008; Schupp _et al._, 2010). The former determines the number of seeds that are transported and delivered unharmed, in conditions adequate for germination; the latter defines the potential range of microsites that will intercept delivered seeds. The aim of this final section is to review how the fruit and frugivore characteristics previously considered interact and result in seed deposition patterns with implications for differential seed and seedling survival.

The spatial pattern of seed fall (i.e. the seed shadow) is a function of the species of frugivore eating the fruit, its movement rates, and seed throughput rates (Hoppes, 1987; Murray, 1988; Chapman and Russo, 2005; Russo _et al._, 2006). Note that two of the factors, namely the species identity and the seed throughput rates, can be expected to remain more or less invariant in their effect on the seed shadow independently of the particular ecological context (e.g. fruit handling patterns, defecation rates, fruit capture behaviours and other characteristics of the frugivore). In contrast, movement rates that depend on movements between foraging locations and the distances between these locations are much more 'context sensitive' and dependent on the particular ecological situation.

Fruit processing and seed deposition

Fruit processing by frugivores determines how many seeds are delivered to potential safe sites in an unharmed condition. Two important components of fruit processing are the number of fruits handled and the probability that seeds survive the fruit handling by the frugivore. If the number of safe sites increases with distance from parent plants or, if the probability of seed and early seedling survival increases with distance, then an important component of seed processing will be how fast seeds are delivered after fruit capture.

A typical feeding bout for most frugivores, especially small-sized temperate and tropical birds and phyllostomid bats, includes consumption of one or a few fruits during discrete visits to individual plants that occur along foraging sequences (Herrera and Jordano, 1981; Fleming, 1988; Snow and Snow, 1988; Green, 1993; Sun and Moermond, 1997; Jordano and Schupp, 2000; Russo _et al._, 2006). The resulting pattern of seed delivery will differ markedly between species that process fruits through the digestive tract and defecate seeds and those that process seeds orally by spitting, regurgitating or mashing prior to ingestion. These two general types of seed processing behaviours are present in most communities and differ in their immediate consequences for seed delivery. I must emphasize here that they do not represent a dichotomy of frugivore strategies but rather, a continuum gradient of seed processing rate (e.g. the number of viable seeds delivered per unit foraging time). Even the same frugivore species can be ranked in different positions along this gradient (e.g. Fig. 2.5) when interacting with different plant species.

Rapid processing of seeds by frugivores that mash or spit out seeds involves mastication and slow mandibulation of the fruit to separate the pulp from the seeds prior to ingestion and this usually results in increased risk of seed damage by cracking of the coat, excessive mechanical scarification, etc. (Hylander, 1975; Levey, 1987; Corlett and Lucas, 1990). Short-distance delivery of seeds, usually below the parent plant, is the likely result of oral fruit processing, resulting in highly clumped seed distributions irrespective of how many seeds are dispersed. In addition, low mixing of different seed species is expected since fruits are processed individually. Frugivores that process fruits orally either expectorate seeds while foraging on the same plant for more fruits (e.g. birds that mash fruits, some neotropical primates) or temporarily exit to nearby perches to process the fruit and then return to the same foraging patch. Highly clumped seed distributions have been reported as a result of the activity of phyllostomid bats that mash fruits (e.g. _Carollia_) or expectorate a food bolus with seeds (*Artibeus*), and territorial birds that regurgitate seeds within a close range of the feeding plant or display perches (Pratt and Stiles, 1983; Snow and Snow, 1984; Pratt, 1984; Bonaccorso and Gush, 1987; Fleming, 1988). The same applies to territorial birds that regurgitate seeds within a close range of the feeding plant or display perches (Pratt and Stiles, 1983; Pratt, 1984; Snow and Snow, 1984; Théry and Larpin, 1993; Kinnaird, 1998; Wenny and Levey, 1998) and tapirs and large primates using recurrent movement patterns (Fragoso, 1997; Julliot, 1997). Clumped seed distributions are not caused by a high number of seeds being processed, since the longer times to handle fruits (birds that regurgitate seeds are an exception) result in slower feeding rates, but are caused by the recurrent use of the same perches for fruit handling, resting, etc.

In contrast, digestive seed processing involves a longer retention time for seeds and increases the probability that the seed will be moved away from the parent plant. This might result in more scattered seed delivery unless postforaging movements concentrate seeds at traditional roosts, latrines, pathways, etc. Also, the degree of scattering depends on frugivore size. Blackcaps scatter 1–3 seeds in single droppings at no particular locations in Mediterranean shrubland (Jordano, 1988; Debussche and Isenmann, 1994), but large ungulates concentrate hundreds of seeds in single droppings (Dinerstein and Wemmer, 1988; Howe, 1989; Fragoso, 1997; Julliot, 1997; Bueno *et al.*, 2013). The longer retention times of seeds within the gut obviously increase the probability of seed delivery to longer distances. Fruit handling prior to ingestion is minimal, but there is a greater risk of digestive seed damage especially in frugivores with long retention times such as ungulates, parrots, some pigeons and terrestrial birds, and some finches (Janzen, 1981a, 1982; Gautier-Hion, 1984; Erard and Sabatier, 1988; Murray, 1988; Bodmer, 1989a; Lambert, 1989b; Lambert, 2011). Finally, seed clumping in faeces is strongly dependent on frugivore size (Howe, 1989; White and Stiles, 1990; Bueno *et al.*, 2013)

and this has important implications for seed survival, germination and seedling competition. Few studies, however, have documented how these patterns translate into positive net effects of non-random ('directed') seed dispersal by frugivores (Reid, 1989; Ladley and Kelly, 1996; Wenny and Levey, 1998).

Proximate consequences of seed deposition patterns

Frugivory influences on plant fitness and recruitment do not end up with seed delivery. For every dispersal episode, it matters how many and where seeds reach the ground and the particular mixing of seed species delivered. There are a number of detailed studies on the ranging behaviour of frugivores and I will not attempt to consider them in detail here (Gautier-Hion *et al.*, 1981; Hladik, 1981; Terborgh, 1983; Fleming, 1988; Murray, 1988, among others). This is probably the aspect of zoochory that is most 'context sensitive'. Most of the animal-orientated studies of frugivore movements and ranging behaviour have emphasized the patchy nature of the movements and foraging effort and the influences of external factors, such as seasonality, between-year variations in the fruit supply and numbers of other frugivores, habitat structure and abundance of alternative fruit sources and other food resources. These factors influence the 'where' component of seed deposition patterns but I wish to concentrate on the 'how' component and point out some recent research and promising directions.

The greater probability of seed mixing for internally processed seeds has far-reaching implications for postdispersal seed and seedling survival that have only recently been considered in detail in explicit relation to frugivore activity. Studies by Lieberman and Lieberman (1980), Herrera (1984b,c), Jordano (1988), Loiselle (1990), White and Stiles (1990), Théry and Larpin (1993) and Julliot (1997) strongly support the hypothesis that frugivorous animals can have determinant effects on plant community composition by differentially dispersing particular combinations of seed species. Detailed studies are needed to obtain experimental support for this hypothesis, but recent analyses (Clark *et al.*, 2004; Chapman and Russo, 2005; Carlo and Morales, 2008; Damschen *et al.*, 2008) show compelling evidence for large-scale effects of dissemination by frugivores.

Observational evidence indicates that particular combinations of seed species in faeces of dispersal agents are not the result of a process of random assortment of the available fruits in the diet, but rather indicate the presence of consistent choice patterns. Preliminary correlative evidence comes from studies of hemiparasitic and parasitic plants that need highly directed dispersal to particular hosts (Herrera, 1988; Reid, 1989; Ladley and Kelly, 1996), but a similar effect can be important for vines. Additional evidence has been obtained from detailed studies of individual diet variation in frugivore populations (Jordano, 1988; Loiselle, 1990; White and Stiles, 1990) and seed fall studies (Stiles and White, 1986; Clark *et al.*, 2004; Carlo and Morales, 2008). Loiselle (1990) has demonstrated experimentally that specific combinations of dispersed seeds in faeces of tropical frugivorous birds have direct influence on seed germination and early seedling vigour and survival.

Studies of germination rates in deposited seeds, early seedling survival and variations in seedling biomass, adequately linked with detailed information of frugivory patterns such as those described above, are the necessary tools for exploring the potential consequences of the fruit–frugivory interface in plant demography. The main influences arise from limitation of the two processes that determine animal-mediated seed dispersal: arrival of seeds to dissemination sites and postdispersal survival and establishment. These limitation effects have several components that can be directly linked to the activity of the frugivorous animals (Fig. 2.7; Nathan and Muller-Landau, 2000). While reproductive output – directly related to reproductive success after flowering – limits the amount of seeds available for dispersal (source limitation), multiple influences determine whether or not frugivore activity limits recruitment (Fig. 2.7) through the combined effects of quantity and quality components of effectiveness (Schupp *et al.*, 2010).

Concluding Remarks: An Agenda for the Fruit–Frugivory Interface

Seed dispersal is a central demographic process in plant populations. The interaction of fruits and frugivores determines the net result of the whole predispersal reproductive phase, being its last step. However, events occurring during this fruit removal–seed delivery transition have a direct influence on later-occurring stages such as germination and early seedling establishment. The studies of fruit–frugivore interactions considered in this chapter have documented what could be designated as largely 'invariant' fruit and frugivory patterns that characterize each interacting species in a particular scenario where the interaction occurs (e.g. fruit and seed size, design, nutrient configuration, fruiting display; and body size, ecomorphology, fruit handling behaviour and digestive processing of food, etc.). Description of these patterns has enabled us in the last 35 years to elaborate predictions about the outcomes of particular combinations of characteristics and test them by evaluating the associated costs in terms of seed losses for the plants or as foraging costs for the frugivorous animals.

But we need to translate the effects of these interactions into a demographic and evolutionary context to assess the relative contributions of the derived selection pressures in shaping the patterns we are observing. In this context, the net outcomes of the interactions may or may not have evolutionary consequences if their effects are 'screened off' by factors external to the interaction itself. The same can be said for the potential of frugivores to impose 'dispersal limitation' on the recruitment of their food plants (Fig. 2.7; Jordano and Herrera, 1995; Clark *et al.*, 1999; Nathan and Muller-Landau, 2000). Thus, the outcome of the invariant

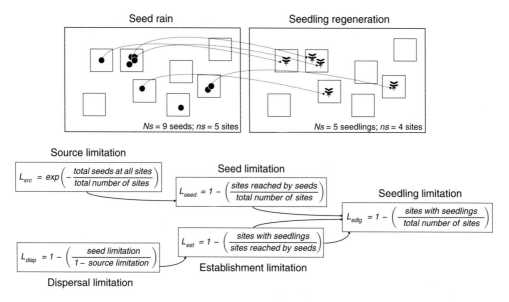

Fig. 2.7. Decomposing recruitment limitation (Nathan and Muller-Landau, 2000). Total failure to recruit at a given site or recruitment at less than maximum density can be the result of failure of seeds to arrive and/or the lack of suitable conditions for seedling and sapling establishment. Indices of dispersal and recruitment limitation can be estimated from seed trap data (seed rain) and seedling recruitment plots, from basic data on number of seeds sampled, number of sites (traps) with at least one seed, number of seedlings emerging, and number of sites (plots) with at least one seedling recruited. These indices indicate the influence of dispersal (arrival) and recruitment (survival) factors by calculating how many sites would be won if that factor were not limiting, but all other limitations were still present and thus the proportion of those sites lost because it is limiting.

patterns described above depends in addition on 'context-sensitive' effects that represent a largely stochastic component of the fruit removal–seed dispersal phase. Among them, plant spacing patterns, neighbourhood structure, site-specific habitat heterogeneity, density of alternative resources, temporal variations in fruit production and frugivore numbers, etc., produce effects that shape the result of the 'invariant' fruit–frugivore patterns.

A future avenue of research would assess the net demographic outcome of the fruit–frugivory interface by associating probabilities of seed delivery, resulting from a given interaction, with probabilities of seed and seedling survival in different microhabitats (e.g. see Chapters 3 and 4 in this volume). This is central to assessing the role of frugivore activity in limiting colonization and, potentially, the success of vegetation restoration efforts (García *et al.*,

2010). In this way, the relative roles of seed dispersal limitation and recruitment limitation in determining abundance and genetic structure could be gauged (Dalling *et al.*, 1998; García and Grivet, 2011). The preliminary protocols have been developed (e.g. Heithaus *et al.*, 1982; Jordano, 1989) for incorporating the consequences of the predispersal events and the deferred consequences for the postdispersal phase (McDonnell and Stiles, 1983; Howe *et al.*, 1985; Fleming, 1988; Katusic-Malmborg and Willson, 1988; Murray, 1988; Schupp, 1988, 1993; Herrera *et al.*, 1994; Jordano and Herrera, 1995; Schupp and Fuentes, 1995; Wenny and Levey, 1998; Clark *et al.*, 1999; Jordano and Schupp, 2000). These studies emphasize the need for measures of the net outcome of interactions with frugivores for individual plants in natural populations and consider whether the effects of frugivores are offset by events in subsequent stages of recruitment. In addition, it is necessary to consider how demographic processes (especially seed germination and seedling establishment) are influenced by variation in traits relevant to the plant–frugivore interaction.

In 1591, Italian painter Giuseppe Arcimboldo finished *Vertumnus*, an oil painting on wood depicting a portrait of Emperor Rudolf II in a frontal view of head and shoulders. When admired from a distance, this image of Vertumnus, a Roman deity responsible for vegetation and metamorphosis, appears as a neat, brightly coloured and meticulously elaborated picture. On approaching the painting, one discovers that Arcimboldo illustrated at least 34 species of fleshy fruits, which, carefully assembled, served as natural models to produce Vertumnus' image. Grapes, cherries, pears, figs, blackberries, peaches and plums, among many others, serve as the eyes, ears, lips, nose, etc. of this incredible fruit dish. What I admire about this intriguing funny face is the painter's ability to produce an ordered image from such a chaotic ensemble of fruits and plant parts. I think that the last three decades of research on the fruit–frugivory interface have yielded many fruits that, like Arcimboldo's model objects, need an elaborate assembly to produce a neat image. The efforts to bridge the consequences of frugivory and seed dispersal with demographic and evolutionary processes in plant and frugivore populations are a first sketch of that picture.

Acknowledgements

This research was supported by the Spanish Ministry of Economy and Innovation and by funds of the Consejería de Educación y Ciencia (Junta de Andalucía, Excellence Grant RNM5731).

Appendix

Summary statistics (sample size, mean and standard error of the mean for each family and variable) of fruit characteristics and pulp constituents of vertebrate-dispersed plants, by families. The FRUBASE dataset (Jordano, 1995a) is accessible at the DRYAD Digital Repository: DOI:10.5061/dryad.9tb73.

Family	Fruit diameter (mm)	Pulp dry mass (g)	Seed dry mass (g)	Relative yield	Kcal/g dry mass	Kcal/ fruit	Percent water	Lipids	Protein	Carbo-hydrates	Ash
Anacardiaceae N=12											
	5	6	5	6	10	5	9	10	10	10	7
	7.6	0.047	0.117	21.25	5.410	0.122	57.12	0.240	0.054	0.638	0.033
	2.3	0.029	0.093	6.90	0.473	0.051	7.69	0.080	0.005	0.090	0.007
Annonaceae N=11											
	3	5	3	4	5	3	7	8	9	7	5
	15.1	0.374	0.405	16.28	3.043	1.458	71.67	0.114	0.042	0.636	0.022
	1.8	0.156	0.233	4.28	0.629	1.181	6.53	0.039	0.009	0.093	0.008
Apocynaceae N=10											
	2	3	3	3	8	2	7	9	9	9	6
	6.1	0.313	0.147	15.80	4.734	2.026	79.09	0.143	0.047	0.762	0.032
	2.4	0.290	0.099	6.05	0.412	1.904	3.47	0.069	0.014	0.094	0.011
Caprifoliaceae N=26											
	16	17	16	17	21	14	25	17	21	17	15
	6.6	0.088	0.127	15.97	4.175	0.426	71.60	0.057	0.060	0.756	0.060
	0.4	0.057	0.104	1.80	0.086	0.284	3.27	0.016	0.010	0.049	0.007
Ericaceae N=10											
	8	8	8	8	6	4	10	6	6	6	6
	9.9	0.199	0.026	17.25	4.200	1.275	78.61	0.047	0.034	0.899	0.024
	1.4	0.129	0.007	2.70	0.029	1.091	2.85	0.006	0.002	0.012	0.006
Lauraceae N= 46											
	36	39	26	39	27	21	41	39	40	28	4
	15.6	0.510	0.680	14.32	4.337	1.956	68.03	0.271	0.061	0.274	0.032
	0.9	0.089	0.134	0.93	0.360	0.396	2.05	0.021	0.007	0.044	0.004
Liliaceae N=13											
	11	13	12	13	8	8	13	8	8	8	10
	9.3	0.055	0.091	14.18	4.056	0.243	69.06	0.030	0.046	0.782	0.061
	0.6	0.008	0.022	1.94	0.078	0.049	2.88	0.008	0.006	0.067	0.008
Melastomataceae N=7											
	2	3	3	3	6	2	7	4	6	6	3
	4.9	0.035	0.009	22.03	3.407	0.202	75.11	0.044	0.035	0.738	0.057
	0.4	0.027	0.006	8.30	0.386	0.176	4.85	0.016	0.009	0.080	0.012
Meliaceae N=19											
	4	7	4	7	15	4	9	17	18	15	8
	12.4	0.237	0.120	20.96	5.627	1.232	53.88	0.305	0.075	0.588	0.032
	2.7	0.052	0.015	4.19	0.346	0.283	7.16	0.059	0.016	0.071	0.008
Moraceae N=39											
	14	8	7	7	20	6	18	19	25	18	12
	13.4	0.588	0.286	10.77	3.462	2.997	79.67	0.044	0.055	0.653	0.071
	2.0	0.254	0.177	1.19	0.238	1.378	1.50	0.008	0.007	0.057	0.008

Continued

Continued.

Family	Fruit diameter (mm)	Pulp dry mass (g)	Seed dry mass (g)	Relative yield	Kcal/g dry mass	Kcal/fruit	Percent water	Lipids	Protein	Carbohydrates	Ash
Myrsinaceae N= 4											
	3	4	3	4	3	3	4	4	4	3	2
	8.9	0.029	0.030	11.98	3.376	0.126	82.45	0.062	0.041	0.629	0.066
	2.4	0.009	0.013	1.39	0.942	0.052	2.94	0.021	0.019	0.165	0.013
Myrtaceae N=18											
	8	8	4	8	11	3	14	14	16	12	9
	15.5	0.730	0.477	10.86	3.265	0.805	82.29	0.022	0.040	0.722	0.037
	3.1	0.433	0.313	1.85	0.347	0.374	2.02	0.004	0.003	0.077	0.005
Oleaceae N= 9											
	7	6	5	6	8	5	6	8	9	7	7
	7.4	0.123	0.072	15.62	4.254	0.207	62.98	0.079	0.046	0.796	0.029
	0.7	0.084	0.038	1.13	0.334	0.094	4.79	0.049	0.005	0.060	0.005
Palmae N=17											
	6	7	6	7	13	3	11	14	14	13	6
	13.7	0.582	1.436	12.34	4.356	5.396	54.30	0.181	0.061	0.592	0.079
	1.2	0.412	1.015	4.35	0.361	3.999	9.00	0.048	0.012	0.069	0.021
Piperaceae N=11											
	1	2	1	2	10	1	11	11	11	10	1
	5.1	0.118	0.170	13.55	2.468	0.964	83.27	0.057	0.074	0.389	0.125
	–	0.103	–	5.75	0.285	–	2.24	0.014	0.007	0.044	–
Rhamnaceae N=13											
	7	7	7	7	10	6	11	10	11	10	7
	8.2	0.110	0.090	16.20	3.785	0.494	66.50	0.014	0.053	0.839	0.051
	1.2	0.084	0.044	3.12	0.120	0.389	3.44	0.004	0.011	0.031	0.013
Rosaceae N=47											
	37	34	31	34	36	26	40	31	38	30	26
	12.3	0.390	0.120	21.83	3.928	1.757	66.78	0.023	0.044	0.787	0.044
	0.9	0.116	0.025	1.75	0.109	0.594	2.08	0.002	0.004	0.045	0.004
Rubiaceae N=23											
	8	15	9	15	10	5	19	10	16	10	7
	7.8	0.019	0.013	11.31	3.875	0.035	81.99	0.047	0.045	0.728	0.043
	1.6	0.006	0.005	2.22	0.171	0.007	2.82	0.016	0.011	0.052	0.010
Rutaceae N=6											
	3	3	2	3	4	2	3	5	4	4	4
	16.5	0.503	0.862	4.29	4.285	2.178	72.50	0.104	0.100	0.650	0.066
	6.8	0.276	0.826	4.47	0.371	1.931	8.88	0.030	0.007	0.043	0.011
Sapotaceae N=10											
	2	4	3	4	7	2	6	9	9	8	7
	16.2	0.477	0.145	21.13	3.761	1.327	74.08	0.073	0.063	0.742	0.045
	4.4	0.228	0.065	3.64	0.309	0.902	3.59	0.016	0.013	0.066	0.011
Smilacaceae N=4											
	3	4	2	4	4	4	4	4	4	4	2
	7.4	0.036	0.051	12.55	4.215	0.153	77.45	0.011	0.050	0.488	0.069
	0.1	0.005	0.012	0.62	0.214	0.026	3.67	0.004	0.006	0.215	0.019
Solanaceae N=25											
	13	13	10	13	19	8	24	21	22	19	7
	11.2	0.099	0.085	10.54	3.019	0.522	81.33	0.044	0.093	0.487	0.056
	0.8	0.021	0.021	1.04	0.345	0.123	1.43	0.022	0.009	0.063	0.012

Continued

Continued.

Family	Fruit diameter (mm)	Pulp dry mass (g)	Seed dry mass (g)	Relative yield	Kcal/g dry mass	Kcal/ fruit	Percent water	Lipids	Protein	Carbo- hydrates	Ash
Tiliaceae N=6											
	0	0	0	0	6	0	0	6	6	6	6
	–	–	–	–	2.945	–	–	0.010	0.064	0.650	0.039
	–	–	–	–	0.249	–	–	0.003	0.012	0.054	0.006
Ulmaceae N=5											
	3	3	2	3	5	3	3	4	5	4	3
	8.9	0.118	0.068	33.67	5.044	0.494	44.87	0.241	0.084	0.380	0.082
	0.5	0.063	0.066	5.02	0.628	0.243	11.20	0.136	0.027	0.183	0.017
Viscaceae N=9											
	6	7	4	6	6	3	7	6	7	5	3
	5.6	0.041	0.010	15.55	4.847	0.161	74.13	0.163	0.084	0.671	0.040
	0.5	0.021	0.003	3.08	0.430	0.049	5.12	0.075	0.023	0.081	0.003
Vitaceae N=8											
	6	5	2	5	5	3	7	5	5	4	2
	9.2	0.071	0.050	13.72	4.528	0.279	81.86	0.138	0.041	0.509	0.016
	0.5	0.022	0.000	2.20	0.286	0.127	3.16	0.060	0.017	0.227	0.010

Only families with >4 species sampled have been included. Figures for pulp constituents are proportions relative to pulp dry mass.

References used: Snow (1962c); Sherburne (1972); White (1974); Crome (1975); McDiarmid *et al.* (1977); Nagy and Milton (1979); Snow (1979); Frost (1980); Morrison (1980); Howe (1981); Howe and Vande Kerckhove (1981); Beehler (1983); Foster and McDiarmid (1983); Jordano (1983, 1995a); Viljoen (1983); Estrada *et al.* (1984); Wheelwright *et al.* (1984); Johnson *et al.* (1985); Moermond and Denslow (1985); Dinerstein (1986); Piper (1986b); Sourd and Gautier-Hion (1986); Debussche *et al.* (1987); Herrera (1987); Pannell and Koziol (1987); Atramentowicz (1988); Dowsett-Lemaire (1988); Abrahamson and Abrahamson (1989); Izhaki and Safriel (1989); Worthington (1989); F.H.J. Crome, personal communication; C.M. Herrera and P. Jordano, unpublished data.

References

Abrahamson, W.G. and Abrahamson, C.R. (1989) Nutritional quality of animal dispersed fruits in Florida sandridge habitats. *Bulletin of the Torrey Botanical Club* 116, 215–228.

Afik, D. and Karasov, W.H. (1995) The trade-offs between digestion rate and efficiency in warblers and their ecological implications. *Ecology* 76, 2247–2257.

Alexandre, D.Y. (1978) Le rôle disseminateur des éléphants en forêt de Tai, Cote-d'Ivoire. *La Terre et la Vie* 32, 47–72.

Alexandre, D.Y. (1980) Caractère seasonier de la fructification dans une forêt hygrophile de Côte-d'Ivoire. *Revue d'Ecologie (Terre Vie)* 34, 335–350.

Atramentowicz, M. (1988) La frugivorie opportuniste de trois marsupiaux didelphidés de Guyane. *Revue d'Ecologie (Terre Vie)* 43, 47–57.

Baird, J.W. (1980) The selection and use of fruit by birds in an Eastern forest. *Wilson Bulletin* 92, 63–73.

Barquín, E. and Wildpret, W. (1975) Diseminación de plantas canarias. Datos iniciales. *Vieraea* 5, 38–60.

Beehler, B. (1983) Frugivory and polygamy in birds of paradise. *Auk* 100, 1–12.

Berthold, P. (1976) Animalische und vegetabilische Ernährung omnivorer Singvogelarten: Nahrungsbevorzugung, Jahresperiodik der Nahrungswahl, physiologische und ökologische Bedeutung. *Journal für Ornithologie* 117, 145–209.

Berthold, P. (1977) Proteinmangel als Ursache der schädigenden Wirkung rein vegetabilischer Ernhärung omnivorer Singvogelarten. *Journal für Ornithologie* 118, 202–205.

Berthold, P. and Moggingen, S. (1976) Der Seidenschwanz *Bombycilla garrulus* als frugivorer Ernährungsspezialist. *Experientia* 32, 1445.

Bhide, S.A. (1980) Observations on the stomach of the Indian fruit bat, *Roussetus leschenaulti* (Desmarest). *Mammalia* 44, 571–579.

Blake, J.G. and Hoppes, W.G. (1986) Influence of resource abundance on use of tree-fall gaps by birds in an isolated woodlot. *Auk* 103, 328–340.

Blake, J.G., Loiselle, B.A., Moermond, T.C., Levey, D.J. and Denslow, J.S. (1990) Quantifying abundance of fruits for birds in tropical habitats. *Studies in Avian Biology* 13, 73–79.

Bodmer, R.E. (1989a) Frugivory in Amazonian Artiodactyla: evidence for the evolution of the ruminant stomach. *Journal of Zoology, London* 219, 457–467.

Bodmer, R.E. (1989b) Ungulate biomass in relation to feeding strategy within Amazonian forests. *Oecologia (Berlin)* 81, 547–550.

Bodmer, R.E. (1990) Ungulate frugivores and browser-grazer continuum. *Oikos* 57, 319–325.

Bonaccorso, F.J. (1979) Foraging and reproductive ecology in a Panamanian bat community. *Bulletin of the Florida State Museum* 24, 359–408.

Bonaccorso, F.J. and Gush, T.J. (1987) Feeding behaviour and foraging strategies of captive phyllostomid fruit bats: an experimental study. *Journal of Animal Ecology* 56, 907–920.

Boojh, R. and Ramakrishnan, P.S. (1981) Phenology of trees in a sub-tropical evergreen montane forest in North-east India. *Geo-Eco-Trop* 5, 189–209.

Borchert, R. (1983) Phenology and control of flowering in tropical trees. *Biotropica* 15, 81–89.

Bowen, S.H., Lutz, EN. and Ahlgren, M.O. (1995) Dietary protein and energy as determinants of food quality: trophic strategies compared. *Ecology* 76, 899–907.

Bueno, R.S., Guevara, R., Ribeiro, M.C., Culot, L., Bufalo, F.S. and Galetti, M. (2013) Functional redundancy and complementarities of seed dispersal by the last neotropical megafrugivores. *PLoS ONE* 8, e56252.

Bullock, S.H. (1978) Plant abundance and distribution in relation to types of seed dispersal. *Madroño* 25, 104–105.

Burrows, C.I. (1994) Fruits, seeds, birds and the forests of Banks Peninsula. *New Zealand Natural Sciences* 21, 87–108.

Cadow, A. (1933) Magen und Darm der Fruchttauben. *Journal für Ornithologie* 81, 236–252.

Carlo, T.A. and Morales, J.M. (2008) Inequalities in fruit-removal and seed dispersal: consequences of bird behaviour, neighbourhood density and landscape aggregation. *Journal of Ecology* 96, 609–618.

Chapman, C. and Russo, S. (2005) Primate seed dispersal linking behavioral ecology with forest community structure. In: Campbell, C.J., Fuentes, A.F., MacKinnon, K.C., Panger, M. and Bearder, S. (eds) *Primates in Perspective*. Oxford University Press, Oxford, UK, pp. 510–525.

Chapman, L.I., Chapman, C.I. and Wrangham, R W. (1992a) *Balanites wilsoniana*: elephant dependent dispersal. *Journal of Tropical Ecology* 8, 275–283.

Chapman, C.A., Wrangham, R.W. and Chapman, L.J. (1992b) Estimators of fruit abundance of tropical trees. *Biotropica* 24, 527–531.

Chapman, C.A., Wrangham, R. and Chapman, L.I. (1994) Indices of habitat-wide fruit abundance in tropical forests. *Biotropica* 26, 160–171.

Charles-Dominique, P., Atramentowicz, M., Charles-Dominique, M., Gerard, H., Hladik, A., Hladik, C.M. and Prevost, M.F. (1981) Les mammiféres frugivores arboricoles nocturnes d'une forêt Guyanaise: inter-relations plantes-animaux. *Revue d'Ecologie (Terre et Vie)* 35, 341–435.

Cipollini, M.L. and Levey, D.J. (1992) Relative risks of microbial rot for fleshy fruits: significance with respect to dispersal and selection for secondary defense. *Advances in Ecological Research* 23, 35–91.

Cipollini, M.L. and Levey, D.J. (1997) Secondary metabolites of fleshy vertebrate-dispersed fruits: adaptive hypotheses and implications for seed dispersal. *American Naturalist* 150, 346–372.

Clark, C., Poulsen, J., Connor, E. and Parker, V. (2004) Fruiting trees as dispersal foci in a semi-deciduous tropical forest. *Oecologia* 139, 66–75.

Clark, J.S., Beckage, B., Camill, P., Cleveland, B., HilleRisLambers, J., Lichter, J., McLachlan, J., Mohan, J. and Wyckoff, P. (1999) Interpreting recruitment limitation in forests. *American Journal of Botany* 86, 1–16.

Conklin-Brittain, Wrangham, R.W. and Hunt, K.D. (1998) Dietary responses of chimpanzees and cercopithecines to seasonal variation in fruit abundance. I. Antifeedants. *International Journal of Primatology* 19, 949–970.

Corlett, R.T. (1996) Characteristics of vertebrate-dispersed fruits in Hong Kong. *Journal of Tropical Ecology* 12, 819–833.

Corlett, R.T. (1998) Frugivory and seed dispersal by vertebrates in the Oriental (Indomalayan) region. *Biological Review* 73, 413–448.

Corlett, R.T. (2011) How to be a frugivore (in a changing world). *Acta Oecologica* 37, 674–681.

Corlett, R.T. and Lucas, P.W. (1990) Alternative seed-handling strategies in primates: seed-spitting by long-tailed macaques (*Macaca fascicularis*). *Oecologia (Berlin)* 82, 166–171.

Courts, S.E. (1998) Dietary strategies of Old World fruit bats (Megachiroptera, Pteropodidae): how do they obtain sufficient protein? *Mammal Review* 28, 185–194.

Croat, T.B. (1978) *Flora of Barro Colorado Island*, Stanford University Press, Stanford, USA.

Crome, F.H.J. (1975) The ecology of fruit pigeons in tropical Northern Queensland. *Australian Wildlife Research* 2, 155–185.

Cruz, A. (1981) Bird activity and seed dispersal of a montane forest tree (*Dunalia arborescens*) in Jamaica. *Biotropica, Supplement* pp. 34–44.

Cvitanic, A. (1970) The relationships between intestine and body length and nutrition in several bird species. *Larus* 21–22, 181–190.

Dalling, L.W., Hubbell, S.P. and Silvera, K. (1998) Seed dispersal, seedling establishment and gap partitioning among tropical pioneer trees. *Journal of Ecology* 86, 674–689.

Damschen, E.I., Brudvig, L.A., Haddad, N.M., Levey, D.J., Orrock, J.L. and Tewksbury, J.J. (2008) The movement ecology and dynamics of plant communities in fragmented landscapes. *Proceedings of the National Academy of Sciences USA* 105, 19078–19083.

Daubenmire, R. (1972) Phenology and other characteristics of tropical semi-deciduous forest in north-western Costa Rica. *Journal of Ecology* 60, 147–170.

Davies, A.G., Bennett, E.L. and Waterman, P.G. (1988) Food selection by two south-east Asian colobine monkeys (*Presbytis rubicunda* and *Presbytis melalophos*) in relation to plant chemistry. *Biological Journal of the Linnean Society* 34 B, 33–56.

Davies, S.J.J.F. (1976) Studies on the flowering season and fruit production of some arid zone shrubs and trees in Western Australia. *Journal of Ecology* 64, 665–687.

Davis, D.E. (1945) The annual cycle of plants, mosquitoes, birds, and mammals in two Brazilian forests. *Ecological Monographs* 15, 245–295.

Debussche, M. (1988) La diversité morphologique des fruits charnus en Languedoc méditerranéen: relations avec les caracteristiques biologiques et la distribution des plantes, et avec les dissemi-nateurs. *Acta Oecologica, Oecologia Plantarum* 9, 37–52.

Debussche, M. and Isenmann, P. (1989) Fleshy fruit characters and the choices of bird and mammal seed dispersers in a Mediterranean region. *Oikos* 56, 327–338.

Debussche, M. and Isenmann, P. (1992) A Mediterranean bird disperser assemblage: composition and phenology in relation to fruit availability. *Revue d'Ecologie* 47, 411–432.

Debussche, M. and Isenmann, P. (1994) Bird-dispersed seed rain and seedling establishment in patchy Mediterranean vegetation. *Oikos* 69, 414–426.

Debussche, M., Escarré, J. and Lepart, J. (1982) Ornithochory and plant succession in Mediterranean abandoned orchards. *Vegetatio* 48, 255–266.

Debussche, M., Cortez, J. and Rimbault, I. (1987) Variation in fleshy fruit composition in the Mediterranean region: the importance of ripening season, life-form, fruit type, and geographical distribution. *Oikos* 49, 244–252.

Decoux, J.P. (1976) Régime, comportement alimentaire et regulation écologique du Metabolisme chez Colius striatus. *La Terre et la Vie* 30, 395–420.

De Foresta, H., Charles-Dominique, P. and Erard, C. (1984) Zoochorie et premières stades de la régénération naturelle aprés coupe en forêt guyanaise. *Revue d'Ecologie (Terre Vie)* 39, 369–400.

Dennis, A., Schupp, E.W., Green, R. and Westcott, D. (eds) (2007) *Seed Dispersal Theory and its Application in a Changing World*. CAB International, Wallingford, UK.

Denslow, J.S. (1987) Fruit removal rates from aggregated and isolated bushes of the red elderberry, *Sambucus pubens*. *Canadian Journal of Botany* 65, 1229–1235.

Denslow, J.S. and Moermond, T.C. (1982) The effect of fruit accessibility on rates of fruit removal from tropical shrubs: an experimental study. *Oecologia (Berlin)* 54, 170–176.

Desselberger, H. (1931) Der verdauungskanal der Dicaeiden nach gestalt und funktion. *Journal für Ornithologie* 79, 353–374.

Dinerstein, E. (1986) Reproductive ecology of fruit bats and the seasonality of fruit production in a Costa Rican cloud forest. *Biotropica* 18, 307–318.

Dinerstein, E. and Wemmer, C.M. (1988) Fruits rhinoceros eat: dispersal of *Trewia nudiflora* (Euphorbiaceae) in lowland Nepal. *Ecology* 69, 1768–1774.

Docters van Leeuwen, W.M. (1954) On the biology of some Loranthaceae and the role birds play in their life-history. *Beaufortia* 4, 105–208.

Donoghue, M.J. (1989) Phylogenies and the analysis of evolutionary sequences, with examples from seed plants. *Evolution* 43, 1137–1156.

Dowsett-Lemaire, F. (1988) Fruit choice and seed dissemination by birds and mammals in the evergreen forests of Upland Malawi. *Revue d'Ecologie (Terre Vie)* 43, 251–286.

Dubost, G. (1984) Comparison of the diets of frugivorous forest ruminants of Gabon. *Journal of Mammalogy* 65, 298–316.

Erard, C. and Sabatier, D. (1988) Rôle des oiseaux frugivores terrestres dans la dinamique forestière en Guyane française. In: Ouellet, H. (ed.) *Acta XIX Congressus Internationalis Ornithologici*, Ottawa, Canada, pp. 803–815.

Erard, C., Théry, M. and Sabatier, D. (1989) Régime alimentaire de *Rupicola rupicola* (Cotingidae) en Guyane française. Relations avec la frugivorie et la zoochorie. *Revue d'Ecologie (Terre Vie)* 44, 47–74.

Eriksson, K. and Nummi, H. (1982) Alcohol accumulation from ingested berries and alcohol metabolism in passerine birds. *Ornis Fennica* 60, 2–9.

Eriksson, O. and Ehrlén, J. (1991) Phenological variation in fruit characteristics in vertebrate-dispersed plants. *Oecologia* 86, 463–470.

Estrada, A. and Fleming, T.H. (eds) (1986) *Frugivores and Seed Dispersal*, Dr. W. Junk Publishers, Dordrecht, the Netherlands.

Estrada, A., Coates-Estrada, R. and Vázquez-Yanes, C. (1984) Observations on fruiting and dispersers of *Cecropia obtusifolia* at Los Tuxtlas, Mexico. *Biotropica* 16, 315–318.

Fenner, M. (1998) The phenology of growth and reproduction in plants. *Perspectives in Plant Ecology, Evolution and Systematics* 1, 78–91.

Fitzpatrick, J.W. (1980) Foraging behavior of neotropical tyrant flycatchers. *Condor* 82, 43–57.

Fleming, T.H. (1982) Foraging strategies of plant-visiting bats. In: Kunz, T.H. (ed.) *Ecology of Bats*. Plenum Press, New York, pp. 287–325.

Fleming, T.H. (1986) Opportunism versus specialization: the evolution of feeding strategies in frugivorous bats. In: Estrada, A. and Fleming, T.H. (eds) *Frugivores and Seed Dispersal*. Dr W. Junk Publishers, Dordrecht, the Netherlands, pp. 105–118.

Fleming, T.H. (1988) *The Short-tailed Fruit Bat. A Study in Plant-Animal Interactions*. University of Chicago Press, Chicago.

Fleming, T.H. and Estrada, A. (eds) (1993) *Frugivory and Seed Dispersal: Ecological and Evolutionary Aspects*. Kluwer Academic Publ., Dordrecht, the Netherlands.

Forbes, W.A. (1880) Contributions to the anatomy of passerine birds. Part I. On the structure of the stomach in certain genera of tanagers. *Proceedings of the Zoological Society, London* 188, 143–147.

Foster, M.S. (1977) Ecological and nutritional effects of food scarcity on a tropical frugivorous bird and its fruit source. *Ecology* 58, 73–85.

Foster, M.S. (1978) Total frugivory in tropical passerines: a reappraisal. *Tropical Ecology* 19, 131–154.

Foster, M.S. (1987) Feeding methods and efficiencies of selected frugivorous birds. *Condor* 89, 566–580.

Foster, M.S. and McDiarmid, R.W. (1983) Nutritional value of the aril of *Trichilia cuneata*, a bird-dispersed fruit. *Biotropica* 15, 26–31.

Foster, R.B. (1982) Famine on Barro Colorado Island. In: Leigh Jr., E.G., Rand, E.S. and Windsor, D. (eds) *The Ecology of a Tropical Forest*. Smithsonian Institution Press, Washington, D.C., pp. 201–212.

Fragoso, J.M.V. (1997) Tapir-generated seed shadows: scale-dependent patchiness in the Amazon rain forest. *Journal of Ecology* 85, 519–529.

Frankie, G.W., Baker, H.G. and Opler, P.A. (1974a) Comparative phenological studies in tropical wet and dry forests in the lowlands of Costa Rica. *Journal of Ecology* 62, 881–919.

Frankie, G.W., Baker, H.G. and Opler, P.A. (1974b) Tropical plant phenology: applications for studies in community ecology. In: Lieth, H. (ed.) *Phenology and Seasonality Modelling*. Ecological Studies vol. 8. Springer Verlag, Berlin, Germany, pp. 287–296.

Franklin, J.F., Maeda, T., Ohsumi, Y., Matsui, M., Yagi, H. and Hawk, G.M. (1979) Subalpine coniferous forests of central Houshu, Japan. *Ecological Monographs* 49, 311–334.

French, K. (1992) Phenology of fleshy fruits in a wet sclerophyll forest in Southeastern Australia: are birds an important influence? *Oecologia* 90, 366–373.

Frost, P.G.H. (1980) Fruit-frugivore interactions in a South African coastal dune forest. In: Noring, R. (ed.) *Acta XVII Congressus Internationalis Ornithologici Deutsche Ornithologen Gesellschaft*, Berlin, pp. 1179–1184.

Fuentes, M. (1992) Latitudinal and elevational variation in fruiting phenology among western European bird-dispersed plants. *Ecography* 15, 177–183.

Funakoshi, K., Watanabe, H. and Kunisaki, T. (1993) Feeding ecology of the northern Ryukyu fruit bat, *Pteropus dasymallus dasymallus*, in a warm-temperate region. *Journal of Zoology* 230, 221–230.

Galetti, M., Pizo, M.A. and Morellato, P. (2011) Diversity of functional traits of fleshy fruits in a species-rich Atlantic rain forest. *Biota Neotropica* 11, 1–14.

Garber, P.A. (1986) The ecology of seed dispersal in two species of callitrichid primates (*Sanguinus mystax* and *Sanguinus fuscicollis*). *American Journal of Primatology* 10, 155–170.

García, C. and Grivet, D. (2011) Molecular insights into seed dispersal mutualisms driving plant population recruitment. *Acta Oecologica* 37, 632–640.

García, D. and Ortiz-Pulido, R. (2004) Patterns of resource tracking by avian frugivores at multiple spatial scales: two case studies on discordance among scales. *Ecography* 27, 187–196.

García, D., Zamora, R. and Amico, G.C. (2010) Birds as suppliers of seed dispersal in temperate ecosystems: Conservation guidelines from real-world landscapes. *Conservation Biology* 24, 1070–1079.

García, D., Zamora, R. and Amico, G.C. (2011) The spatial scale of plant-animal interactions: effects of resource availability and habitat structure. *Ecological Monographs* 81, 103–121.

Gardner, A.L. (1977) Feeding habits. In: Baker, R.J., Knox-Jones, J. and Carter, D.C. (eds) Biology of bats in the New World family Phyllostomatidae. Part II. *Special Publication, Museum Texas Tech. University* No.13, Texas, pp. 295–328.

Garwood, N.C. (1983) Seed germination in a seasonal tropical forest in Panama: a community study. *Ecological Monographs* 53, 159–181.

Gautier-Hion, A. (1984) La dissémination des graines par les cercopithecidés forestiers africains. *Revue d'Ecologie (Terre Vie)* 39, 159–165.

Gautier-Hion, A., Gautier, J.P. and Quris, R. (1981) Forest structure and fruit availability as complementary factors influencing habitat use of monkeys (*Cercopithecus cephus*). *Revue d'Ecologie (Terre Vie)* 35, 511–536.

Gautier-Hion, A., Duplantier, J.M., Emmons, L., Feer, F., Heckestweiler, P., Moungazi, A., Quris, R. and Sourd, C. (1985a) Coadaptation entre rythmes de fructification et frugivorie en forêt tropicale humide du Gabon: mythe ou realité. *Revue d'Ecologie (Terre Vie)* 40, 405–434.

Gautier-Hion, A., Duplantier, J.M., Quris, R., Feer, F., Sourd, C., Decoux, J.P., Dubost, G., Emmons, L., Erard, C. and Hecketsweiler, P. (1985b) Fruit characters as a basis of fruit choice and seed dispersal in a tropical forest vertebrate community. *Oecologia (Berlin)* 65, 324–337.

Gentry, A.H. (1982) Patterns of neotropical plant species diversity. In: Hecht, M.K., Wallace, B. and Prance, G.T. (eds) *Evolutionary Biology*, vol. 15. Plenum Press, New York, USA, pp. 1–84.

Givnish, T.J. (1980) Ecological constraints on the evolution of breeding systems in seed plants: dioecy and dispersal in gymnosperms. *Evolution* 34, 959–972.

Gorchov, D.L. (1990) Pattern, adaptation, and constraint in fruiting synchrony within vertebrate-dispersed woody plants. *Oikos* 58, 169–180.

Green, R.J. (1993) Avian seed dispersal in and near subtropical rainforests. *Wildlife Research* 20, 535–537.

Guimarães Jr., P.R., Galetti, M. and Jordano, P. (2008) Seed dispersal anachronisms: Rethinking the fruits extinct megafauna ate. *PLoS ONE* 3, e1745.

Guitián, J. (1984) Ecología de una comunidad de Passeriformes en un bosque montano de la Cordillera Cantábrica Occidental. Unpublished Ph. D. Thesis, Universidad de Santiago, Santiago.

Halls, L.K. (1973) Flowering and fruiting of southern browse species. *US Department of Agriculture Forest Service Research Paper* SO-90, 10 pp.

Harding, R.S.O. (1981) An order of omnivores: nonhuman primate diets in the wild. In: Harding, R.S.O. and Teleki, G. (eds) *Omnivorous Primates. Gathering and Hunting in Human Evolution*. Columbia University Press, New York, USA, pp. 191–214.

Heiduck, S. (1997) Food choice in masked titi monkeys (*Callicebus personatus melanochir*): selectivity or opportunism? *International Journal of Primatology* 18, 487–502.

Heithaus, E.R., Stashko, E. and Anderson, P.K. (1982) Cumulative effects of plant-animal interactions on seed production by *Bauhinia ungulata*, a neotropical legume. *Ecology* 63, 1294–1302.

Herbst, L.H. (1986) The role of nitrogen from fruit pulp in the nutrition of the frugivorous bat *Carollia perspicillata*. *Biotropica* 18, 39–44.

Herrera, C.M. (1981a) Are tropical fruits more rewarding to dispersers than temperate ones? *American Naturalist* 118, 896–907.

Herrera, C.M. (1981b) Fruit variation and competition for dispersers in natural populations of *Smilax aspera*. *Oikos* 36, 51–58.

Herrera, C.M. (1982) Seasonal variation in the quality of fruits and diffuse coevolution between plants and avian dispersers. *Ecology* 63, 773–785.

Herrera, C.M. (1984a) Adaptation to frugivory of Mediterranean avian seed dispersers. *Ecology* 65, 609–617.

Herrera, C.M. (1984b) Habitat-consumer interactions in frugivorous birds. In: Cody, M. L. (ed.) *Habitat Selection in Birds*. Academic Press, New York, USA, pp. 341–365.

Herrera, C.M. (1984c) A study of avian frugivores, bird-dispersed plants, and their interaction in Mediterranean scrublands. *Ecological Monographs* 54, 1–23.

Herrera, C.M. (1984d) Tipos morfológicos y funcionales en plantas del matorral mediterráneo del sur de España. *Studia Oecologica* 5, 7–34.

Herrera, C.M. (1987) Vertebrate-dispersed plants of the Iberian Peninsula: a study of fruit characteristics. *Ecological Monographs* 57, 305–331.

Herrera, C.M. (1988) The fruiting ecology of *Osyris quadripartita*: individual variation and evolutionary potential. *Ecology* 69, 233–249.

Herrera, C.M. (1992) Interspecific variation in fruit shape: allometry, phylogeny, and adaptation to dispersal agents. *Ecology* 73, 1832–1841.

Herrera, C.M. (1998) Long-term dynamics of Mediterranean fingivorous birds and fleshy fruits: a 12-year study. *Ecological Monographs* 68, 511–538.

Herrera, C.M. and Jordano, P. (1981) *Prunus mahaleb* and birds: the high efficiency seed dispersal system of a temperate fruiting tree. *Ecological Monographs* 51, 203–221.

Herrera, C., Jordano, P., Lopez-Soria, L. and Amat, J. (1994) Recruitment of a mast-fruiting, bird-dispersed tree: bridging frugivore activity and seedling establishment. *Ecological Monographs* 64, 315–344.

Hilty, S.L. (1977) Food supply in a tropical frugivorous bird community. Unpublished Ph. D. Thesis, University of Arizona, USA.

Hilty, S.L. (1980) Flowering and fruiting periodicity in a premontane rain forest in pacific Colombia. *Biotropica* 12, 292–306.

Hladik, A. (1978) Phenology of leaf production in rain forest of Gabon: distribution and composition of food for folivores. In: Montgomery, G. G. (ed.) *The Ecology of Arboreal Folivores*. Smithsonian Institution Press, Washington, USA, pp. 51–71.

Hladik, C.M. (1981) Diet and the evolution of feeding strategies among forest primates. In: Harding, R.S.O. and Teleki, G. (eds) *Omnivorous Primates*. Columbia University Press, New York, USA, pp. 215–254.

Hladik, C.M. and Hladik, A. (1967) Observations sur le rôle des primates dans la dissémination des vegetaux de la forêt gabonaise. *Biologia Gabonica* 3, 43–58.

Hladik, C.M., Hladik, A., Bousset, J., Valdebouze, P., Viroben, G. and Delrot-Laval, J. (1971) Le régime alimentaire des primates de l'île de Barro-Colorado (Panama). *Folia Primatologica* 16, 85–122.

Hoffmann, A.J. and Armesto, J.J. (1995) Modes of seed dispersal in the Mediterranean regions in Chile, California, and Australia. In: Arroyo, M.T.K., Zedler, P.R. and Fox, M.D. (eds) *Ecology and Biogeography of Mediterranean Ecosystems in Chile, California and Australia*. Springer-Verlag, New York, USA, pp. 289–310.

Holl, K.D. (1998) Do bird perching structures elevate seed rain and seedling establishment in abandoned tropical pasture? *Restoration Ecology* 6, 253–261.

Hopkins, M.S. and Graham, A.W. (1989) Community phenological patterns of a lowland tropical rainforest in north-eastern Australia. *Australian Journal of Ecology* 14, 399–413.

Hoppes, W.G. (1987) Pre- and post-foraging movements of frugivorous birds in an eastern deciduous forest woodland, USA. *Oikos* 49, 281–290.

Hoppes, W.G. (1988) Seedfall pattern of several species of bird-dispersed plants in an Illinois woodland. *Ecology* 69, 320–329.

Horn, M.H., Correa, S.B., Parolin, P., Pollux, B.J.A., Anderson, J.T., Lucas, C., Widmann, P., Tjiu, A., Galetti, M. and Goulding, M. (2011) Seed dispersal by fishes in tropical and temperate fresh waters: The growing evidence. *Acta Oecologica* 37, 561–577.

Houssard, C., Escarre, J. and Romane, F. (1980) Development of species diversity in some Mediterranean plant communities. *Vegetatio* 43, 59–72.

Howe, H.F. (1980) Monkey dispersal and waste of a neotropical fruit. *Ecology* 61, 944–959.

Howe, H.F. (1981) Dispersal of neotropical nutmeg (*Virola sebifera*) by birds. *Auk* 98, 88–98.

Howe, H.F. (1983) Annual variation in a neotropical seed-dispersal system. In: Sutton, S.L., Whitmore, T.C. and Chadwick, A.C. (eds) *Tropical Rainforest: Ecology and Management.* Blackwell Scientific Publications, London, pp. 211–227.

Howe, H.F. (1986) Seed dispersal by fruit-eating birds and mammals. In: Murray, D.R. (ed.) *Seed Dispersal.* Academic Press, North Ryde, Australia, pp. 123–190.

Howe, H.F. (1989) Scatter- and clump-dispersal and seedling demography: Hypothesis and implications. *Oecologia (Berlin)* 79, 417–426.

Howe, H.F. (1993) Specialized and generalized dispersal systems: where does 'the paradigm' stand? In: Fleming, T.H. and Estrada, A. (eds) *Frugivory and Seed Dispersal: Ecological and Evolutionary Aspects.* Kluwer Academic Publishers, Dordrecht, the Netherlands, pp. 3–13.

Howe, H.F. and Estabrook, G.F. (1977) On intraspecific competition for avian dispersers in tropical trees. *American Naturalist* 111, 817–832.

Howe, H.F. and Richter, W.M. (1982) Effects of seed size on seedling size in *Virola surinamensis*. A within and between tree analysis. *Oecologia (Berlin)* 53, 347–351.

Howe, H.F. and Smallwood, J. (1982) Ecology of seed dispersal. *Annual Review of Ecology and Systematics* 13, 201–228.

Howe, H.F. and Vande Kerckhove, G.A. (1981) Removal of wild nutmeg (*Virola surinamensis*) crops by birds. *Ecology* 62, 1093–1106.

Howe, H.F., Schupp, E.W. and Westley, L.C. (1985) Early consequences of seed dispersal for a neotropical tree (*Virola surinamensis*). *Ecology* 66, 781–791.

Hudson, D.A., Levin, R.J. and Smith, D.H. (1971) Absorption from the alimentary tract. In: Bell, D.J. and Freeman, B.M. (eds) *Physiology and Biochemistry of the Domestic Fowl*, Vol. I. Academic Press, London, pp. 51–71.

Hughes, L., Westoby, M. and Johnson, A.D. (1993) Nutrient costs of vertebrate-dispersed and ant-dispersed fruits. *Functional Ecology* 7, 54–62.

Hume, I.D. (1989) Optimal digestive strategies in mammalian herbivores. *Physiological Zoology* 62, 1145–1163.

Hylander, W.L. (1975) Incisor size and diet in anthropoids with special reference to Cercopithecidae. *Science* 189, 1095–1098.

Idani, G. (1986) Seed dispersal by pygmy chimpanzees (*Pan paniscus*): a preliminary report. *Primates* 27, 441–447.

Innis, G.J. (1989) Feeding ecology of fruit pigeons in subtropical rainforests of south-eastern Queensland. *Australian Wildlife Research* 16, 365–394.

Izhaki, I. (2002) Emodin - a secondary metabolite with multiple ecological functions in higher plants. *New Phytologist* 155, 205–217.

Izhaki, I. and Safriel, U.N. (1989) Why are there so few exclusively frugivorous birds? Experiments on fruit digestibility. *Oikos* 54, 23–32.

Izhaki, I., Walton, P.B. and Safriel, U.N. (1991) Seed shadows generated by frugivorous birds in an eastern Mediterranean scrub. *Journal of Ecology* 79, 575–590.

Janson, C.H. (1983) Adaptation of fruit morphology to dispersal agents in a neotropical forest. *Science* 219, 187–189.

Janson, C.H., Stiles, E.W. and White, D.W. (1986) Selection on plant fruiting traits by brown capuchin monkeys: a multivariate approach. In: Estrada, A. and Fleming, T.H. (eds) *Frugivores and Seed Dispersal.* Dr. W. Junk Publishers, Dordrecht, the Netherlands, pp. 83–92.

Janzen, D.H. (1967) Synchronization of sexual reproduction of trees within the dry season in Central America. *Evolution* 21, 620–637.

Janzen, D.H. (1981a) Digestive seed predation by a Costa Rican Baird's tapir. *Biotropica Supplement*, 59–63.

Janzen, D.H. (1981b) *Enterolobium cyclocarpum* seed passage rate and survival in horses, Costa Rican Pleistocene seed dispersal agents. *Ecology* 62, 593–601.

Janzen, D.H. (1981c) Guanacaste tree seed-swallowing by Costa Rican range horses. *Ecology* 62, 587–592.

Janzen, D.H. (1982) Differential seed survival and passage rates in cows and horses, surrogate pleistocene dispersal agents. *Oikos* 38, 150–156.

Janzen, D.H. (1983) Dispersal of seeds by vertebrate guts. In: Futuyma, D.J. and Slatkin, M. (eds) *Coevolution.* Sinauer Associates, Sunderland, MA, USA, pp. 232–262.

Janzen, D.H. and Martin, P.S. (1982) Neotropical anachronisms: the fruits the Gomphoteres ate. *Science* 215, 19–27.

Janzen, D.H., Miller, G.A., Hackforth-Jones, J., Pond, C.M., Hooper, K. and Janos, D.P. (1976) Two Costa Rican bat-generated seed shadows of *Andira inermis* (Leguminosae). *Ecology* 57, 1068–1075.

Johnson, A.S. and Landers, J.L. (1978) Fruit production in slash pine plantations in Georgia. *Journal of Wildlife Management* 42, 606–613.

Johnson, R.A., Willson, M.F., Thompson, J.N. and Bertin, R.I. (1985) Nutritional values of wild fruits and consumption by migrant frugivorous birds. *Ecology* 66, 819–827.

Jordano, P. (1981) Alimentación y relaciones tróficas entre los paseriformes en paso otoñal por una localidad de Andalucía central. *Doñana Acta Vertebrata* 8, 103–124.

Jordano, P. (1982) Migrant birds are the main seed dispersers of blackberries in southern Spain. *Oikos* 38, 183–193.

Jordano, P. (1983) Fig-seed predation and dispersal by birds. *Biotropica* 15, 38–41.

Jordano, P. (1984) Relaciones entre plantas y aves frugívoras en el matorral mediterráneo del área de Doñana. Unpublished Ph. D. Thesis, Universidad de Sevilla, Sevilla.

Jordano, P. (1985) El ciclo anual de los paseriformes frugívoros en el matorral mediterráneo del sur de España: importancia de su invernada y variaciones interanuales. *Ardeola* 32, 69–94.

Jordano, P. (1987a) Avian fruit removal: effects of fruit variation, crop size, and insect damage. *Ecology* 68, 1711–1723.

Jordano, P. (1987b) Frugivory, external morphology and digestive system in Mediterranean sylviid warblers *Sylvia* spp. *Ibis* 129, 175–189.

Jordano, P. (1987c) Notas sobre la dieta no-insectívora de algunos Muscicapidae. *Ardeola* 34, 89–98.

Jordano, P. (1988) Diet, fruit choice and variation in body condition of frugivorous warblers in Mediterranan scrubland. *Ardea* 76, 193–209.

Jordano, P. (1989) Pre-dispersal biology of *Pistacia lentiscus* (Anacardiaceae): cumulative effects on seed removal by birds. *Oikos* 55, 375–386.

Jordano, P. (1993) Geographical ecology and variation of plant-seed disperser interactions: southern Spanish junipers and frugivorous thrushes. In: Fleming, T.H. and Estrada, A. (eds) *Frugivory and Seed Dispersal: Ecological and Evolutionary Aspects.* Kluwer Academic Publisher, Dordrecht, the Netherlands, pp. 85–104.

Jordano, P. (1995a) Frugivore-mediated selection on fruit and seed size: birds and St Lucie's cherry, *Prunus mahaleb. Ecology* 76, 2627–2639.

Jordano, P. (1995b) Angiosperm fleshy fruits and seed dispersers: a comparative analysis of adaptation and constraints in plant-animal interactions. *American Naturalist* 145, 163–191.

Jordano, P. and Herrera, C.M. (1995) Shuffling the offspring: uncoupling and spatial discordance of multiple stages in vertebrate seed dispersal. *Écoscience* 2, 230–237.

Jordano, P. and Schupp, E. (2000) Seed disperser effectiveness: The quantity component and patterns of seed rain for *Prunus mahaleb. Ecological Monographs* 70, 591–615.

Julliot, C. (1997) Impact of seed dispersal of red howler monkeys *Alouatta seniculus* on the seedling population in the understorey of tropical rain forest. *Journal of Ecology* 85, 431–440.

Kalko, E.K.V., Herre, E.A. and Handley, C.O. (1996) Relation of fig fruit characteristics to fruit-eating bats in the New and Old World tropics. *Journal of Biogeography* 23, 565–576.

Kantak, G.E. (1979) Observations on some fruit-eating birds in Mexico. *Auk* 96, 183–186.

Kaplin, B.A. and Moermond, T.C. (1998) Variation in seed handling by two species of forest monkeys in Rwanda. *American Journal of Primatology* 45, 83–101.

Karasov, W.H. (1996) Digestive plasticity in avian energetics and feeding ecology. In: Carey, C. (ed.) *Avian Energetics and Nutritional Ecology.* Chapman and Hall, New York, pp. 61–84.

Karasov, W.H. and Levey, D.J. (1990) Digestive system trade-offs and adaptations of frugivorous passerine birds. *Physiological Zoology* 63, 1248–1270.

Karr, J.R. and James, F.C. (1975) Eco-morphological configurations and convergent evolution in species and communities. In: Cody, M.L. and Diamond, J.M. (eds) *Ecology and Evolution of Communities.* Belknap Press, Cambridge, Massachusetts, USA, pp. 258–291.

Katusic-Malmborg, P. and Willson, M.F. (1988) Foraging ecology of avian frugivores and some consequences for seed dispersal in an Illinois woodlot. *Condor* 90, 173–186.

Kenward, R.E. and Sibly, R.M. (1977) A woodpigeon (*Columba palumbus*) feeding preference explained by a digestive bottle-neck. *Journal of Applied Ecology* 14, 815–826.

Kinnaird, M.F. (1998) Evidence for effective seed dispersal by the Sulawesi red-knobbed hornbill, *Aceros cassidix*. *Biotropica* 30, 50–55.

Kitamura, S., Yumoto, T., Poonswad, P., Chuailua, P., Plongmai, K., Maruhashi, T. and Noma, N. (2002) Interactions between fleshy fruits and frugivores in a tropical seasonal forest in Thailand. *Oecologia* 133, 559–572.

Ko, M.P., Corlett, R.T. and Xu, R.J. (1998) Sugar composition of wild fruits in Hong Kong, China. *Journal of Tropical Ecology* 14, 381–387.

Kochmer, J.P. and Handel, S.N. (1986) Constraints and competition in the evolution of flowering phenology. *Ecological Monographs* 56, 303–325.

Koelmeyer, K.O. (1959) The periodicity of leaf change and flowering in the principal forest communities of Ceylon. *Ceylon Forester* 4, 157–189.

Kollmann, J. and Poschlod, P. (1997) Population processes at the grassland–scrub interface. *Phytocoenologia* 27, 235–256.

Korine, C., Arad, Z. and Arieli, A. (1996) Nitrogen and energy balance of the fruit bat *Rousettus aegyptiacus* on natural fruit diets. *Physiological Zoology* 69, 618–634.

Kunz, T.H. and Diaz, C.A. (1995) Folivory in fruit-eating bats, with new evidence from *Artibeus jamaicensis* (Chiroptera: Phyllostomidae). *Biotropica* 27, 106–120.

Kunz, T.H., Braun de Torrez, E., Bauer, D., Lobova, T. and Fleming, T.H. (2011) Ecosystem services provided by bats. *Annals of the New York Academy of Sciences*, 1223, 1–38.

Ladley, J.J. and Kelly, D. (1996) Dispersal, germination and survival of New Zealand mistletoes (Loranthaceae): dependence on birds. *New Zealand Journal of Ecology* 20, 69–79.

Lambert, F. (1989a) Fig-eating by birds in a Malaysian lowland rain forest. *Journal of Tropical Ecology* 5, 401–412.

Lambert, F. (1989b) Pigeons as seed predators and dispersers of figs in a Malaysian lowland forest. *Ibis* 131, 521–527.

Lambert, J.E. (2011) Primate nutritional ecology: feeding biology and diet at ecological and evolutionary scales. In: Campbell, C., Fuentes, A., MacKinnon, K., Bearder, S. & Stumpf, R. (eds) *Primates in Perspective*. Oxford University Press, New York, USA, pp. 512–522.

Lambert, J.E. and Garber, P.A. (1998) Evolutionary and ecological implications of primate seed dispersal. *American Journal of Primatology* 45, 9–28.

Langer, P. (1986) Large mammalian herbivores in tropical forests with either hindgut- or forestomach-fermentation. *Zeitschrift für Saugetierkunde* 51, 173–187.

Lee, W.G., Grubb, P.J. and Wilson, J.B. (1991) Patterns of resource allocation in fleshy fruits of nine European tall-shrub species. *Oikos* 61, 307–315.

Leigh Jr., E.G. (1975) Structure and climate in tropical rain forest. *Annual Review of Ecology and Systematics* 6, 67–86.

Leighton, M. and Leighton, D.R. (1984) Vertebrate responses to fruiting seasonality within a Bornean rainforest. In: Sutton, S. L., Whitmore, T.C. and Chadwick, A.C. (eds) *Tropical Rainforests: Ecology and Management*. Blackwell Scientific Publications, Oxford, UK, pp. 181–209.

Levey, D.J. (1986) Methods of seed processing by birds and seed deposition patterns. In: Estrada, A. and Fleming, T.H. (eds) *Frugivores and Seed Dispersal*. Dr. W. Junk Publishers, Dordrecht, the Netherlands, pp. 147–158.

Levey, D.J. (1987) Seed size and fruit-handling techniques of avian frugivores. *American Naturalist* 129, 471–485.

Levey, D.J. (1988a) Spatial and temporal variation in Costa Rican fruit and fruit-eating bird abundance. *Ecological Monographs* 58, 251–269.

Levey, D.J. (1988b) Tropical wet forest treefall gaps and distributions of understory birds and plants. *Ecology* 69, 1076–1089.

Levey, D.J. (1990) Habitat-dependent fruiting behaviour of an understorey tree, *Miconia centrodesma*, and tropical treefall gaps as keystone habitats for frugivores in Costa Rica. *Journal of Tropical Ecology* 6, 409–420.

Levey, D.J. and Duke, G.E. (1992) How do frugivores process fruit: gastrointestinal transit and glucose absorption in cedar waxwings (*Bombycilla cedrorum*). *Auk* 109, 722–730.

Levey, D.J. and Grajal, A. (1991) Evolutionary implications of fruit-processing limitations in cedar waxwings. *American Naturalist* 138, 171–189.

Levey, D.J. and Stiles, F.G. (1992) Evolutionary precursors of long-distance migration: resource availability and movement patterns in neotropical landbirds. *American Naturalist* 140, 447–476.

Levey, D.J., Moermond, T.C. and Denslow, J.S. (1984) Fruit choice in neotropical birds: the effect of distance between fruits on preference patterns. *Ecology* 65, 844–850.

Levey, D.J., Silva, W.R. and Galletti, M. (2002) *Seed Dispersal and Frugivory: Ecology, Evolution and Conservation.* CAB International, Wallingford, UK.

Levey, D.J., Tewksbury, J.J. and Bolker, B.M. (2008) Modelling long-distance seed dispersal in heterogeneous landscapes. *Journal of Ecology* 96, 599–608.

Lieberman, D. (1982) Seasonality and phenology in a dry tropical forest in Ghana. *Journal of Ecology* 70, 791–806.

Lieberman, D., Lieberman, M. and Martin, C. (1987) Notes on seeds in elephant dung from Bia National Park, Ghana. *Biotropica* 19, 365–369.

Lieberman, M. and Lieberman, D. (1980) The origin of gardening as an extension of infra-human seed dispersal. *Biotropica* 12, 316.

Loiselle, B.A. (1990) Seeds in droppings of tropical fruit-eating birds: importance of considering seed composition. *Oecologia (Berlin)* 82, 494–500.

Loiselle, B.A. and Blake, I.G. (1991) Temporal variation in birds and fruits along an elevational gradient in Costa Rica. *Ecology* 72, 180–193.

Losos, J.B. and Greene, H.W. (1988) Ecological and evolutionary implications of diet in monitor lizards. *Biological Journal of the Linnean Society* 35, 379–407.

Mack, A.L. (1990) Is frugivory limited by secondary compounds in fruits? *Oikos* 57, 135–138.

Mack, A.L. (1993) The sizes of vertebrate-dispersed fruits: a neotropical-paleotropical comparison. *American Naturalist* 142, 840–856.

Magnan, A. (1912) Essai de morphologie stomacal en fonction du régime alimentaire chez les oiseaux. Annales des Sciences Naturelles, *Zoologie*, 9e série 15, 1–41.

Manasse, R.S. and Howe, H.F. (1983) Competition for dispersal agents among tropical trees: influences of neighbors. *Oecologia (Berlin)* 59, 185–190.

Marks, P.L. (1974) The role of pin cherry (*Prunus pennsylvanica* L.) in the maintenance of stability in northern hardwood ecosystems. *Ecological Monographs* 44, 73–88.

Marks, P.L. and Harcombe, P.A. (1981) Forest vegetation of the Big Thicket, Southeast Texas. *Ecological Monographs* 51, 287–305.

Marshall, A.G. (1983) Bats, flowers and fruit: evolutionary relationships in the Old World. *Biological Journal of the Linnean Society* 20, 115–135.

Marshall, A.G. and McWilliam, A.N. (1982) Ecological observations on Epomorphorinae fruit-bats (Megachiroptera) in West African savanna woodland. *Journal of Zoology, London* 198, 53–67.

Martin, T.E. (1985a) Resource selection by tropical frugivorous birds: integrating multiple interactions. *Oecologia (Berlin)* 66, 563–573.

Martin, T.E. (1985b) Selection of second-growth woodlands by frugivorous migrating birds in Panama: an effect of fruit size and plant density? *Journal of Tropical Ecology* 1, 157–170.

Martin, T.E. and Karr, J.R. (1986) Temporal dynamics of neotropical birds with special reference to frugivores in second-growth woods. *Wilson Bulletin* 98, 38–60.

Martínez del Rio, C. and Karasov, W.H. (1990) Digestion strategies in nectar- and fruit-eating birds and the sugar composition of plant rewards. *American Naturalist* 136, 618–637.

Martínez del Rio, C. and Restrepo, C. (1993) Ecological and behavioral consequences of digestion in frugivorous animals In: Fleming, T.H. and Estrada, A. (eds) *Frugivory and Seed Dispersal: Ecological and Evolutionary Aspects.* Kluwer Academic Publisher, Dordrecht, the Netherlands, pp. 205–216.

Mazer, S.J. and Wheelwright, N.T. (1993) Fruit size and shape: allometry at different taxonomic levels in bird-dispersed plants. *Evolutionary Ecology* 7, 556–575.

McDiarmid, R.W., Ricklefs, R.E. and Foster, M.S. (1977) Dispersal of *Stemmadennia donnell-smithii* (Apocyanaceae) by birds. *Biotropica* 9, 9–25.

McDonnell, M.J. and Stiles, E.W. (1983) The structural complexity of old field vegetation and the recruitment of bird-dispersed plant species. *Oecologia (Berlin)* 56, 109–116.

McKey, D. (1975) The ecology of coevolved seed dispersal systems. In: Gilbert, L.E. and Raven, P.H. (eds) *Coevolution of Animals and Plants.* University of Texas Press, Austin, USA, pp. 159–191.

McKey, D.B., Gartlan, J.S., Waterman, P.G. and Choo, G.M. (1981) Food selection by black colobus monkeys (*Colobus satanas*) in relation to plant chemistry. *Biological Journal of the Linnean Society* 16, 115–146.

McWilliams, S.R. and Karasov, W.H. (1998) Test of a digestion optimization model: effect of variable-reward feeding schedules on digestive performance of a migratory bird. *Oecologia* 114, 160–169.

Medway, L. (1972) Phenology of a tropical rain forest in Malaya. *Biological Journal of the Linnean Society* 4, 117–146.

Merz, G. (1981) Recherches sur la biologie de nutrition et les habitats preferés de l'éléphant de forêt, *Loxodonta africana cyclotis* Matschie. *Mammalia* 45, 299–312.

Milewski, A.V. (1982) The occurrence of seeds and fruits taken by ants versus birds in mediterranean Australia and Southern Africa, in relation to the availability of soil potassium. *Journal of Biogeography* 9, 505–516.

Milewski, A.V. and Bond, W.J. (1982) Convergence in myrmecochory in mediterranean Australia and South Africa. In: Buckley, R.C. (ed.) *Ant-plant Interactions in Australia.* Dr. W. Junk Publishers, The Hague, the Netherlands, pp. 89–98.

Milton, K.L. (1981) Food choice and digestive strategies of two sympatric primate species. *American Naturalist* 117, 496–505.

Moermond, T.C. and Denslow, J.S. (1985) Neotropical avian frugivores: patterns of behavior, morphology, and nutrition, with consequences for fruit selection. In: Buckley, P.A., Foster, M.S., Morton, E.S., Ridgely, R.S. and Buckley, F.G. (eds) *Neotropical Ornithology.* Ornithological Monographs No. 36. American Ornithologist Union, Washington, USA, pp. 865–897.

Moermond, T.C., Denslow, J.S., Levey, D.J. and Santana C.E. (1986) The influence of morphology on fruit choice in neotropical birds. In: Estrada, A. and Fleming, T.H. (eds) *Frugivores and Seed Dispersal.* Dr. W. Junk Publishers, Dordrecht, the Netherlands, pp. 137–146.

Mooney, H.A., Kummerov, J., Johnson, A.W., Parsons, D.J., Keeley, S.A., Hoffmann, A., Hays, R.I., Gilberto, J. and Chu, C. (1977) The producers – their resources and adaptive responses. In: Mooney, H.A. (ed.) *Convergent Evolution in Chile and California.* Dowden, Hutchinson and Ross, Stroudsburg, Pennsylvania, USA, pp. 85–143.

Morel, G. and Morel, M.Y. (1972) Recherches écologiques sur une savane sahelienne du Ferlo septentrional, Sénégal: l'avifaune et son cycle annuel. *La Terre et la Vie* 26, 410–439.

Morellato, L.P.C., Talora, D.C., Takahasi, A., Bencke, C.C., Romera, E.C. and Zipparro, V.B. (2000) Phenology of atlantic rain forest trees: a comparative study. *Biotropica* 32, 811–823.

Morris, J.G. and Rogers, Q.R. (1983) Nutritionally related metabolic adaptations of carnivores and ruminants. In: Margaris, N.S., Arianoutsou-Faraggitaki, M. and Reiter, R.J. (eds) *Plant, Animal and Microbial Adaptations to Terrestrial Environment.* Plenum Publishing Corporation, New York, USA, pp. 165–180.

Morrison, D.W. (1980) Efficiency of food utilization by fruit bats. *Oecologia (Berlin)* 45, 270–273.

Morton, E.S. (1973) On the evolutionary advantages and disadvantages of fruit eating in tropical birds. *American Naturalist* 107, 8–22.

Murray, K.G. (1988) Avian seed dispersal of three neotropical gap-dependent plants. *Ecological Monographs* 58, 271–298.

Muscarella, R. and Fleming, T.H. (2007) The role of frugivorous bats in tropical forest succession. *Biological Reviews* 82, 573–590.

Nagy, K.A. and Milton, K. (1979) Aspects of dietary quality, nutrient assimilation and water balance in wild howler monkeys (*Alouatta palliata*). *Oecologia (Berlin)* 39, 249–258.

Nathan, R. and Muller-Landau, H.C. (2000) Spatial patterns of seed dispersal, their determinants and consequences for recruitment. *Trends in Ecology and Evolution* 15, 278–285.

Nathan, R., Getz, W.M., Revilla, E., Holyoak, M., Kadmon, R., Saltz, D. and Smouse, P.E. (2008) A movement ecology paradigm for unifying organismal movement research. *Proceedings of the National Academy of Sciences, USA* 105, 19052–19059.

Noma, N. and Yumoto, T. (1997) Fruiting phenology of animal-dispersed plants in response to winter migration of frugivores in a warm temperate forest on Yakushima Island, Japan. *Ecological Research* 12, 119–129.

Oates, J.F. (1978) Water-plant and soil consumption by guereza monkeys (*Colobus guereza*): a relationship with minerals and toxins in the diet? *Biotropica* 10, 241–253.

Oates, J.F., Waterman, P.G. and Choo, G.M. (1980) Food selection by the South Indian leaf-monkey, *Presbytis johnii*, in relation to food chemistry. *Oecologia (Berlin)* 45, 45–56.

O'Dowd, D.J. and Gill, A.M. (1986) Seed dispersal syndromes in Australian *Acacia*. In: Murray, D.R. (ed.) *Seed Dispersal*. Academic Press, North Ryde, Australia, pp. 87–121.

Olesen, J. and Valido, A. (2004) Lizards and birds as generalized pollinators and seed dispersers of island plants. In: Fernández-Palacios, J.M. and Morici, C. (eds) *Island Ecology*. AEET-Cabildo Insular de La Palma, Tenerife, pp. 229–249.

Opler, P.A., Frankie, G.W. and Baker, H.G. (1980) Comparative phenological studies of treelet and shrub species in tropical wet and dry forests in the lowlands of Costa Rica. *Journal of Ecology* 68, 167–188.

Pannell, C.M. and Koziol, M.J. (1987) Ecological and phytochemical diversity of arillate seeds in *Aglaia* (Meliaceae): a study of vertebrate dispersal in tropical trees. *Philosophical Transactions of the Royal Society, London, Ser. B* 316, 303–333.

Parrish, J.D. (1997) Patterns of frugivory and energetic condition in Nearctic landbirds during autumn migration. *Condor* 99, 681–697.

Perea, R., Delibes, M., Polko, M., Suárez-Esteban, A. and Fedriani, J.M. (2013) Context-dependent fruit-frugivore interactions: partner identities and spatio-temporal variations. *Oikos* 122: 943–951.

Piper, J.K. (1986a) Effects of habitat and size of fruit display on removal of *Smilacina stellata* (Liliaceae) fruits. *Canadian Journal of Botany* 64, 1050–1054.

Piper, J.K. (1986b) Seasonality of fruit characters and seed removal by birds. *Oikos* 46, 303–310.

Poupon, H. and Bille, J.C. (1974) Recherches écologiques sur une savane sahelienne du Ferlo septentrional, Sénégal: influence de la secheresse de l'année 1972-1973 sur la strate ligneuse. *La Terre et la Vie* 28, 49–75.

Pratt, T.K. (1983) Diet of the dwarf cassowary *Casuarius bennetti picticollis* at Wau, Papua New Guinea. *Emu* 82, 283–285.

Pratt, T.K. (1984) Examples of tropical frugivores defending fruit-bearing plants. *Condor* 86, 123–129.

Pratt, T.K. and Stiles, E.W. (1983) How long fruit-eating birds stay in the plants where they feed: implications for seed dispersal. *American Naturalist* 122, 797–805.

Pratt, T.K. and Stiles, E.W. (1985) The influence of fruit size and structure on composition of frugivore assemblages in New Guinea. *Biotropica* 17, 314–321.

Primack, R.B. (1987) Relationships among flowers, fruits, and seeds. *Annual Review of Ecology and Systematics* 18, 409–430.

Pulliainen, E., Helle, P. and Tunkkari, P. (1981) Adaptive radiation of the digestive system, heart and wings of *Turdus pilaris, Bombycilla garrulus, Sturnus vulgaris, Pyrrhula pyrrhula, Pinicola enucleator* and *Loxia pyttyopsittacus*. *Ornis Fennica* 58, 21–28.

Putz, F.E. (1979) Aseasonality in Malaysian tree phenology. *Malaysian Forester* 42, 1–24.

Reid, N. (1989) Dispersal of mistletoes by honeyeaters and flowerpeckers: components of seed dispersal quality. *Ecology* 70, 137–145.

Reid, N. (1990) Mutualistic interdependence between mistletoes (*Amyema quandang*), and spiny-cheeked honeyeaters and mistletoebirds in an arid woodland. *Australian Journal of Ecology* 15, 175–190.

Restrepo, C. and Gómez, N. (1998) Responses of understory birds to anthropogenic edges in a neotropical montane forest. *Ecological Applications* 8, 170–183.

Rey, P.J. and Gutiérrez, J.E. (1996) Pecking of olives by frugivorous birds: a shift in feeding behaviour to overcome gape limitation. *Journal of Avian Biology* 27, 327–333.

Rey, P.J., Gutierrez, L.E., Alcantara, J. and Valera, F. (1997) Fruit size in wild olives: implications for avian seed dispersal. *Functional Ecology* 11, 611–618.

Robbins, C.T. (1983) *Wildlife Feeding and Nutrition*. Academic Press, New York, USA.

Rogers, M.E., Maisels, F., Williamson, E.A., Fernández, M. and Tutin, C.E.G. (1990) Gorilla diet in the Lopé Reserve, Gabon: a nutritional analysis. *Oecologia (Berlin)* 84, 326–339.

Russo, S.E., Portnoy, S. and Augspurger, C.K. (2006) Incorporating animal behavior into seed dispersal models: Implications for seed shadows. *Ecology* 87, 3160–3174.

Sallabanks, R. (1992) Fruit fate, frugivory, and fruit characteristics: a study of the hawthorn, *Crataegus monogyna* (Rosaceae). *Oecologia* 91, 296–304.

Santana, E. and Milligan, B.G. (1984) Behavior of toucanets, bellbirds, and quetzals feeding on lauraceous fruits. *Biotropica* 16, 152–154.

Schlesinger, W.H. (1978) Community structure, dynamics and nutrient cycling in the Okefenokee Cypress swamp forest. *Ecological Monographs* 48, 43–65.

Schupp, E.W. (1988) Seed and early seedling predation in the forest understory and in treefall gaps. *Oikos* 51, 71–78.

Schupp, E.W. (1993) Quantity, quality, and the effectiveness of seed dispersal by animals. In: Fleming, T.H. and Estrada, A. (eds) *Frugivory and Seed Dispersal: Ecological and Evolutionary Aspects.* Kluwer Academic Publishers, Dordrecht, the Netherlands, pp. 15–29.

Schupp, E.W. and Fuentes, M. (1995) Spatial patterns of seed dispersal and the unification of plant population ecology. *Écoscience* 2, 267–275.

Schupp, E.W., Jordano, P. and Gómez, J.M. (2010) Seed dispersal effectiveness revisited: a conceptual review. *New Phytologist* 188, 333–353.

Sherburne, J.A. (1972) Effects of seasonal changes in the abundance and chemistry of the fleshy fruits of northern woody shrubs on patterns of exploitation by frugivorous birds. Unpublished Ph. D. Thesis, Cornell University, USA.

Short, J. (1981) Diet and feeding behaviour of the forest elephant. *Mammalia* 45, 177–185.

Sibly, R.M. (1981) Strategies of digestion and defecation. In: Townsend, C.R. and Calow, P. (eds) *Physiological Ecology: an Evolutionary Approach to Resource Use.* Blackwell, London, pp. 109–138.

Simons, D. and Bairlein, F. (1990) The significance of seasonal frugivory in migratory garden warblers *Sylvia borin. Journal für Ornithologie* 131, 381–401.

Smith, A.J. (1975) Invasion and ecesis of bird-disseminated woody plants in a temperate forest sere. *Ecology* 56, 19–34.

Snow, B.K. (1979) The oilbirds of Los Tayos. *Wilson Bulletin* 91, 457–461.

Snow, B.K. and Snow, D.W. (1984) Long-term defence of fruit by mistle thrushes *Turdus viscivorus. Ibis* 126, 39–49.

Snow, B.K. and Snow, D.W. (1988) *Birds and Berries.* T. and A.D. Poyser, Calton, UK.

Snow, D.W. (1962a) A field study of the black and white manakin, *Manacus manacus,* in Trinidad. *Zoologica* 47, 65–109.

Snow, D.W. (1962b) A field study of the golden-headed manakin, *Pipra erythrocephala,* in Trinidad, W.I. *Zoologica* 47, 183–198.

Snow, D.W. (1962c) The natural histroy of the oilbird, *Steatornis caripensis,* in Trinidad, W.I. Part 2. Population, breeding ecology and food. *Zoologica* 47, 199–221.

Snow, D.W. (1971) Evolutionary aspects of fruit-eating by birds. *Ibis* 113, 194–202.

Snow, D.W. (1981) Tropical frugivorous birds and their food plants: a world survey. *Biotropica* 13, 1–14.

Sorensen, A.E. (1981) Interactions between birds and fruit in a temperate woodland. *Oecologia (Berlin)* 50, 242–249.

Sorensen, A.E. (1983) Taste aversion and frugivore preference. *Oecologia (Berlin)* 56, 117–120.

Sorensen, A.E. (1984) Nutrition, energy and passage time: experiments with fruit preference in European blackbirds (*Turdus merula*). *Journal of Animal Ecology* 53, 545–557.

Sourd, C. and Gautier-Hion, A. (1986) Fruit selection by a forest guenon. *Journal of Animal Ecology* 55, 235–244.

Staggemeier, V.G., Diniz-Filho, J.A.F. and Morellato, L.P.C. (2010) The shared influence of phylogeny and ecology on the reproductive patterns of Myrteae (Myrtaceae). *Journal of Ecology* 98, 1409–1421.

Stiles, E.W. (1980) Patterns of fruit presentation and seed dispersal in bird-disseminated woody plants in the eastern deciduous forest. *American Naturalist* 116, 670–688.

Stiles, E.W. and White, D.W. (1986) Seed deposition patterns: influence of season, nutrients, and vegetation structure. In: Estrada, A. and Fleming, T.H. (eds) *Frugivores and Seed Dispersal.* Dr. W. Junk Publishers, Dordrecht, the Netherlands, pp. 45–54.

Stocker, G.C. and Irvine, A.K. (1983) Seed dispersal by cassowaries (*Casuarius casuarius*) in North Queensland's rainforests. *Biotropica* 15, 170–176.

Stransky, J.J. and Halls, L.K. (1980) Fruiting of woody plants affected by site preparation and prior land use. *Journal of Wildlife Management* 44, 258–263.

Studier, E.H., Szuch, E.J., Thompkins, T.M. and Cope, V.M. (1988) Nutritional budgets in free flying birds: cedar waxwings (*Bombycilla cedrorum*) feeding on Washington hawthorn fruit (*Crataegus phaenopyrum*). *Comparative Biochemistry and Physiology* 89A, 471–474.

Sukumar, R. (1990) Ecology of the Asian elephant in southern India. II. Feeding habits and crop raiding patterns. *Journal of Tropical Ecology* 6, 33–53.

Sun, C. and Moermond, T.C. (1997) Foraging ecology of three sympatric turacos in a montane forest in Rwanda. *Auk* 114, 396–404.

Swain, T. (1979) Tannins and lignins. In: Rosenthal, G.A. and Janzen, D.H. (eds) *Herbivores. Their Interaction with Secondary Plant Metabolites.* Academic Press, New York, USA, pp. 657–682.

Tanner, E.V.J. (1982) Species diversity and reproductive mechanisms in Jamaican trees. *Biological Journal of the Linnean Society* 18, 263–278.

Terborgh, J. (1983) *Five New World Primates. A Study in Comparative Ecology.* Princeton University Press, Princeton, NJ, USA.

Tester, M., Paton, D., Reid, N. and Lange, R.T. (1987) Seed dispersal by birds and densities of shrubs under trees in arid south Australia. *Transactions of the Royal Society of South Australia* 111, 1–5.

Théry, M. and Larpin, D. (1993) Seed dispersal and vegetation dynamics at a cock-of-the-rock's lek in the tropical forest of French-Guiana. *Journal of Tropical Ecology* 9, 109–116.

Thomas, D.W. (1984) Fruit intake and energy budgets of frugivorous bats. *Physiological Zoology* 57, 457–467.

Thompson, J.N. and Willson, M.F. (1978) Disturbance and the dispersal of fleshy fruits. *Science* 200, 1161–1163.

Traveset, A. (1994) Influence of type of avian frugivory on the fitness of *Pistacia terebinthus. Evolutionary Ecology* 8, 618–627.

Traveset, A., Willson, M. and Verdú, M. (2004) Characteristics of fleshy fruits in southeast Alaska: phylogenetic comparison with fruits from Illinois. *Ecography* 27, 41–48.

Turcek, F.J. (1961) *Okologische Beziehungen der Vögel und Gehölze.* Slowakische Akademie der Wiesenschaften, Bratislava.

Tutin, C.E.G., Williamson, R.A., Rogers, M.E. and Fernandez, M. (1991) A case study of a plant-animal relationship: *Cola lizae* and lowland gorillas in the Lope Reserve, Gabon. *Journal of Tropical Ecology* 7, 181–199.

Tutin, C.E.G., Parnell, R.J. and White, F. (1996) Protecting seeds from primates: examples from *Diospyros* spp. in the Lope Reserve, Gabon. *Journal of Tropical Ecology* 12, 371–384.

van der Pijl, L. (1982) *Principles of Dispersal in Higher Plants.* Springer-Verlag, Berlin, Germany.

van Schaik, C.T., Terborgh, J.W. and Wright, S.J. (1993) The phenology of tropical forests: adaptive significance and consequences for primary consumers. *Annual Review of Ecology and Systematics* 24, 353–377.

Viljoen, S. (1983) Feeding habits and comparative feeding rates of three Southern African arboreal squirrels. *South African Journal of Zoology* 18, 378–387.

Voronov, H.R. and Voronov, P.H. (1978) Morphometric study of the digestive system of the Waxwing (*Bombycilla garrulus* L.) (Aves, Bombycillidae). *Vestny k Zoology* 5, 28–31.

Walsberg, G.E. (1975) Digestive adaptations of *Phainopepla nitens* associated with the eating of mistletoe berries. *Condor* 77, 169–174.

Wang, B. and Smith, T. (2002) Closing the seed dispersal loop. *Trends in Ecology and Evolution* 17, 379–385.

Waterman, P.G., Mbi, C.N., Mckey, D.B. and Gartlan, J.S. (1980) African rainforest vegetation and rumen microbes: phenolic compounds and nutrients as correlates of digestibility. *Oecologia (Berlin)* 47, 22–33.

Wendeln, M.C., Runkle, J.R. and Kalko, E.K.V. (2000) Nutritional values of 14 fig species and bat feeding preferences in Panama. *Biotropica* 32, 489–501.

Wenny, D.G. and Levey, D.J. (1998) Directed seed dispersal by bellbirds in a tropical cloud forest. *Proceedings of the National Academy of Sciences USA* 95, 6204–6207.

Wetmore, A. (1914) The development of stomach in the euphonias. *Auk* 31, 458–461.

Wheelwright, N.T. (1983) Fruits and the ecology of resplendent quetzals. *Auk* 100, 286–301.

Wheelwright, N.T. (1985) Fruit size, gape width, and the diets of fruit-eating birds. *Ecology* 66, 808–818.

Wheelwright, N.T. (1986) The diet of American robins: an analysis of U.S. Biological Survey records. *Auk* 103, 710–725.

Wheelwright, N.T. (1988) Seasonal changes in food preferences of American robins in captivity. *Auk* 105, 374–378.

Wheelwright, N.T. (1993) Fruit size in a tropical tree species: variation, preference by birds, and heritability. In: Fleming, T.H. and Estrada, A. (eds) *Frugivory and Seed Dispersal: Ecological and Evolutionary Aspects.* Kluwer Academic Publisher, Dordrecht, the Netherlands, pp. 163–174.

Wheelwright, N.T., Haber, W.A., Murray, K.G. and Guindon, C. (1984) Tropical fruit-eating birds and their food plants: a survey of a Costa Rican lower montane forest. *Biotropica* 16, 173–192.

White, D.W. and Stiles, E.W. (1990) Co-occurrences of foods in stomachs and feces of fruit-eating birds. *Condor* 92, 291–303.

White, D.W. and Stiles, E.W. (1991) Fruit harvesting by American robins: influence of fruit size. *Wilson Bulletin* 103, 690–692.

White, S.C. (1974) Ecological aspects of growth and nutrition in tropical fruit-eating birds. Unpublished Ph.D. Thesis, University of Pennsylvania, USA.

Whitney, K.D. and Smith, T.B. (1998) Habitat use and resource tracking by African *Ceratogymna* hornbills: implications for seed dispersal and forest conservation. *Animal Conservation* 1, 107–117.

Williamson, E.A., Tutin, C.E.G., Rogers, M.E. and Fernandez, M. (1990) Composition of the diet of lowland gorillas at Lope in Gabon. *American Journal of Primatology* 21, 265–277.

Willson, M.F. (1986) Avian frugivory and seed dispersal in eastern North America. In: Johnston, R.F. (ed.) *Current Ornithology*. Vol. 3, Plenum Press, New York, USA, pp. 223–279.

Willson, M.F. (1988) Spatial heterogeneity of post-dispersal survivorship of Queensland rainforest seeds. *Australian Journal of Ecology* 13, 137–145.

Willson, M.F., Porter, E.A. and Condit, R.S. (1982) Avian frugivore activity in relation to forest light gaps. *Caribbean Journal of Science* 18, 1–4.

Willson, M.F., Irvine, A.K. and Walsh, N.G. (1989) Vertebrate dispersal syndromes in some Australian and New Zealand plant communities, with geographic comparisons. *Biotropica* 21, 133–147.

Witmer, M.C. (1996) Annual diet of cedar waxwings based on US Biological Survey records (1885-1950) compared to diet of American robins: contrasts in dietary patterns and natural history. *Auk* 113, 414–430.

Witmer, M.C. (1998a) Ecological and evolutionary implications of energy and protein requirements of avian frugivores eating sugary diets. *Physiological Zoology* 71, 599–610.

Witmer, M.C. (1998b) Do seeds hinder digestive processing of fruit pulp? Implications for plant/frugivore mutualisms. *Auk* 115, 319–326.

Witmer, M.C. and van Soest, P.J. (1998) Contrasting digestive strategies of fruit-eating birds. *Functional Ecology* 12, 728–741.

Wood, C.A. (1924) The Polynesian fruit pigeon, *Globicera pacifica*, its food and digestive apparatus. *Auk* 41, 433–438.

Worthington, A.H. (1982) Population sizes and breeding rhythms of two species of manakins in relation to food supply. In: Leigh Jr., E.G., Rand, A.S. and Windsor, D.M. (eds) *The Ecology of a Tropical Forest. Seasonal Rhythms and Long-Term Changes*. Smithsonian Institution Press, Washington, USA, pp. 213–225.

Worthington, A.H. (1989) Adaptations for avian frugivory: assimilation efficiency and gut transit time of *Manacus vitellinus* and *Pipra mentalis*. *Oecologia (Berlin)* 80, 381–389.

Wrangham, K.W., Chapman, C.A. and Chapman, L.J. (1994) Seed dispersal by forest chimpanzees in Uganda. *Journal of Tropical Ecology* 10, 355–368.

Wrangham, R.W., Conklin-Brittain, N.L. and Hunt, K.D. (1998) Dietary responses of chimpanzees and cercopithecines to seasonal variation in fruit abundance. I. Antifeedants. *International Journal of Primatology* 19, 949–970.

Zhang, S.Y. and Wang, L.X. (1995) Comparison of three fruit census methods in French Guiana. *Journal of Tropical Ecology* 11, 281–294.

3 The Ecology of Seed Dispersal

Anna Traveset,[1]* Ruben Heleno[2] and Manuel Nogales[3]
[1]*Mediterranean Institute of Advanced Studies (CSIC-UIB), Terrestrial Ecology Group, Mallorca, Balearic Islands, Spain;* [2]*Centre for Functional Ecology (CEF-UC), Department of Life Sciences, University of Coimbra, Coimbra, Portugal;* [3]*Island Ecology and Evolution Research Group (CSIC-IPNA), Tenerife, Canary Islands, Spain*

Introduction

Seed dispersal is one of the key phases in the regeneration process of plant populations. It determines the potential area of recruitment and, simultaneously, acts as a template for the subsequent stages of plant growth. Seed dispersal is the most common means for plants to colonize new areas and to avoid sibling competition and natural enemies such as herbivores or pathogens. Seeds can be dispersed by wind, water, gravity and by a wide assemblage of animals (including those that consume fruits and/or seeds as well as those that move seeds via their fur, plumage or feet). By directly dispersing seeds to favourable recruitment sites (Wenny and Levey, 1998) or by virtue of the treatment offered to ingested seeds (Verdú and Traveset, 2004; Traveset *et al.*, 2007), animals actually play an important role as seed dispersers for most (60–80%) plant species (Levey *et al.*, 2002; Dennis *et al.*, 2007). Moreover, seed dispersers are crucial in plant-community dynamics in many ecosystems around the globe and contribute to numerous ecosystem services offered by forests, including fruit, wood and non-timber products, carbon sequestration and forest cover – at no cost to humans (Forget *et al.*, 2011).

Seed dispersal is currently a very active research area that includes both ecological and evolutionary aspects. Studies have diversified in the last few decades mainly towards the study of landscape ecology (movement patterns), plant genetic diversity and structure (gene flow through pollen and/or seeds), community ecology (e.g. mutualistic interaction networks), dispersal adaptations of both plants and seed dispersers, conservation biology (effects of different types of disturbances such as habitat fragmentation, defaunation and biological invasions) and ecological restoration. Research on seed dispersal has actually shifted from being organism oriented towards being currently more mechanism oriented in order to unravel the mechanistic processes behind seed dispersal (Jordano *et al.*, 2011). Thus, for instance, by means of increasingly precise tools, such as radio or satellite tracking, researchers can now study the movement of animal seed dispersers across habitats (Blake *et al.*, 2012), or even across continents (Kays *et al.*, 2011), and evaluate plant seed dispersal kernels (i.e. the probability density function of the dispersal distance for an individual or population). Physiological, chemical and morphological mechanisms,

---

* E-mail: atraveset@imedea.uib-csic.es

of both fruits/seeds (e.g. odour, colour, shape, presence of secondary metabolites) and frugivores (e.g. digestive physiology, foraging behaviour) are also being examined in order to provide insights into the evolution of fruit displays and animal frugivory. One of the most promising fields in seed dispersal research is the estimation of dispersal kernels by using a multidisciplinary approach that combines high-resolution tracking of frugivore movements, mechanistic models of fruit processing and seed dissemination with genetic data (DNA polymorphisms) (see Jordano *et al.*, 2007; Jones and Muller-Landau, 2008).

Although most information available is from continental systems, an increasing number of studies are being carried out on islands. Insular ecosystems can be considered as natural laboratories for the evolution of dispersal adaptations. Here, frugivore assemblages are less diverse than on the continents and often unique, especially on oceanic islands, where reptiles (lizards, iguanas and tortoises), bats and pigeons play an important role as seed dispersers (Olesen and Valido, 2004; Nogales *et al.*, 2012). However, islands are highly vulnerable to disturbances and most mutualistic disruptions have actually been reported in such insular systems (Traveset and Richardson, 2006). Current studies are being performed to assess how alien species can replace extinct disperser taxa and restore lost ecosystem functions (Foster and Robinson, 2007; Hansen *et al.*, 2010).

The goal of this chapter is to review the existing literature on seed dispersal, highlighting the essential findings so far, as well as the missing gaps of information. We first review the determinants of seed dispersal, examining the main dispersal vectors, and what determines when and where seeds are moved. Next we deal with the consequences of seed dispersal right from plant genetic structure to global biogeography, passing by the population and community levels. In the following section, we focus on the causes and risks of seed dispersal disruptions for conservation and their implications for ecological restoration. Finally, we suggest directions for future studies that we believe will contribute to deepening the understanding of this crucial phase in plant

regeneration and to its integration with other disciplines in the quest to comprehend ecosystem functioning.

The Determinants of Seed Dispersal

Seeds are highly specialized structures, which greatly increase plant success by conferring their offspring a higher probability of finding suitable recruiting sites (Howe and Smallwood, 1982). This selective pressure favours seeds that are particularly well equipped for dispersal, resulting in the vast array of elaborate solutions we see today, including plumes, hairs, wings, floating devices, nutritive tissues, sticky surfaces, hooks, etc. Nevertheless, the fate of seeds is far from certain, and the journey of seeds from when they leave the mother plant until recruitment, if achieved, can be highly eventful. Such a journey is dependent on a wide variety of factors which together will determine how, when, where, by whom and if seeds will ever be dispersed and given the chance to become established adults.

How are seeds dispersed?

The presence of numerous mechanisms of seed dispersal in nature suggests that these are well adapted to enhance dispersal (Ridley, 1930; Van der Pijl, 1982). Thus, different structures are easily interpretable in terms of favouring a certain dispersal mode (i.e. dispersal syndromes), such as the presence of flotation devices to travel by oceanic currents or rivers, diaspores with wings or plumes capable of being dispersed by air, explosive opening of fruits that project their own seeds, seeds with hooks or sticky surfaces able to hitchhike on the fur of mammals, or seeds with a nutritive reward that attract foraging animals which in turn transport the seeds in their gut. Regarding animals, most seeds are dispersed by three classes of vertebrates: birds, mammals (including humans) and reptiles. Other dispersing agents can however be quite active seed dispersers, such as ants, fish, amphibians, grasshoppers, beetles, slugs, bees and even earthworms (see Chapter 2, this volume). The relative importance of all these dispersers varies greatly according to the species of seed, but also with the local distribution, abundance and behaviour of the disperser (Fig. 3.1). More globally, the importance of the different disperser guilds varies according to the latitude and type of ecosystem; birds and mammals being crucial in temperate environments (Herrera, 1995), while reptiles take on a more important role in subtropical and tropical islands (Olesen and Valido, 2003; Nogales *et al.*, 2005) where large mammals are usually absent (Williamson, 1983; Whittaker and Fernández-Palacios, 2007) (see Box 3.1). Distinct habitats may also be related to certain dispersal mechanisms such as the high proportion of vertebrate-dispersed seeds in tropical wet forests (Willson *et al.*, 1989), and the high frequency of ant-dispersed species in sclerophyllous biomes in the southern hemisphere (Rice and Westoby, 1986). Furthermore, external dispersal on vertebrates seems to be more common along riparian corridors in arid parts of southern Africa (Sorensen, 1986).

Although morphological adaptations unequivocally increase the probability of dispersal by a certain mechanism, this does not mean that seeds cannot be dispersed by other vectors for which they are not particularly adapted (Higgins *et al.*, 2003b). Similarly, many species have simple seeds that have not developed any specific adaptation for dispersal – 'unassisted seeds' – which does not preclude these seeds being successfully dispersed (Cain *et al.*, 1998). An illustrative example of this potential is the proportion of plants with such 'unassisted' diaspores that arrived by natural means and colonized even highly remote archipelagos such as the Galápagos (Vargas *et al.*, 2012).

The different modes of dispersal generate a great variation in the dispersal potential of plants and in the relationship between dispersal mode and seed mass. This is especially evident in the genus *Pinus*; pine seeds weighing less than *c.*100 mg tend to be wind dispersed whereas heavier seeds usually have developed adaptations for bird dispersal (Benkman, 1995). Also, the size and chemical composition of the edible

appendage on ant-dispersed seeds influences the rate of seed removal, the array of dispersing ant species, and consequently the pattern of dispersal and the eventual fate of the seeds (Gorb and Gorb, 1995; Mark and Olesen, 1996). Even closely related species can display very different adaptations for dispersal. One of the most illustrative examples is found in the widely spread genus *Acacia*, with Australian species exhibiting morphological adaptations for dispersal by ants or birds (Willson *et al.*, 1989), American species adapted to dispersal by birds and/or large mammals (O'Dowd and Gill, 1986), and African species mostly adapted to dispersal by large mammals (Dudley, 1999).

Seed size is another important factor that influences the seed dispersal mode (Westoby *et al.*, 1996). Intuitively, larger seeds are generally more limited in terms of dispersal than smaller ones, because they cannot be dispersed by mechanisms such as ballistochory or by adhering to the animal exterior, and must be disseminated by larger animals, or less commonly by water currents and gravity (Foster and Janson, 1985; Wheelwright, 1985).

Plant habit also differentially affects the efficiency of particular dispersal mechanisms and therefore seed dispersal syndromes are frequently related to plant growth form. For example, the seeds of high trees are more frequently adapted to wind dispersal, while small plants may more often use ballistic mechanisms (Thompson and Rabinowitz, 1989; Willson *et al.*, 1990).

So far we have considered seed dispersal as a one-way ticket, however one seed can be transported sequentially by several processes, which can greatly affect the final outcome – a phenomenon named diplochory or secondary seed dispersal (reviewed in Vander Wall and Longland, 2004) (Fig. 3.2). Although diplochory is considerably less well studied than 'traditional' single-vector dispersal, due to the understandable logistical challenges, it is likely to be essential in the reproduction of some plants (Chambers and Macmahon, 1994). Secondary seed dispersal systems are widely variable because they depend on the potential combinations

Fig. 3.1. Different stages of the seed dispersal process. (a) A seed of *Ipomoea pes-caprae* recruiting in a Galápagos beach after having been dispersed by thalassochory (oceanic drift). (b) The anemochorous seeds of *Asclepias curassavica* ready for being dispersed by wind. (c) The Canarian lizard (*Gallotia galloti*) dispersing the seeds of *Rubia fruticosa* in the Canary Islands. (d) A Southern grey shrike (*Lanius meridionalis*) preying upon a frugivorous lizard (*G. galloti*) and acting as a secondary dispersal for the seeds in its gut. (e) Seeds of the tree *Hippomane mancinella* recruiting in a dung pile of the Galápagos giant tortoise (*Chelonoidis nigra*). (f) The blue tit (*Cyanistes teneriffae*), an essentially insectivorous bird, consuming and dispersing the seeds of *Opuntia maxima*. (g) The granivorous large beak ground finch (*Geospiza magnirostris*) handling and dispersing the hard seed of *Cordia lutea*. Photo credits: (a), (b), (e), (g) – Ruben Heleno; (c) – Beneharo Rodríguez; (d) – Gustavo Peña; (f) – José Juan Hernández.

Box 3.1. Seed dispersal on islands: main differences with mainland environments

The composition of island biota is the reflection of biogeographical, ecological and evolutionary processes that have simultaneously occurred since their formation (see Whittaker and Fernández-Palacios, 2007 and references therein). These insular communities are frequently considered disharmonic in relation to the species composition of the mainland source(s) due to the differential colonization capacity of organisms, often expressed in the different composition of plant families (Gulick, 1932). They are characterized by a low diversity of plants and animals, a high proportion of endemic species – especially oceanic islands – and the presence of relict species (i.e. ancient evolutionary lineages) (Carlquist, 1974). Therefore, the new species that reach and become established on islands clearly force a new order of interactions, which are particular with respect to mainland environments.

    While birds and mammals are the most common dispersers of fleshy-fruited plants in mainland ecosystems, in many tropical and subtropical oceanic islands (the most biodiverse), birds and reptiles play a more determinant role (Nogales et al., 2005). The community of seed dispersers on islands is often characterized by the absence of medium and large mammals, except for some large bats, and the existence of 'unusual' dispersers, which include a large array of reptiles such as tortoises, iguanas and lizards (Olesen and Valido, 2003; McConkey and Drake, 2006, 2007; Whittaker and Fernández-Palacios, 2007; Blake et al., 2012).

    At the biogeographical and ecological levels, this insular scenario promotes the emergence of unusual interactions, which are more frequent on islands than in mainland areas. Some of these interactions involve: (i) large omnivorous birds (e.g. gulls or corvids), not typically frugivorous, which frequently broaden their diet – a phenomenon known as niche expansion on islands (Wright, 1980) – and include fleshy fruits, acting as potentially long-distance dispersers (Nogales et al., 1999, 2001); (ii) 'seed predator birds' which are often found to be legitimate seed dispersers (Guerrero and Tye, 2009; Heleno et al., 2011; Young et al., 2012); and (iii) predatory birds which may frequently act as secondary seed dispersers when preying upon frugivorous lizards and birds (Nogales et al., 1998, 2007; Padilla et al., 2012). When fruits and seeds are dispersed by a mechanism other than that to which they are particularly well adapted, i.e. a non-standard dispersal mechanism (Higgins et al., 2003b), such events have often been classified as stochastic (e.g. Clark et al., 2001). However, dismissing such means of dispersal as attributable to chance alone might oversimplify the importance of deterministic, but poorly understood, processes. Such 'stochastic' mechanisms might actually acquire a major biogeographical and ecological importance if they occurred regularly and in a number of oceanic archipelagos (Nogales et al., 2012). However, despite many of these dispersal mechanisms involving regular events repeated year after year, our knowledge to evaluate their real importance in the context of islands worldwide is still incomplete.

    In recent years, the use of interaction networks has grown popular as a framework to compare community structure and function (see Box 3.2, this chapter). Some general differences have been identified between mainland and island pollination networks (Olesen and Jordano, 2002), but sufficient data are so far lacking to permit generalizations about differences in seed dispersal networks. The first study comparing the structure of seed dispersal networks between island and mainland habitats has been made by González-Castro et al. (2012). This compares the same types of habitat (Mediterranean shrublands) in the Canary and Balearic Islands with those on the European continent (Southern Iberian Peninsula); studying the same types of habitat better allows us to infer differences due to the insularity effect. Results show that island networks are smaller (less diverse), more highly connected, and have a more asymmetric proportion of plant/animal species and a lower relative nestedness. The lower species richness and higher specialization on islands appear to promote the prevalence of more symmetric interaction frequencies than those found on the mainland (see also Box 3.2, this chapter).

of many possible abiotic and biotic dispersal agents. In some cases, the first dispersal phase is mediated by an abiotic mechanism (e.g. wind or water) whilst the second is mediated by scatter-hoarding vertebrates (Vander Wall, 2002) or invertebrates (Pizo et al., 2005). In other cases, primary seed dispersal includes a first process of endozoochory

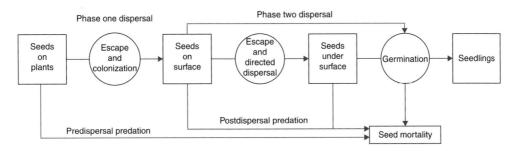

Fig. 3.2. The potential benefits of diplochory and the possible fate of seeds from when they are produced until they die or otherwise germinate and establish as seedlings. Rectangles represent physical states whereas circles symbolize the most important advantages resulting from seed dispersal. (From Vander Wall and Longland, 2004.)

(involving mammals or birds) followed by secondary dispersal by invertebrates such as ants (Christianini and Oliveira, 2010) or beetles (Santos-Heredia *et al.*, 2010).

A special type of diplochory occurs when predatory vertebrates prey upon frugivores that have recently consumed fruits, ingesting the entire seed load contained within the digestive tract of their prey. Although this phenomenon has been described for some time (Damstra, 1986), only a few studies have assessed its importance in an ecological context (Nogales *et al.*, 1998, 2007). In the Canary archipelago, for instance, this process appears to be highly important, involving a total of 78 plant species found in shrike (*Lanius meridionalis*) and kestrel (*Falco tinnunculus*) pellets (Padilla *et al.*, 2012). The large number of seeds achieving dispersal by this secondary mechanism suggests that even if representing a non-standard dispersal mechanism, it can be ecologically significant for increasing the dispersal distance of many native and endemic insular plants. While such dispersal by two vertebrate systems is likely to also occur on continents, it is not as generalized as the secondary dispersal due to a conjunction of abiotic and biotic vectors.

When are seeds dispersed?

Dispersal phenology is influenced by several ecological factors. From the plant's viewpoint, fruit development and seed maturation should be timed to match with the seasonal availability of legitimate dispersal agents and suitable environmental conditions for dispersal and plant establishment. Some authors argue that differences in flowering phenology of tropical forests are primarily caused by abiotic factors (e.g. climatic: water or light), whereas differences in fruiting phenology are mostly influenced by biotic ones (e.g. presence of seed dispersers) (Thies and Kalko, 2004); however, few clear general patterns have emerged so far. Furthermore, wind-dispersed neotropical trees often release their seeds during the dry season, when trade winds are stronger and trees are leafless (Foster, 1982; Morellato and Leitão, 1996), while zoochorous dispersal usually occurs during the rainy season (Griz and Machado, 2001). Fleshy fruits in northern temperate latitudes mature during the late summer and autumn, coinciding with a high abundance of frugivorous birds (Willson and Thompson, 1982; Snow and Snow, 1988). On the other hand, in southern temperate latitudes fleshy fruits mature in winter, coinciding with the presence of foraging flocks of migrant birds (Herrera, 1995). While in many areas fruit and bird phenology tend to coincide (de Castro *et al.*, 2012), wide-ranging plant species show no latitudinal shift in fruiting times, as would be expected if their fruiting seasons were 'adapted' to disperser timing (Willson and Whelan, 1993); thus, patterns may not always be entirely interpretable.

Where are seeds dispersed?

The spatial distribution of seeds with respect to their source plant is called the seed shadow (Janzen, 1971); it marks the end of the dispersal stage and sets the template for future processes that might eventually lead to plant recruitment. This simple descriptor of shape is generally augmented by information on seed density per distance class, forming the dispersal kernel. Although directionality is frequently asymmetric in respect to the source and clearly significant for many ecological questions (Cain *et al.*, 2000), it is usual to discuss seed distributions chiefly in terms of the relationship between seed density and distance from the source (i.e. dispersal curve). The majority of studies of dispersal curves fit a unimodal leptokurtic distribution (with a higher peak and a longer 'fatter' tail than a normal distribution) (Willson, 1993; Kot *et al.*, 1996). However, dispersal curves can form any kind of distribution, including multimodal distributions (Schupp, 1993). Different species of dispersers generate characteristic seed shadows, depending on foraging behaviour, seed retention times, secondary seed dispersal, patterns of fruit selection and disperser responses to vegetation structure and other ecological cues (Herrera, 1995; Borges *et al.*, 2011; Guttal *et al.*, 2011). Yet other factors which determine the final shape of the seed shadow are climatic, e.g. wind, rainfall or humidity, and intrinsic characteristics of the mother plant, e.g. height, ballistic mechanisms and of course diaspore morphology.

While wind-dispersed seeds can travel remarkable distances (Nathan *et al.*, 2002), on average it seems that animal-dispersed seeds travel longer distances (Clark *et al.*, 2005). Regarding animal-dispersed seeds, different vertebrates (e.g. birds and bats), produce different seed shadows; for example, Thomas *et al.* (1988) described that birds generated a strongly skewed seed shadow towards open savanna, while bats produced a similar-shaped shadow but oriented to the nearest forest edge.

A critical component of the seed shadow is its size, which is determined by the maximum travelling distance of seeds. Because such long-distance dispersal events tend to be rare, they are increasingly difficult to study. Different methods have been applied to study long-distance dispersal such as mechanistic models, often used in wind-dispersal systems (e.g. Nathan and Muller-Landau, 2000; Katul *et al.*, 2005; Nathan *et al.*, 2011), or more recently, genetic methods have become a highly active and diverse field of research including assignment likelihood, genealogical and demographic methods (e.g. Godoy and Jordano, 2001; Jordano *et al.*, 2007; Jones and Muller-Landau, 2008) (see Fig. 3.3). Such methods point to a close relationship between the frequency of long-distance dispersal events and the amount of genetic diversity preserved during colonization (Bialozyt *et al.*, 2006).

Mechanistic models in wind dispersal applied in open habitats indicate that autocorrelated turbulent fluctuations in vertical wind speed are key factors for long-distance dispersal (LDD) (Tackenberg, 2003; Soons *et al.*, 2004). However, abscission is often controlled by horizontal wind speed and therefore this is another important factor that needs to be successfully modelled (Greene, 2005). With regards to the plant biotype, windy and stormy weather conditions are of much less overall importance in non-tree species (Tackenberg, 2003). However, according to this last author, the number of seeds dispersed on a long-distance scale is not correlated to horizontal wind speed but to frequency of updrafts (especially thermals). Wind dispersal distances are increased for lighter seeds and probably more importantly by plant height (Thomson *et al.*, 2011).

Animal seed-dispersal events are often grouped into loose categories regarding the distance travelled by seeds, such as short-, medium- and long-distance seed dispersal. However, no general scale has yet been proposed that can be applied more consistently across studies. While it is clear that each system requires its own ecological interpretation of dispersal distances, we consider that it is useful to define some guiding classes in order to facilitate inter-study comparisons. According to published information we suggest the following categories: (i) short-distance dispersal

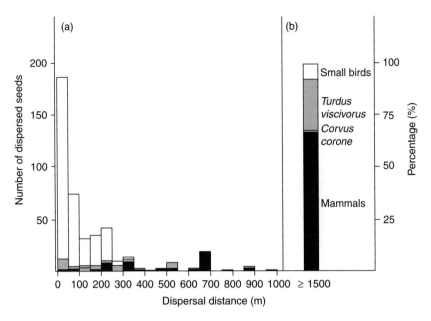

Fig. 3.3. Seed dispersal kernel of the tree *Prunus mahaleb* in Southern Spain. (a) Shows the relative contribution of major disperser guilds to different distance classes. Open bars, small- to medium-sized frugivorous birds; light grey, *Turdus viscivorus*; dark grey, *Corvus corone*; black, carnivorous mammals. (b) Shows the weighted contribution of each dispersal guild to seed immigration to the study population (dispersal distances ≥1,500 m). (From Jordano *et al.*, 2007.)

(SDD) when seeds travel less than 25 m; (ii) medium-distance dispersal (MDD), from 25 m to 250 m; (iii) long-distance dispersal (LDD), from 250 m to 10,000 m; and (iv) very long-distance dispersal (VLDD), for distances over 10,000 m. Although these categories are useful as a first approximation, we envisage that a careful analysis of dispersal distances by different animal groups will help to increase precision in the limits of each category. Precise distances currently obtained by means of molecular analysis and radio and satellite tracking will help this purpose. Such an analysis will allow a better understanding of the spatial scale at which different disperser guilds disperse seeds, and evaluate the relationship between dispersers' body size and dispersal distances.

Most precise assessments of long-distance dispersal, including considerations on animal movement and retention times, have been measured in birds and mammals (Guttal *et al.*, 2011). Levey *et al.* (2008) modelled seed dispersal by terrestrial birds and showed that the dispersal kernel was uniform in homogeneous landscapes and irregular in heterogeneous ones. In both environments, dispersal distances >150 m made up *c.*50% of all dispersal events. Gómez (2003) recorded dispersal distances of over 250 m, and up to 1 km for *Quercus ilex* acorns dispersed by Eurasian jays. In this respect, some of the long-distance movements for a large avian frugivore were up to 290 km recorded for rainforest hornbills in Cameroon (Holbrook *et al.*, 2002). Based on DNA-genotyping techniques on *Prunus mahaleb*, Jordano *et al.* (2007) recorded that small passerines dispersed most seeds over short distances (50% dispersed <51 m from source trees), while mammals and medium-sized birds dispersed seeds over long distances (50% of mammals: <495 m, and 50% of medium-sized birds: <110 m) (Fig. 3.3). In some Amazonian fish (*Colossoma macropomum*), at least 5% of seeds are estimated to be dispersed around 1700–2110 m (Anderson *et al.*, 2011).

Most of our knowledge on animal seed dispersal is focused on the internal dispersal of fruits (endozoochory) by birds and mammals or the external movement of seeds with elaiosomes by ants. The vast majority of external transport of seeds with adhesion structures (i.e. epizoochory) has been described in large vertebrates (Bullock *et al.*, 2011). An interesting study described the transport of seeds on the fur of 'transhumant' migrating sheep for several hundred kilometres (Manzano and Malo, 2006). Long-distance dispersal by epizoochory in birds is frequently mentioned as a mechanism with likely important ecological and biogeographical consequences (such as island colonization); however, there is almost no information on this phenomenon which might be largely confined to water birds (Figuerola and Green, 2002; Nogales *et al.*, 2012). Recent evidence suggests that in migrating passerines in Portugal, endozoochory is at least two orders of magnitude more common than epizoochory (Costa *et al.*, in press).

Consequences of Seed Dispersal

Although the advantages of seed dispersal for plant reproductive fitness and vegetation structure have been largely accepted, quantifying such effects has proven methodologically challenging (Wang and Smith, 2002; Levine and Murrell, 2003; García *et al.*, 2005). This is due to (i) the complexity of the process, which involves the simultaneous dispersal of many different seeds by different vectors to different distances and in all directions; (ii) a practical difficulty in measuring seed movement, particularly at longer distances; (iii) the difficulty in correctly monitoring the variability of each step of the process in both space and time; and (iv) problems in tracking the effects of seed dispersal up to the establishment of long-lived adult plants (Herrera, 1998; Levey and Benkman, 1999; Wang and Smith, 2002). Despite such challenges, a growing interest in seed dispersal and its consequences for natural communities has resulted in significant advances in the field towards providing answers to old and new hypotheses, which are constantly being put to the test with rapidly emerging empirical and theoretical evidence.

Seed rain and dispersal effectiveness

The output of seed dispersal sets the spatial framework upon which seeds have to survive, germinate and recruit in order to become established as reproductive adults incorporated into local communities. Therefore, the seed shadow, i.e. the spatial distribution of dispersed seeds around their parent plant, and seed rain, i.e. the sum of all conspecific seed shadows in a certain area, have important consequences for the subsequent processes of plant establishment (Janzen, 1971; Willson, 1993).

Although dispersal is a key parameter in models that seek to understand species distributions (Chisholm and Lichstein, 2009), not many studies have yet linked seed dispersal with its demographic consequences, or provided reliable estimates of seed dispersal effectiveness (Côrtes *et al.*, 2009; Rodríguez-Pérez and Traveset, 2012). Furthermore, studies which have followed the complete sequence of plant regeneration show that focusing on a single stage can lead to misleading conclusions, strongly suggesting that a thorough analysis of seed dispersal effectiveness is vital for solid inferences about the consequences of dispersal (Wright *et al.*, 2000; Rodríguez-Pérez and Traveset, 2007; Côrtes *et al.*, 2009; Figueroa-Esquivel *et al.*, 2009; Calviño-Cancela, 2011).

Disperser effectiveness is the contribution of a disperser to plant fitness (Schupp, 1993), or more practically, the number of plant recruits resulting from the activity of each disperser relative to other dispersers, abiotic disperser or non-dispersed seeds (Calviño-Cancela and Martín-Herrero, 2009). Such a contribution is dependent on a sequence of steps, which can be divided into two main groups: (i) those related to the quantity of seeds dispersed (quantitative component); and (ii) the probability of each dispersed seed to survive, germinate and be recruited as a new adult (qualitative component) (Schupp, 1993; Schupp *et al.*, 2010). Thus, although the initial quantitative component varies greatly according to the number of animal visits, feeding rate and handling technique, the treatment and deposition site also varies greatly between dispersers, which frequently have determinant consequences for the final pattern of recruitment (Jordano and Schupp, 2000; Calviño-Cancela and Martín-Herrero, 2009). An extreme example of the importance of the quality of seed deposition is the disproportionate arrival of seeds to favoured establishment sites, known as directed dispersal (Howe and Smallwood, 1982; Wenny, 2001). On the other hand, low quality seed dispersal can in fact reduce plant fitness whenever fruits are a limiting factor for recruitment (Jordano and Schupp, 2000). The net result is that different dispersers can have very different dispersal effectiveness based on their quantitative or qualitative effects on germination (Nogales *et al.*, 2005; Padrón *et al.*, 2011) and although merging dispersers into functional groups can be useful as a first approach to their effectiveness (e.g. seed predators, pulp feeders, gulpers, etc; Jordano and Schupp, 2000), it seems that often dispersers lay along a continuum from poor to good dispersers (Heleno *et al.*, 2011). Two factors that largely determine the quality of seed dispersal are the deposition site and the effect of seed ingestion (in the case of endozoochory) on seed germination rate, which can be either accelerated or slowed down depending on both the seed and the disperser (Traveset, 1998; Traveset *et al.*, 2001). The germination asynchrony resulting from the heterogeneity of interactions might have an important adaptive value in unpredictable environments, leading to changes in at least some seeds so that they germinate during more favourable periods (Moore, 2001; Mooney *et al.*, 2005).

The advantages of seed dispersal

The advantages of seed dispersal can be divided into three non-exclusive categories of hypotheses: (i) escape from density-dependent mortality in which dispersed seeds have increased fitness by escaping disproportional mortality near the parent plants, due to high predation rate, acquisition of pathogens or conspecific competition (Connell, 1971; Janzen, 1971; Harms *et al.*, 2000); (ii) colonization of newly available sites ensures that the main advantage of dispersal is the chance occupation of favourable sites that are unpre-

dictable in time and space (Holmes and Wilson, 1998); and (iii) directed dispersal where dispersed seeds benefit from non-random deposition in sites appropriate for establishment and growth (Wenny, 2001).

The debate about recruitment probability in relation to distance from the parent plant has been one of the most exciting and controversial debates in plant ecology (Nathan and Casagrandi, 2004). The Janzen–Connell hypothesis (Connell, 1971; Janzen, 1971) proposes that the density of dispersed seeds decreases with increasing distance from the parent plant, but because of a reduced impact of distance- and density-responsive seed and seedling enemies, propagule survival increases, resulting in peak recruitment at some distance from the parent and little recruitment near adult conspecifics. This pattern has received considerable empirical support (Silander, 1978; Wills *et al.*, 1997; Harms *et al.*, 2000; Jansen *et al.*, 2008; Choo *et al.*, 2012), yet it has resisted broad generalization as the effect seems to be largely species specific (Hyatt *et al.*, 2003). Moreover, definitive tests of this model require a demonstration that effects carry over to recruitment of new reproductive adults, however few studies have gone beyond early development stages (Schupp and Jordano, 2011; but see Steinitz *et al.*, 2011).

While the advantages of seed dispersal for plants can be categorized according to the scale on which they operate (e.g. populations, communities, etc.), such divisions are mostly for our convenience, as all these levels continuously and reciprocally affect each other. Consequences on other levels, such as evolutionary (Riera *et al.*, 2002; Chen and Chen, 2011; Schaefer, 2011; Pickup and Barrett, 2012), economic (Fujita and Tuttle, 1991; Hougner *et al.*, 2006; García *et al.*, 2010), or for the dispersers (Chapter 2, this volume) would require a differentiated approach, which is beyond the scope of this chapter. Next, we briefly point out the main consequences of seed dispersal at five levels of biological organization, from genes up to global biogeography.

Genes

Seed and pollen dispersal are the two main processes available for plant gene movement, and particularly seed dispersal is the only method available for moving genes of self-fertilized flowers and maternally transmitted genes in outcrossing plants. When seed dispersal is low, genetic structure can give rise to 'genetic neighbourhoods' on a relatively small scale (Gibson and Wheelwright, 1995; Nagy and Rice, 1997). However, dispersal by animals can impose distinct spatial signatures in gene flow, resulting from directionality or clustered seed dispersal even at long distances, for example by seed-caching animals or those which use dormitories (García et al., 2007; Pringle et al., 2011).

Therefore, the dispersal pattern of seeds contributes to the genetic structure of populations and the genetic relatedness between seedlings, in turn affecting plant fitness as a result of competition, facilitation, genetic drift and natural selection (Donohue, 2003; Koelewijn, 2004; Crawford and Whitney, 2010; García and Grivet, 2011). Furthermore, at least occasional passage of genes out of a local neighbourhood or between populations is important in maintaining the genetic diversity of the recipient population, which is particularly valuable in a scenario of fragmented populations (Hanski and Gilpin, 1997; Figueroa-Esquivel et al., 2009; Calviño-Cancela et al., 2012).

Because there are so many factors affecting the genetic structure of populations, it remains a challenge to determine the exact importance of seed dispersal to plant population genetic structure, but several studies unequivocally show that such effects exist (Duminil et al., 2007; Wang et al., 2007; Voigt et al., 2009; Zhou and Chen, 2010). Long-distance dispersal, in particular, increases genetic diversity within populations and reduces genetic diversity among populations (Ray and Excoffier, 2010), thus slowing down the rate of population differentiation (Linhart and Grant, 1996).

Populations

Seed dispersal is well documented to influence colonization rate, population spread and persistence and metapopulation dynamics (Levin et al., 2003; García et al., 2005; Mendoza et al., 2009; Schupp et al.,

2010). In fact, dispersal is inherent to the very concept of plant metapopulations, broadly defined as spatially disjoint populations linked by dispersal (Hanski and Gilpin, 1997; Cain et al., 2000). In unpredictable and highly disturbed environments, dispersal can allow metapopulation persistence through dispersal from one temporally favourable site to another, even if the growth rate in all local populations is negative (Metz et al., 1983), thus providing a rescue effect for small populations in sink habitats (Brown and Kodricbrown, 1977).

Seed dispersal seems to be particularly important in early successional communities and in the case of expanding populations, such as plant invasions (Hovestadt et al., 2000; Traveset and Richardson, 2006; Iponga et al., 2008). However, all species that cannot be competitive in climax communities are destined to local extinction and are in this sense 'fugitive' species largely dependent on dispersal for long-term survival (Holmes and Wilson, 1998).

Despite its potential demographic importance, intra- and inter-population variation in the assemblages of dispersers has not been well documented (Jordano, 1994; Traveset, 1994; Loiselle and Blake, 1999; Padrón et al., 2011). However, when such studies have been possible, reduced recruitment has been reported as a likely response to lack of appropriate dispersal (Voigt et al., 2009; Rodríguez-Pérez and Traveset, 2010; Traveset et al., 2012).

The search for a relationship between dispersal distances and plant abundance in equilibrium populations has led to ambiguous results (Eriksson and Jakobsson, 1998; Bolker and Pacala, 1999; Thompson et al., 1999), with little evidence of a general pattern (Levine and Murrell, 2003). The causes of this spatial 'uncoupling' are mainly attributed to the spatiotemporal variation in the relative importance of mortality factors (e.g. predation, pathogens, competition) for seeds and seedlings (Houle, 1998). Plant population dynamics in patchy environments depend on patch suitability across all stages (Schupp and Fuentes, 1995; Aguiar and Sala, 1997; Forget, 1997). For example seed–seedling conflicts may exist when a microhabitat offers high probability of seed survival but low seedling survival or vice versa (Jordano and Herrera, 1995; Rodríguez-Pérez and Traveset, 2010).

Community

Superior competitors and colonizers are thought to be able to coexist because the superior competitor cannot rapidly fill all available niches, allowing some free space for the weaker competitor yet better colonizer species (Horn and MacArthur, 1972; Tilman, 1994; Holmes and Wilson, 1998). Colonization ability is a function of both fecundity and dispersal ability (Holmes and Wilson, 1998; Levine and Murrell, 2003; Murrell and Law, 2003; Muller-Landau *et al.*, 2008). While competition and dispersal/fecundity trade-offs might enable equilibrium coexistence in spatially variable habitats (Yu and Wilson, 2001), clear empirical support for the importance of dispersal in such trade-offs is still scant (Levine and Murrell, 2003). Supporting this idea, Seidler and Plotkin (2006) showed that tree-species spatial distribution patterns in tropical forests are highly correlated with their dispersal mode (Muller-Landau and Hardesty, 2005).

Recent studies suggest that local community diversity is largely limited by the regional species pool (Srivastava, 1999; Turnbull *et al.*, 2000). Although the influence of dispersal on community structure is only beginning to be rigorously examined (particularly in the context of habitat fragmentation, plant invasions and global climate change) (Bacles *et al.*, 2006; Wright *et al.*, 2009; Hampe, 2011; Traveset and Richardson, 2011), its importance seems intuitive as dispersal is known to influence population spread and persistence (Howe and Smallwood, 1982; Howe and Miriti, 2000). However, such variables are very different from most measures of community structure, which include patterns of abundance, distribution and coexistence in climax communities (Levine and Murrell, 2003).

An interesting, yet largely unexplored, question in seed dispersal at the community level is how interactions between seeds and animals other than legitimate seed dispersal, e.g. parasitism, disease, herbivory, plant competition, root symbiosis, etc., affect plant recruitment patterns. Evidence suggests that many such processes can greatly influence the outcome of seed dispersal leading to highly complex and unexpected outcomes. Some of these interferences to seed dispersal occur before seeds are even removed from the mother plant, such as fruit and seed predation and pathogen attack (Beckman and Muller-Landau, 2011), while others occur after seeds are dispersed, such as secondary dispersal (Nogales *et al.*, 2007), postdispersal seed predation (Orrock *et al.*, 2006) or indirect interactions such as scatter hoarder predation (Steele *et al.*, 2011).

Ecosystem

The final destination of a seed depends not only on seed morphology and the dispersal vectors, but also on the interaction between these and habitat structure or topography, which traps or sorts propagules at different scales – a less appreciated and more difficult to predict determinant of dispersal kernels (Levine and Murrell, 2003; García *et al.*, 2011). Because plants, particularly large trees, can physically shape landscape characteristics and the distribution of those plants simultaneously affects and is affected by seed deposition patterns, feedback loops between species composition, seed dispersal and habitat structure are to be expected (Purves *et al.*, 2007). For example by attracting seed dispersers, isolated plants can act as dispersal foci for many forest plants, as seeds accumulate under their canopies after visits by frugivores (Zahawi and Augspurger, 2006; Kelm *et al.*, 2008; Herrera and García, 2009; Morales *et al.*, 2012).

The signature of animal seed dispersal on habitat features has been shown in several systems, including bird-mediated dispersal of invasive trees driving savanna nucleation processes (Milton *et al.*, 2007), bat dispersal-assisted tropical forest recovery (Kelm *et al.*, 2008), elephant-driven spatial distribution of trees (Campos-Arceiz and Blake, 2011), and ant-selected community composition (Christian, 2001). Even on a geological time-frame, ecosystems depend on the ability of plants to adapt to changing climate, which is largely dependent on their long-distance dispersal potential (Clark *et al.*, 1998; Higgins *et al.*, 2003a; Hampe, 2011).

Biogeography

The way that seeds are moved over long distances can have consequences that go beyond the landscape level and influence the global distribution of plants, i.e. biogeography. At this level, seed dispersal is particularly important in determining plant 'migration' rates, for example tracking favourable climatic conditions such as after glaciations (Clark *et al.*, 2003; Powell and Zimmermann, 2004; Corlett, 2009), or as a consequence of recent anthropogenic climate change (Pitelka *et al.*, 1997; Higgins and Richardson, 1999). Other global processes that are largely influenced by the facility of seeds to travel over long distances are the spread of invasive species (Bartuszevige and Gorchov, 2006; Brochet *et al.*, 2009) and the colonization of oceanic islands by plants (Nogales *et al.*, 2012; Vargas *et al.*, 2012). In the latter case, seed dispersal is determinant as all native vegetation of oceanic islands, i.e. those that emerged from the sea floor and were never connected to a continent, had to initially cross the sea barrier, usually in the form of seeds dispersed by sea currents, wind or animals (see section entitled 'The Determinants of Seed Dispersal', p. 63, this chapter).

Seed Dispersal and Conservation Implications

Seed dispersal is universally considered important for biodiversity conservation. Seed dispersal by animals, in particular, is considered a pivotal ecosystem function that drives plant-community dynamics in natural habitats and vegetation recovery in human-altered landscapes. Nevertheless, there is still a lack of suitable ecological knowledge to develop basic conservation and management guidelines for this ecosystem service (Gosper *et al.*, 2005; García *et al.*, 2010).

The structure of the landscape has strong effects on the distances travelled by seeds, regardless of the dispersal mode. Hence, any type of disturbance such as habitat fragmentation or modification by an invasive plant species is likely to change the patterns of seed movement and recruitment, as well as the genetic structure of plant populations. Wind-dispersed seeds travel much further in open landscapes than in dense forest, due to differences in the shape of the wind profile. By contrast, plants depending on animals for seed movement are susceptible to dispersal failure when their seed vectors become rare or extinct (Traveset and Richardson, 2006).

Disruption of seed dispersal mutualistic interactions can have serious consequences for the population maintenance of the organisms involved, but especially for the plant. An increasing number of studies are showing how the populations of seed dispersers are being decimated, both in the tropics and in the temperate zones, and how this translates into a lower dispersal success of the plants depending upon their services (Wotton and Kelly, 2011; Rodríguez-Pérez and Traveset, 2012; Traveset *et al.*, 2012; Young *et al.*, 2012). In tropical areas in particular, the widespread decline of dispersers by overhunting, selective logging and fragmentation is expected to have long-term negative consequences for the maintenance of tree species diversity (Wright *et al.*, 2000; Markl *et al.*, 2012). A recent meta-analysis indicates that disrupted plant–frugivore interactions could actually trigger a homogenization of seed traits in tree communities of disturbed tropical forests, as hunting and logging show a differential effect on the dispersal of large versus small-seeded tree species (Markl *et al.*, 2012).

Species response to habitat loss and fragmentation

The response of plant species with different dispersal modes to habitat loss is highly variable. A study by Montoya *et al.* (2008) examining the responses of 34 tree species found that animal-dispersed species were less vulnerable to forest loss than those dispersed by other vectors, thus suggesting that plant–animal interactions can help to prevent the collapse of forest communities after habitat degradation.

One question which has only recently started to be addressed is whether a diverse assemblage of dispersers is important to sustain a high quality of seed dispersal services in a community, as reported for the pollination service (Biesmeijer *et al.*, 2006; Fontaine *et al.*, 2006). The first results suggest that despite frugivore abundance being a good surrogate of landscape-scale seed dispersal and an indicator of patch quality for the dispersal function (García *et al.*, 2010), the identity of dispersers might also play an important role (García and Martínez, 2012). More data from different systems are needed to assess how generalized these results are. In the agricultural matrix in Costa Rica, for example, bird abundance rather than diversity best predicted the richness of bird-dispersed seeds (Pejchar *et al.*, 2008). In Central Europe, the ecosystem function is apparently unaltered in areas of high human land use mainly because birds have increased their foraging flying distances to locate fruits (Breitbach *et al.*, 2010). Other systems have been shown to be more vulnerable to forest degradation, such as the effective dispersal of seeds from many bushes and trees in a vast number of tropical freshwater systems (Horn *et al.*, 2011).

The emergence of novel communities

Throughout the world, alien invasive species rank among the most serious threats to native biodiversity and are considered a major factor of global change (Millennium Ecosystem Assessment, 2005). A recent review on the dispersal mode of 622 alien plant species has shown that birds are the most important agent of dispersal for invasive alien trees (*c.*43%) and shrubs (*c.*61%), with wind being the second most important vector (Richardson and Rejmánek, 2011). Furthermore, increasing evidence is showing that propagule pressure, determined by seed dispersal, is important in invasive plant

establishment and spread (Milton *et al.*, 2007; Simberloff, 2009). Nowadays, we also know that seed dispersal has direct consequences on vegetation structure and may be particularly decisive in a scenario of plant invasions, in which frugivores include the fruit of invasive plants into their diets and consequently facilitate their establishment and spread (Nathan and Muller-Landau, 2000). An illustrative example of how a plant–frugivore interaction may promote an invasion that changes the entire community structure was provided by Milton *et al.* (2007). These authors reported that birds facilitate the invasion of arid savannas in South Africa, where alien fleshy-fruited plants infiltrate prevailing seed dispersal networks. Once infiltrated, the natural dispersal network is disrupted because some invasive plants transform the savannas by overtopping and suppressing native trees that act as crucial perching sites and foci for directed dispersal (Iponga *et al.*, 2008).

Although the role of seed dispersal processes in biological invasions has still received relatively little attention (see a review in Westcott and Fletcher, 2011), an increasing number of studies report how invasive species are integrated into natural dispersal communities (Milton *et al.*, 2007; Padrón *et al.*, 2011; Heleno *et al.*, 2013a,b) as well as how intruders affect native dispersal interactions (reviewed in Traveset and Richardson, 2011; López-Bao and González-Varo, 2011; Rodríguez-Cabal *et al.*, 2012). Clear examples of dispersal disruptions have been documented mainly in oceanic archipelagos such as Hawaii (Chimera and Drake, 2010) and the Canary Islands (Nogales *et al.*, 2005; López-Darias and Nogales, 2008). Competition between natives and invaders for mutualistic partners is often being reported, although it is not yet clear to what extent such competition affects native populations rather than just facilitating the spread of invaders. However, it seems likely that plant–seed disperser (as well as pollinator) interaction webs will be irreversibly adjusted in these novel communities in response to the spread of invaders (Ghazoul, 2005). Further work is needed across different species and ecosystems to better understand the overall cost of native mutualistic disruptions, but the evidence so far indicates that this can be quite high (Traveset et al., 2012).

Long-distance dispersal of seeds has caused surprisingly fast invasion rates in many species (Nathan, 2006). It is somehow paradoxical that long-distance dispersal, which is disproportionally important for plant biodiversity and conservation, is, at another level, the root of one of the great threats to biodiversity in the form of human-increased long dispersal of alien invasive species (Trakhtenbrot et al., 2005).

Another paradox can be found in those cases in which native plants currently rely exclusively on alien seed dispersers. This has been reported, for instance, for most common understorey native plants in Hawaiian rainforests, dispersed by alien birds (Foster and Robinson, 2007). Another case can be found in the Balearic Islands where alien pine martens are the main dispersers of plants that used to be dispersed by currently extinct native lizards (Traveset et al., 2012). More research in other systems is necessary to assess how common such replacements are and to what extent the ecosystem function is maintained with these new dispersers. Fortunately, most plant–disperser interactions are generalized and a tight specialization of dispersers does not seem to be required in order to achieve highly effective dispersal (Calviño-Cancela, 2002, 2004). However, seed dispersal by simplified fauna composed of abundant and generalist species is likely to accelerate species shuffling according to their abundance and thus accelerate the homogenization of biota, even if resulting in highly connected and highly nested (i.e. robust) communities (McKinney and Lockwood, 1999; Heleno et al., 2012). If both mutualists are alien, and facilitating each other's spread, as is often the case, then we are faced with an 'invasional meltdown' (Simberloff and Holle, 1999; Simberloff, 2006).

Effect of climate change

Climate change is another major driver of global change that presents a potentially severe threat to biodiversity. Climatic models show that species will be required to disperse rapidly through fragmented landscapes, across both latitudinal and altitudinal gradients in order to keep pace with the changing climate (Pearson and Dawson, 2005). Hampe (2011) reviews the empirical evidence for the role of long-distance seed dispersal in past and ongoing expansions, and examines how some major ecological determinants of seed dispersal and colonization processes might be altered by a rapidly changing climate. Relating dispersal processes and pathways with the establishment of pioneer populations ahead of the continuous species range remains a real challenge. An in-depth treatment of the relationship between seed dispersal and climate change is given in Chapter 9.

Restoration of seed dispersal processes

Evidence is growing that focusing only on species conservation is not enough, and that in order to preserve and restore biodiversity, we need to maintain and re-establish the integrity of interactions between species (Memmott et al., 2007; Heleno et al., 2010; Kaiser-Bunbury et al., 2010; Tylianakis et al., 2010). The restoration of the seed dispersal function is crucial for the long-term stability of restored communities (Handel, 1997). Seed dispersal to a great extent determines vegetation structure (Wang and Smith, 2002), such that the incomplete restoration of such an important ecosystem service may actually result in a community failing to be self-sustaining (Kremen and Hall, 2005). Moreover, even if restoration results in the recovery of species diversity this does not guarantee that the processes in which those species are involved, such as seed dispersal, are re-established (Palmer et al., 1997; Forup et al., 2007).

The preservation of the seed dispersal process and its mobile agents should be considered a tool for passive and hence low-cost ecological restoration (Howe and Miriti, 2004). A land-management approach trying to maintain and increase the presence of scattered remnant trees over a deforested matrix might accelerate secondary succession through a process of facilitation mediated by seed dispersers (Valladares and Gianoli, 2007; Herrera and García, 2009). A recent study carried out in southern Spain has shown that the seed rain produced by frugivorous birds on a tree plantation is strongly determined by the nature of the surrounding vegetation (Zamora *et al.*, 2010). Recent simulation studies have also found that fruit removal rates and mean dispersal distances are strongly affected by fruiting plant neighbourhoods (Morales *et al.*, 2012). Plants in denser neighbourhoods had greater fruit removal and shorter mean dispersal distances than more isolated plants. The interplay between frugivore behavioural decisions and the spatial distribution of plants could thus have important consequences for plant spatial dynamics and should be taken into account in any restoration programme.

Birds have often been documented as aiding alien plant spread (Bartuszevige and Gorchov, 2006; Williams, 2006; Gosper and Vivian-Smith, 2009). Whenever native and exotic plants do not share the same physical space or the same dispersers, natural or artificial bird perching structures can conceivably be used to direct the dispersal of exotic seeds to unsuitable or easily controlled areas (sink habitats), or to direct the dispersal of native seeds to suitable recruitment sites. Several studies have conducted perch manipulations in the field in order to direct seed dispersal (e.g. Holl *et al.*, 2000; Robinson and Handel, 2000), usually with a clear increase in seed deposition under perches (Wenny, 2001; Shiels and Walker, 2003). Even if the effectiveness of such methods remains a matter of debate (Holl *et al.*, 2000; Robinson and Handel, 2000; Shiels and Walker, 2003), it stands as a practical example of conservation measures that can be greatly enhanced by an in-depth knowledge of seed dispersal systems. Several other more general guidelines based on seed dispersal are frequently implemented by practitioners in conservation projects such as initiating the control

of weeds starting from elevated areas and working downhill (as most seeds will be dispersed in this direction by wind and gravity), or trying to tackle plant invasions before the fruiting season.

Ecological networks can provide a holistic approach to ecosystem management, with benefits for both restoration science and restoration practice (Heleno *et al.*, 2010). In a scenario of ecological restoration, seed dispersal networks can be useful in both the planning stage and the monitoring of restoration effectiveness as they provide a valuable tool to look beyond species composition and into ecosystem functioning. For example, they can be important to predict the potential consequences (positive and negative, direct and indirect) of species introductions or species eradication programmes (Memmott, 2009; Padrón *et al.*, 2009; Hansen *et al.*, 2010). Such an exercise can be useful to identify keystone species, or in the context of seed dispersal, keystone mutualists, which should then be the focus of conservation, management or restoration efforts (Kaiser-Bunbury *et al.*, 2010).

Conservation and restoration priorities are inherently idiosyncratic to each area and it may not always be possible or important to address seed dispersal when restoring a habitat. In the medium to long term, restoration has to tackle seed dispersal interactions but also pollinator interactions to avoid genetic deterioration of the plant community. Small populations of restored plant species will maintain relatively little genetic diversity, which reduces the chance of successful adaptation to small- or large-scale alterations such as climate change. Hence, in a second phase, large-scale, long-term restoration needs to be carried out to account for seed dispersal interactions to eventually result in reproductively self-sustaining communities (Kaiser-Bunbury *et al.*, 2010). In order to predict which ecological interactions are at risk from loss of critical species, empirical comparisons of the population-level impacts of mutualist animals are also necessary (Brodie *et al.*, 2009).

Conclusions and Future Avenues of Research

The study of the dispersal of plants has rapidly advanced in the last two decades, as essential elements of the evolutionary and ecological causes and consequences of dispersal have been examined. Specifically in the last decade, there has been an increasing effort to link the process of seed dispersal and its demographic and genetic consequences. This has been mainly due to the development of several technological tools which allow following seed fate via radio-tracking (Pons and Pausas, 2007), rare isotope enrichment (Morales *et al.*, 2012) or fluorescent markers (Levey and Sargent, 2000), following the movements of dispersers with satellite tracking of long-range seed dispersers such as seabirds, the development of increasingly robust mathematical modelling (Levey *et al.*, 2005), and particularly with the advent of widely available and cost-effective genetic markers (Terakawa *et al.*, 2009; Choo *et al.*, 2012). For example, it is now possible to distinguish between maternally and paternally inherited genes by comparing DNA from the nucleus with that from organelles (chloroplasts and mitochondria) that are inherited from the mother plant (Ouborg *et al.*, 1999). Furthermore, the analysis of several microsatellite loci of the seed endocarp allows matching each seed to its individual mother plant, which in turn permits the study of long-distance dispersal (Godoy and Jordano, 2001; Ziegenhagen *et al.*, 2003; Grivet *et al.*, 2005; Jordano *et al.*, 2007). Such techniques, for instance, allow estimating seed dispersal kernels, characterizing animal foraging behaviour, and understanding the colonization history of oceanic islands by plants (e.g. Rumeu *et al.*, 2011; Vargas *et al.*, 2012). The possibility to identify the disperser species based on surface DNA from droppings and pellets collected in the field has also only come about with molecular tools (Marrero *et al.*, 2009). Much has been advanced towards the integration of multiscales of dispersal (García, 2002), and in following the seed dispersal signature up to the establishment of adult trees (Calviño-Cancela and Martín-Herrero, 2009). Quantifying seed dispersal patterns at increasingly large scales is a particular challenge due to the uncertain consequences of rare events (Higgins *et al.*, 2003b; Nogales *et al.*, 2012); despite different types of seed traps proving to be highly useful in the study of short-distance dispersal, they are impractical for detecting rare long-distance dispersal events (Greene and Calogeropoulos, 2002). Another important information gap is the lack of long-term studies, which are crucial if we are to link seed dispersal and the distribution of adult trees in natural habitats. A short-cut to address this issue might be to start with the distribution of adult trees and try to predict previous dispersal based on population age structure using dendrochronology and kinship analysis via molecular methods.

It is now widely accepted that animal seed dispersal represents a key process in the functioning of many different ecosystems, from deserts to rainforests, and for the maintenance of their biodiversity (Forget *et al.*, 2011). We also know that this mutualistic interaction is seriously threatened by human activities, which promote fast changes in ecological conditions. New conceptual frameworks such as movement ecology, complex networks of interactions among species (see Box 3.2, this chapter) and their associated ecological functions and services, and landscape-level analyses combined with models of range shifts due to climate change are likely to provide important advances in this field (Jordano *et al.*, 2011). The integration of different disciplines related to seed dispersal that involve different methodologies (including ecophysiology, landscape ecology, population genetics, biogeography, conservation biology, evolutionary ecology and phylogenetics, and climate change biology) is increasingly essential to build mechanistic models and frame robust predictions about the consequences of the loss of plant–frugivore interactions for natural and human-modified habitats. This, in turn, will permit foreseeing critical conservation risks, developing early-warning signals of seed dispersal disruptions, and designing better management plans for

Box 3.2. Seed dispersal networks

Species do not exist in isolation, and therefore the interactions they establish with one another play a determinant role in defining the community structure, the functions performed by ecosystems and the services they deliver to humans (Duffy *et al.*, 2007; Tylianakis *et al.*, 2010). Seed dispersal by animals, in particular, is one such function that greatly illustrates the importance of studying ecological interactions at a community level. Such interactions have been frequently depicted in the form of food webs or more generally as interaction networks composed of nodes (i.e. plants and animals), and links (i.e. dispersal events). The consolidation of this network approach has allowed scientists to simultaneously focus on the effects of conservation efforts on focal species and on the overall community, and contributed to improving the focus of conservation efforts for a more integrated viewpoint (Millennium Ecosystem Assessment, 2005). Seed dispersal is inherently a complex process at many levels. The use of a network approach offers a theoretical framework to look for patterns and general rules without losing information due to clustering species into functional groups.

Although first networks represented antagonistic interactions (who eats whom), rapidly the approach has been translated into mutualistic interactions, stemming from the growing perception that facilitation, like competition, is an important process shaping natural communities (Bruno *et al.*, 2003). Within this context, pollination has received the most attention followed by seed dispersal and more recently by root symbionts (Memmott, 1999; Bascompte and Jordano, 2007; Montesinos-Navarro *et al.*, 2012).

Although in his seminal book, Ridley (1930) compiled information on the dispersal vectors of a large number of plants, only in the 1970s did ecologists begin to empirically document all dispersal events within a single community (e.g. Snow and Snow, 1971; Herrera, 1984). The work by Jordano (1987) formally applied the network theory to the study of seed dispersal, and more recently others have followed (e.g. Donatti *et al.*, 2011; Mello *et al.*, 2011b; Heleno *et al.*, 2013a,b).

Seed dispersal networks can be especially useful for understanding the process of plant invasions, particularly those with fleshy fruits, and there has been an all-out effort to document the dispersers of invasive plants and the plants dispersed by invasive animals (e.g. Milton *et al.*, 2007; Kelly *et al.*, 2010; Linnebjerg *et al.*, 2010; Padrón *et al.*, 2011; Heleno *et al.*, 2013b). However, rigorous information on entire species assemblages of plants, seed dispersers and their interactions is still rarely available (Buckley *et al.*, 2006), which hinders informed decisions in conservation planning (Gosper *et al.*, 2005).

Networks allow us to escape descriptive studies and look for patterns across different communities by means of standard network parameters, and to link those patterns to community stability and functioning (Bascompte and Jordano, 2007), for example between islands and continents (Box 3.1, this chapter).

Although the application of network theory to seed dispersal is still in its youth, some common patterns have been identified such as: (i) a great heterogeneity in the strength of interactions among species; (ii) rarity of strong inter-specific co-evolution between plants and dispersers (Jordano, 1987); and (iii) a high level of asymmetry in the interactions, such that specialist plants tend to depend on generalist dispersers while specialist dispersers depend on generalist plants (Bascompte *et al.*, 2003). Such a characteristic results from a nested pattern between plants and animals, which has been shown to protect communities from cascading extinctions (Bascompte and Jordano, 2007). Furthermore, recent work revealed seed dispersal interactions are arranged in a modular pattern, i.e. with some clusters of species interacting more with each other than with species outside that cluster—module (Donatti *et al.*, 2011). Such a pattern, emerging through a combination of phylogenetic history and trait convergence (Donatti *et al.*, 2011) suggests that despite being a less specialized interaction than pollination (Howe and Smallwood, 1982), species within different modules may follow different co-evolutionary pathways driving adaptations among plants and their dispersers (Mello *et al.*, 2011b).

Regardless of the proclaimed potential of seed dispersal networks for the advance of ecological theory and practice, broad generalizations and application of this approach are still meagre (Carlo and Yang, 2011). We briefly consider four challenges that seed dispersal networks have to overcome in order to meet their full potential.

Continued

Box 3.2. Continued.

More precise networks – Seed dispersal is a multistage process, where each part contributes to the overall outcome (Fig. 3.4). Although the emergence of quantitative networks, which incorporate measures of interaction strength, has been an important advance from qualitative networks, a most necessary step is to include more precise assessments of dispersal effectiveness in seed dispersal networks. Despite effectiveness being the best parameter to quantify interaction strength, it has seldom been used in a network context (Carlo and Yang, 2011). Most often, interaction frequency is used instead as a surrogate of dispersal effectiveness (Vázquez *et al.*, 2005), which ignores the quality of seed dispersal, focusing on its quantity (Schupp *et al.*, 2010). Conclusions based on quantitative proxies might then be misleading (Calviño-Cancela and Martín-Herrero, 2009) and we need to try to incorporate dispersal effectiveness into seed dispersal networks in order to validate how network properties predict ecological outcomes (Tylianakis *et al.*, 2010; Carlo and Yang, 2011).

Bigger networks – Networks are assembled under multiple decisions taken by researchers either for logistical reasons or following their own 'comfort zones' (Memmott *et al.*, 2007), which might be taxonomic groups or habitats. Even if seed dispersal does not stop at habitat interfaces, most seed dispersal networks are confined to one functional group of dispersers, frequently birds (Mello *et al.*, 2011a), and to physical or perceived borders in the habitat. In addition, seed dispersal by animals is only one of many interactions that both organisms establish in nature. Seed dispersal networks can thus be included in a wider context of other interactions, in effect a network of networks (Pocock *et al.*, 2012), as these simultaneous interactions can greatly affect the outcome of seed dispersal (e.g. Wright *et al.*, 2000; Steele *et al.*, 2011).

Node characterization – The soundness of the insights gained from analysing seed dispersal networks is dependent on the quality of data and the correct interpretation of results, rather than on sophisticated analytical methods. Thus, while for network analysis all nodes within a level are considered equivalent, all species are distinct in many ways (e.g. behaviour, abundance, conservation value, etc.) which constitutes important information for interpreting the context-specific outputs (Carlo and Yang, 2011).

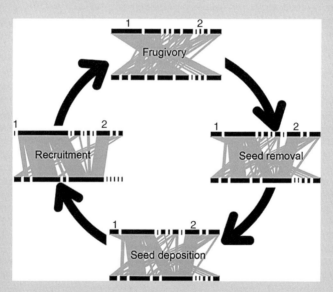

Fig. 3.4. Each network represents successive stages in the whole process of seed dispersal: frugivory, seed removal, frugivore movement and seed recruitment. Although the same species are represented in each network: dispersers on the top row and plants on the bottom row, their relative contribution is largely dependent upon the information used to quantify the interaction strength in each network. (From Carlo and Yang, 2011.)

Continued

Box 3.2. Continued.

For example, one should not dismiss concepts of species conservation value when translating network results for a conservation scenario (Heleno *et al.*, 2012). For instance, while high specialization can contribute to increasing the vulnerability to disruption (Aizen *et al.*, 2012), to protect vulnerable specialist species might be the appropriate conservation target as high network robustness can be a consequence of the previous extinction of specialists or the accumulation of generalist species (Menke *et al.*, 2012), both of which can represent environmental degradation and not a direct indicator of conservation value (Heleno *et al.*, 2012).

Better experimental design – The progress made in the development of mathematical tools to describe interaction networks contrasts with the scarcity of empirical data to feed them, which is vital to validate network model predictions (Carlo and Yang, 2011; Heleno *et al.*, 2013a). A major difficulty has been the *a posteriori* comparison of networks assembled by different researchers for different ends and which vary greatly in the sampling protocols, sampling effort and taxonomical resolution. In order to increase the statistical power and the likelihood of detecting emerging patterns, it is important to plan robust experimental designs or wait for enough high-quality studies to become available for meaningful meta-analysis.

In conclusion, network theory provides ecologists with an important tool to examine the intricate web of interactions between plants and their dispersers; however, whether networks become fully informative will depend on our ability to put theory to the test with more and better datasets and appropriate experimental designs explicitly set to test network-driven predictions.

efficient restoration of ecological functions (Forget *et al.*, 2011; Jordano *et al.*, 2011). We also need to consider that the effects of different types of disturbance (e.g. frugivore hunting, fragmentation, invasions, climate change, etc.) are not only ecological (e.g. truncating dispersal kernels, changing density-dependent plant mortality at different stages of the life cycle), but also evolutionary. Direct anthropogenic impacts on the community of dispersers may translate into rapid evolutionary shifts in seed and fruit traits (Wotton and Kelly, 2011; Markl *et al.*, 2012).

The field of dispersal ecology also needs to be more integrated with that of invasion ecology (and vice versa), as Westcott and Fletcher (2011) have pointed out. Invasions represent 'natural' experiments that allow testing models related to dispersal processes and their influence on population and community structure. Moreover, they can contribute to the understanding of the evolution of dispersal-relevant traits such as seed size (Muller-Landau, 2010). A detailed and integrated understanding of dispersal processes is also important to effectively manage biological invasions (Traveset and Richardson, 2011).

When considering the conservation and restoration of ecosystem functions, such as that provided by seed dispersers, it is increasingly recognized that network theory is a particularly valuable tool in providing a structural and functional approach to frame the whole community and the full variety of inter-specific interactions that hold communities together (Carlo and Yang, 2011; Cruz *et al.*, in press). Such an approach may be particularly useful on islands due to the simplicity of ecosystems (Kaiser-Bunbury *et al.*, 2010), although it should also be valid for continental habitats. Insights gained through an increasing number of mutualistic networks encompass similar interaction patterns at the guild or ecosystem level between communities (Bascompte *et al.*, 2006; Olesen *et al.*, 2007). One major challenge, however, is to adequately capture and describe the spatial and temporal dynamics of mutualistic networks (Bascompte and Jordano, 2007; Olesen *et al.*, 2008; Tylianakis, 2008). Such network analyses appear to be promising as well, to detect geographical patterns, colonization and extinction dynamics of native species, and appear particularly useful to study the functioning of the novel ecosystems emerging due to the movement of species across the planet.

In short, dispersal ecology is a rapidly developing field that offers a wealth of investigative opportunity at many different levels, ranging from good natural history to sophisticated modelling and conceptual synthesis.

Acknowledgements

We dedicate this chapter to Mary F. Willson, whose review chapter in the two previous editions of this book has been an important baseline when preparing this updated version. We also thank all our colleagues and students who have shared with us their interest for mutualistic interactions and whose experience has been greatly valuable to us when reviewing information on seed dispersal. During the preparation of the chapter we have benefitted from the financial support of several organizations to which we are thankful, namely: the BBVA Foundation, the Spanish Ministry of Science and Innovation, National Park Organism – Spain, Government of the Canary Islands and Fundação para a Ciência e Tecnologia – Portugal. Lastly, we thank Robert Gallagher for inviting us to write this chapter and for the opportunity to share our interest in the fascinating process of seed dispersal with others.

References

Aguiar, M.R. and Sala, O.E. (1997) Seed distribution constrains the dynamics of the Patagonian steppe. *Ecology* 78, 93–100.

Aizen, M.A., Sabatino, M. and Tylianakis, J.M. (2012) Specialization and rarity predict nonrandom loss of interactions from mutualist networks. *Science* 335, 1486–1489.

Anderson, J.T., Nuttle, T., Rojas, J.S.S., Pendergast, T.H. and Flecker, A.S. (2011) Extremely long-distance seed dispersal by an overfished Amazonian frugivore. *Proceedings of the Royal Society B: Biological Sciences* 278, 3329–3335.

Bacles, C.F.E., Lowe, A.J. and Ennos, R.A. (2006) Effective seed dispersal across a fragmented landscape. *Science* 311, 628.

Bartuszevige, A.M. and Gorchov, D.L. (2006) Avian seed dispersal of an invasive shrub. *Biological Invasions* 8, 1013–1022.

Bascompte, J. and Jordano, P. (2007) Plant-animal mutualistic networks: The architecture of biodiversity. *Annual Review of Ecology Evolution and Systematics* 38, 567–593.

Bascompte, J., Jordano, P., Melian, C.J. and Olesen, J.M. (2003) The nested assembly of plant-animal mutualistic networks. *Proceedings of the National Academy of Sciences of the United States of America* 100, 9383–9387.

Bascompte, J., Jordano, P. and Olesen, J.M. (2006) Asymmetric coevolutionary networks facilitate biodiversity maintenance. *Science* 312, 431–433.

Beckman, N.G. and Muller-Landau, H.C. (2011) Linking fruit traits to variation in predispersal vertebrate seed predation, insect seed predation, and pathogen attack. *Ecology* 92, 2131–2140.

Benkman, C.W. (1995) Wind dispersal capacity of pine seeds and the evolution of different seed dispersal modes in pines. *Oikos* 73, 221–224.

Bialozyt, R., Ziegenhagen, B. and Petit, R.J. (2006) Contrasting effects of long distance seed dispersal on genetic diversity during range expansion. *Journal of Evolutionary Biology* 19, 12–20.

Biesmeijer, J.C., Roberts, S.P.M., Reemer, M., Ohlemuller, R., Edwards, M., Peeters, T., Schaffers, A.P., Potts, S.G., Kleukers, R., Thomas, C.D., Settele, J. and Kunin, W.E. (2006) Parallel declines in pollinators and insect-pollinated plants in Britain and the Netherlands. *Science* 313, 351–354.

Blake, S., Wikelski, M., Cabrera, F., Guezou, A., Silva, M., Sadeghayobi, E., Yackulic, C.B. and Jaramillo, P. (2012) Seed dispersal by Galápagos tortoises. *Journal of Biogeography* 39, 1961–1972.

Bolker, B.M. and Pacala, S.W. (1999) Spatial moment equations for plant competition: Understanding spatial strategies and the advantages of short dispersal. *American Naturalist* 153, 575–602.

Borges, R.M., Ranganathan, Y., Krishnan, A., Ghara, M. and Pramanik, G. (2011) When should fig fruit produce volatiles? Pattern in a ripening process. *Acta Oecologica* 37, 611–618.

Breitbach, N., Laube, I., Steffan-Dewenter, I. and Bohning-Gaese, K. (2010) Bird diversity and seed dispersal along a human land-use gradient: high seed removal in structurally simple farmland. *Oecologia* 162, 965–976.

Brochet, A.L., Guillemain, M., Fritz, H., Gauthier-Clerc, M. and Green, A.J. (2009) The role of migratory ducks in the long-distance dispersal of native plants and the spread of exotic plants in Europe. *Ecography* 32, 919–928.

Brodie, J.F., Helmy, O.E., Brockelman, W.Y. and Maron, J.L. (2009) Functional differences within a guild of tropical mammalian frugivores. *Ecology* 90, 688–698.

Brown, J.H. and Kodricbrown, A. (1977) Turnover rates in insular biogeography: Effect of immigration on extinction. *Ecology* 58, 445–449.

Bruno, J.F., Stachowicz, J.J. and Bertness, M.D. (2003) Inclusion of facilitation into ecological theory. *Trends in Ecology & Evolution* 18, 119–125.

Buckley, Y.M., Anderson, S., Catterall, C.P., Corlett, R.T., Engel, T., Gosper, C.R., Nathan, R., Richardson, D.M., Setter, M., Spiegel, O., Vivian-Smith, G., Voigt, F.A., Weir, J.E.S. and Westcott, D.A. (2006) Management of plant invasions mediated by frugivore interactions. *Journal of Applied Ecology* 43, 848–857.

Bullock, J.M., Galsworthy, S.J., Manzano, P., Poschlod, P., Eichberg, C., Walker, K. and Wichmann, M.C. (2011) Process-based functions for seed retention on animals: a test of improved descriptions of dispersal using multiple data sets. *Oikos* 120, 1201–1208.

Cain, M.L., Damman, H. and Muir, A. (1998) Seed dispersal and the Holocene migration of woodland herbs. *Ecological Monographs* 68, 325–347.

Cain, M.L., Milligan, B.G. and Strand, A.E. (2000) Long-distance seed dispersal in plant populations. *American Journal of Botany* 87, 1217–1227.

Calviño-Cancela, M. (2002) Spatial patterns of seed dispersal and seedling recruitment in *Corema album* (Empetraceae): the importance of unspecialized dispersers for regeneration. *Journal of Ecology* 90, 775–784.

Calviño-Cancela, M. (2004) Ingestion and dispersal: direct and indirect effects of frugivores on seed viability and germination of *Corema album* (Empetraceae). *Acta Oecologica* 26, 55–64.

Calviño-Cancela, M. (2011) Simplifying methods to assess site suitability for plant recruitment. *Plant Ecology* 212, 1375–1383.

Calviño-Cancela, M. and Martín-Herrero, J. (2009) Effectiveness of a varied assemblage of seed dispersers of a fleshy-fruited plant. *Ecology* 90, 3503–3515.

Calviño-Cancela, M., Escudero, M., Rodríguez-Pérez, J., Cano, E., Vargas, P., Velo-Antón, G. and Traveset, A. (2012) The role of seed dispersal, pollination and historical effects on genetic patterns of an insular plant that has lost its only seed disperser. *Journal of Biogeography* 39, 1996–2006.

Campos-Arceiz, A. and Blake, S. (2011) Megagardeners of the forest – the role of elephants in seed dispersal. *Acta Oecologica* 37, 542–553.

Carlo, T.A. and Yang, S. (2011) Network models of frugivory and seed dispersal: Challenges and opportunities. *Acta Oecologica* 37, 619–624.

Carlquist, S. (1974) *Island Biology*. Columbia University Press, Columbia.

Chambers, J.C. and Macmahon, J.A. (1994) A day in the life of a seed - movements and fates of seeds and their implications for natural and managed systems. *Annual Review of Ecology and Systematics* 25, 263–292.

Chen, F. and Chen, J. (2011) Dispersal syndrome differentiation of *Pinus armandii* in Southwest China: Key elements of a potential selection mosaic. *Acta Oecologica* 37, 587–593.

Chimera, C.G. and Drake, D.R. (2010) Patterns of seed dispersal and dispersal failure in a Hawaiian dry forest having only introduced birds. *Biotropica* 42, 493–502.

Chisholm, R.A. and Lichstein, J.W. (2009) Linking dispersal, immigration and scale in the neutral theory of biodiversity. *Ecology Letters* 12, 1385–1393.

Choo, J., Juenger, T.E. and Simpson, B.B. (2012) Consequences of frugivore-mediated seed dispersal for the spatial and genetic structures of a neotropical palm. *Molecular Ecology* 21, 1019–1031.

Christian, C.E. (2001) Consequences of a biological invasion reveal the importance of mutualism for plant communities. *Nature* 413, 635–639.

Christianini, A.V. and Oliveira, P.S. (2010) Birds and ants provide complementary seed dispersal in a neotropical savanna. *Journal of Ecology* 98, 573–582.

Clark, C.J., Poulsen, J.R. and Parker, V.T. (2001) The role of arboreal seed dispersal groups on the seed rain of a lowland tropical forest. *Biotropica* 33, 606–620.

Clark, C.J., Poulsen, J.R., Bolker, B.M., Connor, E.F. and Parker, V.T. (2005) Comparative seed shadows of bird-, monkey-, and wind-dispersed trees. *Ecology* 86, 2684–2694.

Clark, J.S., Fastie, C., Hurtt, G., Jackson, S.T., Johnson, C., King, G.A., Lewis, M., Lynch, J., Pacala, S., Prentice, C., Schupp, E.W., Webb, T. and Wyckoff, P. (1998) Reid's paradox of rapid plant migration – Dispersal theory and interpretation of paleoecological records. *Bioscience* 48, 13–24.

Clark, J.S., Lewis, M., McLachlan, J.S. and HilleRisLambers, J. (2003) Estimating population spread: What can we forecast and how well? *Ecology* 84, 1979–1988.

Connell, J.H. (1971) On the role of natural enemies in preventing competitive exclusion in some marine mammals and in rain forest trees. In: Boer, P.J. and Gradwell, G.R. (eds) *Dynamics of Populations.* Centre for Agricultural Publishing and Documentation, Wageningen, the Netherlands, pp. 298–310.

Corlett, R.T. (2009) Seed dispersal distances and plant migration potential in tropical East Asia. *Biotropica* 41, 592–598.

Côrtes, M.C., Cazetta, E., Staggemeier, V.G. and Galetti, M. (2009) Linking frugivore activity to early recruitment of a bird dispersed tree, *Eugenia umbelliflora* (Myrtaceae) in the Atlantic rainforest. *Austral Ecology* 34, 249–258.

Costa, J., Ramos, J., Silva, L., Timoteo, S., Araujo, P., Felgueiras, M., Rosa, A., Matos, C., Encarnação, P., Tenreiro, P. and Heleno, R. (in press) Endozoochory largely outweighs epizoochory in migrating passerines. *Journal of Avian Biology*, DOI: 10.1111/j.1600-048x.2013.00271.x

Crawford, K.M. and Whitney, K.D. (2010) Population genetic diversity influences colonization success. *Molecular Ecology* 19, 1253–1263.

Cruz, J., Ramos, J., Silva, L., Tenreiro, P. and Heleno, P. (in press) Seed dispersal networks in an urban novel ecosystem. *European Journal of Research*, DOI: 10.1007/s10342-013-0722-1.

Damstra, K. (1986) Editorial (notes by George Hall). *Tree Life* 71, 5.

de Castro, E.R., Cortes, M.C., Navarro, L., Galetti, M. and Morellato, L.P.C. (2012) Temporal variation in the abundance of two species of thrushes in relation to fruiting phenology in the Atlantic rainforest. *Emu* 112, 137–148.

Dennis, A.J., Schupp, E.W., Green, A.J. and Westcott, D.A. (eds) (2007) *Seed Dispersal: Theory and its Application in a Changing World.* CAB International, Wallingford.

Donatti, C.I., Guimarães, P.R., Galetti, M., Pizo, M.A., Marquitti, F.M.D. and Dirzo, R. (2011) Analysis of a hyper-diverse seed dispersal network: modularity and underlying mechanisms. *Ecology Letters* 14, 773–781.

Donohue, K. (2003) The influence of neighbor relatedness on multilevel selection in the Great Lakes sea rocket. *American Naturalist* 162, 77–92.

Dudley, J.P. (1999) Seed dispersal of *Acacia erioloba* by African bush elephants in Hwange National Park, Zimbabwe. *African Journal of Ecology* 37, 375–385.

Duffy, J.E., Carinale, B.J., France, K.E., McIntyre, P.B., Thébault, E. and Loreau, M. (2007) The functional role of biodiversity in ecosystems: incorporating trophic complexity. *Ecology Letters* 10, 522–538.

Duminil, J., Fineschi, S., Hampe, A., Jordano, P., Salvini, D., Vendramin, G.G. and Petit, R.J. (2007) Can population genetic structure be predicted from life-history traits? *American Naturalist* 169, 662–672.

Eriksson, O. and Jakobsson, A. (1998) Abundance, distribution and life histories of grassland plants: a comparative study of 81 species. *Journal of Ecology* 86, 922–933.

Figueroa-Esquivel, E., Puebla-Olivares, F., Godinez-Álvarez, H. and Núñez-Farfán, J. (2009) Seed dispersal effectiveness by understory birds on *Dendropanax arboreus* in a fragmented landscape. *Biodiversity and Conservation* 18, 3357–3365.

Figuerola, J. and Green, A. (2002) How frequent is external transport of seeds and invertebrate eggs by waterbirds? A study in Doñana, SW Spain. *Archiv fur Hydrobiologie* 155, 557–565.

Fontaine, C., Dajoz, I., Meriguet, J. and Loreau, M. (2006) Functional diversity of plant-pollinator inter-action webs enhances the persistence of plant communities. *PLOS Biology* 4, 129–135.

Forget, P.M. (1997) Effect of microhabitat on seed fate and seedling performance in two rodent-dispersed tree species in rain forest in French Guiana. *Journal of Ecology* 85, 693–703.

Forget, P.-M., Jordano, P., Lambert, J.E., Böhning-Gaese, K., Traveset, A. and Wright, S.J. (2011) Frugivores and seed dispersal (1985 - 2010); the 'seeds' dispersed, established and matured. *Acta Oecologica* 37, 517–520.

Forup, M.L., Henson, K.S.E., Craze, P.G. and Memmott, J. (2007) The restoration of ecological interactions: plant-pollinator networks on ancient and restored heathlands. *Journal of Applied Ecology* 45, 742–752.

Foster, J.T. and Robinson, S.K. (2007) Introduced birds and the fate of Hawaiian rainforests. *Conservation Biology* 21, 1248–1257.

Foster, R.B. (1982) The seasonal rhythm of fruitfall on Barro Colorado Island. In: Leigh, E.G., Rand, A.S. and Windsor, D.M. (eds) *The Ecology of a Tropical Forest.* Smithsonian, Washington DC, pp. 151–172.

Foster, S.A. and Janson, C.H. (1985) The relationship between seed size and establishment conditions in tropical woody-plants. *Ecology* 66, 773–780.

Fujita, M.S. and Tuttle, M.D. (1991) Flying foxes (*Chiroptera pteropodidae*) threatened animals of key ecological and economic importance. *Conservation Biology* 5, 455–463.

García, C. and Grivet, D. (2011) Molecular insights into seed dispersal mutualisms driving plant population recruitment. *Acta Oecologica* 37, 632–640.

García, C., Jordano, P. and Godoy, J.A. (2007) Contemporary pollen and seed dispersal in a *Prunus mahaleb* population: patterns in distance and direction. *Molecular Ecology* 16, 1947–1955.

García, D. and Martínez, D. (2012) Species richness matters for the quality of ecosystem services: a test using seed dispersal by frugivorous birds. *Proceedings of the Royal Society B: Biological Sciences* 279, 3106–3113.

García, D., Obeso, J.R. and Martínez, I. (2005) Spatial concordance between seed rain and seedling establishment in bird-dispersed trees: does scale matter? *Journal of Ecology* 93, 693–704.

García, D., Zamora, R. and Amico, G.C. (2010) Birds as suppliers of seed dispersal in temperate ecosystems: conservation guidelines from real-world landscapes. *Conservation Biology* 24, 1070–1079.

García, D., Zamora, R. and Amico, G.C. (2011) The spatial scale of plant-animal interactions: effects of resource availability and habitat structure. *Ecological Monographs* 81, 103–121.

García, J.D. (2002) Interaction between introduced rats and a frugivore bird-plant system in a relict island forest. *Journal of Natural History* 36, 1247–1258.

Ghazoul, J. (2005) Pollen and seed dispersal among dispersed plants. *Biological Reviews* 80, 413–443.

Gibson, J.P. and Wheelwright, N.T. (1995) Genetic structure in a population of a tropical tree *Ocotea tenera* (Lauraceae): influence of avian seed dispersal. *Oecologia* 103, 49–54.

Godoy, J.A. and Jordano, P. (2001) Seed dispersal by animals: exact identification of source trees with endocarp DNA microsatellites. *Molecular Ecology* 10, 2275–2283.

Gómez, J.M. (2003) Spatial patterns in long-distance dispersal of *Quercus ilex* acorns by jays in a heterogeneous landscape. *Ecography* 26, 573–584.

González-Castro, A., Traveset, A. and Nogales, M. (2012) Seed dispersal interactions in the Mediterranean Region: contrasting patterns between islands and mainland. *Journal of Biogeography* 39, 1938–1947.

Gorb, S.N. and Gorb, E.V. (1995) Removal rates of seeds of five myrmecochorous plants by the ant *Formica polyctena* (Hymenoptera: Formicidae). *Oikos* 73, 367–374.

Gosper, C.R. and Vivian-Smith, G. (2009) The role of fruit traits of bird-dispersed plants in invasiveness and weed risk assessment. *Diversity and Distributions* 15, 1037–1046.

Gosper, C.R., Stansbury, C.D. and Vivian-Smith, G. (2005) Seed dispersal of fleshy-fruited invasive plants by birds: contributing factors and management options. *Diversity and Distributions* 11, 549–558.

Greene, D.F. (2005) The role of abscission in long-distance seed dispersal by the wind. *Ecology* 86, 3105–3110.

Greene, D.F. and Calogeropoulos, C. (2002) Measuring and modelling seed dispersal of terrestrial plants. In: Bullock, J.M., Kenward, R.E. and Hails, R.S. (eds) *Dispersal Ecology*. Blackwell, Oxford, pp. 3–23.

Grivet, D., Smouse, P.E. and Sork, V.L. (2005) A novel approach to an old problem: tracking dispersed seeds. *Molecular Ecology* 14, 3585–3595.

Griz, L.M.S. and Machado, I.C.S. (2001) Fruiting phenology and seed dispersal syndromes in caatinga, a tropical dry forest in the northeast of Brazil. *Journal of Tropical Ecology* 17, 303–321.

Guerrero, A.M. and Tye, A. (2009) Darwin's finches as seed predators and dispersers. *The Wilson Journal of Ornithology* 121, 752–764.

Gulick, A. (1932) Biological peculiarities of oceanic islands. *The Quarterly Review of Biology* 7, 405–427.

Guttal, V., Bartumeus, F., Hartvigsen, G. and Nevai, A.L. (2011) Retention time variability as a mechanism for animal mediated long-distance dispersal. *PLOS ONE* 6, e28447.

Hampe, A. (2011) Plants on the move: The role of seed dispersal and initial population establishment for climate-driven range expansions. *Acta Oecologica* 37, 666–673.

Handel, S.N. (1997) *The Role of Plant-Animal Mutualisms in the Design and Restoration of Natural Communities*. Cambridge University Press, Cambridge.

Hansen, D.M., Donlan, C.J., Griffiths, C.J. and Campbell, K.J. (2010) Ecological history and latent conservation potential: large and giant tortoises as a model for taxon substitutions. *Ecography* 33, 272–284.

Hanski, I.A. and Gilpin, M.E. (1997) *Metapopulation Biology*. Academic, San Diego.

Harms, K.E., Wright, S.J., Calderon, O., Hernández, A. and Herre, E.A. (2000) Pervasive density-dependent recruitment enhances seedling diversity in a tropical forest. *Nature* 404, 493–495.

Heleno, R.H., Lacerda, I., Ramos, J.A. and Memmott, J. (2010) Evaluation of restoration effectiveness: community response to the removal of alien plants. *Ecological Applications* 20, 1191–1203.

Heleno, R.H., Ross, G., Everard, A., Ramos, J.A. and Memmott, J. (2011) On the role of avian seed predators as seed dispersers. *Ibis* 153, 199–203.

Heleno, R., Devoto, M. and Pocock, M. (2012) Connectance of species interaction networks and conservation value: is it any good to be well connected? *Ecological Indicators* 14, 7–10.

Heleno, R., Ramos, R. and Memmott, J. (2013a) Integration of exotic seeds into an Azorean seed dispersal network. *Biological Invasions* 15, 1143–1154.

Heleno, R., Olesen, J., Nogales, M., Vargas, P. and Traveset, A. (2013b) Seed dispersal networks in the Galápagos and the consequences of alien plant invasions. *Proceedings of the Royal Society* B280, 2012–2112.

Herrera, C.M. (1984) A study of avian frugivores, bird-dispersed plants, and their interaction in Mediterranean scrublands. *Ecological Monographs* 54, 1–23.

Herrera, C.M. (1995) Plant-vertebrate seed dispersal systems in the Mediterranean: ecological, evolutionary, and historical determinants. *Annual Review of Ecology and Systematics* 26, 705–727.

Herrera, C.M. (1998) Long-term dynamics of Mediterranean frugivorous birds and fleshy fruits: A 12-year study. *Ecological Monographs* 68, 511–538.

Herrera, J.M. and García, D. (2009) The role of remnant trees in seed dispersal through the matrix: Being alone is not always so sad. *Biological Conservation* 142, 149–158.

Higgins, S. and Richardson, D. (1999) Predicting plant migration rates in a changing world: the role of long-distance dispersal. *American Naturalist* 153, 464–475.

Higgins, S.I., Lavorel, S. and Revilla, E. (2003a) Estimating plant migration rates under habitat loss and fragmentation. *Oikos* 101, 354–366.

Higgins, S.I., Nathan, R. and Cain, M.L. (2003b) Are long-distance dispersal events in plants usually caused by nonstandard means of dispersal? *Ecology* 84, 1945–1956.

Holbrook, K.M., Smith, T.B. and Hardesty, B.D. (2002) Implications of long-distance movements of frugivorous rain forest hornbills. *Ecography* 25, 745–749.

Holl, K.D., Loik, M.E., Lin, E.H.V. and Samuels, I.A. (2000) Tropical montane forest restoration in Costa Rica: Overcoming barriers to dispersal and establishment. *Restoration Ecology* 8, 339–349.

Holmes, E.E. and Wilson, H.B. (1998) Running from trouble: Long-distance dispersal and the competitive coexistence of inferior species. *American Naturalist* 151, 578–586.

Horn, H.S. and MacArthur, R.H. (1972) Competition among fugitive species in a harlequin environment. *Ecology* 53, 749–752.

Horn, M.H., Correa, S.B., Parolin, P., Pollux, B.J.A., Anderson, J.T., Lucas, C., Widmann, P., Tjiu, A., Galetti, M. and Goulding, M. (2011) Seed dispersal by fishes in tropical and temperate fresh waters: The growing evidence. *Acta Oecologica* 37, 561–577.

Hougner, C., Colding, J. and Soderqvist, T. (2006) Economic valuation of a seed dispersal service in the Stockholm National Urban Park, Sweden. *Ecological Economics* 59, 364–374.

Houle, G. (1998) Seed dispersal and seedling recruitment of *Betula alleghaniensis*: Spatial inconsistency in time. *Ecology* 79, 807–818.

Hovestadt, T., Poethke, H.J. and Messner, S. (2000) Variability in dispersal distances generates typical successional patterns: a simple simulation model. *Oikos* 90, 612–619.

Howe, H.F. and Miriti, M.N. (2000) No question: seed dispersal matters. *Trends in Ecology & Evolution* 15, 434–436.

Howe, H.F. and Miriti, M.N. (2004) When seed dispersal matters. *Bioscience* 54, 651–660.

Howe, H.F. and Smallwood, J. (1982) Ecology of seed dispersal. *Annual Review of Ecology and Systematics* 13, 201–228.

Hyatt, L.A., Rosenberg, M.S., Howard, T.G., Bole, G., Fang, W., Anastasia, J., Brown, K., Grella, R., Hinman, K., Kurdziel, J.P. and Gurevitch, J. (2003) The distance dependence prediction of the Janzen-Connell hypothesis: a meta-analysis. *Oikos* 103, 590–602.

Iponga, D.M., Milton, S.J. and Richardson, D.M. (2008) Superiority in competition for light: A crucial attribute defining the impact of the invasive alien tree *Schinus molle* (Anacardiaceae) in South African savanna. *Journal of Arid Environments* 72, 612–623.

Jansen, P.A., Bongers, F. and van der Meer, P.J. (2008) Is farther seed dispersal better? Spatial patterns of offspring mortality in three rainforest tree species with different dispersal abilities. *Ecography* 31, 43–52.

Janzen, D.H. (1971) Seed predation by animals. *Annual Review of Ecology and Systematics* 2, 465–492.

Jones, F.A. and Muller-Landau, H.C. (2008) Measuring long-distance seed dispersal in complex natural environments: an evaluation and integration of classical and genetic methods. *Journal of Ecology* 96, 642–652.

Jordano, P. (1987) Patterns of mutualistic interactions in pollination and seed dispersal – connectance, dependence asymmetries, and coevolution. *American Naturalist* 129, 657–677.

Jordano, P. (1994) Spatial and temporal variation in the avian-frugivore assemblage of *Prunus mahaleb* – patterns and consequences. *Oikos* 71, 479–491.

Jordano, P. and Herrera, C.M. (1995) Shuffling the offspring: Uncoupling and spatial discordance of multiple stages in vertebrate seed dispersal. *Écoscience* 2, 230–237.

Jordano, P. and Schupp, E.W. (2000) Seed disperser effectiveness: The quantity component and patterns of seed rain for *Prunus mahaleb. Ecological Monographs* 70, 591–615.

Jordano, P., García, C., Godoy, J.A. and García-Castano, J.L. (2007) Differential contribution of frugivores to complex seed dispersal patterns. *Proceedings of the National Academy of Sciences of the United States of America* 104, 3278–3282.

Jordano, P., Forget, P.M., Lambert, J.E., Bohning-Gaese, K., Traveset, A. and Wright, S.J. (2011) Frugivores and seed dispersal: mechanisms and consequences for biodiversity of a key ecological interaction. *Biology Letters* 7, 321–323.

Kaiser-Bunbury, C.N., Traveset, A. and Hansen, D.M. (2010) Conservation and restoration of plant-animal mutualisms on oceanic islands. *Perspectives in Plant Ecology, Evolution and Systematics* 12, 131–143.

Katul, G.G., Porporato, A., Nathan, R., Siqueira, M., Soons, M.B., Poggi, D., Horn, H.S. and Levin, S.A. (2005) Mechanistic analytical models for long-distance seed dispersal by wind. *American Naturalist* 166, 368–381.

Kays, R., Jansen, P.A., Knecht, E.M.H., Vohwinkel, R. and Wikelski, M. (2011) The effect of feeding time on dispersal of *Virola* seeds by toucans determined from GPS tracking and accelerometers. *Acta Oecologica* 37, 625–631.

Kelly, D., Ladley, J.J., Robertson, A.W., Anderson, S.H., Wotton, D.M. and Wiser, S.K. (2010) Mutualisms with the wreckage of an avifauna: the status of bird pollination and fruit-dispersal in New Zealand. *New Zealand Journal of Ecology* 34, 66–85.

Kelm, D.H., Wiesner, K.R. and von Helversen, O. (2008) Effects of artificial roosts for frugivorous bats on seed dispersal in a Neotropical forest pasture mosaic. *Conservation Biology* 22, 733–741.

Koelewijn, H.P. (2004) Sibling competition, size variation and frequency-dependent outcrossing advantage in *Plantago coronopus. Evolutionary Ecology* 18, 51–74.

Kot, M., Lewis, M.A. and vandenDriessche, P. (1996) Dispersal data and the spread of invading organisms. *Ecology* 77, 2027–2042.

Kremen, C. and Hall, G. (2005) Managing ecosystem services: what do we need to know about their ecology? *Ecology Letters* 8, 468–479.

Levey, D.J. and Benkman, C.W. (1999) Fruit-seed disperser interactions: timely insights from a long-term perspective. *Trends in Ecology & Evolution* 14, 41–43.

Levey, D.J. and Sargent, S. (2000) A simple method for tracking vertebrate-dispersed seeds. *Ecology* 81, 267–274.

Levey, D.J., Silva, W.R. and Galetti, M. (2002) *Seed Dispersal and Frugivory: Ecology, Evolution and Conservation.* CAB International, Wallingford, UK.

Levey, D.J., Bolker, B.M., Tewksbury, J.J., Sargent, S. and Haddad, N.M. (2005) Effects of landscape corridors on seed dispersal by birds. *Science* 309, 146–148.

Levey, D.J., Tewksbury, J.J. and Bolker, B.M. (2008) Modelling long-distance seed dispersal in heterogeneous landscapes. *Journal of Ecology* 96, 599–608.

Levin, S.A., Muller-Landau, H.C., Nathan, R. and Chave, J. (2003) The ecology and evolution of seed dispersal: A theoretical perspective. *Annual Review of Ecology Evolution and Systematics* 34, 575–604.

Levine, J.M. and Murrell, D.J. (2003) The community-level consequences of seed dispersal patterns. *Annual Review of Ecology Evolution and Systematics* 34, 549–574.

Linhart, Y.B. and Grant, M.C. (1996) Evolutionary significance of local genetic differentiation in plants. *Annual Review of Ecology and Systematics* 27, 237–277.

Linnebjerg, J.F., Hansen, D.M., Bunbury, N. and Olesen, J.M. (2010) Diet composition of the invasive red-whiskered bulbul *Pycnonotus jocosus* in Mauritius. *Journal of Tropical Ecology* 26, 347–350.

Loiselle, B.A. and Blake, J.G. (1999) Dispersal of melastome seeds by fruit-eating birds of tropical forest understory. *Ecology* 80, 330–336.

López-Bao, J.V. and González-Varo, J.P. (2011) Frugivory and spatial patterns of seed deposition by carnivorous mammals in anthropogenic landscapes: A multi-scale approach. *PLOS ONE* 6, e14569.

López-Darias, M. and Nogales, M. (2008) Effects of the invasive Barbary ground squirrel (*Atlantoxerus getulus*) on seed dispersal systems of insular xeric environments. *Journal of Arid Environments* 72, 926–939.

Manzano, P. and Malo, J.E. (2006) Extreme long-distance seed dispersal via sheep. *Frontiers in Ecology and the Environment* 4, 244–248.

Mark, S. and Olesen, J.M. (1996) Importance of elaiosome size to removal of ant-dispersed seeds. *Oecologia* 107, 95–101.

Markl, J.S., Schleuning, M., Forget, P.M., Jordano, P., Lambert, J.E., Traveset, A., Wright, S.J. and Böhning-Gaese, K. (2012) Meta-analysis of the effects of human disturbance on seed dispersal by animals. *Conservation Biology* 26, 1072–1081.

Marrero, P., Fregel, R., Cabrera, V.M. and Nogales, M. (2009) Extraction of high-quality host DNA from feces and regurgitated seeds: a useful tool for vertebrate ecological studies. *Biological Research* 42, 147–151.

McConkey, K.R. and Drake, D.R. (2006) Flying foxes cease to function as seed dispersers long before they become rare. *Ecology* 87, 271–276.

McConkey, K.R. and Drake, D.R. (2007) Indirect evidence that flying foxes track food resources among islands in a Pacific Archipelago. *Biotropica* 39, 436–440.

McKinney, M.L. and Lockwood, J.L. (1999) Biotic homogenization: a few winners replacing many losers in the next mass extinction. *Trends in Ecology & Evolution* 14, 450–453.

Mello, M., Marquitti, F., Guimarães, P.R., Kalko, E., Jordano, P. and Aguiar, M.d. (2011a) The missing part of seed dispersal networks: structure and robustness of bat-fruit interactions. *PLOS ONE* 6, e17395.

Mello, M., Marquitti, F., Guimarães, P.R., Kalko, E., Jordano, P. and Aguiar, M.d. (2011b) The modularity of seed dispersal: differences in structure and robustness between bat– and bird–fruit networks. *Oecologia* 167, 131–140.

Memmott, J. (1999) The structure of a plant-pollinator food web. *Ecology Letters* 2, 276–280.

Memmott, J. (2009) Food webs: a ladder for picking strawberries or a practical tool for practical problems? *Philosophical Transactions of the Royal Society B: Biological Sciences* 364, 1693–1699.

Memmott, J., Gibson, R., Carvalheiro, L., Henson, K., Heleno, R., Lopezaraiza, M. and Pearce, S. (2007) The conservation of ecological interactions. In: Stewart, A.A., New, T.R. and Lewis, O.T. (eds) *Insect Conservation Biology*. The Royal Entomological Society, London, pp. 226–244.

Mendoza, I., Gomez-Aparicio, L., Zamora, R. and Matias, L. (2009) Recruitment limitation of forest communities in a degraded Mediterranean landscape. *Journal of Vegetation Science* 20, 367–376.

Menke, S., Böhning-Gaese, K. and Schleuning, M. (2012) Plant-frugivore networks are less specialized and more robust at forest-farmland edges than in the interior of a tropical forest. *Oikos* 121, 1553–1566.

Metz, J.A.J., Dejong, T.J. and Klinkhamer, P.G.L. (1983) What are the advantages of dispersing: a paper by Kuno explained and extended. *Oecologia* 57, 166–169.

Millennium Ecosystem Assessment (2005) *Millennium Ecosystem Assessment – Ecosystems and Human Well-being: A Synthesis*. Island Press, Washington D.C.

Milton, S.J., Wilson, J.R.U., Richardson, D.M., Seymour, C.L., Dean, W.R.J., Iponga, D.M. and Proches, S. (2007) Invasive alien plants infiltrate bird-mediated shrub nucleation processes in arid savanna. *Journal of Ecology* 95, 648–661.

Montesinos-Navarro, A., Segarra-Moragues, J.G., Valiente-Banuet, A. and Verdú, M. (2012) The network structure of plant–arbuscular mycorrhizal fungi. *New Phytologist* 194, 536–547.

Montoya, D., Zavala, M.A., Rodriguez, M.A. and Purves, D.W. (2008) Animal versus wind dispersal and the robustness of tree species to deforestation. *Science* 320, 1502–1504.

Mooney, H.A., Mack, R.N., McNeely, J.A., Neville, L.E., Schei, P.J. and Waage, J.K. (2005) *Invasive Alien Species: A New Synthesis*. Island Press, Washington.

Moore, P.D. (2001) The guts of seed dispersal. *Nature* 414, 406–407.

Morales, J.M., Rivarola, M.D., Amico, G. and Carlo, T.A. (2012) Neighborhood effects on seed dispersal by frugivores: testing theory with a mistletoe-marsupial system in Patagonia. *Ecology* 93, 741–748.

Morellato, P.C. and Leitão, H.F. (1996) Reproductive phenology of climbers in a Southeastern Brazilian forest. *Biotropica* 28, 180–191.

Muller-Landau, H.C. (2010) The tolerance-fecundity trade-off and the maintenance of diversity in seed size. *Proceedings of the National Academy of Sciences of the United States of America* 107, 4242–4247.

Muller-Landau, H.C. and Hardesty, B.D. (2005) Seed dispersal of woody plants in tropical forests: concepts, examples and future directions. In: Burslem, D., Pinard, M. and Hartley, S. (eds) *Biotic Interactions in the Tropics: Their Role in the Maintenance of Species Diversity.* Cambridge University Press, New York, pp. 267–309.

Muller-Landau, H.C., Wright, S.J., Calderon, O., Condit, R. and Hubbell, S.P. (2008) Interspecific variation in primary seed dispersal in a tropical forest. *Journal of Ecology* 96, 653–667.

Murrell, D.J. and Law, R. (2003) Heteromyopia and the spatial coexistence of similar competitors. *Ecology Letters* 6, 48–59.

Nagy, E.S. and Rice, K.J. (1997) Local adaptation in two subspecies of an annual plant: Implications for migration and gene flow. *Evolution* 51, 1079–1089.

Nathan, R. (2006) Long-distance dispersal of plants. *Science* 313, 78.

Nathan, R. and Casagrandi, R. (2004) A simple mechanistic model of seed dispersal, predation and plant establishment: Janzen-Connell and beyond. *Journal of Ecology* 92, 733–746.

Nathan, R. and Muller-Landau, H.C. (2000) Spatial patterns of seed dispersal, their determinants and consequences for recruitment. *Trends in Ecology & Evolution* 15, 278–285.

Nathan, R., Katul, G.G., Horn, H.S., Thomas, S.M., Oren, R., Avissar, R., Pacala, S.W. and Levin, S.A. (2002) Mechanisms of long-distance dispersal of seeds by wind. *Nature* 418, 409–413.

Nathan, R., Katul, G.G., Bohrer, G., Kuparinen, A., Soons, M.B., Thompson, S.E., Trakhtenbrot, A. and Horn, H.S. (2011) Mechanistic models of seed dispersal by wind. *Theoretical Ecology* 4, 113–132.

Nogales, M., Delgado, J. and Medina, F. (1998) Shrikes, lizards and *Lycium intricatum* (Solanaceae) fruits: a case of indirect seed dispersal on an oceanic island. *Journal of Ecology* 86, 866–871.

Nogales, M., Hernández, E.C. and Valdés, F. (1999) Seed dispersal by common ravens *Corvus corax* among island habitats (Canarian Archipelago). *Écoscience* 6, 56–61.

Nogales, M., Medina, F., Quilis, V. and González-Rodríguez, M. (2001) Ecological and biogeographical implications of Yellow-Legged Gulls (*Larus cachinnans* Pallas) as seed dispersers of *Rubia fruticosa* Ait. (Rubiaceae) in the Canary Islands. *Journal of Biogeography* 28, 1137–1145.

Nogales, M., Nieves, C., Illera, J.C., Padilla, D.P. and Traveset, A. (2005) Effect of native and alien vertebrate frugivores on seed viability and germination patterns of *Rubia fruticosa* (Rubiaceae) in the eastern Canary Islands. *Functional Ecology* 19, 429–436.

Nogales, M., Padilla, D., Nieves, C., Illera, J. and Traveset, A. (2007) Secondary seed dispersal systems, frugivorous lizards and predatory birds in insular volcanic badlands. *Journal of Ecology* 95, 1394–1403.

Nogales, M., Heleno, R., Traveset, A. and Vargas, P. (2012) Evidence for overlooked mechanisms of long-distance seed dispersal to and between oceanic islands. *New Phytologist* 194, 313–317.

O'Dowd, D.J. and Gill, A.M. (1986) Seed dispersal syndromes in Australian Acacia. In: Murray, D.R. (ed.) *Seed Dispersal.* Academic Press, Sydney, pp. 87–121.

Olesen, J.M. and Jordano, P. (2002) Geographic patterns in plant-pollinator mutualistic networks. *Ecology* 83, 2416–2424.

Olesen, J. and Valido, A. (2003) Lizards as pollinators and seed dispersers: an island phenomenon. *Trends in Ecology & Evolution* 18, 177–181.

Olesen, J. and Valido, A. (2004) Lizards and birds as generalized pollinators and seed dispersers of island plants. In: Fernández-Palacios, J.M. and Morici, C. (eds) *Island Ecology.* Asociación Española de Ecología Terrestre & Cabildo Insular de La Palma, Palma de Mallorca, Spain, pp. 229–249.

Olesen, J.M., Bascompte, J., Dupont, Y.L. and Jordano, P. (2007) The modularity of pollination networks. *Proceedings of the National Academy of Sciences of the United States of America* 104, 19891–19896.

Olesen, J.M., Bascompte, J., Elberling, H. and Jordano, P. (2008) Temporal dynamics in a pollination network. *Ecology* 89, 1573–1582.

Orrock, J.L., Levey, D.J., Danielson, B.J. and Damschen, E.I. (2006) Seed predation, not seed dispersal, explains the landscape-level abundance of an early-successional plant. *Journal of Ecology* 94, 838–845.

Ouborg, N.J., Piquot, Y. and Van Groenendael, J.M. (1999) Population genetics, molecular markers and the study of dispersal in plants. *Journal of Ecology* 87, 551–568.

Padilla, D.P., González-Castro, A. and Nogales, M. (2012) Significance and extent of secondary seed dispersal by predatory birds on oceanic islands: the case of the Canary archipelago. *Journal of Ecology* 100, 416–427.

Padrón, B., Traveset, A., Biedenweg, T., Díaz, D., Nogales, M. and Olesen, J.M. (2009) Impact of alien plant invaders on pollination networks in two archipelagos. *Plos One* 4, e6275.

Padrón, B., Nogales, M., Traveset, A., Vilà, M., Martínez-Abraín, A., Padilla, D.P. and Marrero, P. (2011) Integration of invasive *Opuntia* spp. by native and alien seed dispersers in the Mediterranean area and the Canary Islands. *Biological Invasions* 13, 831–844.

Palmer, M.A., Ambrose, R.F. and Poff, N.L. (1997) Ecological theory and community restoration ecology. *Restoration Ecology* 5, 291–300.

Pearson, R.G. and Dawson, T.P. (2005) Long-distance plant dispersal and habitat fragmentation: identifying conservation targets for spatial landscape planning under climate change. *Biological Conservation* 123, 389–401.

Pejchar, L., Pringle, R.M., Ranganathan, J., Zook, J.R., Duran, G., Oviedo, F. and Daily, G.C. (2008) Birds as agents of seed dispersal in a human-dominated landscape in southern Costa Rica. *Biological Conservation* 141, 536–544.

Pickup, M. and Barrett, S.C.H. (2012) Reversal of height dimorphism promotes pollen and seed dispersal in a wind-pollinated dioecious plant. *Biology Letters* 8, 245–248.

Pitelka, L.F., Gardner, R.H., Ash, J., Berry, S., Gitay, H., Noble, I.R., Saunders, A., Bradshaw, R.H.W., Brubaker, L., Clark, J.S., Davis, M.B., Sugita, S., Dyer, J.M., Hengeveld, R., Hope, G., Huntley, B., King, G.A., Lavorel, S., Mack, R.N., Malanson, G.P., McGlone, M., Prentice, I.C. and Rejmanek, M. (1997) Plant migration and climate change. *American Scientist* 85, 464–473.

Pizo, M.A., Guimaraes, P.R. and Oliveira, P.S. (2005) Seed removal by ants from faeces produced by different vertebrate species. *Écoscience* 12, 136–140.

Pocock, M.J.O., Evans, D.M. and Memmott, J. (2012) The robustness and restoration of a network of ecological networks. *Science* 335, 973–977.

Pons, J. and Pausas, J.G. (2007) Acorn dispersal estimated by radio-tracking. *Oecologia* 153, 903–911.

Powell, J.A. and Zimmermann, N.E. (2004) Multiscale analysis of active seed dispersal contributes to resolving Reid's paradox. *Ecology* 85, 490–506.

Pringle, J.M., Blakeslee, A.M.H., Byers, J.E. and Roman, J. (2011) Asymmetric dispersal allows an upstream region to control population structure throughout a species' range. *Proceedings of the National Academy of Sciences of the United States of America* 108, 15288–15293.

Purves, D.W., Zavala, M.A., Ogle, K., Prieto, F. and Benayas, J.M.R. (2007) Environmental heterogeneity, bird-mediated directed dispersal, and oak woodland dynamics in Mediterranean Spain. *Ecological Monographs* 77, 77–97.

Ray, N. and Excoffier, L. (2010) A first step towards inferring levels of long-distance dispersal during past expansions. *Molecular Ecology Resources* 10, 902–914.

Rice, B. and Westoby, M. (1986) Evidence against the hypothesis that ant-dispersed seeds reach nutrient-enriched microsites. *Ecology* 67, 1270–1274.

Richardson, D.M. and Rejmánek, M. (2011) Trees and shrubs as invasive alien species – a global review. *Diversity and Distributions* 17, 788–809.

Ridley, H.N. (1930) *The Dispersal of Plants Throughout the World.* L. Reeve & Co., Kent, UK.

Riera, N., Traveset, A. and García, O. (2002) Breakage of mutualisms by exotic species: the case of *Cneorum tricoccon* L. in the Balearic Islands (Western Mediterranean Sea). *Journal of Biogeography* 29, 713–719.

Robinson, G.R. and Handel, S.N. (2000) Directing spatial patterns of recruitment during an experimental urban woodland reclamation. *Ecological Applications* 10, 174–188.

Rodríguez-Cabal, M.A., Stuble, K.L., Guenard, B., Dunn, R.R. and Sanders, N.J. (2012) Disruption of ant-seed dispersal mutualisms by the invasive Asian needle ant (*Pachycondyla chinensis*). *Biological Invasions* 14, 557–565.

Rodríguez-Pérez, J. and Traveset, A. (2007) A multi-scale approach in the study of plant regeneration: Finding bottlenecks is not enough. *Perspectives in Plant Ecology, Evolution and Systematics* 9, 1–13.

Rodríguez-Pérez, J. and Traveset, A. (2010) Seed dispersal effectiveness in a plant-lizard interaction and its consequences for plant regeneration after disperser loss. *Plant Ecology* 207, 269–280.

Rodríguez-Pérez, J. and Traveset, A. (2012) Demographic consequences for a threatened plant after the loss of its only disperser. Habitat suitability buffers limited seed dispersal. *Oikos* 121, 835–847.

Rumeu, B., Elias, R.B., Padilla, D.P., Melo, C. and Nogales, M. (2011) Differential seed dispersal systems of endemic junipers in two oceanic Macaronesian archipelagos: the influence of biogeographic and biological characteristics. *Plant Ecology* 212, 911–921.

Santos-Heredia, C., Andresen, E. and Zárate, D.A. (2010) Secondary seed dispersal by dung beetles in a Colombian rain forest: effects of dung type and defecation pattern on seed fate. *Journal of Tropical Ecology* 26, 355–364.

Schaefer, H.M. (2011) Why fruits go to the dark side. *Acta Oecologica* 37, 604–610.

Schupp, E.W. (1993) Quantity, quality and the effectiveness of seed dispersal by animals. *Vegetatio* 108, 15–29.

Schupp, E.W. and Fuentes, M. (1995) Spatial patterns of seed dispersal and the unification of plant population ecology. *Écoscience* 2, 267–275.

Schupp, E.W. and Jordano, P. (2011) The full path of Janzen-Connell effects: genetic tracking of seeds to adult plant recruitment. *Molecular Ecology* 20, 3953–3955.

Schupp, E.W., Jordano, P. and Gómez, J.M. (2010) Seed dispersal effectiveness revisited: a conceptual review. *New Phytologist* 188, 333–353.

Seidler, T.G. and Plotkin, J.B. (2006) Seed dispersal and spatial pattern in tropical trees. *Plos Biology* 4, 2132–2137.

Shiels, A.B. and Walker, L.R. (2003) Bird perches increase forest seeds on Puerto Rican landslides. *Restoration Ecology* 11, 457–465.

Silander, J.A. (1978) Density-dependent control of reproductive success in *Cassia biflora*. *Biotropica* 10, 292–296.

Simberloff, D. (2006) Invasional meltdown 6 years later: Important phenomenon, unfortunate metaphor, or both? *Ecology Letters* 9, 912–919.

Simberloff, D. (2009) The role of propagule pressure in biological invasions. *Annual Review of Ecology Evolution and Systematics* 40, 81–102.

Simberloff, D. and Holle, B.V. (1999) Positive interactions of nonindigenous species: Invasional meltdown? *Biological Invasions* 1, 21–32.

Snow, B.K. and Snow, D.W. (1971) The feeding ecology of tanagers and honeycreepers in Trinidad. *Auk* 88, 291–322.

Snow, B.K. and Snow, D.W. (1988) *Birds and Berries: A Study of an Ecological Interaction*. Poyser, London.

Soons, M.B., Nathan, R. and Katul, G.G. (2004) Human effects on long-distance wind dispersal and colonization by grassland plants. *Ecology* 85, 3069–3079.

Sorensen, A.E. (1986) Seed dispersal by adhesion. *Annual Review of Ecology and Systematics* 17, 443–463.

Srivastava, D.S. (1999) Using local-regional richness plots to test for species saturation: pitfalls and potentials. *Journal of Animal Ecology* 68, 1–16.

Steele, M.A., Bugdal, M., Yuan, A., Bartlow, A., Buzalewski, J., Lichti, N. and Swihart, R. (2011) Cache placement, pilfering, and a recovery advantage in a seed-dispersing rodent: Could predation of scatter hoarders contribute to seedling establishment? *Acta Oecologica* 37, 554–560.

Steinitz, O., Troupin, D., Vendramin, G.G. and Nathan, R. (2011) Genetic evidence for a Janzen-Connell recruitment pattern in reproductive offspring of *Pinus halepensis* trees. *Molecular Ecology* 20, 4152–4164.

Tackenberg, O. (2003) Modeling long-distance dispersal of plant diaspores by wind. *Ecological Monographs* 73, 173–189.

Terakawa, M., Isagi, Y., Matsui, K. and Yumoto, T. (2009) Microsatellite analysis of the maternal origin of *Myrica rubra* seeds in the feces of Japanese macaques. *Ecological Research* 24, 663–670.

Thies, W. and Kalko, E.K.V. (2004) Phenology of neotropical pepper plants (Piperaceae) and their association with their main dispersers, two short-tailed fruit bats, *Carollia perspicillata* and *C. castanea* (Phyllostomidae). *Oikos* 104, 362–376.

Thomas, D.W., Cloutier, D., Provencher, M. and Houle, C. (1988) The shape of bird- and bat-generated seed shadows around a tropical fruiting tree. *Biotropica* 20, 347–348.

Thompson, K. and Rabinowitz, D. (1989) Do big plants have big seeds? *American Naturalist* 133, 722–728.

Thompson, K., Gaston, K.J. and Band, S.R. (1999) Range size, dispersal and niche breadth in the herbaceous flora of central England. *Journal of Ecology* 87, 150–155.

Thomson, F.J., Moles, A.T., Auld, T.D. and Kingsford, R.T. (2011) Seed dispersal distance is more strongly correlated with plant height than with seed mass. *Journal of Ecology* 99, 1299–1307.

Tilman, D. (1994) Competition and biodiversity in spatially structured habitats. *Ecology* 75, 2–16.

Trakhtenbrot, A., Nathan, R., Perry, G. and Richardson, D.M. (2005) The importance of long-distance dispersal in biodiversity conservation. *Diversity and Distributions* 11, 173–181.

Traveset, A. (1994) Influence of type of avian frugivory on the fitness of *Pistacia terebinthus* L. *Evolutionary Ecology* 8, 618–627.

Traveset, A. (1998) Effect of seed passage through vertebrate frugivores' guts on germination: a review. *Perspectives in Plant Ecology, Evolution and Systematics* 1, 151–190.

Traveset, A. and Richardson, D.M. (2006) Biological invasions as disruptors of plant reproductive mutualisms. *Trends in Ecology & Evolution* 21, 208–216.

Traveset, A. and Richardson, D.M. (2011) *Mutualisms: Key Drivers of Invasions… Key Casualties of Invasions.* Wiley-Blackwell, Oxford, UK.

Traveset, A., Riera, N. and Mas, R.E. (2001) Passage through bird guts causes interspecific differences in seed germination characteristics. *Functional Ecology* 15, 669–675.

Traveset, A., Robertson, A.W. and Rodríguez, J. (2007) A review on the role of endozoochory on seed germination. In: Dennis, A.J., Schupp, E.W., Green, A.J. and Westcott, D.A. (eds) *Seed Dispersal: Theory and its Application in a Changing World.* CAB International, Wallingford, UK, pp. 78–103.

Traveset, A., González-Varo, J.P. and Valido, A. (2012) Long-term demographic consequences of a seed dispersal disruption. *Proceedings of the Royal Society B-Biological Sciences* 279, 3298–3303.

Turnbull, L.A., Crawley, M.J. and Rees, M. (2000) Are plant populations seed-limited? A review of seed sowing experiments. *Oikos* 88, 225–238.

Tylianakis, J.M. (2008) Understanding the web of life: The birds, the bees, and sex with aliens. *Plos Biology* 6, e47.

Tylianakis, J.M., Laliberté, E., Nielsen, A. and Bascompte, J. (2010) Conservation of species interaction networks. *Biological Conservation* 143, 2270–2279.

Valladares, F. and Gianoli, E. (2007) How much ecology do we need to know to restore Mediterranean ecosystems? *Restoration Ecology* 15, 363–368.

Van der Pijl, L. (1982) *Principles of Dispersal in Higher Plants.* Springer-Verlag, Berlin.

Vander Wall, S.B. (2002) Masting in animal-dispersed pines facilitates seed dispersal. *Ecology* 83, 3508–3516.

Vander Wall, S.B. and Longland, W.S. (2004) Diplochory: are two seed dispersers better than one? *Trends in Ecology & Evolution* 19, 155–161.

Vargas, P., Heleno, R., Traveset, A. and Nogales, M. (2012) Colonization of the Galápagos Islands by plants with no specific syndromes for long-distance dispersal: a new perspective. *Ecography* 35, 33–43.

Vázquez, D.P., Morris, W.F. and Jordano, P. (2005) Interaction frequency as a surrogate for the total effect of animal mutualists on plants. *Ecology Letters* 8, 1088–1094.

Verdú, M. and Traveset, A. (2004) Bridging meta-analysis and the comparative method: a test of seed size effect on germination after frugivores' gut passage. *Oecologia* 138, 414–418.

Voigt, F.A., Arafeh, R., Farwig, N., Griebeler, E.M. and Bohning-Gaese, K. (2009) Linking seed dispersal and genetic structure of trees: a biogeographical approach. *Journal of Biogeography* 36, 242–254.

Wang, B.C. and Smith, T.B. (2002) Closing the seed dispersal loop. *Trends in Ecology & Evolution* 17, 379–386.

Wang, B.C., Sork, V.L., Leong, M.T. and Smith, T.B. (2007) Hunting of mammals reduces seed removal and dispersal of the afrotropical tree *Antrocaryon klaineanum* (Anacardiaceae). *Biotropica* 39, 340–347.

Wenny, D.G. (2001) Advantages of seed dispersal: A re-evaluation of directed dispersal. *Evolutionary Ecology Research* 3, 51–74.

Wenny, D.G. and Levey, D.J. (1998) Directed seed dispersal by bellbirds in a tropical cloud forest. *Proceedings of the National Academy of Sciences of the United States of America* 95, 6204–6207.

Westcott, D.A. and Fletcher, C.S. (2011) Biological invasions and the study of vertebrate dispersal of plants: Opportunities and integration. *Acta Oecologica* 37, 650–656.

Westoby, M., Leishman, M. and Lord, J. (1996) Comparative ecology of seed size and dispersal. *Philosophical Transactions of the Royal Society B: Biological Sciences* 351, 1309–1317.

Wheelwright, N.T. (1985) Fruit-size, gape width, and the diets of fruit-eating birds. *Ecology* 66, 808–818.

Whittaker, R.J. and Fernández-Palacios, J.M. (2007) *Island Biogeography: Ecology, Evolution, and Conservation.* Oxford University Press, Oxford, UK.

Williams, P.A. (2006) The role of blackbirds (*Turdus merula*) in weed invasion in New Zealand. *New Zealand Journal of Ecology* 30, 285–291.

Williamson, M. (1983) *Island Populations.* Oxford University Press, Oxford.

Wills, C., Condit, R., Foster, R.B. and Hubbell, S.P. (1997) Strong density- and diversity-related effects help to maintain tree species diversity in a neotropical forest. *Proceedings of the National Academy of Sciences of the United States of America* 94, 1252–1257.

Willson, M.F. (1993) Dispersal mode, seed shadows, and colonization patterns. *Vegetatio* 108, 261–280.

Willson, M.F. and Thompson, J.N. (1982) Phenology and ecology of color in bird-dispersed fruits, or why some fruits are red when they are 'green'. *Canadian Journal of Botany* 60, 701–713.

Willson, M.F. and Whelan, C.J. (1993) Variation of dispersal phenology in a bird-dispersed shrub, *Cornus drummondii. Ecological Monographs* 63, 151–172.

Willson, M.F., Irvine, A.K. and Walsh, N.G. (1989) Vertebrate dispersal syndromes in some Australian and New Zealand plant communities, with geographical comparisons. *Biotropica* 21, 133–147.

Willson, M.F., Michaels, H.J., Bertin, R.I., Benner, B., Rice, S., Lee, T.D. and Hartgerink, A.P. (1990) Intraspecific variation in seed packaging. *American Midland Naturalist* 123, 179–185.

Wotton, D.M. and Kelly, D. (2011) Frugivore loss limits recruitment of large-seeded trees. *Proceedings of the Royal Society B-Biological Sciences* 278, 3345–3354.

Wright, S.J. (1980) Density compensation in island avifaunas. *Oecologia* 45, 385–389.

Wright, S.J., Zeballos, H., Domínguez, I., Gallardo, M.M., Moreno, M.C. and Ibáñez, R. (2000) Poachers alter mammal abundance, seed dispersal, and seed predation in a neotropical forest. *Conservation Biology* 14, 227–239.

Wright, S.J., Muller-Landau, H.C. and Schipper, J. (2009) The future of tropical species on a warmer planet. *Conservation Biology* 23, 1418–1426.

Young, L.M., Kelly, D. and Nelson, X.J. (2012) Alpine flora may depend on declining frugivorous parrot for seed dispersal. *Biological Conservation* 147, 133–142.

Yu, D.W. and Wilson, H.B. (2001) The competition-colonization trade-off is dead: Long live the competition-colonization trade-off. *American Naturalist* 158, 49–63.

Zahawi, R.A. and Augspurger, C.K. (2006) Tropical forest restoration: Tree islands as recruitment foci in degraded lands of Honduras. *Ecological Applications* 16, 464–478.

Zamora, R., Antonio Hodar, J., Matias, L. and Mendoza, I. (2010) Positive adjacency effects mediated by seed disperser birds in pine plantations. *Ecological Applications* 20, 1053–1060.

Zhou, H.P. and Chen, J. (2010) Spatial genetic structure in an understorey dioecious fig species: the roles of seed rain, seed and pollen-mediated gene flow, and local selection. *Journal of Ecology* 98, 1168–1177.

Ziegenhagen, B., Liepelt, S., Kuhlenkamp, V. and Fladung, M. (2003) Molecular identification of individual oak and fir trees from maternal tissues of their fruits or seeds. *Trees, Structure and Function* 17, 345–350.

4 Seed Predators and Plant Population Dynamics

Michael J. Crawley*

*Department of Biology, Imperial College London,
Ascot, Berkshire, UK*

Introduction

The enormous seed production of most plants, coupled with the general paucity of seedlings and saplings, is vivid testimony to the intensity of seed mortality. The degree to which this mortality results from seed predation (the consumption and killing of seeds by granivorous animals) is the subject of the present review (for earlier references, see Crawley, 2000). To people who are unfamiliar with Darwinist thinking, it appears obvious that seed mortality is so high, because 'plants need only to leave one surviving offspring in a lifetime'. In fact, of course, every individual plant is struggling to ensure that its own offspring make up as big a fraction as possible of the plants in the next generation, and for each individual plant there is a huge evolutionary gain to be achieved by leaving more surviving seedlings, i.e. more copies of its genes, in the next generation. The mass mortality of seeds is part of natural selection in action, and the group selectionist argument that plants only need to replace themselves is simply wrong. Since the numbers of seeds produced are so large, it only takes a small percentage change in seed mortality to make

a massive difference to the number of seedlings that survives. The second point that needs to be emphasized is the difference between plant population consequences of seed feeding, which need to be evaluated over the entire lifespan (effects on seedling numbers need not necessarily translate to effects on adult plant numbers or total seed production in the next generation), and plant evolutionary consequences (which need have no impact at all on plant population size, if selective seed predation leads to changes in the frequency of fitness-affecting genes in the next generation).

Estimates of the magnitude of seed predation are common in agricultural folklore. For example, the traditional Northumbrian recipe for sowing cereal seeds runs like this: 'One for the rook, One for the crow, One for the pigeon, And one to grow', affording an estimate of 75% seed predation that agrees remarkably closely with results from carefully controlled experiments.

It is useful to deal with predispersal and postdispersal seed predators separately. Perhaps the most important reason is that the cast of characters is so different. Most of the species involved in predispersal seed predation are small, sedentary, specialist

* E-mail: m.crawley@imperial.ac.uk

feeders belonging to the insect orders Diptera, Lepidoptera, Coleoptera and Hymenoptera. In contrast, the postdispersal seed predators tend to be larger, more mobile, generalist herbivores such as rodents and granivorous birds (Vander Wall *et al.*, 2005). Some insects such as ants, of course, are important postdispersal seed predators, especially in deserts and nutrient-poor communities (Inouye *et al.*, 1980; Hulme, 1998; Wilby and Shachak, 2000; Ines Pirk and Lopez de Casenave, 2010; Ferreira *et al.*, 2011; Edelman, 2012). Other invertebrate herbivores such as slugs and snails can be important both as pre- and postdispersal predators (Ehrlen, 2002; Allan and Crawley, 2011).

From the seed's perspective it is also clear that pre- and postdispersal predation are different, if only because the costs of predispersal seed defence can be drawn from the resources of the parent plant, whereas the costs of postdispersal defence must all be borne by the independent seed. This has major implications for the chemistry and resource costs of pre- and postdispersal defence (Janzen, 1969).

Finally, the spatial distributions of the seeds and the implications for predator foraging are quite distinct (Munoz and Bonal, 2011; Beckman *et al.*, 2012). Predispersal seed predators exploit a spatially and temporally aggregated resource and can use searching cues based on a conspicuous parent plant. Postdispersal seed predators usually have no such cues, and must search for inconspicuous items scattered or buried, often at low density, against a cryptic background.

Seed predation differs from other kinds of herbivory in several ways (Janzen, 1969; Crawley, 1983). Even the most fervent advocates of the unimportance of competition among herbivores such as Hairston (Hairston, 1997) have to admit that granivores are different. Indeed, some of the most convincing demonstrations of inter-specific competition for resources have come from studies on granivorous desert rodents and ants. Notable is the classic series of experiments carried out by Jim Brown and his colleagues (Brown and Heske, 1990); other more recent field studies confirm the importance of

seed predation for plant community structure (Ferreira *et al.*, 2011, Maron and Pearson, 2011). Among the reasons that granivory differs from other kinds of herbivory are: (i) food quality is relatively high, e.g. protein nitrogen levels are higher in seeds than in many other plant tissues; (ii) the food is parcelled into discrete packets that are often too small to allow the full development of an insect herbivore in a single seed; (iii) the seeds are available on the plant for only a brief period; and (iv) the production of seeds in any one year is typically less predictable than the production of other plant resources such as new leaves.

Patterns of Seed Production

Seed production varies with the weather, with plant density, the size structure of the plant population, the proportion of plants of flowering size, pollination rates, the level of defoliation and, in masting or alternate-bearing species, with the recent history of seed production. As perceived by seed-feeding animals, therefore, the annual seed crop is a highly variable and unpredictable resource.

Most of the classic long-term studies of annual plant demography have paid little or no attention to herbivory of any kind, let alone to the impact of seed predation (Symonides *et al.*, 1986, Rees *et al.*, 1996). Little is known about the degree to which seed predators influence the population dynamics of annual plants over the long term, but a number of studies point to important short-term effects (Kuang and Chesson, 2008). These models involve generalist, postdispersal granivores, and it remains to be demonstrated that specialist predispersal seed feeders have important effects on the population dynamics of annual plants. For many annual plant populations, the impact of seed predation is buffered by recruitment from the bank of dormant seeds in the soil (Bohan *et al.*, 2011, Dalling *et al.*, 2011), or by the immigration of wind-borne propagules from elsewhere (Aguilar *et al.*, 2006). Thus, quite large changes in the seed predation rate may have no measurable impact on plant recruitment, at least in the short term.

Herbaceous perennials often show wide variation in the proportion of individuals producing seeds in any year (Eriksson, 1989), and in the proportion of ramets that flower within a given individual (Maron and Simms, 1997). The size structure of the plant population is important, because fluctuations in the proportion of plants in the reproductive size classes can be substantial. Thus, the size of the seed crop per plant can vary by several orders of magnitude from one year to the next (Kelly, 1994). In some extreme cases where seed set is pollinator limited, the production of any seed at all may be the exception as in the lady's slipper orchids studied by Gill (1989) where in only 1 year out of 10 did more than 5% of the plants produce seed, and in 4 years out of 10 there was complete reproductive failure, despite the fact that hand-pollinated individuals set good seed crops every year. Such extreme examples of pollinator limitation are uncommon, and the consensus is that pollination limitation of seed production is the exception rather than the rule (Burd, 1994; Kearns *et al.*, 1998; Ashman *et al.*, 2004; Klein *et al.*, 2007).

We have rather few long-term records on the fecundity of individual trees, and records are skimpy even for forest seed orchards. Data on commercial fruit production in orchards are of limited value in understanding tree fecundity, because it is not clear how changes in the numbers of seeds are related to changes in the yield of fruit. The few data that are available confirm the view obtained from shorter-lived plants, that seed production is highly variable from year to year and from individual plant to plant (Crone *et al.*, 2011; Visser *et al.*, 2011). Thus, while it is clear that there are good years and bad years for seed production, individual trees are consistently found to be more fecund than others, suggesting both genetic and microenvironmental causes (Archibald *et al.*, 2012).

The variation in seed crop is exacerbated in trees such as ash *Fraxinus excelsior* (Tapper, 1996) and beech *Fagus sylvatica* (Nilsson and Wastljung, 1987) by the propensity of the plants to produce large quantities of empty fruits, often in the year before producing a large crop of viable seeds. Thus, what look superficially to be good seed years, turn out on close inspection of the fruits to be years of low seed production. The cause of this failure to fill the seeds may involve low pollination success, but whether or not the behaviour is adaptive as an anti-predator strategy (e.g. through effects on reducing reproductive success of the granivores in the year before a mast crop) is not yet known.

Predispersal Seed Predation

A great many studies have documented the impact of specialist seed feeders on plant fecundity. The references in Crawley (2000) were selected to illustrate the range of impacts attributable to different kinds of seed predators in different habitats, and to demonstrate the substantial variation in seed predation rates in the same system from year to year and place to place.

An example of the issues involved in predispersal seed predation is afforded by the alien cynipid gall former *Andricus quercuscalicis* (the Knopper gall) which attacks the acorns of *Quercus robur* in its agamic generation, killing between 50% and 95% of the acorns in any year and between 0% and 100% of the acorns on any individual tree. The mechanism of alternate bearing is clearly effective as a means of predator satiation for this insect (Crawley and Long, 1995). The problem is that we need to consider the impact of the guild of vertebrate postdispersal acorn predators/ seed dispersers (jays (*Garrulus glandarius*), pigeons (*Columba palumbus*), woodmice (*Apodemus sylvaticus*), pigs (*Sus scrofa*), etc.) in attempting to get a longer-term perspective on tree recruitment and lifetime fitness in *Quercus robur*.

Differential seed predation may reverse the relative production of viable seed in closely related coexisting plant species, so that the species with the higher potential seed production produces fewer seeds because it suffers a disproportionately high rate of seed predation. For example, in the two *Astragalus* spp. studied by Green and Palmbald (1975), the more fecund *A. cibarius* producing 1400 seeds per plant suffered total seed predation of 93% over 3 years: 74% from bruchids and 19% from chalcid wasps and hemipterans, whereas the less fecund *A. utahensis* producing 1100 seeds suffered an average 60% bruchid predation. This meant that an average individual of *A. cibarius* dispersed only 19 intact seeds compared with 327 for *A. utahensis*. This is just the kind of predator-induced switch in demographic performance that can reduce the likelihood of competitive exclusion and lead to enhanced species richness.

An important problem with many of these studies is the lack of adequate spatial replication and their extremely short-term nature. This means that they provide little more than snapshots of what is clearly a highly dynamic process (Edwards and Crawley, 1999a). In particular, it has only rarely been established that there is a positive relationship between seed production and the recruitment of juvenile plants, which would suggest that predispersal seed predation might have an important role in plant population dynamics (Crawley and Long, 1995; Crawley, 1997). Even less often has the per-seed probability of recruitment been estimated for the same plant species in a variety of habitats over a representative sample of years (Louda and Potvin, 1995).

Postdispersal Seed Predation

Once the seeds have been scattered to the winds, they fall prey to a different suite of seed predators. Generalist herbivores such as small mammals become extremely important at this stage, but larger insects such as ants, lygaeid bugs and carabid beetles are significant postdispersal seed predators in many habitats (Velho *et al.*, 2012). As with predispersal seed predators, the literature on the impact of these seed predators is voluminous. The protocols of some of the studies are rather suspect, because they involve commercially available seeds

that are unfamiliar to the resident seed predators. These might be expected to suffer either untypically high rates of predation if the seeds were especially attractive or untypically low rates if their unfamiliarity led to their being avoided. This difficulty aside, it is clear that postdispersal seed predation rates can be very high and that they are extremely variable in space and time (Crawley, 2000). Notice, in particular, how often 100% postdispersal seed losses are recorded.

Seed burial is a major determinant of the predation rate, and loss rates are substantially lower for buried seed than for seed exposed on the surface (Crawley and Long, 1995). For buried seed, the probability of predation increases rapidly with seed size. Once buried, small seeds are relatively secure from small mammal predation, but rodents will dig up larger seeds from considerable depths (Hulme, 1996). The nature of the soil in which the seeds are buried is also important, and seed recovery rates by rodents have been shown to depend on the texture of the soil and the particle size distribution relative to the size of the seeds in question (Vander Wall, 1998).

Seed-bank Predation

There is a great scarcity of data on mortality attributable to predators of dormant or quiescent seeds in the soil (Bohan *et al.*, 2011; Dalling *et al.*, 2011). Protecting buried seeds in cages with a variety of mesh sizes should enable data on subterranean seed predation rates to be gathered without much difficulty, but so far as I know, such experiments have not been carried out.

Given the tremendous longevity of many species' seeds, especially the seeds of ruderal plants (Rees, 1994), we would predict that the protracted exposure to the risk of predation would select for seeds that were relatively unpalatable to generalist seed predators and too small for specialist, arthropod seed feeders to complete their larval development within a single seed. The data appear to conform to these predictions (Bohan *et al.*, 2011), but much more work is needed on this component of seed predation (Edwards and Crawley, 1999b).

Seed dormancy provides an extremely powerful buffer against the ravages of seed predation, and the existence of a large bank of dormant seeds may mean that wide fluctuations in predispersal seed predation have little or no impact on plant dynamics (Crawley, 1990). Thus, the old gardening proverb 'One year's seeding is seven years' weeding' illustrates the degree to which heavy mortality inflicted by seed predators in one year can be compensated by recruitment from the seed bank in the following years. The failure of many apparently successful establishments of biological weed control agents is often attributed to the existence of a seed bank, so that even seed-feeding insects that destroy over 90% of the annual seed crop may have no long-term impact in reducing the rate of plant recruitment. For example, gorse *Ulex europaeus* in New Zealand loses most of its seeds to the introduced weevil *Apion ulicis*, but it remains a serious pest of pasture land because its seeds can survive for a quarter of a century or more in the seed bank (Miller, 1970).

Mast Fruiting

Masting is a traditional forester's term that has taken on a rather precise ecological meaning (Crawley and Long, 1995). The core of the idea is simply that synchronous plant reproduction over wide geographical areas leads to satiation of both the specialist and generalist seed predators, with the added bonus that specialist seed predators are kept scarce during the inter-mast period (Crone *et al.*, 2011; Visser *et al.*, 2011). The rationale for the evolution of the masting habit is based on five putative selective advantages (Smith *et al.*, 1990): (i) populations of seed predators are reduced during years of small seed crops, and hence unable to exploit a high percentage of the seeds during mast years; (ii) larger mast crops cause seeds to be dispersed over greater distances; (iii) the use of weather cues for the timing of mast crops may also provide optimal conditions for reproductive growth; (iv) weather cues may predict optimum future conditions for seed germination and seedling growth; and (v) concentrations of pollen production in mast years increase the probability of pollination for wind-pollinated species. Taking these conditions together leads to the prediction that masting species should be long-lived, wind-pollinated plants with large, edible seeds that would otherwise suffer high rates of seed predation. While many masting species fit comfortably into this classification, there are a sufficiently large number of exceptions to caution against complacency. It is not at all obvious, for example, why insect-pollinated plants with small seeds and fleshy, animal-dispersed fruits should be mast producers. Also, while predator satiation usually works reasonably well in most mast-fruiting species, this is not always the case, as for example in cycads (Ballardie and Whelan, 1986).

Alternate bearing is distinct from mast fruiting. Not only is the period shorter, but the interval between seed crops is relatively predictable. This has important evolutionary implications for the life history evolution of the insects that specialize on the seeds of alternate-bearing and masting plants, e.g. their investment in protracted dormancy. Also, weather cues are substantially more important in masting than in alternate-beating species (Crawley and Long, 1995; Tapper 1996). It seems useful, therefore, to treat these two categories of fruiting behaviour separately, defining them on the basis of the duration of the inter-crop interval as follows: (i) alternate bearing – 2 sometimes 3 years; and (ii) masting – 3 or more years but with the interval unpredictable.

There is certainly plenty of evidence for the effectiveness of predator satiation in heavy seed crops (Kelly, 1994; Crone *et al.*, 2011; Visser *et al.*, 2011). Both pre- and postdispersal seed predators tend to inflict substantially lower percentage mortality during the mast year than in either the preceding or subsequent years. What is less clear, however, is the relationship between plant recruitment and masting. Rather few studies have shown that successful recruits are significantly more likely to come from a mast crop. It is plausible that they should do so, but largely unproven (see Crawley and Long, 1995 for an exception in which recruitment peaked following peak years of acorn production in oak *Quercus robur*).

Variability in Seed Mortality

A recurrent theme in the literature on seed mortality is the variability in the risk of death from predation to which each seed is exposed. The probability of death varies with plant density, seed crop size, within-season phenology, pollination rate, spatial location, weather conditions, predator density and the availability of alternative foods for generalist seed predators. This whole suite of probabilities varies from year to year with changes in the weather, and the same change in weather may produce a different change in seed death rate in different habitats and at different stages of succession (Desteven, 1991; Ostfeld *et al.*, 1997; Bishop, 2002; Benkman and Siepielski, 2004). The mortality suffered by dormant seeds in the soil is likely also to vary in response to events that occurred during ripening on the parent plant (e.g. phenotypic variation in seed size and dormancy state), as well as to events occurring after addition to the pool of seeds in the soil (e.g. patterns of wetting and drying, degree of chill experienced and the extent of seed coat scarification (Edwards and Crawley, 1999b)).

Numerous studies have documented the extent of phenotypic variation in seed predation rates suffered by different individual plants in the same year (Thompson and Baster, 1992; Ehrlen, 1995; Ollerton and Lack, 1996; Manson *et al.*, 1998; Russell and Schupp, 1998). Rather few studies have investigated whether there is a genetic basis to these differences, and so we have no clear picture of the relative importance of microhabitat heterogeneity and genetic polymorphism in generating the observed patterns of heterogeneity in seed predation. Nevertheless, this variation in the risk of death can be a potent force in promoting coexistence and enhancing plant diversity,

because it allows the inferior competitors a refuge from competitive exclusion (Pacala and Crawley, 1992; Clark *et al.*, 2012).

The Impact of Seed Predation on Plant Population Dynamics

The impact of seed predation on plant recruitment is summarized in Fig. 4.1. When seed densities are low, as they are when plant population density is low, or when a site is a long way from patchily distributed parent plants, then seed predation will inevitably reduce plant recruitment. This is because at low seed densities there is

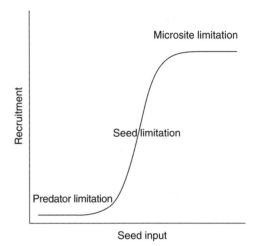

Fig. 4.1. Recruitment limitation in plants. Recruits are defined as juvenile plants that are sufficiently well established that they have a reasonable chance of surviving to reproductive maturity (in distinction to seedlings, most or all of which will die within a few months of germination). There are three distinct circumstances. At low rate of seed input, seed predators may consume all of the seed input, with the result that recruitment is prevented. This is the region of predator-limited recruitment. At intermediate seed densities recruitment will be proportional to seed input. This region of seed-limited recruitment is where predispersal seed-feeding herbivores might be expected to influence plant dynamics. At high rates of seed input, recruitment is limited by microsite availability, and substantial reductions in seed production by predispersal or postdispersal seed predators might have no measurable impact on plant dynamics.

no opportunity for compensatory reductions in other mortality factors, i.e. compensation is only possible when recruitment is subject to density-dependent control.

When seed densities are high (Fig. 4.1), as in monocultures or close to parent plants in species-rich habitats, then there may be intense competition for access to suitable recruitment microsites. Under these circumstances, large reductions in seed density resulting from seed predation may lead to no measurable reduction in the number of plants that recruit to the juvenile population. Thus, when recruitment is microsite limited, seed predation is unlikely to have any impact on mature plant density.

Whether or not reduced seedling recruitment leads to reduced densities of adult plants depends upon the spacing of recruitment microsites and on the various density-dependent interactions of the adult plants (see Harper (1977) and Crawley (1990) for detailed discussion). The essential issues involved are beautifully encapsulated in the Parable of The Sower (Matthew 13:4):

> Some seeds fell by the wayside, and the fowls came and devoured them up: Some fell upon stony places, where they had not much earth: and forthwith they sprung up, because they had no deepness of earth: And when the sun was up they were scorched; and because they had no root they withered away. And some fell among thorns; and the thorns sprung up and choked them: But others fell into good ground, and brought forth fruit, some an hundredfold, some sixtyfold, some thirtyfold.

Note that as well as cataloguing the hazards facing the seed (predation, unsuitable microsites and plant competition), this provides the first published estimates of the intrinsic rate of increase of an annual plant ($3.4 < r < 4.6$).

The first issue, therefore, is to establish whether plant recruitment is seed limited, and if it is not, to understand why not (microsites or predators, for instance; Fig. 4.1). The traditional view is that annual plants are seed limited because they need to reproduce each year, and that herbaceous perennials are not seed limited, because

seedlings are so rarely found in undisturbed perennial vegetation. Both these views are oversimplified. For many annual plants, the evidence suggests that recruitment is not seed limited, either because the seed bank is sufficiently large to make good any temporary shortfall in seed input caused by seed predation (Sheppard et al., 2002; Dalling et al., 2011), or because of microsite limitation. Again, a more detailed study of the literature on seedling recruitment in perennial communities shows that seedlings are present in about 40% of cases (Turnbull et al., 2000). Of course, the mere presence of seedlings does not mean that recruitment is seed limited, but it does indicate that seedling recruitment is possible.

The simple experiment of sowing extra seeds and measuring whether there is any extra plant recruitment is still worth doing, especially if seed addition is crossed with granivore exclusion and soil disturbance. In one unpublished study, it was found that of 20 species of native grassland plants sown into undisturbed mesic grassland at a rate of 1000 extra seeds m^{-2}, only two species, *Lotus corniculatus* and *Rumex acetosella*, showed enhanced recruitment. More recently, we have discovered that *Arrhenatherum elatius, Heracleum sphondylium* and *Centaurea nigra* are seed limited in Nash's Field, but that 60 other species of herbs and grasses are not seed limited (Edwards and Crawley, 1999a). In former times, prior to major human impact, it may well have been that bigger, more abundant herbivores caused higher rates of disturbance than are currently observed in grassland communities, and that seed limitation was consequently more frequent than it is today.

The balance of this meagre evidence suggests that seed-limited recruitment is the exception rather than the rule, and that the expectation, therefore, is that seed predation will tend to have rather little impact on plant recruitment (Buckley et al., 2005). This prediction is borne out by the weed biocontrol literature, which shows that seed-feeding agents have a relatively low rate of success (Sheppard, 1992; Paynter et al., 1996).

Convincing experimental demonstrations for the role of seed predators on plant dynamics have tended to come from arid and semiarid communities where the open vegetation means that microsite limitation is less likely to be important (Louda and Potvin, 1995). Most of these studies were relatively short term, and it is possible that the observed dynamics were merely transitory rather than equilibrium responses to changes in seed predation rate. Long-term exclusion experiments in the south-western deserts of the USA leave no doubt, however, about the potential importance of seed-feeding animals in plant community structure (Brown and Heske, 1990).

It is important to contrast the evolutionary impact and the demographic impact of seed predation. If there is genetic variation for traits affecting the rate of seed predation, and there is differential seed predation from one plant individual to another, then the potential exists for the evolution of seed-defending mechanisms. This evolution can occur whether or not the seed predators have any impact on plant population dynamics.

There is abundant evidence of differential seed predation between individual plants (see above) but the genetic basis of these differences has rarely been established, and it is not possible to disentangle the effects of spatial variations in seed quality and predator density from any genetic component. A review of the impact of predation on seed evolution is beyond the remit of the present chapter, but a useful summary is to be found in Estrada and Fleming (1986).

Compensation

Plants often produce many more flowers than they could ever turn into ripened fruits packed with proteins and carbohydrates. This allows substantial scope for compensating for predispersal seed predation through the differential abortion of damaged fruits prior to seed fill (Crawley, 1997). This kind of compensation requires that fruit production is not pollinator limited, and compensation for predispersal seed

predation is not likely to be important in habitats or in years when pollinators are scarce (Nilsson and Wastljung, 1987; Gill, 1989; Brody and Mitchell, 1997).

There have been rather few carefully controlled, long-term manipulative experiments on compensation for predispersal seed predation; see review in Crawley (1997). There is tremendous scope for further experimentation using well-replicated pairs of plants, one of which is kept free of seed predators by cage exclosure or with systemic insecticides and the other is infested by natural densities of seed predators.

What of the potential for postdispersal compensation? For the plant population as a whole, there may be considerable scope for compensation through predator satiation (see Fig. 4.1). But in terms of individual plant fitness, the options are dramatically reduced as soon as the seed leaves the vicinity of the parent plant. In order to benefit from competitor release, the seedling would need to obtain an advantage from growing among a lower density of conspecific seedlings of the same parentage. At a distance from the parent plant, any given genotype of seed is almost certain to be at the left-hand end of the x axis in Fig. 4.1, where reductions in density lead to reduced probability of recruitment. Thus, what is good for the species (e.g. predator satiation) is not likely to be good for the particular genotypes that are eaten.

Mutualist Seed Predators

The natural history literature is rich in examples of animals that increase plant fitness by dispersing their seeds. When dispersal is achieved by seed predators, the plant pays the price of a certain degree of seed mortality for the evident benefits of removal from the parent plant and the escape from competitors and natural enemies that this normally entails (Zwolak and Crone, 2012). A parallel may be drawn with the mutualism that involves pollinators, as in the classic examples of fig wasps, yucca moths, or the dipteran pollinators of globe flowers where ovules are sacrificed to obtain cross-pollination. See Crawley (1997) for references.

In the case of seed-dispersing predators, we know something about the distances that seeds are moved (jays are recorded as taking the acorns of *Quercus robur* over 20 km and mice can move acorns into Danish heathlands at a rate of about 35 m yr^{-1}), but rather little about the mortality that is inflicted. The best information comes from seed caching species, and periodic sampling of rodent caches provides accurate data on seed survival (Vander Wall *et al.*, 2005). In the case of scatter-hoarding vertebrates like jays this kind of work is more difficult. Thus, while it is reasonable to suppose that seed-dispersing seed predators do indeed increase the fitness of the plants from which they feed, we have no convincing data to demonstrate it.

Case Histories

Small mammals in grassland

There is no doubt that small mammals are important postdispersal seed predators (Ferreira *et al.*, 2011; Maron, 2011; Munoz and Bonal, 2011). The question is whether or not they have an important impact on plant community structure and dynamics, or whether they simply kill seeds that would have died in any case for other reasons. Some long-term studies have provided unequivocal evidence of the impact of granivorous rodents on desert plant communities (Brown and Heske, 1990), but their role in mesic grasslands is less well known.

Hulme (1996) carried out a variety of seed addition and rodent exclosure studies in two grasslands in Silwood Park, Berkshire. He investigated seed and seedling predation by *Apodemus sylvaticus* and *Microtus agrestis* on a wide range of native and exotic plant species whose seeds varied in size between 0.11 mg (*Agrostis capillaris*) and 2.53 mg (*Festuca pratensis*). Small mammals removed an average of 43% of the seeds from the dry grassland site and 37%

from the meadow. Native plant species 'familiar' seeds were taken at a significantly higher rate than alien seeds in the dry grassland (49 versus 35%) but both were taken equally in the meadow habitat. Between plant species, predation rates varied from 80% for *Festuca ovina* (not present in either community) and *Dactylis glomerata* (present in both communities) to 8% for *Lotus corniculatus* and *Plantago lanceolata* both present in both communities. As in other studies of small mammal predation, there was tremendous spatial variation in predation rates.

Experimental seed burial reduced predation rates in both sites and at all times of year; in the dry grassland burial reduced seed losses by over 98% and in the meadow by between 40 and 90%. The larger the buried seed, the more likely it was to be eaten by small mammals. *Festuca pratensis*, *Lolium perenne* and *Trifolium pratense* suffered the highest rates of postburial predation and all had relatively heavy seeds, approximately 2 mg.

Small mammal exclosure increased seedling numbers but increased seedling densities mapped through to increased adult plant densities in only two cases (*Arrhenatherum elatius* and *Centaurea nigra*). In most cases, seedling densities were typically low, so there appears to be little scope for compensatory reductions in other mortality factors, and it is likely that small mammal predation does, indeed, reduce the density, at least of large-seeded plant species in these Silwood grasslands. It is interesting that none of the dominant species in these sites suffered significant seedling density reductions under small mammal predation, e.g. *Holcus lanatus*, *Festuca rubra* and *Plantago lanceolata*, whereas the rare and absent species suffered heavy predation on both seeds and seedlings (Edwards and Crawley, 1999c).

Acorns of *Quercus robur*

The seed production and seed mortality of pedunculate oak *Quercus robur* has been studied at Silwood Park for more than 30 years.

The pattern of acorn production is alternate bearing, with high and low crops alternating in strict sequence with only a few phase shifts during this time, i.e. there were two successive low acorn crops in 1983 and 1984 (see Fig. 4.2). A severe air frost in late April can completely defoliate early-flushing trees and kill their entire crop of female flowers for the year. The regrowth shoots that are produced in May and June do not bear any flowers of either sex. In a given year, it is unlikely that the total seed crop would be destroyed by spring frosts, because of the wide phenological span of the oak. Some of the latest-leafing trees will always tend to escape unscathed, because frosts do not damage flowers that are still protected within the bud.

The flowers that survive frost defoliation may be either wind or insect pollinated. Hand pollination, using pollen from a variety of male parent trees, and adding the pollen on different occasions and to different flowers of different ages, had no effect on final seed production, so there was no evidence of pollen limitation. Between pollination in May and acorn fill in August, a great many of the female flowers are aborted. Whole peduncles may wilt and fall off, and many of the flowers on each peduncle may fail to develop. Thus out of the maximum four peduncles per shoot and six flowers per peduncle (a potential of 24 acorns per shoot), perhaps only one acorn is filled. The modal number of acorns per shoot is zero, even in years of high acorn production. The rate of peduncle abortion varies from tree to tree in a consistent pattern across years, and certain individual trees always abort a higher proportion of their female flowers than other trees in the same microhabitat (Crawley and Long, 1995). The cycle of alternate bearing is therefore a cycle of differential peduncle abortion, rather than a 2-year cycle in flower production.

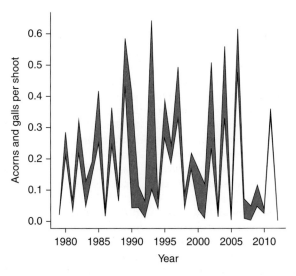

Fig. 4.2. Time series of acorn (white) and Knopper gall (grey) as mean numbers per shoot on *Quercus robur* in Silwood Park over the period 1979–2012. The broad pattern of alternate bearing has been interrupted by late spring frosts in years that would otherwise have been expected to have high acorn crops in 1984, 1991, 2001 and 2008. Only in 1993 did the Knopper gall destroy most of a peak acorn crop; in other peak acorn years, predator satiation seems to have been an effective defence against this seed feeder. A very severe winter in 2010/11 killed most of the sexual generation of the Knopper gall on their over-wintering host *Quercus cerris*, with the result that galling rates were exceptionally low in the high acorn year of 2011. The acorn crop in 2012 (extreme right-hand point) was one of the lowest ever recorded but showed signs of a recovery in Knopper gall numbers over 2011.

The young acorns are prey to two important predispersal seed predators: a native weevil *Curculio glandium* and an alien cynipid gall wasp *Andricus quercuscalicis*. These two insects exhibit strongly asymmetric inter-specific competition, because the gall wasp attacks the female flower early in the summer before the acorn has begun to fill. The Knopper gall, which develops inside the female flower, competes for resources with the developing acorn, and usually outgrows it, so that by the time the female *Curculio* are laying their eggs in late July the galled acorns are completely destroyed. This pre-emptive exploitation competition means that in years when acorns are scarce, the weevils are faced with greatly reduced resource availability as a result of gall formation. The numbers of both insects appear to be resource limited, so in years when acorns are abundant, both weevils and gall formers inflict much lower rates of acorn loss. Despite this, their population densities increase because they are taking a smaller proportion of a vastly greater total crop. In the next year of the alternate-bearing cycle, there are high insect numbers and a low acorn crop leading to high percentage attack by the gall former and intense competition between the weevils for access to sound acorns. And so the cycle goes on.

Insects are not the only predispersal seed predators of oak. A guild of vertebrates, again, curiously, one native and one alien species, takes the ripe acorns from the tree in late summer: the jay *Garrulus garrulus glandarius* and the grey squirrel *Sciurus carolinensis*. The jay is interesting because it is believed to have a positive impact on oak fitness, through the long-distance dispersal of acorns (see above). The impact of grey squirrels is strongly negative. Apart from ring barking and killing mature trees, the squirrels kill large quantities of acorns. Their acorn storage behaviour is of no benefit to the oak because they nip the embryo out of the seed before caching in order to prevent germination. It is notable that the squirrels do not nip out the embryo of red oak acorns like *Quercus borealis*; this is a spring-germinating species whose acorns are readily stored in caches through the winter. Estimates of acorn loss from the canopy for the combined guild of vertebrate acorn feeders range between less than 10% in high-seed crops and more than 40% in low years. As with insect attack rates, these predispersal losses differ enormously from tree to tree.

Once the surviving acorns have fallen to the ground they are fed upon by many of the large vertebrate herbivores (rabbits, pigs, deer, etc.). In low acorn years, these animals (especially rabbits in Silwood Park) can inflict almost 100% mortality on the seeds. Acorns placed in the field in experimental piles of different sizes and at different distances from cover are almost always removed during the first night. For example, out of ten piles of 1000 acorns, only two piles had any acorns left after 24 h, and in both these piles, the few acorns left behind were all weevily (M.J. Crawley, 2013, unpublished data). In high acorn years, the vertebrate herbivores appear to be satiated, and large numbers of sound acorns survive through the winter to produce their first seedling shoots in the following spring.

In summary, the total rate of acorn predation varies in a roughly 2-year cycle, with close to 100% seed loss in the low years, and about 50% seed predation in the years of high acorn abundance. Recruitment is proportional to acorn production in some habitats, but not in all (Crawley and Long, 1995). In many cases, there is no seedling recruitment, except after the very largest peaks of acorn production, e.g. after the massive crop of autumn of 1995 (Fig. 4.2). It must be stressed, however, that these averages hide large, and potentially important, differences in average predation rates from tree to tree and from location to location.

Seed predators of ragwort, *Senecio jacobaea*

Ragwort *Senecio jacobaea* is toxic to most vertebrate herbivores, and is a serious weed in semiarid grazing land in both northern and southern hemispheres (Cameron, 1935; Dempster, 1971). The plant becomes progressively more competitive as the grasses are eaten back and gaps in the sward are opened up in which seedling recruitment can occur. The two major predispersal seed predators are the cinnabar moth *Tyria jacobaeae* and the seed-head fly *Pegohylemyia seneciella*, both of which have been established as potential weed biocontrol agents. As with the gall wasp and the weevil on oaks, their interaction is highly asymmetric because the cinnabar moth feeds on the growing ragwort shoots, and the colonially feeding larvae can strip a plant of its flower heads (capitula) before the seed-head fly has completed its development. The density of the seed-head fly is significantly increased in areas that are kept free of cinnabar moth experimentally, but there is no reciprocal effect of the seed-head fly on the cinnabar moth (Crawley and Gillman, 1989).

Flowering in ragwort varies greatly from year to year. This is due partly to the weather, partly to cinnabar moth defoliation and partly to the recent history of flowering. Autumn and spring rains promote flowering in the following summer by ensuring that rosette growth is rapid so that a high proportion of rosettes are above the threshold size for flowering. Cinnabar moths affect flowering directly by eating flowers and flower buds and indirectly by reducing rosette size, although their impact on rosettes is great only in years when they have stripped the entire crop of flowers and buds. The history of flowering is important because plants that have produced a large seed crop die back to the base. Many of these plants die, and those that survive have to grow back from the small basal shoots. Since it is unusual for rosettes to grow large enough to flower in their first year, a mass flowering is almost always followed by a more or less protracted run of years in which flowering is sparse. At first, this is due to mass postreproductive dieback of the larger reproductive plants. Later on, cinnabar moth defoliation can lead to 100% seed losses for several successive years. Cinnabar moth feeding tends to reduce ragwort death rates because defoliation is less of a drain on resources than seed set. For a plant of a given size, there is a clear negative relationship between the probability of surviving the winter and the size of the seed crop produced the previous summer (Gillman and Crawley, 1990). In mass-flowering years like 1985 and 1990 (Fig. 4.3) the cinnabar moth population is satiated and there is mass seed production, wide dispersal and, presumably, replenishment of the seed bank. These years, however, are not inevitably associated with peaks of ragwort seedling recruitment, and it appears that in many habitats ragwort recruitment is not seed limited (Crawley and Gillman, 1989).

Rather little is known about postdispersal seed predation of ragwort. Crawley and Nachapong (1985) sowed ragwort seed into ragwort-infested grasslands without observing any increase in seedling recruitment, so it appears that recruitment was not seed limited. However, recruitment is more likely to be seed limited in open, sandy habitats like coastal dunes or the Breckland (Bonsall *et al.*, 2003).

Discussion

A cynic might argue that the only pattern to emerge from this review is that seed predation rates tend to lie somewhere between 0 and 100%. It is true that the search for simple patterns to describe the causes and consequences of variation in seed predation rates has been relatively unsuccessful. Neither seed density (Connell, 1971) nor distance from parent plants (Janzen, 1970) has proved to have the degree of explanatory power once hoped for (although recent work does show that spatial density dependence is very widespread, at least for trees (Johnson *et al.*, 2012)). On the other hand, it has become plain that variation in the rate of seed predation is the norm. Thus, we find pronounced variation from individual to individual, with some plants exhibiting consistently high rates of seed predation and others appearing to be more or less immune. There is pronounced spatial variation in predation rates, although the amount of this variation that can be explained by differences in seed density or by isolation

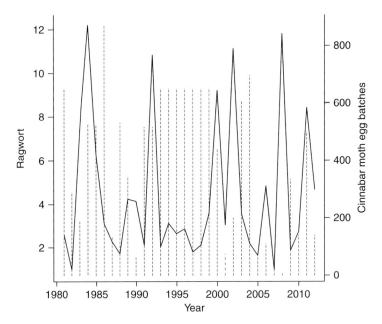

Fig. 4.3. The population dynamics of ragwort (*Senecio jacobaeae*, Asteraceae; solid line, time series showing mean flowering plant density m^{-2}) and cinnabar moth (*Tyris jacobaeae*, Lepidoptera: Arctiidae; vertical dashed lines showing mean egg batches per 50m square) in Silwood Park from 1981 to 2012. There was mass flowering ('ragwort years' > 10 flowering plants m^{-2}) on four occasions, each followed by mass mortality of the flowering plants (the individuals are typically monocarpic unless they are so severely defoliated that they set little if any seed). Flowering stems in the next year come from last year's rosettes (not shown) and last year's perennating individuals. Cinnabar moth caterpillars act as predispersal seed predators, consuming the unripe flower heads (capitulae) and are capable of causing 100% seed loss to defoliated individuals. There is no significant cross-correlation between the two time series, and no significant partial autocorrelations in the separate series. The number of flowering ragworts in year t+2 is not correlated with the number of flowering plants (a rough surrogate for seed production) in year t (p = 0.82). The number of cinnabar moths in year t+1 is not related to ragwort flowering plant density in year t (p = 0.45). Despite the occasionally very severe predispersal seed predation (well over 90% seed loss), the two series are effectively uncoupled, presumably because ragwort recruitment is more often microsite limited than seed limited. Also, when predispersal seed predation rates are very high, the plants may perennate and set seed in the following year (cinnabar moth is one of those herbivores that reduces the death rate of its host plant, which presumably explains why it has been so spectacularly poor as a weed biocontrol agent).

from seed parents may be rather low in most cases. Year-to-year differences in seed production may run to several orders of magnitude. This, in turn, leads to predator satiation in the high-seed years, and is the mechanism that underlies the typical pattern of negative density dependence found in time series data on seed predation. This variation in the risk of predation can have important effects in promoting species richness by reducing the probability of competitive exclusion (Chesson and Warner, 1985; Pacala and Crawley, 1992; also see below).

Even taking full account of the statistical shortcomings of the data, there does appear to be an interesting difference between the patterns of predispersal and postdispersal predation. While the mean rates of reported seed losses are not markedly different (the mean is a little higher in studies of postdispersal predation at 50% than in predispersal studies at 45%), the shapes of the distributions of loss rates are characteristically different. For postdispersal predation, the distribution is distinctly U-shaped with strong classes at both the 0 and 100% ends of the spectrum. The predispersal predation data show no such tendency, and their strongest classes are in the middle range, with relatively few reports of 100 or 0% losses. These figures have not changed significantly since the earlier review (Crawley, 2000).

This is consistent with what we know about the natural history of seed predation and the degree of variation in loss rates from year to year and from plant to plant. The predispersal predators tend to be invertebrate herbivores with narrow host ranges, whose numbers are rather closely coupled to seed production. In a number of cases, their populations can be reasonably well predicted from knowledge of seed production and percentage seed destruction in the previous year. Because of these time-lagged numerical responses to changes in seed density, their impact on seed production tends to fluctuate in an inverse density dependent manner, with high proportions of small seed crops destroyed, and low proportions of large crops. Many of these specialist predispersal predators also suffer relatively high rates of attack by insect parasitoids, which may explain why the range of fluctuation in seed attack rates is relatively small compared with the more generalist, postdispersal predators. This might also help to explain why alien predispersal seed predators, having been freed from their natural enemies, tend to inflict a wider range of seed mortalities than their native counterparts, e.g. the Knopper gall insect on oaks in England (see Fig. 4.2).

The consequences of these patterns for plant population dynamics are potentially great. It is now becoming clear from an expanding body of theoretical work that spatial and temporal variation in death rates can have important consequences for plant coexistence (Chesson and Warner, 1985; Chesson and Kuang, 2010). Variation in seed predation rates of the sort reported here is more than sufficient to promote species diversity in model communities. All that is required is that there is a sufficiently large number of places within a spatially heterogeneous environment where the rates of seed predation suffered by the inferior competitor happen, by chance alone, to be low at the same time as those suffered by the superior competitor happen to be high. This stochastic variation in predation rate can prevent the inherent competitive superiority of the dominant species from being expressed to the full, and hence can permit coexistence. Notice that this mechanism does not require the existence of any clear-cut deterministic processes affecting variation in seed mortality, e.g. density dependence (Connell, 1971) or frequency dependence (Janzen, 1970); although, of course, frequency dependence in seed predation would be a powerful additional force in promoting plant species coexistence (Clark *et al.*, 2012). All that is required is that differences exist in the probabilities of death in different places, and that these probabilities have a sufficiently low covariance that a refuge from competition is created. Given this view of the world, the variation in predation rates becomes the focus of interest. Instead of being a nuisance that tends to blur the estimation of crisp, deterministic parameter values, the variation becomes an important mechanism of population dynamics in its own right.

References

Aguilar, R., Ashworth, L., Galetto, L. and Aizen, M.A. (2006) Plant reproductive susceptibility to habitat fragmentation: review and synthesis through a meta-analysis. *Ecology Letters* 9, 968–980.

Allan, E. and Crawley, M.J. (2011) Contrasting effects of insect and molluscan herbivores on plant diversity in a long-term field experiment. *Ecology Letters* 14, 1246–1253.

Archibald, D.W., McAdam, A.G., Boutin, S., Fletcher, Q.E. and Humphries, M.M. (2012) Within-season synchrony of a masting conifer enhances seed escape. *American Naturalist* 179, 536–544.

Ashman, T.L., Knight, T.M., Steets, J.A., Amarasekare, P., Burd, M., Campbell, D.R., Dudash, M.R., Johnston, M.O., Mazer, S.J., Mitchell, R.J., Morgan, M.T. and Wilson, W.G. (2004) Pollen limitation of plant reproduction: Ecological and evolutionary causes and consequences. *Ecology* 85, 2408–2421.

Ballardie, R.T. and Whelan, R.J. (1986) Masting, seed dispersal and seed predation in the cycad *Macrozamia communis. Oecologia* 70, 100–105.

Beckman, N.G., Neuhauser, C. and Muller-Landau, H.C. (2012) The interacting effects of clumped seed dispersal and distance- and density-dependent mortality on seedling recruitment patterns. *Journal of Ecology* 100, 862–873.

Benkman, C.W. and Siepielski, A.M. (2004) A keystone selective agent? Pine squirrels and the frequency of serotiny in lodgepole pine. *Ecology* 85, 2082–2087.

Bishop, J.G. (2002) Early primary succession on Mount St. Helens: Impact of insect herbivores on colonizing lupines. *Ecology* 83, 191–202.

Bohan, D.A., Boursault, A., Brooks, D.R. and Petit S. (2011) National-scale regulation of the weed seedbank by carabid predators. *Journal of Applied Ecology* 48, 888–898.

Bonsall, M.B., van der Meijden, E. and Crawley, M.J. (2003) Contrasting dynamics in the same plant-herbivore interaction. *Proceedings of the National Academy of Sciences of the United States of America* 100, 14932–14936.

Brody, A.K. and Mitchell, R.J. (1997) Effects of experimental manipulation of inflorescence size on pollination and pre-dispersal seed predation in the hummingbird-pollinated plant *Ipomopsis aggregata. Oecologia* 110, 86–93.

Brown, J.H. and Heske, E.J. (1990) Control of a desert-grassland transition by a keystone rodent guild. *Science* 250, 1705–1707.

Buckley, Y.M., Rees, M., Sheppard, A.W. and Smyth, M.J. (2005) Stable coexistence of an invasive plant and biocontrol agent: a parameterized coupled plant-herbivore model. *Journal of Applied Ecology* 42, 70–79.

Burd, M. (1994) Bateman Principle and plant reproduction - the role of pollen limitation in fruit and seed set. *Botanical Review* 60, 83–139.

Cameron, E. (1935) A study of the natural control of ragwort (*Senecio jacobaea* L.). *Journal of Ecology* 23, 265–322.

Chesson, P. and Kuang, J.J. (2010) The storage effect due to frequency-dependent predation in multi-species plant communities. *Theoretical Population Biology* 78, 148–164.

Chesson, P.L. and Warner, R.R. (1985) Coexistence mediated by recruitment fluctuations: a field guide to the storage effect. *American Naturalist* 125, 769–787.

Clark, C.J., Poulsen, J.R. and Levey, D.J. (2012) Vertebrate herbivory impacts seedling recruitment more than niche partitioning or density-dependent mortality. *Ecology* 93, 554–564.

Connell, J.H. (1971) On the role of natural enemies in preventing competitive exclusion in some marine animals and in rain forest trees. In: den Boer, P.J. and Gradwell, G. (eds) *Dynamics of Populations.* PUDOC, Wageningen, pp. 298–312.

Crawley, M.J. (1983) *Herbivory: the Dynamics of Animal-Plant Interactions.* Blackwell Scientific Publications, Oxford.

Crawley, M.J. (1990) The population-dynamics of plants. *Philosophical Transactions of the Royal Society of London Series B: Biological Sciences* 330, 125–140.

Crawley, M.J. (1997) Plant-herbivore dynamics. In: Crawley, M.J. (ed.) *Plant Ecology.* Blackwell Science, Oxford, pp. 401–474.

Crawley, M.J. (2000) Seed predators and plant population dynamics. In: Fenner, M. (ed.) *Seeds: The Ecology of Regeneration in Plant Communities,* 2nd edn. CAB International, Wallingford, UK, pp. 157–191.

Crawley, M.J. and Gillman, M.P. (1989) Population dynamics of cinnabar moth and ragwort in grassland. *Journal of Animal Ecology* 58, 1035–1050.

Crawley, M.J. and Long, C.R. (1995) Alternate bearing, predator satiation and seedling recruitment in *Quercus robur* L. *Journal of Ecology* 83, 683–696.

Crawley, M.J. and Nachapong, M. (1985) The establishment of seedlings from primary and regrowth seeds of ragwort (*Senecio jacobaea*). *Journal of Ecology* 73, 255–261.

Crone, E.E., McIntire, E.J.B. and Brodie, J. (2011) What defines mast seeding? Spatio-temporal patterns of cone production by whitebark pine. *Journal of Ecology* 99, 438–444.

Dalling, J.W., Davis, A.S., Schutte, B.J. and Arnold, A.E. (2011) Seed survival in soil: interacting effects of predation, dormancy and the soil microbial community. *Journal of Ecology* 99, 89–95.

Dempster, J.P. (1971) The population ecology of the cinnabar moth, *Tyria jacobaeae* L. (Lepidoptra: Arctiidae). *Oecologia* 7, 26–67.

Desteven, D. (1991) Experiments on mechanisms of tree establishment in old-field succession – seedling emergence. *Ecology* 72, 1066–1075.

Edelman, A.J. (2012) Positive interactions between desert granivores: localized facilitation of harvester ants by kangaroo rats. *PLOS ONE* 7, e30914.

Edwards, G.R. and Crawley, M.J. (1999a) Effects of disturbance and rabbit grazing on seedling recruitment of six mesic grassland species. *Seed Science Research* 9, 145–156.

Edwards, G.R. and Crawley, M.J. (1999b) Herbivores, seed banks and seedling recruitment in mesic grassland. *Journal of Ecology* 87, 423–435.

Edwards, G.R. and Crawley, M.J. (1999c) Rodent seed predation and seedling recruitment in mesic grassland. *Oecologia* 118, 288–296.

Ehrlen, J. (1995) Demography of the perennial herb *Lathyrus vernus*. 2. Herbivory and population dynamics. *Journal of Ecology* 83, 297–308.

Ehrlen, J. (2002) Assessing the lifetime consequences of plant-animal interactions for the perennial herb *Lathyrus vernus* (Fabaceae). *Perspectives in Plant Ecology Evolution and Systematics* 5, 145–163.

Eriksson, O. (1989) Seedling dynamics and life histories in clonal plants. *Oikos* 55, 231–238.

Estrada, A. and Fleming, T.H. (1986) *Frugivores and Seed Dispersal*. Junk, Dordrecht, the Netherlands.

Ferreira, A.V., Bruna, E.M. and Vasconcelos, H.L. (2011) Seed predators limit plant recruitment in Neotropical savannas. *Oikos* 120, 1013–1022.

Gill, D.E. (1989) Fruiting failure, pollinator inefficiency, and speciation in orchids. In: Otte, D. and Endler, J.A. (eds) *Speciation and its Consequences*, Sunderland, MA, pp. 458–481.

Gillman, M.P. and Crawley, M.J. (1990) The cost of sexual reproduction in ragwort (*Senecio jacobaea* L). *Functional Ecology* 4, 585–589.

Green, T.W. and Palmbald, I.G. (1975) Effects of insect seed predators on *Astragalus cibarius* and *Astragalus utahensis* (Leguminosae). *Ecology* 56, 1435–1440.

Hairston, N.G. (1997) Does food web complexity eliminate trophic-level dynamics? *American Naturalist* 149, 1001–1007.

Harper, J.L. (1977) *Population Biology of Plants*. Academic Press, London.

Hulme, P.E. (1996) Herbivory, plant regeneration, and species coexistence. *Journal of Ecology* 84, 609–615.

Hulme, P.E. (1998) Post-dispersal seed predation: Consequences for plant demography and evolution. *Perspectives in Plant Ecology Evolution and Systematics* 1, 32–46.

Ines Pirk, G. and Lopez de Casenave, J. (2010) Influence of seed size on feeding preferences and diet composition of three sympatric harvester ants in the central Monte Desert, Argentina. *Ecological Research* 25, 439–445.

Inouye, R.S., Byers, G.S. and Brown, J.H. (1980) Effects of predation and competition on survivorship, fecundity, and community structure of desert annuals. *Ecology* 61, 1344–1351.

Janzen, D.H. (1969) Seed-eaters versus seed size, number, toxicity and dispersal. *Evolution* 23, 1–27.

Janzen, D.H. (1970) Herbivores and the number of tree species in tropical forests. *American Naturalist* 104, 501–528.

Johnson, D.J., Beaulieu, W.T., Bever, J.D. and Clay, K. (2012) Conspecific negative density dependence and forest diversity. *Science* 336, 904–907.

Kearns, C.A., Inouye, D.W. and Waser, N.M. (1998) Endangered mutualisms: The conservation of plant-pollinator interactions. *Annual Review of Ecology and Systematics* 29, 83–112.

Kelly, D. (1994) The evolutionary ecology of mast seeding. *Trends in Ecology & Evolution* 9, 465–470.

Klein, A.M., Vaissiere, B.E., Cane, J.H., Steffan-Dewenter, I., Cunningham, S.A., Kremen, C., and Tscharntke, T. (2007) Importance of pollinators in changing landscapes for world crops. *Proceedings of the Royal Society B: Biological Sciences* 274, 303–313.

Kuang, J.J. and Chesson, P. (2008) Predation-competition interactions for seasonally recruiting species. *American Naturalist* 171, E119–E133.

Louda, S.M. and Potvin, M.A. (1995) Effect of infloresence-feeding insects on the demography and lifetime fitness of a native plant. *Ecology* 76, 229–245.

Manson, R.H., Ostfeld, R.S. and Canham, C.D. (1998) The effects of tree seed and seedling density on predation rates by rodents in old fields. *Ecoscience* 5, 183–190.

Maron, J.L. (2011) Vertebrate predators have minimal cascading effects on plant production or seed predation in an intact grassland ecosystem. *Ecology Letters* 14, 661–669.

Maron, J.L. and Pearson, D.E. (2011) Vertebrate predators have minimal cascading effects on plant production or seed predation in an intact grassland ecosystem. *Ecology Letters* 14, 661–669.

Maron, J.L. and Simms, E.L. (1997) Effect of seed predation on seed bank size and seedling recruitment of bush lupine (*Lupinus arboreus*). *Oecologia* 111, 76–83.

Miller, D. (1970) *Biological Control of Weeds in New Zealand 1927-48*. New Zealand Department of Scientific and Industrial Research Information Series.

Munoz, A. and Bonal, R. (2011) Linking seed dispersal to cache protection strategies. *Journal of Ecology* 99, 1016–1025.

Nilsson, S.G. and Wastljung, U. (1987) Seed predation and cross pollination in mast seeding beech (*Fagus sylvatica*) patches. *Ecology* 68, 260–265.

Ollerton, J. and Lack, A. (1996) Partial predispersal seed predation in *Lotus corniculatus* L (Fabaceae). *Seed Science Research* 6, 65–69.

Ostfeld, R.S., Manson, R.H. and Canham, C.D. (1997) Effects of rodents on survival of tree seeds and seedlings invading old fields. *Ecology* 78, 1531–1542.

Pacala, S.W. and Crawley, M.J. (1992) Herbivores and plant diversity. *American Naturalist* 140, 243–260.

Paynter, Q., Fowler, S.V., Hinz, H.L., Memmott, J., Shaw, R., Sheppard, A.W. and Syrett, P. (1996) Are seed-feeding insects of use for the biological control of broom? *Proceedings of the 9th International Symposium on Biological Control of Weeds*, Stellenbosch, South Africa, pp. 495–501.

Rees, M. (1994) Delayed germination of seeds: a look at the effects of adult longevity, the timing of reproduction, and population age/stage structure. *American Naturalist* 144, 43–64.

Rees, M., Grubb, P.J. and Kelly, D. (1996) Quantifying the impact of competition and spatial heterogeneity on the structure and dynamics of a four-species guild of winter annuals. *American Naturalist* 147, 1–32.

Russell, S.K. and Schupp, E.W. (1998) Effects of microhabitat patchiness on patterns of seed dispersal and seed predation of *Cercocarpus ledifolius* (Rosaceae). *Oikos* 81, 434–443.

Sheppard, A.W. (1992) Predicting biological weed-control. *Trends in Ecology & Evolution* 7, 290–291.

Sheppard, A.W., Hodge, P., Paynter, Q. and Rees, M. (2002) Factors affecting invasion and persistence of broom Cytisus scoparius in Australia. *Journal of Applied Ecology* 39, 721–734.

Smith, C.C., Hamrick, J.L. and Kramer, C.L. (1990) The advantage of mast years for wind pollination. *American Naturalist* 136, 154–166.

Symonides, E., Silvertown, J. and Andreasen, V. (1986) Population-cycles caused by overcompensating density-dependence. *Oecologia* 71, 156–158.

Tapper, P.G. (1996) Long-term patterns of mast fruiting in *Fraxinus excelsior*. *Ecology* 77, 2567–2572.

Thompson, K. and Baster, K. (1992) Establishment from seed of selected Umbelliferae in unmanaged grassland. *Functional Ecology* 6, 346–352.

Turnbull, L.A., Crawley, M.J. and Rees, M. (2000) Are plant populations seed-limited? A review of seed sowing experiments. *Oikos* 88, 225–238.

Vander Wall, S.B. (1998) Foraging success of granivorous rodents: effects of variation in seed and soil water on olfaction. *Ecology* 79, 233–241.

Vander Wall, S.B., Kuhn, K.M. and Beck, M.J. (2005) Seed removal, seed predation, and secondary dispersal. *Ecology* 86, 801–806.

Velho, N., Isvaran, K. and Datta, A. (2012) Rodent seed predation: effects on seed survival, recruitment, abundance, and dispersion of bird-dispersed tropical trees. *Oecologia* 169, 995–1004.

Visser, M.D., Jongejans, E., van Breugel, M., Zuidema, P.A., Chen, Y.-Y., Kassim, A.R. and de Kroon, H. (2011) Strict mast fruiting for a tropical dipterocarp tree: a demographic cost-benefit analysis of delayed reproduction and seed predation. *Journal of Ecology* 99, 1033–1044.

Wilby, A. and Shachak, M. (2000) Harvester ant response to spatial and temporal heterogeneity in seed availability: pattern in the process of granivory. *Oecologia* 125, 495–503.

Zwolak, R. and Crone, E.E. (2012) Quantifying the outcome of plant–granivore interactions. *Oikos* 121, 20–27.

5   Light-mediated Germination

Thijs L. Pons*

*Department of Plant Ecophysiology, Institute of Environmental
Biology, Utrecht University, Utrecht, the Netherlands*

Introduction

The light response of seeds can control the
time and place of germination of a seed, a
crucial factor in the survival of the resulting
seedlings, and the growth and fitness in
subsequent developmental stages. The ultimate effect of light on seeds depends on
genotype, and on environmental factors
during ripening of the seeds, during dormancy and during germination itself. These
environmental factors may include light, or
factors other than light such as soil temperatures and soil chemical factors (see Chapter 6
of this volume). The picture is further complicated by the fact that the light climate
itself has various aspects that have different
effects on seeds, such as irradiance, spectral
composition and duration of exposure of
the seeds. All the above-mentioned factors
can interact in one way or another in their
effect on seeds. Moreover, the factors are
not constant in time and are difficult to
characterize at the seed's position in the
soil, thus complicating further the analysis
of what is actually happening with a seed
in a natural situation and the interpretation
of a possible ecological significance of light
responses.

Not all seeds are sensitive to light (photoblastic). Large-seeded species are generally not sensitive or only to a limited extent,
whereas photoblastism is very common among
small-seeded species (Grime *et al.*, 1981).
The responsiveness to light helps them to
avoid germination in the soil and to trigger
germination and emergence after some form
of disturbance. In these seeds, light serves
as a signal rather than as a resource. Only
sufficient water and oxygen (in most species)
and a temperature within the range that permits growth are prerequisites for growth of the
heterotrophic embryo. Following Bewley
and Black (1982), light responses in seed are
better referred to as evidence of the involvement of light as a signal for the enforcement or
termination of dormancy rather than as a direct
control over the germination process itself.

In the first part of this chapter, the
seed's light climate is described and the
various light responses are dealt with in
physiological terms, including the interaction with environmental factors other than
light. Mechanisms underlying these light
responses are considered next. They can
help to bring order to the enormous variety
of phenomena observed. In the second part
an attempt is made to interpret the ecological

* E-mail: t.l.pons@uu.nl

© CAB International 2014. *Seeds: The Ecology of Regeneration in Plant Communities,*
3rd Edition (ed. R.S. Gallagher)

significance of the various light responses in different (micro-) habitats. A possible adaptive value of the response to the various aspects of the light climate is considered for seeds in the soil or for those on the soil surface where they may or may not be shaded by litter or a leaf canopy. A number of case studies from different climatic regions are treated in the third part. Available data on light responses for selected species are reviewed, and their importance in the life cycle of the plant is discussed.

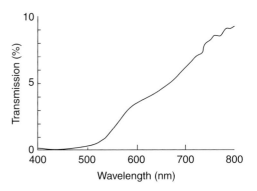

Fig. 5.1. Spectral transmission percentages in the 400 to 800 nm wavelength region under 3 mm of moist sand. (From Bliss and Smith, 1985.)

The Light Climate of Seeds

The light climate of a seed depends on its position in the soil or on the surface, and whether it is exposed to full daylight or under litter or a leaf canopy. As a result, when exposed to light, there can be variation in spectral composition, irradiance and in duration of exposure. The light conditions within a seed, where phytochrome, the pigment system that perceives light, is located, furthermore depend on the absorption characteristics of the seed coat and any other surrounding structures.

Light is strongly attenuated by soil. Hence, seeds in deeper soil layers are in perpetual darkness. Measurable quantities of light do not penetrate to greater depths than a few millimetres or centimetres at the most (Tester and Morris, 1987). Sandy soils tend to have higher light transmission than loam, and humic substances are efficient absorbers of light. The picture is complicated when cracks and soil aggregates reduce the homogeneity of the soil, which is more the rule than the exception. Transmission spectra indicate that the shorter wavelengths are more strongly absorbed than longer ones in the spectral region of interest (Fig. 5.1). The steep gradient of irradiance in the upper soil layers and the horizontal inhomogeneity means that the actual light climate of the seed greatly depends on its exact position. As the surface is approached from below, light conditions may change from ineffective to stimulating and further to inhibiting for germination across only a few millimetres.

Exposures of seeds to light of short duration are commonly used in physiological studies. The only occasion when short exposure can occur under field conditions is when soil is disturbed. Seeds may be brought to the surface and become buried again immediately afterwards (Sauer and Struik, 1964; Scopel et al., 1994) or on a later occasion (Pons, 1984). Once seeds are at or close to the soil surface, day length determines further variation in duration of exposure.

The presence of a leaf canopy above a seed reduces irradiance of all wavelengths relative to full daylight, but much more in the photosynthetically active part of the spectrum (400–700 nm) than in the near infra-red (700–1000 nm) due to strong absorption by chlorophyll (Fig. 5.2). Hence, canopy shade light is relatively rich in far-red (FR) and poor in red (R). A light climate is conveniently characterized by the R:FR ratio with regard to phytochrome functioning, which is the principal pigment involved in the light responses of seeds (Smith, 1982). The R:FR is defined as the ratio in irradiance in 10 nm-wide bands centred at 660 nm and 730 nm, the absorption maxima of the R and FR absorbing forms of phytochrome respectively (P_r and P_{fr}). Other wavelengths are also absorbed by phytochrome; hence, the ultimate effect of broad spectrum light on phytochrome depends also to some extent on the presence of other wavelengths. Unfiltered daylight typically has R:FR of c.1.2 and leaf canopies may reduce this value to 0.2

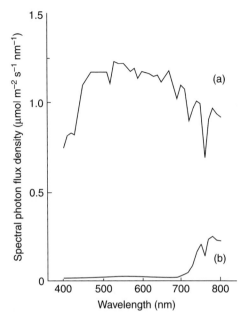

Fig. 5.2. The spectral photon distribution in the 400 to 800 nm wavelength region of (a) daylight and (b) leaf canopy-filtered light. Measurements were made in overcast weather conditions under an ash canopy. Red:far-red photon ratio of daylight was 1.2, and under the canopy 0.18 (T.L. Pons, 1975, unpublished results).

depending on leaf area index (Federer and Tanner, 1966; Stoutjesdijk, 1972; Holmes and Smith, 1977; Pons, 1983) (Fig. 5.2).

Phytochrome is located in the embryo. Hence, when interpreting light responses of seeds, the optical properties of the surrounding structures must be taken into account. Seed coats can be intensely pigmented, which reduces the irradiance and alters the spectral composition of the light in the seed (Widell and Vogelmann, 1988). The blue part of the spectrum can be particularly strongly absorbed.

Physiological Aspects

Types of light responses

There is a wide variety of light responses in seeds. Two main types can be distinguished: (i) effects of short duration

exposures – whether germination is stimulated or inhibited depends on wavelength and light dose (fluence) up to a threshold value; and (ii) effects of long duration exposures – the stimulating or inhibiting effect depends also on spectral composition, but in this case on irradiance (fluence rate).

When light is given in short exposures (e.g. less than 1 h), red light of wavelengths around 660 nm (R) tends to break dormancy and far-red radiation of wavelengths around 730 nm (FR) tends to impose it (Flint and McAllister, 1937). The discovery of the reversibility of these responses contributed to the discovery of phytochrome (Borthwick *et al.*, 1954). The earlier experiments were largely carried out with lettuce seeds, but similar responses are now known for many other species. The reversibility is limited to a certain period of time (escape time) (Fig. 5.3). Rather low photon doses are required for this response (c.100 μmol m^{-2}) which obeys the law of reciprocity (at the same fluence, the response is independent of duration and irradiance). This response is known as the low fluence response (LFR; Fig. 5.4). After preconditioning at low or high temperature, light-sensitive seeds may even respond to extremely low photon doses (down to 0.01 μmol m^{-2}). This is known as the very low fluence response (VLFR; Fig. 5.4) (Blaauw-Jansen and Blaauw, 1975; Blaauw-Jansen, 1983; Cone and Kendrick, 1986). As a consequence of the extreme light sensitivity, seeds may respond to green light, which has been used for recording germination in darkness (Baskin and Baskin, 1979; Grime *et al.*, 1981). This complicates the interpretation of the results of several studies where a green 'safe' light was used.

Some seeds do not respond to short exposures of light but require long exposure times for dormancy breaking (e.g. *Plantago major* in Fig. 5.3). In such cases repeated short exposures are often equally effective and the response is considered essentially a LFR. However, in many species long exposure times tend to inhibit germination, particularly when irradiance is high or the R:FR is low (Górski and Górska, 1979; Frankland, 1986; Fig. 5.4). The inhibiting effect of continuous illumination (or light interrupted by a night period) on germination

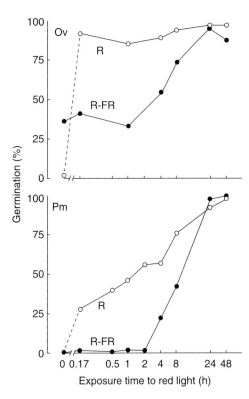

Fig. 5.3. Effect of exposure time to red light (R) on germination of *Origanum vulgare* (Ov) and *Plantago major* (Pm), and escape from phytochrome control as indicated by the effect of a subsequent exposure to far-red (R-FR). Seeds were exposed to R for the period indicated. The exposure time for FR was 10 min. After the exposure(s) the seeds were tested for final germination percentage in darkness at 22°C (Ov) or 27°C (Pm). (From Pons, 1991a.)

is known as the high irradiance response (HIR). This response does not obey the law of reciprocity and is only irradiance dependent. Wavelengths of 710–720 nm tend to be most effective (FR-HIR). Species that are characterized as negatively photoblastic are inhibited at a low irradiance in continuous light, whereas short exposures to white or red light may be stimulating in these seeds (Frankland, 1976). The seeds thus show a strong inhibition by HIR, whereas dormancy breaking through the LFR operates as well. Exposure of seeds to light in the field is generally of long duration. Hence, the HIR is potentially involved in most light responses under natural conditions.

The actual response of seeds to light varies greatly between species and seed lots, and further depends on pretreatment and conditions during the germination test. The three responses as described above (VLFR, LFR, HIR) can sometimes be demonstrated in one species. For example in lettuce seeds the LFR can be easily demonstrated, but when the seeds are pretreated at high temperatures the VLFR becomes effective (Blaauw-Jansen and Blaauw, 1975). When exposed for long periods to daylight, germination of lettuce is negatively affected by high irradiance (Fig. 5.4) (Górski and Górska, 1979). Similar phenomena have been described for *Plantago major* and *Sinapis arvensis* (Frankland and Poo, 1980). Unfortunately, a reciprocity test is not always performed when allocating a light response to the LFR or the HIR (Casal and Sánchez, 1998), which can make the distinction uncertain.

The different types of light responses as described above enable the seed to perceive different aspects of its light environment. Since preconditioning at specific temperatures can influence the light response mode and the sensitivity of the seeds, dormancy breaking by light is also intrinsically involved in seasonally dependent germination (Derkx and Karssen, 1993b). In a natural setting, the LFR and the VLFR are involved in the perception of the absence of light, which enforces dormancy as long as the situation persists. The LFR and the HIR are involved in the perception of spectral composition and intensity of the light to which the seeds are exposed.

Phytochrome action

The physiological mechanisms underlying the various seed responses to light are summarized shortly. For more extensive information and references the reader is referred to reviews by Shinomura (1997), Casal and Sánchez (1998) and Franklin and Quail (2010). The light environment is principally perceived by the photoreceptor phytochrome, which absorbs primarily in the R and FR regions of the spectrum. Other

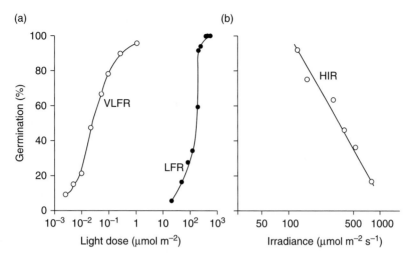

Fig. 5.4. The three main types of light responses of seeds demonstrated in one species (lettuce; *Lactuca sativa*). (a) Response to a dose of red light of seeds pretreated in two ways, at 37°C which induces the very low fluence response (VLFR) and with exposure to far-red which demonstrates the low fluence response (LFR) (Blaauw-Jansen and Blaauw, 1975). (b) Irradiance response to daylight showing inhibition of germination with increasing irradiance as evidence of the high-irradiance response (HIR) (Górki and Górska, 1979).

photoreceptors may also be involved, but they will not be considered here. Phytochrome is the collective name for a small family of photoreceptors consisting of a specific apoprotein covalently bound to a common chromophore (phytochromobilin), which exist as dimers. Phytochrome is synthesized in the inactive R absorbing form (P_r). Absorption of R causes photoconversion to the FR absorbing form (P_{fr}), which is the biologically active form. P_{fr} is translocated to the nucleus where it interacts with transcription factors (phytochrome interacting factors; PIFs). These regulate gene expression that facilitates the responses to specific light conditions.

Photoconversion of P_{fr} can also go back to P_r, which occurs as the result of absorption of FR. Since the two forms of phytochrome have partly overlapping absorption spectra, irradiation with R and FR does not result in pure pools of P_{fr} and P_r respectively. In daylight, R, FR and other wavelengths that are absorbed by phytochrome occur. Light establishes a certain photostationary state of phytochrome ($P_{fr}:P_{tot}$, the ratio of P_{fr} to total phytochrome). A saturating dose of daylight results in *c*.65% of phytochrome in P_{fr} form; in canopy-filtered light,

this percentage can drop to 20% (Fig. 5.5). Pure R and FR result in *c*.85% and 2% P_{fr} respectively. A low photon dose can be too low to saturate the photoconversion of phytochrome, resulting in a different $P_{fr}:P_{tot}$ ratio than expected from the spectral composition of the light. A particular $P_{fr}:P_{tot}$ can thus be established by irradiation with light of a specific spectral composition (R:FR), or by a sub-saturating light dose of R or FR. In the LFR, and probably also the VLFR, a threshold value of $P_{fr}:P_{tot}$ is required for dormancy breaking, which can be achieved by either one of the above pathways.

The presence of P_{fr} is required for a certain period of time, several hours or more, for the completion of its dormancy-breaking action, the escape time as referred to above. When P_{fr} is converted back to P_r within this period, enforcement of dormancy is re-established. When the escape time has elapsed, P_{fr} is no longer necessary and germination cannot be prevented any more by a short exposure to FR (Fig. 5.3). P_{fr} can also be converted to P_r in darkness (dark reversion). When that happens before the escape time has passed, germination can also not occur. Dark reversion P_{fr} to P_r ultimately

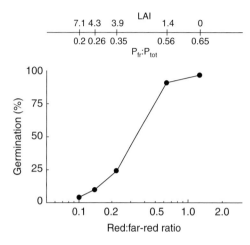

Fig. 5.5. Germination of *Plantago major* in daylight in stands of *Sinapis alba* of different densities resulting in different red:far-red photon ratios (R:FR) of the transmitted light under the canopy. Corresponding leaf area index (LAI) and phytochrome photoequilibria (P_{fr}:P_{tot} ratios) are indicated. (From Frankland and Poo, 1980.)

leads to a pure P_r phytochrome pool after a prolonged period in darkness at a sufficiently high temperature. The process is strongly temperature dependent and proceeds slowly at low temperatures (Schäfer and Schmidt, 1974). In seeds that have a high P_{fr}:P_{tot} threshold and possibly a fast dark reversion, P_{fr} can drop to below threshold values before it has completed its action or in other words within the escape time. Such seeds may require long exposure times or repeated short exposures (e.g. *Plantago major* in Fig. 5.3). Another consequence is that in continuous light of low irradiance, dark reversion of P_{fr} to P_r can reduce the P_{fr}:P_{tot} ratio below that expected from photoconversions only (Frankland and Taylorson, 1983). A stimulating effect of increasing irradiance on germination is found where this phenomenon plays a role.

Phytochrome may be pre-existent in the seeds as P_{fr}, which originates from before ripening of the seed. Phytochrome is roughly arrested at the P_{fr}:P_{tot} ratio as it was when the seed dried, which thus depends on the light environment of the mother plant and on chlorophyll content of the seed coat and other structures surrounding the embryo

(Cresswell and Grime, 1981). The P_{fr} in the dry seed may result in germination in darkness upon imbibition. Repeated exposure to FR is sometimes required to prevent germination, since P_{fr} can be formed in the imbibing seed from intermediates trapped in the drying seed (Kendrick and Spruit, 1977).

Five members of the phytochrome family have been identified in *Arabidopsis thaliana* (phyA–E) (Mathews and Sharrock, 1997). Other species may have slightly different forms (e.g. tomato; Appenroth *et al.*, 2006). Before the different phytochromes were characterized, two types were distinguished on the basis of physiological experiments, a labile type I and a more stable type II. PhyA appeared to be the labile form, and phyB–E that have much slower turnover rates are equivalent to type II. The phytochromes have distinct characteristics associated with their functions as evident from studies with mutants deficient in particular phytochromes (Koornneef and Kendrick, 1994; Smith, 1995). This has been most extensively studied in *Arabidopsis thaliana*, to which most of the following allocation of functions refers. The spectrum of variation in the plant kingdom has yet to be investigated.

PhyB is the most abundant pre-existent phytochrome in light-sensitive seed. The light responses of seeds pertaining to the classical LFR as described above can for a large part be ascribed to the action of this phytochrome. However, mutants deficient in phyB do still exhibit a light sensitivity that can be ascribed to the other phytochromes (Koornneef and Kendrick, 1994; Smith, 1995). PhyA is involved in the VLFR. It is synthesized in the P_r form after imbibition and is unstable once transformed to P_{fr}. The threshold level of P_{fr} required for dormancy breaking by phyA is much lower than by phyB. The very low photon doses that are effective in the VLFR are a manifestation of this very low requirement of phyA in P_{fr} form. PhyA is synthesized *de novo* after imbibition. Dormancy breaking by FR that establishes very low levels of P_{fr} can also be interpreted as an indication of the involvement of the VLFR (*Origanum vulgare* in Fig. 5.3). The response of photosensitized lettuce seeds to moonlight or even starlight is another example of the extreme light sensitivity caused by phyA (Hartmann *et al.*, 1998). The R-FR reversibility in phyA-phyB double mutants indicates the activity of additional phytochromes in this process (Poppe and Schäfer, 1997). Particularly phyE appeared to be important for photoreversibility and also for germination in continuous light (Hennig *et al.*, 2002) and for germination at low temperatures (Heschel *et al.*, 2007). PhyC and phyD appeared to be less important for light responses in *Arabidopsis* seeds, but that can be different for other species.

During exposure to light, phytochrome is constantly transformed from P_{fr} to P_r and back; the rate of cycling depends on irradiance and spectral composition. Evidence indicates that this phytochrome cycling may somehow be responsible for the inhibiting effect on germination in continuous light in the HIR (Frankland, 1986; Hennig *et al.*, 2000). The HIR is still effective after expiration of the escape time in the LFR, which indicates that it interferes with a later step in the cascade of reactions leading to germination than the LFR. PhyA is involved in the HIR in seedlings. PhyA has also been identified as the phytochrome species involved in the HIR in tomato seeds (Shichijo *et al.*, 2001). Arabidopsis does not exhibit the classical HIR as known from many other species. However, phyA appeared to be involved in the stimulation of germination in continuous light in this species (Hennig *et al.*, 2002), which may also be considered a form of HIR (Franklin and Quail, 2010).

The phytochrome system is a multi-functional pigment system. It is involved in the perception of various aspects of the light climate (Smith and Whitelam, 1990) and interacts with other environmental factors. The differences in these light responses between species are large, pointing to a large diversity in modes of phytochrome signalling (Maloof *et al.*, 2000). This complicates the interpretation of the ecological significance of a light response as studied in the laboratory, if there is one at all.

The pathway from perception of light to dormancy breaking and germination

The structures surrounding the embryo can restrict radicle growth and thus prevent germination. Hard seed coats do that passively by preventing water uptake. However, light is not involved in breaking this physical dormancy (see Chapter 7 of this volume; Baskin and Baskin, 2001). Living structures surrounding the embryo such as the endosperm can restrict its growth and thus keep seeds in a dormant state. This physiological dormancy is then controlled by the balance between the growth potential of the embryo, particularly the radicle on the one hand and the constraint exerted by the surrounding structures on the other (Bewley and Black, 1982). Light can influence both these factors, as evident from many studies done with different species (Casal and Sánchez, 1998).

The morphology of seeds varies between species and so does the pathway leading to dormancy breaking and germination after the perception of light (Bewley, 1997; Koornneef *et al.*, 2002). Phytohormones play an important role in this process. Abscisic acid (ABA) keeps seeds dormant by reducing the extensibility of the radicle and thus its growth potential. It also keeps the endosperm rigid. ABA levels are high in dormant seeds, and its synthesis is reduced after R and induced after FR (Seo *et al.*, 2009). Gibberellic acid (GA) produced in the embryo has an antagonistic effect. It promotes the growth potential of the radicle and migrates to the endosperm where it induces the hydrolysis of cell wall components, thus loosening this tissue (Seo *et al.*, 2009). This allows the radicle to break through. R and other factors involved in dormancy breaking can induce the synthesis of and/or sensitivity to GA (Karssen *et al.*, 1989; Derkx and Karssen, 1993a).

Interaction with other environmental factors

The actual response of a seed to light depends strongly on other environmental conditions. Both the present and preceding conditions are relevant. These factors, such as temperature, water potential and chemical substances, affect germination in their own right and are treated in other chapters of this book. They are discussed briefly here in relation to their interaction with light responses.

Temperature fluctuation can break dormancy in many seeds, and light can often substitute for this requirement as in *Nicotiana tabacum* (Toole *et al.*, 1955) and *Rumex obtusifolius* (Taylorson and Hendricks, 1972; Totterdel and Roberts, 1980). In other cases, such as *Lycopus europaeus* (Thompson, 1969) and *Fimbristylis littoralis* (Pons and Schröder, 1986), the joint action of temperature fluctuation and light is required. There are indications that temperature changes somehow interfere with P_{fr} action (Takaki *et al.*, 1981; Probert and Smith, 1986).

Temperature itself has dual effects (Bouwmeester and Karssen, 1992). It affects germination and the dormancy status of the seed. A seed may be light requiring at one temperature but not at another one; e.g. lettuce seeds do not require light at low temperatures, but do so at higher ones (Smith, 1975). Other species have a reverse requirement, such as *Betula verrucosa* (Black and Wareing, 1955). These effects can be ascribed to an effect of temperature on the threshold level of P_{fr} for dormancy breaking, on reversion of P_{fr} in darkness, or a combination of these. Temperature also has an effect on the response of seeds to the R:FR (Fig. 5.6) supposedly also by an effect on the P_{fr} threshold (Pons, 1986; Senden *et al.*, 1986; van Tooren and Pons, 1988). A lower P_{fr} threshold at low temperatures may add to the dormancy-enforcing effect of the low R:FR under canopies where cool temperatures prevail (Gallagher and Cardina, 1998a).

There are many examples that illustrate temperature effects on light responses through an influence on the dormancy status of seeds. Chilling or overwintering in soil can replace light in many species, or greatly modifies the light response. Germination in darkness increased after a chilling treatment in e.g. *Lactuca sativa* (VanderWoude and Toole, 1980), *Amaranthus retroflexus* (Taylorson and Hendricks, 1969), *Cirsium palustre* (Pons, 1983, 1984), *Aster pilosus* (Baskin and Baskin, 1985), *Carlina vulgaris* (van Tooren and Pons, 1988) and *Plantago lanceolata* (Pons and van der Toorn, 1988). A lower R:FR was

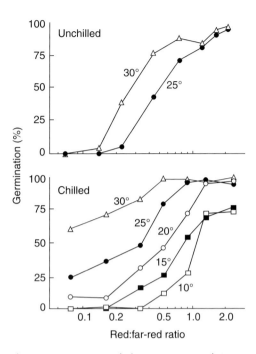

Fig. 5.6. Germination of *Plantago major* seeds in light of different red:far-red photon ratios and at different temperatures. Seeds were either used directly after dry storage (unchilled) or chilled for 8 weeks at 4°C in darkness. Unchilled seeds did not germinate below 25°C. (From Pons, 1986.)

needed to inhibit germination in *Plantago major* after chilling, indicating that a lower P_{fr} level was required for dormancy breaking (Pons, 1986; Fig. 5.6). In buried *Polygonum aviculare* seeds, the VLFR developed during chilling in a fraction of the seed population (Batlla and Benech-Arnold, 2005). This can be ascribed to the synthesis of phyA.

Seasonal changes in dormancy status of seeds are described for many species which are caused by seasonal changes in temperature (Karssen, 1982; Baskin and Baskin, 2001). This can be reflected in seasonal changes in the light requirement for dormancy breaking, as found for *Ambrosia artemisiifolia* (Baskin and Baskin, 1980), *Rumex obtusifolius* (van Assche and Vanlerberghe, 1989), several species in a chalk grassland community (Pons, 1991b) and weed species (Milberg and Andersson, 1997). These species typically pass from light requiring in autumn (in some cases preceded by non-responsiveness to light) to non-light requiring in spring, which is repeated in successive years. Derkx and Karssen (1993b) have studied the change in sensitivity to light during burial in more detail in *Sysimbrium officinale*. Alleviation of primary dormancy was accompanied by the appearance of the VLFR. Indications were also found for an involvement of an increase in phytochrome receptors and GA synthesis.

Another factor that strongly interferes with the seed's light response is water potential. The threshold level of P_{fr} for dormancy breaking is increased at water potentials that do not totally inhibit germination (Scheibe and Lang, 1965; Karssen, 1970). This results in a requirement for higher R:FR ratios at low water potentials as found for *Plantago major* (Pons, 1986). Similarly, the development of the VLFR was hampered in buried *Polygonum aviculare* seeds during burial at low soil moisture content (Batlla *et al.*, 2007).

Nitrate, which is a naturally occurring dormancy breaking agent (see Chapters 6 and 7 of this volume), can change the light response of seeds appreciably. As with temperature fluctuation, nitrate can replace a light requirement, e.g. in *Plantago lanceolata* (Steinbauer and Grigsby, 1957), including at the low naturally occurring concentrations (Pons, 1989a), and in *Sinapis arvensis* (Goudey *et al.*, 1987). In other species nitrate must act together with light to break dormancy, e.g. *Sisymbrium officinale* (Hilhorst *et al.*, 1986; Derkx and Karssen, 1993b) and *Ranunculus sceleratus* (Probert *et al.*, 1987). In *Plantago major*, nitrate lowered the R:FR required for breaking dormancy (Pons, 1986). There are indications, as with temperature fluctuation, that nitrate interferes with P_{fr} action.

These are examples of factors that interfere with the light response of seeds. The situation may become even more complex when more than two factors operate simultaneously. Chilling, nitrate, temperature fluctuation and light in various combinations showed additive effects or positive interactions in a number of weed species studied by Vincent and Roberts (1977). Simultaneous operation of several environmental factors

together is more the rule than the exception under field conditions. The various interactions complicate ecological interpretations on the one hand, but on the other hand they can be viewed as tools available to seeds for the fine tuning of their germination to occur at the time and place best suited to the requirements of the subsequent developmental stages.

Light Responses in Distinct Seed Positions

Three main positions of seeds can be distinguished with respect to the soil–atmosphere interface and the presence of a leaf canopy overhead where light responses play a role. Seeds can be deep in soil where they are excluded from light; they can be at or close to the soil surface where irradiance varies greatly; and they may be exposed at the surface to leaf canopy-filtered light where spectral composition is important. The position of a seed in the soil can change by burial when seeds fall into cracks or by animal activity, or by movement to the surface during soil disturbance such as with tillage. The degree of leaf-canopy shade can change seasonally and with disturbance of the canopy. The light responses in the different seed positions with respect to the soil surface and a leaf canopy are schematically represented in Fig. 5.7.

A light requirement enforces dormancy in soil

The most obvious significance of a light requirement for dormancy breaking is avoidance of germination too deep in the soil for the seedlings to reach the surface on the available nutrient reserves. Only when a seed is brought to the surface is it exposed to light and its dormancy alleviated. The seed can germinate there when other factors are conducive. Considering the above argument it is not surprising that most light-requiring seeds are small. In a survey by Grime *et al.* (1981) mostly using dry stored seeds of 69 species, many of the small-seeded species (<0.1 mg) had a light requirement at least in some of their seeds. The majority of species with a seed mass of 1 mg or more lacked a light requirement (Table 5.1). This relationship with seed mass was also true after a chilling treatment (Milberg *et al.*, 2000). Also other studies indicate such a predominance of a light requirement in small-seeded species (Kinzel, 1926; Toole, 1973; Pearson *et al.*, 2002).

A light requirement is an important trait for the avoidance of germination in the soil and hence, for the formation of a persistent

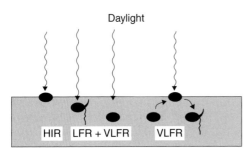

Daylight

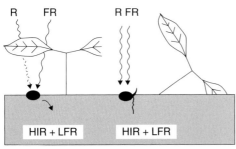

Fig. 5.7. Breaking of seed dormancy by light in seeds in distinct positions in soil and under a leaf canopy. The upper picture shows the involvement of the HIR, LFR and VLFR in seeds at, and close under, the soil surface where a steep gradient in light intensity exists. At the right, the effect of a short exposure to light during soil disturbance in the range where the VLFR is active is demonstrated. The lower picture shows the dormancy-enforcing effect of leaf canopy-filtered light as a result of a shift in the R:FR. The HIR and the LFR are involved in this response. During dormancy enforced by leaf-canopy shade, seeds can become buried in the soil where dormancy is further enforced by absence of light. When remaining at or brought to the soil surface during disturbance, seeds are exposed to unfiltered daylight, which breaks dormancy.

seed bank. However, the germination percentage of freshly harvested seeds in darkness is often not a very good predictor of avoidance of germination in the soil. A light requirement can be induced upon burial as found in several arable weed species by Wesson and Wareing (1969). Or it may have disappeared after overwintering in soil as in *Leontodon hispidus* and several other grassland species (Pons, 1991b).

Germination in the soil may be caused by the presence of P_{fr} in the seed. This can disappear by dark reversion and thus a light requirement can be induced. A prerequisite is that germination is prevented by conditions not conducive to germination. This may be either by forms of dormancy other than those mediated by phytochrome, or by germination-inhibitory conditions such as unsuitable temperature and low water potential (Fig. 5.8). Induction of a light requirement at low water potentials has been demonstrated in *Rumex crispus* (Duke, 1978), *Chenopodium album* (Karssen, 1970), *Chenopodium bonus-henricus* (Khan and Karssen, 1980), and *Plantago major* and *Origanum vulgare* (Pons, 1991a).

Although seeds may have a light requirement when ripe, it is most probably more the rule than the exception that they have absorbed sufficient water for phytochrome transformations to occur before they are excluded from light by burial. Such seeds would then germinate in the soil when further imbibed. Hence, induction of a light requirement is not only important for survival of the seeds in the soil when they

Table 5.1. Light requirement of germination in relation to seed mass. The distribution of 69 species is shown over three seed mass classes and three germination categories. Germination in darkness is expressed as a percentage of germination in light. N.S., germination in darkness not significantly different from germination in light. (From Grime *et al.*, 1981.)

Seed mass (mg)	Germination in darkness		
	<20%	>20%	N.S.
<0.1	8	6	1
0.1–1.0	13	14	12
>1.0	3	2	10

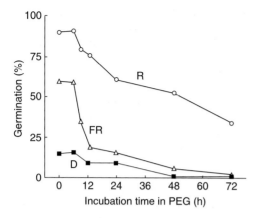

Fig. 5.8. Induction of a light requirement in seeds of *Origanum vulgare* during incubation in polyethylene glycol (PEG-6000, 1.2 MPa). The seeds were irradiated with red (R) or far-red (FR) for 10 min before incubation, or were not exposed to light at all (D). After various incubation periods the seeds were rinsed with water and tested for germination. The experiment was carried out in darkness at 27°C. Controls that were irradiated with R or FR after incubation in PEG indicated that the responsiveness to light had not changed. (From Pons, 1991a.)

are not light requiring when they are ripe (Wesson and Wareing, 1969), but also when they have been exposed to light before burial. Depth of burial of the seeds can influence the type of light response that is induced in the seeds, as in *Datura ferox* (Botto *et al.*, 1998). The species in which a light requirement can be induced are well known to form large seed reserves in the soil. The capacity for induction of dark dormancy appears thus to be of great importance for the formation of a persistent seed bank.

From the temperature dependence of dark reversion (Schäfer and Schmidt, 1974) it can be predicted that induction of a light requirement is much slower at low compared to high temperatures. It takes typically a few days at temperatures of 20°C or higher, but is very slow at 4°C or does not occur at all at low temperature (Khan and Karssen, 1980; Pons, 1984, 1991a). This could be the reason that seeds exposed to light in winter, e.g. with soil disturbance during tree felling, may still respond to that light stimulus in spring when temperatures have become favourable for germination as

shown for *Cirsium palustre* in coppiced woodland (Pons, 1984). Also the stimulation of the emergence of weed seedlings up to nine months after exposure to light during soil cultivation in late autumn but not in summer could be related to the maintenance of high P_{fr} levels at low temperature (Hartmann *et al.*, 2005).

Seasonal changes in dormancy do not necessarily lead to germination in the soil when the seeds are not fully light requiring (Karssen, 1982). In the case of *Ambrosia artemisiifolia* Baskin and Baskin (1980) have shown that the seeds could only germinate in darkness at temperatures that are higher than those prevailing in the soil in spring. Gradually rising soil temperatures later in spring induce a light requirement before they are sufficiently high to allow germination. However, an unknown proportion of the seeds in which this mechanism apparently did not work germinated in the first spring after their burial. Germination in the soil of a considerable proportion of the seeds was also found in the first spring after burial of *Scabiosa columbaria* (50%), *Carlina vulgaris* (50%) and *Leontodon hispidus* (95%) (Pons, 1991b). The remaining seeds showed a seasonal change in light requirement while germination did not occur in the soil, presumably due to the same mechanism as described above for *A. artemisiifolia*.

In buried seeds, a light requirement functions as a gap detection mechanism. Seeds excluded from light in the soil can be brought to the surface where they are exposed to light during disturbance of the soil. Soil disturbance generally coincides with destruction of established vegetation and hence, is a good predictor of reduced competition by established plants. Only short exposures are sufficient to break dormancy of many buried weed seeds (Sauer and Struik, 1964). Hence, seeds can germinate even when they are buried again by continued movement of the soil after being exposed at the surface (Pons, 1984). This may be essential when disturbance occurs infrequently, for example in *Cirsium palustre* growing in coppiced woodland (Pons and During, 1987).

The short millisecond exposure times during soil disturbance when seeds are brought to the surface and reburied again immediately during the same disturbance action result in light doses in the range where the VLFR is active. Scopel *et al.* (1991, 1994) showed that exposure to light during soil cultivation in an arable field was indeed essential for germination of a large part of the seed population. Emergence of seedlings was stimulated more by cultivation during daytime than during the night. Daytime cultivation with the soil surface protected from exposure to light resulted in less emergence than occurred on full exposure (Fig. 5.9). However, soil tillage at night was not considered a reliable approach for the reduction of weed emergence in an agricultural setting (Gallagher and Cardina, 1998b).

Exposure to light in soil

The steep gradient in irradiance near the surface of soils has a profound effect on

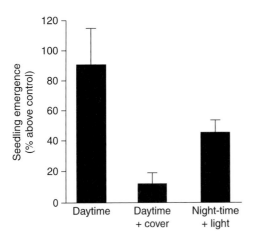

Fig. 5.9. The effect of manipulation of light conditions during soil cultivation on emergence of weed seedlings. Values are expressed as percentage above the control treatment where cultivation was carried out during the night, which resulted in an emergence of 366 seedlings m^{-2}. Soil cultivation during daytime and during the night with supplemental lighting stimulated germination of seeds brought to the surface, which were briefly exposed to light and buried again. Light doses are in the range where the VLFR is active. (From Scopel *et al.*, 1994.)

seeds. Studies indicate that dormancy of seeds can still be broken by the very low irradiance penetrating the upper few millimetres of soil. Woolley and Stoller (1978) found that lettuce seeds germinated at a depth of 2 mm to which *c.*1% of daylight penetrated, but not at 6 mm. Van der Meyden and van der Waals-Kooi (1979) studied the influence of soil cover on germination of *Senecio jacobaea* in two ways. Emergence of seedlings from seeds planted at different depths in dune sand was compared with germination in dishes covered with layers of sand of different thickness. The experiment indicated that germination was stimulated even by light penetrating at 16 mm; no data on irradiance at that depth are given.

The dormancy of *Sinapis arvensis* and *Plantago major* can also be broken by the very low irradiance in compost at 6 and 8 mm depth respectively (Frankland and Poo, 1980). Assuming that germination is under control of the LFR, the authors interpret the decrease of germination with depth from the decrease in P_{fr}:P_{tot} ratio as a result of the decrease in rate of irradiance-dependent P_r to P_{fr} transformation, while the P_{fr} to P_r dark reversion remains constant. Equilibrium P_{fr}:P_{tot} ratios were calculated to be 0.16 at a depth of 3 mm, while this would be 0.48 on the basis of the reduced R:FR in soil alone. Daylight establishes a P_{fr}:P_{tot} ratio of *c.*0.6 (Fig. 5.5). These calculations show that reduction of irradiance can be more important for dormancy breaking than the change in spectral composition of the light in soil provided that the assumptions for the rate of dark reversion of P_{fr} are correct.

The influence of the light climate in soil on germination of a number of species was investigated by Bliss and Smith (1985). They observed widely different responses among the species to the steep light gradient in the upper soil layers (Fig. 5.10). *Chenopodium album* did not germinate under a cover of more than 2 mm sand; *Rumex obtusifolius*, *Cecropia obtusifolius* and *Plantago major* also germinated under 6 mm, but high percentages of *Galium aparine* and particularly *Digitalis purpurea* seeds were stimulated to germinate by light penetrating 10 mm at an

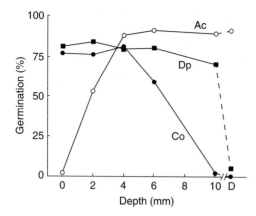

Fig. 5.10. Germination of three species in light transmitted by layers of sand of different thickness. The experiments were carried out at 20°C and an irradiance of 280 µmol m^{-2} s^{-1} at the soil surface. Controls in darkness (D) were included. The species were *Amaranthus caudatus* (Ac), *Digitalis purpurea* (Dp) and *Cecropia obtusifolia* (Co). (From Bliss and Smith, 1985.)

estimated irradiance of only 0.026 µmol m^{-2} s^{-1}. The authors suggest that dormancy breaking at such a low irradiance may involve the absence of dark reversion of P_{fr}, which would allow integration of light over longer periods. However, this would be in conflict with the significance of dark reversion for the formation of a seed bank as argued above, and *Digitalis purpurea* forms large seed banks (van Baalen, 1982). A more likely explanation seems that the VLFR is involved in the seeds that germinate at greater soil depths. This would add another ecologically significant role for this light-response type to the effect of soil disturbance mentioned above.

Germination on the soil surface

Inhibition of germination of seeds on the soil surface as compared with superficially buried seeds has been reported in several cases (Fenner, 1985). This inhibition could be ascribed to reduced moisture availability in the case of *Senecio jacobaea* (van der Meyden and van der Waals-Kooi, 1979) or to the inhibiting effect of oxygen in the case of the rice field weed *Scirpus juncoides* (Pons and Schröder, 1986). However, Bliss

and Smith (1985) could ascribe the inhibition of germination at the soil surface of several species to the high irradiance in continuous light, e.g. *Chenopodium album* and *Galium aparine* (>70 µmol m^{-2} s^{-1}) and *Amaranthus caudatus* (>3 µmol m^{-2} s^{-1}), which is indicative of the involvement of the HIR.

Inhibition of germination due to prolonged exposure to light has been described for several species. It is most obvious in seeds that are reported to be negatively photoblastic, e.g. *Amaranthus caudatus* (Kendrick and Frankland, 1969), *Phacelia tanacetifolia* (Rollin and Maignan, 1967) and *Oldenlandia corymbosa* (Corbineau and Côme, 1982) in which prolonged exposure to rather low irradiance reduces germination already below dark control values. Higher irradiance is generally required for photoinhibition in otherwise positively photoblastic seeds such as *Sinapis arvensis* (20 µmol m^{-2} s^{-1}; Frankland and Poo, 1980) and *Urtica urens* (200 µmol m^{-2} s^{-1}; Grime and Jarvis, 1975). The experiments of Ellis *et al.* (1986a,b, 1989) included species that were inhibited at low daily irradiances (<10 mol m^{-2} day^{-1}) and species that were only inhibited at high irradiance. Alternatively, there are also seeds that are stimulated at high as compared to low irradiance. Several species showed this phenomenon in the survey of Grime *et al.* (1981). Pons (1989b) showed that freshly harvested seeds of *Calluna vulgaris* and *Erica tetralix*, but not previously dark-incubated seeds, showed a strong positive effect on germination of increasing irradiance up to 250 µmol m^{-2} s^{-1}.

The ecological significance of the HIR at the soil surface in the absence of a leaf canopy is not altogether clear. Some authors suggest that it may avoid germination on the soil surface on exposed sites where conditions are not suitable for establishment, since the seedling may suffer desiccation. However, this is not necessarily the case. A dry soil surface itself can inhibit germination and water relations will not be that different between a seedling germinated at the surface or at a few millimetres depth, unless penetration of the radicle is hampered in surface-germinating seeds. No studies are available where the survival value of the

HIR at an exposed soil surface is investigated. This matter awaits critical evaluation.

Spectral composition, a signal for the presence of established plants

After the discovery of the response of seeds to R and FR, and the effect of leaves on the R:FR ratio of daylight (Fig. 5.2), the potential ecological significance of canopy-filtered light for seed dormancy and germination was recognized. Cumming (1963) carried out experiments with artificial light source and filter combinations to produce different R:FR ratios. Germination of the light-requiring *Chenopodium botrys* appeared to be inhibited at an R:FR ratio similar to those found under leaves.

Using leaves as filters, Taylorson and Borthwick (1969) and van der Veen (1970) showed that germination was reduced below unfiltered light control values in several light-requiring seeds (e.g. *Chenopodium album, Rumex obtusifolius, Plantago major, Ageratum conyzoide, Amaranthus retroflexus, Potentilla norvegis*). Germination of some species that germinated in darkness was even reduced below dark control values in the study of van der Veen (1970) (e.g. *Ruellia tuberosa, Calotropa procera*). Other studies have further illustrated the potential significance of leaf-canopy shade light for maintenance of seed dormancy. Stoutjesdijk (1972) found dormancy enforced by leaf-filtered light in five species from widely different habitats, but six were not influenced. King (1975) mentioned three short-lived grassland species that were inhibited by light filtered through leaves. Silvertown (1980) found dormancy enforced by leaf-filtered light in 17 out of 27 grassland species. Fenner (1980a) reported that 16 out of 18 African weed species showed this type of dormancy. The most extensive survey on leaf canopy-induced dormancy was done by Górski and coworkers (Górski, 1975; Górski et al., 1977, 1978). They tested 271 species under a canopy of rhubarb (*Rheum rhabarbarum*) with controls in diffuse white light of similar irradiance. Leaf-canopy light

reduced germination in 90% of the wild species that were tested, but in only 53% of cultivated plants (Table 5.2). Leaf canopy-enforced dormancy was present in all positively photoblastic seeds, in 71% of the negatively photoblastic seeds and in 58% of the light-indifferent seeds.

The effect on seed germination of leaf-filtered daylight is generally similar to the effect of a mixture of artificial light sources giving the same R:FR ratio (Frankland and Letendre, 1978; Pons, 1983; Fig. 5.5 and 5.6). Frankland and Poo (1980) showed clearly that the response of the seeds of *Plantago major* and *Sinapis arvensis* to the R:FR operates through its effect on the photoequilibrium of phytochrome ($P_{fr}:P_{tot}$), thus through the LFR (Fig. 5.5). However, germination was not so strongly reduced when the seeds were exposed to light with a low R:FR for only 1 h as compared to 20 h. The authors attributed the additional reduction in prolonged exposure to photoinhibition by the HIR, which is most effective at long wavelengths. The dormancy enforced by canopy shade light can thus be under control of the LFR and the HIR. The contribution of the two responses to the ultimate effect is likely to be different per species and to vary with environmental conditions.

The significance of the responses of seeds to leaf canopy-filtered light interpreted on the information presented so far can be summarized as follows. With phytochrome as the photoreceptor, seeds have a

Table 5.2. Dormancy enforced by leaf-canopy shade. The distribution of 77 wild species and 75 cultivated species over three categories of germination in leaf canopy-filtered light. The latter was expressed as a percentage of germination in diffuse unfiltered daylight. N.S., germination in canopy-filtered light not significantly different from germination in unfiltered light. (From Górski et al., 1978.)

Species type	Germination in leaf canopy-filtered light		
	<20%	>20%	N.S.
Wild	49	20	8
Cultivated	7	33	35

device for the detection of shading by established plants and can respond by delayed germination. Only when gaps arise in vegetation can the dormancy of a seed be broken by the high $P_{fr}:P_{tot}$ ratio established by unmodified daylight allowing the seed to germinate. Gaps are considered to be important for the regeneration of many plants from seeds (Grubb, 1977), since shading and other competitive effects of established plants can greatly reduce growth and survival of seedlings (Fenner, 1978; van der Toorn and Pons, 1988). However, as Stoutjesdijk (1972) pointed out, this response is found among species from temperate deciduous woodland also, but the mechanism cannot be of much ecological importance in spring in that habitat, since the tree leaf canopy is absent when most seeds germinate. This indicates that a light response must be evaluated critically in its ecological context before conclusions can be drawn on its significance in a plant's life cycle.

Importance of Light Responses in the Life Cycle of Selected Plant Species

The above-mentioned studies provide evidence for a potential ecological significance of a particular light response only. A complicating factor for a sound ecological interpretation of results of seed germination experiments is that the photoreceptor has multiple roles in seeds. Hence, a particular response might just be a consequence of the role that phytochrome plays in another situation. Another complicating factor is the strong interaction of phytochrome-mediated processes with other environmental factors and the preconditioning of the seed. Many experiments have been carried out with freshly harvested or dry stored seeds. Since as shown above, the light response can change considerably after dispersal and further with the seasons, the results may only apply to newly dispersed seeds, and may not be relevant for overwintering seeds or seeds in the seed bank. A third reason that may confound conclusions on the significance of a light response is that different

mechanisms may have potentially similar functions. For instance, seeds have an array of detection mechanisms for disturbance events. Light responses represent only a few of those in the 'tool box' available to a seed. A critical evaluation must decide what the importance of a particular light response is. In this section the available information on light responses of a few selected plant species is critically evaluated with regard to the actual significance of phytochrome functioning for their seeds.

The complexities of deciphering the ecological significance of seed responses to light can be illustrated in *Plantago major*, a weedy short-lived perennial species occurring on open places in grassland and along paths and roadsides. Establishment from naturally dispersed seeds in a dense grass sward generally does not occur, and when the rather small seeds (0.2 mg) are experimentally sown in dense vegetation successful establishment is again absent (Sagar and Harper, 1964; Blom, 1978; van der Toorn and Pons, 1988). The species can form large seed banks (van Altena and Minderhoud, 1972), with the seeds surviving long periods in soil (Toole and Brown, 1946; Kivilaan and Bandurski, 1981). They show a seasonal pattern of emergence (Froud-Williams *et al.*, 1984a), which is probably due to seasonal changes in dormancy (Pons and van der Toorn, 1988). Several studies indicate an almost absolute light requirement for dormancy breaking (Steinbauer and Grigsby, 1957; Frankland and Poo, 1980; Grime *et al.*, 1981; Froud-Williams *et al.*, 1984b; Pons, 1986, 1991a; Fig. 5.3).

Far-red and leaf canopy-filtered light inhibit germination of *P. major* (van der Veen, 1970; Górski *et al.*, 1977; Frankland and Poo, 1980; Froud-Williams *et al.*, 1984b; Pons, 1986; Pons and van der Toorn, 1988). Frankland and Poo (1980) exposed seeds to light under a *Sinapis alba* leaf canopy. Increasing plant density and thus increasing leaf area index resulted in decreasing R:FR ratio of the light. This in turn caused decreasing $P_{fr}:P_{tot}$ ratios in the seeds, resulting in a decline in germination with increasing plant density with complete inhibition at high densities (Fig. 5.5). Similar results were obtained with dry stored seeds sown under grass by Pons and van der Toorn (1988). However, several seeds sown in the field in autumn under grass germinated the following March and April, although germination in gaps was higher.

The experiments carried out with freshly harvested and/or cool and dry stored seeds of *P. major* can be considered as representative of summer conditions in the field. There are two important points of difference in spring as compared to summer conditions in the temperate climate where the species originates: (i) leaf canopies are not so dense in spring as in summer, resulting in higher R:FR ratios; and (ii) lower R:FR ratios are required to enforce dormancy of chilled seeds than of freshly harvested seeds (Pons, 1986; Fig. 5.6). The two factors together permit germination in spring under grass of several seeds that are at or near the soil surface. However, none of the seedlings that germinated under grass survived (Pons and van der Toorn, 1988). Hence, the light climate in spring is not a reliable environmental signal for avoidance of ultimately fatal germination under established plants. However, newly ripened seeds have a high temperature and P_{fr} requirement, they are exposed to summer droughts that further increases the P_{fr} requirement (Pons, 1986), and leaf canopies are denser than in spring. Dormancy enforced by canopy shade light by a combination of the LFR and the HIR is thus likely to play a more important role in summer after seed fall. Germination will be retarded or avoided altogether and the small seeds may have a chance to slip into crevices or become buried in the soil in another way. The P_{fr} that may be present in the seeds as a result of exposure before burial will disappear (Pons, 1991a) rendering the seed obligate light requiring. Soil disturbance will bring seeds to the surface again, where some of them may germinate upon exposure. However, there is no evidence for the engagement of the VLFR in this species. Hence, the very short exposures resulting from quick re-burial may not be effective. Other seeds that do not respond to a short exposure (Frankland and Poo, 1980; Fig. 5.3) avoid possible fatal germination at

depth and await another disturbance. Hence, the significance of leaf canopy-induced dormancy may lie more in the prevention of germination shortly after dispersal of the seeds rather than during the time of normal germination of *P. major* in spring.

The significance of induction of a light requirement by leaf-canopy shade before burial has also been suggested by Fenner (1980b) in the case of the weed *Bidens pilosa*. It was even more pronounced in the biennial *Cirsium palustre*, which emerges in ash coppice after felling (Pons, 1983, 1984). Although germination is inhibited at low R:FR, there is no leaf canopy in spring when germination takes place, which can serve as an environmental signal for reduced competition. Again it was concluded in this case that the significance must be found in summer after seed dispersal. Soil disturbance during felling is a principal trigger for germination, as concluded from the stimulation of emergence by artificial soil disturbance in a non-felled situation and its reduction by carefully avoiding disturbance during felling.

Origanum vulgare is a perennial growing on exposed rocky places, forest edges and chalk grasslands. Large numbers of small seeds ($\approx$0.1 mg) are produced in the autumn. There is some germination in autumn, presumably of newly shed seeds, but most of the emergence of seedlings is concentrated in spring (van Tooren and Pons, 1988). The species forms large persistent seed banks (Thompson and Grime, 1979; Willems, 1995). Widely different percentages of germination in darkness have been reported for *O. vulgare*. Silvertown (1980) and Grime *et al.* (1981) reported high dark germination, but van Tooren and Pons (1988) and Pons (1991a) found lower percentages, which decreased with decreasing temperature (Figs 5.3 and 5.6), and also depended on chilling of the seeds. Short exposures to R effectively broke dormancy in all seeds and even FR stimulated germination (van Tooren and Pons, 1988; Pons, 1991a) (Figs 5.3 and 5.6), which is indicative of the engagement of the VLFR. Seeds buried experimentally in the soil in a chalk grassland did not germinate and quickly lost their germinability in darkness (Pons, 1991b). Induction of dark dormancy was completed in a few days at high temperatures under conditions of osmotic stress (Fig. 5.6), but was slow at 4°C and in seeds that had been exposed to R (Pons, 1991a). There may also be slight seasonal changes in dormancy (Pons, 1991b). There was a strong inhibiting effect of a low R:FR ratio of the light at low temperatures in freshly harvested seeds, but this was virtually absent at high temperatures and in chilled seeds (van Tooren and Pons, 1988). From these studies the picture emerges that *Origanum* seeds can have a variable amount of P_{fr} in ripe seeds, which disappears in the soil at higher temperatures when conditions are not conducive for germination, thus inducing a light requirement. The dormancy-enforcing effect of leaf canopy-filtered light at low temperatures may successfully avoid germination of seeds dispersed in vegetation in autumn, but not on open places (van Tooren and Pons, 1988). The small seeds may easily become buried in the soil and a light requirement will be quickly induced when previously exposed to leaf-filtered light, but will take somewhat longer when previously exposed to unfiltered daylight. The engagement of the VLFR in this species, particularly in chilled seeds, clearly points to the functioning of phytochrome as a light detector in spring rather than as a leaf-canopy shade detector, whereas the latter function is likely to be more prominent directly after dispersal.

Similar results were obtained with the heather species *Calluna vulgaris* and *Erica tetralix* (Pons, 1989b,c). Freshly harvested seeds were inhibited by leaf canopy-filtered light, both due to a low R:FR and to the low irradiance. This may prevent their germination shortly after dispersal. However, germination was independent of light quality or intensity after a pretreatment resembling burial, but no germination occurred in darkness, pointing to phytochrome functioning as a light detector once the seeds are part of the soil seed bank and not as a leaf-canopy detector.

The above-mentioned studies have been carried out in temperate climates where leaf canopies vary seasonally in density or are deciduous. The general picture that arises from them is that induction of dormancy by leaf canopy-filtered light is most important in summer and/or autumn after dispersal of the seeds. The function of phytochrome shifts more towards a role as a light detector in early spring when seeds are less dormant and the leaf canopy is absent or not so dense.

Members of the genus *Cecropia* are common as pioneers in the evergreen tropical rainforest of Central and South America. The short-lived trees produce large numbers of small animal-dispersed seeds. They form persistent seed banks in forest soil (Holthuyzen and Boerboom, 1982; Fornara and Dalling, 2005). Experiments have shown that seeds can remain viable in the soil for periods of at least a year (Vázquez-Yanes and Smith, 1982), and presumably much longer. Seedlings cannot survive under closed forest canopy and establishment from seed occurs mainly when large gaps are formed. The germination of seeds of *Cecropia* is stimulated by exposure to normal daylight, but not by leaf canopy-filtered light (Valio and Joly, 1979; Vázquez-Yanes, 1980; Vázquez-Yanes and Orozco-Segovia, 1994). Prolonged exposure to light is required, which can be replaced by intermittent short exposures (Vázquez-Yanes and Smith, 1982; Vázquez-Yanes and Orozco-Segovia, 1990). Time to germination is long, up to 45 days, particularly at moderately low R:FR where germination can also occur (Pearson *et al.*, 2003). Low irradiance is effective in breaking dormancy. This is evident from the stimulation of germination in an experiment where FR was excluded from the spectrum under a dense forest canopy (Vázquez-Yanes, 1980). Also light at 6 mm depth in sand stimulated germination (Bliss and Smith, 1985). No reports are available on an inhibitory effect of high irradiance. Hence, the HIR is not likely to be active in these species. To break dormancy, the minimum length of a daily exposure to high R:FR white light in the middle of a day that further consisted of FR only was 6 h. The available data suggest a long escape time and a relatively high P_{fr} threshold for dormancy breaking. These characteristics of the regulation of the light response by phytochrome contribute to distinguishing large gaps, which are suitable for establishment from short sunflecks, and small gaps, which are not. This reasoning applies only to newly dispersed seeds exposed to light at or close to the soil surface. It is not known whether changes in the light response occur after dispersal. Nevertheless, the available evidence suggests that as in the above-mentioned examples from temperate regions, canopy shade light-imposed dormancy in newly dispersed seeds is important for the formation of a seed bank under established vegetation. Soil disturbance during gap formation can expose buried seeds to light, which breaks dormancy only when the period of unobstructed daylight is sufficiently long as in large gaps. Phytochrome as a canopy-gap sensor for detecting safe sites for germination thus seems to be more important in evergreen tropical rainforest compared to temperate deciduous forest where the leaf canopy is largely absent during the germination season.

Concluding Remarks With Respect to the Ecological Significance of Light Responses

In this chapter various light responses of seeds have been reviewed and attempts have been made to indicate their possible ecological significance. In some cases the responses appear to have an adaptive function. In other cases this is less clear, and further studies that are specifically designed to critically investigate the ecological significance of a particular light response are required.

There seems to be little doubt about the significance of phytochrome as a detector for the absence of light in soil that maintains dormancy in seed banks. The requirement for P_{fr} in the LFR and the VLFR is the working mechanism. Dark reversion of P_{fr}, pre-existent or established by exposure to light before burial, can induce a light requirement when not present. Also, the role of phytochrome as a light detector with respect to soil disturbance events is straightforward. The LFR plays a role here when seeds are brought to the surface. When seeds are quickly reburied, the VLFR enables the ones that can develop this response to detect the sometimes very short exposures during soil disturbance. The detection of extremely low irradiances just below the soil surface may also require the VLFR.

At the soil surface in unobstructed daylight, the HIR may avoid seedling desiccation by inhibiting germination. However, there is no experimental evidence for such a function. The HIR also causes a further reduction of germination in addition to the reduction caused by the LFR in canopy-filtered light. Hence, the inhibition of germination at high irradiance caused by the HIR at the soil surface in full daylight may also be a consequence of its role for enforcing dormancy in leaf-canopy shade. Critical experimentation is required to resolve this.

The role of enforcement of dormancy in leaf canopy-filtered light, as a result of combined action of LFR and HIR, varies with vegetation type. The available evidence indicates that this light response plays an important role in gap detection in evergreen vegetation such as tropical rainforest. In temperate regions where the vegetation has often a more deciduous character, the role of leaf canopy-enforced dormancy appears to be more related to the prevention of germination shortly after dispersal, thus facilitating the incorporation of the seeds into the soil seed bank.

References

Appenroth, K.J., Lenk, G., Goldau, L. and Sharma, R. (2006) Tomato seed germination: regulation of different response modes by phytochrome B2 and phytochrome A. *Plant Cell and Environment* 29, 701–709.

Baskin, C.C. and Baskin, J.M. (2001) *Seeds – Ecology, Biogeography and Evolution of Dormancy and Germination.* Academic Press, San Diego.

Baskin, J.M. and Baskin, C.C. (1979) Promotion of germination of *Stellaria media* seeds by light from a green safe lamp. *New Phytologist* 82, 381–383.

Baskin, J.M. and Baskin, C.C. (1980) Ecophysiology of secondary dormancy in seeds of *Ambrosia artemisiifolia. Ecology* 61, 475–480.

Baskin, J.M. and Baskin, C.C. (1985) The light requirement for germination of *Aster pilosus* seeds: temporal aspects and ecological consequences. *Journal of Ecology* 73, 765–773.

Batlla, D. and Benech-Arnold, R.L. (2005) Changes in the light sensitivity of buried *Polygonum aviculare* seeds in relation to cold-induced dormancy loss: development of a predictive model. *New Phytologist* 165, 445–452.

Batlla, D., Nicoletta, M. and Benech-Arnold, R.L. (2007) Sensitivity of *Polygonum aviculare* seeds to light as affected by soil moisture conditions. *Annals of Botany* 99, 915–924.

Bewley, J.D. (1997) Breaking down the walls – a role for endo-beta-mannanase in release from seed dormancy? *Trends in Plant Science* 2, 464–469.

Bewley, J.D. and Black, M. (1982) *Physiology and Biochemistry of Seeds.* Vol. 2, Springer, Berlin.

Blaauw-Jansen, G. (1983) Thoughts on the possible role of phytochrome destruction in phytochrome-controlled responses. *Plant Cell and Environment* 6, 173–179.

Blaauw-Jansen, G. and Blaauw, O.H. (1975) A shift of the threshold to red irradiation in dormant lettuce seeds. *Acta Botanica Neerlandica* 24, 199–202.

Black, M. and Wareing, P.F. (1955) Growth studies in woody species. VII. Photoperiodic control of germination in *Betula pubescens. Physiologia Plantarum* 8, 300–316.

Bliss, D. and Smith, H. (1985) Penetration of light into soil and its role in the control of seed germination. *Plant Cell and Environment* 8, 475–483.

Blom, C.W.P.M. (1978) Germination, seedling emergence and establishment of some *Plantago* species under laboratory and field conditions. *Acta Botanica Neerlandica* 27, 257–271.

Borthwick, H.A., Hendricks, S.B., Toole, E.H. and Toole, V.K. (1954) Action of light on lettuce seed germination. *Botanical Gazette* 115, 205–225.

Botto, J.F., Sánchez, R.A. and Casal, J.J. (1998) Burial conditions affect light responses of *Datura ferox* seeds. *Seed Science Research* 8, 423–429.

Bouwmeester, H.J. and Karssen, C.M. (1992) The dual role of temperature in the regulation of the seasonal changes in dormancy and germination of seeds of *Polygonum persicaria* L. *Oecologia* 90, 88–94.

Casal, J.J. and Sánchez, R.A. (1998) Phytochromes and seed germination. *Seed Science Research* 8, 317–329.

Cone, J.W. and Kendrick, R.E. (1986) Photocontrol of seed germination. In: Kendrick, R.E. and Kronenberg, G.H.M. (eds) *Photomorphogenesis in Plants.* Martinus Nijhoff, Dordrecht, pp. 443–465.

Corbineau, F. and Côme, D. (1982) Effect of intensity and duration of light at various temperatures on the germination of *Oldenlandia corymbosa* L. seeds. *Plant Physiology* 70, 1518–1520.

Cresswell, E.G. and Grime, J.P. (1981) Induction of light requirement during seed development and its ecological consequences. *Nature* 291, 583–585.

Cumming, B.G. (1963) The dependence of germination on photoperiod, light quality, and temperature in *Chenopodium* spp. *Canadian Journal of Botany* 41, 1211–1233.

Derkx, M.P.M. and Karssen, C.M. (1993a) Effects of light and temperature on seed dormancy and gibberellin-stimulated germination in *Arabidopsis thaliana*: studies with gibberellin-deficient and insensitive mutants. *Physiologia Plantarum* 89, 360–368.

Derkx, M.P.M. and Karssen, C.M. (1993b) Changing sensitivity to light and nitrate but not to gibberellins regulates seasonal dormancy patterns in *Sisymbrium officinale* seeds. *Plant Cell and Environment* 16, 469–479.

Duke, S.O. (1978) Interactions of seed water content with phytochrome-initiated germination of *Rumex crispus* L. seeds. *Plant Cell Physiology* 19, 1043–1049.

Ellis, R.H., Hong, T.D. and Roberts, E.H. (1986a) Quantal response of seed germination in *Brachiaria humidicola, Echinochloa turnerana, Eragrostis tef* and *Panicum maximum* to photon dose for the low energy reaction and the high irradiance reaction. *Journal of Experimental Botany* 37, 742–753.

Ellis, R.H., Hong, T.D. and Roberts, E.H. (1986b) The response of seeds of *Bromus sterilis* L. and *Bromus mollis* L. to white light of varying photon flux density and photoperiod. *New Phytologist* 104, 485–496.

Ellis, R.H., Hong, T.D. and Roberts, E.H. (1989) Response of seed germination in three genera of *Compositae* to white light of varying photon flux density and photoperiod. *Journal of Experimental Botany* 40, 13–22.

Federer, C.A. and Tanner, C.B. (1966) Spectral distribution of light in the forest. *Ecology* 47, 555–560.

Fenner, M. (1978) A comparison of the abilities of colonizers and closed-turf species to establish from seed in artificial swards. *Journal of Ecology* 66, 953–963.

Fenner, M. (1980a) Germination tests on thirty-two East African weed species. *Weed Research* 20, 135–138.

Fenner, M. (1980b) The induction of a light-requirement in *Bidens pilosa* seeds by leaf canopy shade. *New Phytologist* 84, 103–106.

Fenner, M. (1985) *Seed Ecology*. Chapman and Hall, London.

Flint, L.H. and McAllister, E.D. (1937) Wavelengths of radiation in the visible spectrum promoting the germination of light sensitive lettuce seeds. *Smithsonian Miscellaneous Collections* 96, 1–8.

Fornara, D.A. and Dalling, J.W. (2005) Seed bank dynamics in five Panamanian forests. *Journal of Tropical Ecology* 21, 223–226.

Frankland, B. (1976) Phytochrome control of seed germination in relation to the light environment. In: Smith, H. (ed.) *Light and Plant Development*. Butterworth, London, pp. 477–491.

Frankland, B. (1986) Perception of light quantity. In: Kendrick, R.E. and Kronenberg, G.H.M. (eds) *Photomorphogenesis in Plants*. Martinus Nijhoff Publishers, Dordrecht, pp. 219–235.

Frankland, B. and Letendre, R.J. (1978) Phytochrome and effects of shading on the growth of woodland plants. *Photochemistry and Photobiology* 27, 223–230.

Frankland, B. and Poo, W.K. (1980) Phytochrome control of seed germination in relation to natural shading. In: De Greef, J. (ed.) *Photoreceptors and Plant Development*. University Press, Antwerpen, pp. 357–366.

Frankland, B. and Taylorson, R. (1983) Light control of seed germination. In: Shropshire, W. and Mohr, H. (eds) *Encyclopedia of Plant Physiology*, New Series Vol. 16A, Springer, Berlin, pp. 428–456.

Franklin, K.A. and Quail, P.H. (2010) Phytochrome functions in Arabidopsis development. *Journal of Experimental Botany* 61, 11–24.

Froud-Williams, R.J., Chancellor, R.J. and Drennan, D.S.H. (1984a) The effects of seed burial and soil disturbance on emergence and survival of arable weeds in relation to minimal cultivation. *Journal of Applied Ecology* 21, 629–641.

Froud-Williams, R.J., Drennan, D.S.H. and Chancellor, R.J. (1984b) The influence of burial and dry storage upon cyclic changes in dormancy, germination and response to light in seeds of various arable weeds. *New Phytologist* 96, 473–481.

Gallagher, R.S. and Cardina, J. (1998a) Ecophysiological aspects of phytochrome-mediated germination in soil seed banks. *Aspects of Applied Biology* 51, 165–172.

Gallagher, R.S. and Cardina, J. (1998b) The effect of light environment during tillage on the recruitment of various summer annuals. *Weed Science* 46, 214–216.

Górski, T. (1975) Germination of seeds in the shadow of plants. *Physiologia Plantarum* 34, 342–346.

Górski, T. and Górska, K. (1979) Inhibitory effects of full daylight on the germination of *Lactuca sativa* L. *Planta* 144, 121–124.

Górski, T., Górska, K. and Nowicki, J. (1977) Germination of seeds of various herbaceous species under leaf canopy. *Flora* 166, 249–259.

Górski, T., Górska, K. and Rybicki, J. (1978) Studies on the germination of seeds under leaf canopy. *Flora* 167, 289–299.

Goudey, J.S., Saini, H.S. and Spencer, M.S. (1987) Seed germination of wild mustard (*Sinapis arvensis*): factors required to break primary dormancy. *Canadian Journal of Botany* 65, 849–852.

Grime, J.P. and Jarvis, B.C. (1975) Shade avoidance and shade tolerance in flowering plants II. Effects of light on the germination of contrasted ecology. In: Evans, G.C., Bainbridge, R. and Rackham, O. (eds) *Light as an Ecological Factor: II*. Blackwell, Oxford, pp. 525–532.

Grime, J.P., Mason, G., Curtis, A.V., Rodman, J., Band, S.R., Mowforth, M.A.G., Neal, A.M. and Shaw, S. (1981) A comparative study of germination characteristics in a local flora. *Journal of Ecology* 69, 1017–1059.

Grubb, P.J. (1977) The maintenance of species richness in plant communities: the importance of the regeneration niche. *Biological Reviews* 52, 107–145.

Hartmann, K.M., Mollwo, A. and Tebbe, A. (1998) Photocontrol of germination by moon- and starlight. *Zeitschrift für Pflanzenkrankheiten und Pflanzenschutz, Sonderheft* XVI, 119–127.

Hartmann, K.M., Grundy, A.C. and Market, R. (2005) Phytochrome-mediated long-term memory of seeds. *Protoplasma* 227, 47–52.

Hennig, L., Buche, C. and Schafer, E. (2000) Degradation of phytochrome A and the high irradiance response in Arabidopsis: a kinetic analysis. *Plant Cell and Environment* 23, 727–734.

Hennig, L., Stoddart, W.M., Dieterle, M., Whitelam, G.C. and Schäfer, E. (2002) Phytochrome E controls light-induced germination of Arabidopsis. *Plant Physiology* 128, 194–200.

Heschel, M.S., Selby, J., Butler, C., Whitelam, G.C., Sharrock, R.A. and Donohue, K. (2007) A new role for phytochromes in temperature-dependent germination. *New Phytologist* 174, 735–741.

Hilhorst, H.W.M., Smitt, A.I. and Karssen, C.M. (1986) Gibberellin-biosynthesis and -sensitivity mediated stimulation of seed germination of *Sisymbrium officinale* by red light and nitrate. *Physiologia Plantarum* 67, 285–290.

Holmes, M.G. and Smith, H. (1977) The function of phytochrome in the natural environment. I. The influence of vegetation canopies on the spectral energy distribution of natural daylight. *Photochemistry and Photobiology* 25, 539–545.

Holthuyzen, A.M.A. and Boerboom, J.H.A. (1982) The *Cecropia* seed bank in the Surinam lowland rainforest. *Biotropica* 14, 62–68.

Karssen, C.M. (1970) The light promoted germination of the seeds of *Chenopodium album* L. IV. Effects of red, far-red and white light on non-photoblastic seeds incubated in mannitol. *Acta Botanica Neerlandica* 19, 95–108.

Karssen, C.M. (1982) Seasonal patterns of dormancy in weed seeds. In: Khan, A.A. (ed.) *The Physiology and Biochemistry of Seed Development Dormancy and Germination*. Elsevier Biomedical Press, Amsterdam.

Karssen, C.M., Zagórski, S., Kepczynski, J. and Groot, S.P.C. (1989) Key role for endogenous gibberellins in the control of seed germination. *Annals of Botany* 63, 71–80.

Kendrick, R.E. and Frankland, B. (1969) Photocontrol of germination in *Amaranthus caudatus*. *Planta* 85, 326–339.

Kendrick, R.E. and Spruit, C.J.P. (1977) Phototransformations of phytochrome. *Photochemistry and Photobiology* 26, 201–204.

Khan, A.A. and Karssen, C.M. (1980) Induction of secondary dormancy in *Chenopodium bonus-henricus* L. seeds by osmotic and high temperature treatments and its prevention by light and growth regulators. *Plant Physiology* 66, 175–181.

King, T.J. (1975) Inhibition of seed germination under leaf canopies in *Arenaria serpyllifolia, Veronica arvensis* and *Cerastium holosteoides*. *New Phytologist* 75, 87–90.

Kinzel, W. (1926) *Frost und Licht als beeinflussende Kräfte der Samenkeimung*. Ulmer, Stuttgart.

Kivilaan, A. and Bandurski, R.S. (1981) The one hundred-year period for Dr Beal's seed viability experiment. *American Journal of Botany* 68, 1290–1292.

Koornneef, M. and Kendrick, R.E. (1994) Photomorphogenic mutants of higher plants. In: Kendrick, R.E. and Kronenberg, G.H.M. (eds) *Photomorphogenesis in Plants*, 2nd edn. Kluwer, Dordrecht, pp. 601–630.

Koornneef, M., Bentsink, L. and Hilhorst, H. (2002) Seed dormancy and germination. *Current Opinion in Plant Biology* 5, 33–36.

Maloof, J.N., Borevitz, J.O., Weigel, D. and Chory, J. (2000) Natural variation in phytochrome signaling. *Seminars in Cell and Developmental Biology* 11, 523–530.

Mathews, S. and Sharrock, R.A. (1997) Phytochrome gene diversity. *Plant Cell and Environment* 20, 666–671.

Milberg, P. and Andersson, L. (1997) Seasonal variation in dormancy and light sensitivity in buried seeds of eight annual weed species. *Canadian Journal of Botany* 75, 1998–2004.

Milberg, P., Andersson, L. and Thompson, K. (2000) Large-seeded species are less dependent on light for germination than small-seeded ones. *Seed Science Research* 10, 99–104.

Pearson, T.R.H., Burslem, D.F.R.P., Mullins, C.E. and Dalling, J.W. (2002) Germination ecology of neotropical pioneers: Interacting effects of environmental conditions and seed size. *Ecology* 83, 2798–2807.

Pearson, T.R.H., Burslem, D.F.R.P., Mullins, C.E. and Dalling, J.W. (2003) Functional significance of photoblastic germination in neotropical pioneer trees: a seed's eye view. *Functional Ecology* 17, 394–402.

Pons, T.L. (1983) Significance of inhibition of seed germination under the leaf canopy in ash coppice. *Plant Cell and Environment* 6, 385–392.

Pons, T.L. (1984) Possible significance of changes in the light requirement of *Cirsium palustre* seeds after dispersal in ash coppice. *Plant Cell and Environment* 7, 263–268.

Pons, T.L. (1986) Response of *Plantago major* seeds to the red/far-red ratio as influenced by other environmental factors. *Physiologia Plantarum* 68, 252–258.

Pons, T.L. (1989a) Breaking of seed dormancy by nitrate as a gap detection mechanism. *Annals of Botany* 63, 139–143.

Pons, T.L. (1989b) Dormancy and germination of *Calluna vulgaris* (L.) Hull and *Erica tetralix* L. seeds. *Acta Oecologica: Oecologia Plantarum* 10, 35–43.

Pons, T.L. (1989c) Dormancy, germination and mortality of seeds in heathland and inland sand dunes. *Acta Botanica Neerlandica* 38, 327–335.

Pons, T.L. (1991a) Induction of dark-dormancy in seeds: its importance for the seed bank in the soil. *Functional Ecology* 5, 669–675.

Pons, T.L. (1991b) Dormancy, germination and mortality of seeds in a chalk grassland flora. *Journal of Ecology* 79, 765–780.

Pons, T.L. and During, H.J. (1987) Biennial behaviour of *Cirsium palustre* in ash coppice. *Holarctic Ecology* 10, 40–44.

Pons, T.L. and Schröder, H.F.J.M. (1986) Significance of temperature fluctuation and oxygen concentration of the rice field weeds *Fimbristylis littoralis* and *Scirpus juncoides*. *Oecologia* 68, 315–319.

Pons, T.L. and van der Toorn, J. (1988) Establishment of *Plantago lanceolata* L. and *Plantago major* L. among grass. I Significance of light for germination. *Oecologia* 75, 394–399.

Poppe, C. and Schäfer, E. (1997) Seed germination of *Arabidopsis thaliana* phyA/phyB double mutants is under phytochrome control. *Plant Physiology* 114, 1487–1492.

Probert, R.J. and Smith, R.D. (1986) The joint action of phytochrome and alternating temperatures in the control of seed germination in *Dactylis glomerata*. *Physiologia Plantarum* 67, 299–304.

Probert, R.J., Gajjar, K.H. and Haslam, I.K. (1987) The interactive effects of phytochrome, nitrate and thiourea on the germination response to alternating temperatures in seeds of *Ranunculus sceleratus* L.: a quantal approach. *Journal of Experimental Botany* 38, 1012–1025.

Rollin, P. and Maignan, G. (1967) Phytochrome and the photoinhibition of germination. *Nature* 214, 741–742.

Sagar, G.R. and Harper, J.L. (1964). Biological flora of the British Isles. *Plantago major* L., *Plantago media* L., *Plantago lanceolata* L. *Journal of Ecology* 52, 189–221.

Sauer, J. and Struik, G. (1964) A possible ecological relation between soil disturbance, light-flash, and seed germination. *Ecology* 45, 884–886.

Schäfer, E. and Schmidt, W. (1974) Temperature dependence of phytochrome dark reactions. *Planta* 116, 257–266.

Scheibe, J. and Lang, A. (1965) Lettuce seed germination: evidence for a reversible light-induced increase in growth potential and for phytochrome mediation of the low temperature effect. *Plant Physiology* 40, 485–492.

Scopel, A.L., Ballaré, C.L. and Sánchez, R.A. (1991) Induction of extreme light sensitivity in buried weed seeds and its role in the perception of soil cultivations. *Plant Cell and Environment* 14, 501–508.

Scopel, A.L., Ballaré, C.L. and Radosevitch, S.R. (1994) Photostimulation of seed germination during soil tillage. *New Phytologist* 126, 145–152.

Senden, J.W., Schenkeveld, A.J. and Verkaar, H.J. (1986) The combined effect of temperature and red:far-red ratio on the germination of some short-lived chalk grassland species. *Acta Oecologica: Oecologia Plantarum* 7, 251–259.

Seo, M., Nambara, E., Choi, G. and Yamaguchi, S. (2009) Interaction of light and hormone signals in germinating seeds. *Plant Molecular Biology* 69, 463–472.

Shichijo, C., Katada, K., Tanaka, O. and Hashimoto, T. (2001) Phytochrome A-mediated inhibition of seed germination in tomato. *Planta* 213, 764–769.

Shinomura, T. (1997) Phytochrome regulation of seed germination. *Journal of Plant Research* 110, 151–161.

Silvertown, J. (1980) Leaf-canopy-induced seed dormancy in a grassland flora. *New Phytologist* 85, 109–118.

Smith, H. (1975) *Phytochrome and Photomorphogenesis*. McGraw-Hill, London.

Smith, H. (1982) Light quality, photoperception, and plant strategy. *Annual Review of Plant Physiology* 33, 481–518.

Smith, H. (1995) Physiological and ecological function within the phytochrome family. *Annual Review of Plant Physiology and Plant Molecular Biology* 46, 289–315.

Smith, H. and Whitelam, G.C. (1990) Phytochrome, a family of photoreceptors with multiple physiological roles. *Plant Cell and Environment* 13, 695–707.

Steinbauer, G.P. and Grigsby, B. (1957) Dormancy and germination characteristics of the seeds of four species of *Plantago*. *Proceedings of the Association of Official Seed Analysts of North America* 47, 158–164.

Stoutjesdijk, P. (1972) Spectral transmission curves of some types of leaf canopies with a note on seed germination. *Acta Botanica Neerlandica* 21, 185–191.

Takaki, M., Kendrick, R.E. and Dietrich, S.M.C. (1981) Interaction of light and temperature on the germination of *Rumex obtusifolius*. *Planta* 152, 209–214.

Taylorson, R.B. and Borthwick, H.A. (1969) Light filtration by foliar canopies: significance for light controlled weed seed germination. *Weed Science* 17, 48–51.

Taylorson, R.B. and Hendricks, S.B. (1969) Action of phytochrome during prechilling of *Amaranthus retroflexus* L. seeds. *Plant Physiology* 44, 821–825.

Taylorson, R.B. and Hendricks, S.B. (1972) Interaction of light and a temperature shift in seed germination. *Plant Physiology* 49, 127–130.

Tester, M. and Morris, C. (1987) The penetration of light through soil. *Plant Cell and Environment* 10, 281–286.

Thompson, K. and Grime, J.P. (1979) Seasonal variation in the seed banks of herbaceous species in contrasting habitats. *Journal of Ecology* 67, 893–921.

Thompson, P.A. (1969) Germination of *Lycopus europaeus* in response to fluctuating temperatures and light. *Journal of Experimental Botany* 20, 1–11.

Toole, E.H. and Brown, E. (1946) Final results of the Duvel buried seed experiment. *Journal of Agricultural Research* 72, 201–210.

Toole, E.H., Toole, V.K., Borthwick, H.A. and Hendricks, S.B. (1955) Interaction of temperature and light in germination of seeds. *Plant Physiology* 30, 473–478.

Toole, V.K. (1973) Effects of light, temperature and their interactions on the germination of seeds. *Seed Science and Technology* 1, 339–396.

Totterdel, S. and Roberts, E.H. (1980) Characteristics of alternating temperatures which stimulate loss of dormancy in seeds of *Rumex obtusifolius* L. and *Rumex crispus* L. *Plant Cell and Environment* 3, 3–12.

Valio, I.F.M. and Joly, C.A. (1979) Light sensitivity of the seeds on the distribution of *Cecropia glaziovi* Snethlage (Moraceae). *Zeitschrift für Pflanzenphysiologie* 91, 371–376.

van Altena, S.C. and Minderhoud, J.W. (1972) Keimfähige Samen von Gräsern und Kräutern in der Narbenschicht der Niederländischen Weiden. *Zeitschrift für Acker- und Pflanzenbau* 136, 95–109.

van Assche, J.A. and Vanlerberghe, K.A. (1989) The role of temperature on the dormancy cycle of seeds of *Rumex obtusifolius* L. *Functional Ecology* 3, 107–115.

van Baalen, J. (1982) Germination ecology and seed population dynamics of *Digitalis purpurea*. *Oecologia* 53, 61–67.

van der Meyden, E. and van der Waals-Kooi, R.E. (1979) The population ecology of *Senecio jacobaea* in a sand dune system. I. Reproductive strategy and the biennial habit. *Journal of Ecology* 67, 131–153.

van der Toorn, J. and Pons, T.L. (1988) Establishment of *Plantago lanceolata* L. and *Plantago major* L. among grass. II. Shade tolerance of seedlings and selection on time of germination. *Oecologia* 76, 341–347.

van der Veen, R. (1970) The importance of the red-far red antagonism in photoblastic seeds. *Acta Botanica Neerlandica* 19, 809–812.

VanderWoude, W.J. and Toole, V.K. (1980) Studies of the mechanism of enhancement of phytochrome dependent lettuce seed germination by prechilling. *Plant Physiology* 66, 220–224.

van Tooren, B.F. and Pons, T.L. (1988) Effects of temperature and light on the germination in chalk grassland species. *Functional Ecology* 2, 303–310.

Vázquez-Yanes, C. (1980) Light quality and seed germination in *Cecropia obtusifolia* and *Piper auritum* from a tropical rain forest in Mexico. *Fyton* 38, 33–35.

Vázquez-Yanes, C. and Orozco-Segovia, A. (1990) Ecological significance of light controlled seed germination in two contrasting tropical habitats. *Oecologia* 83, 171–175.

Vázquez-Yanes, C. and Orozco-Segovia, A. (1994) Signals for seeds to sense and respond to gaps. In: Caldwell, M.M. and Pearcy, R.W. (eds) *Exploitation of Environmental Heterogeneity by Plants. Ecophysiological Processes Above and Below-Ground.* Academic Press, New York, pp. 261–318.

Vázquez-Yanes, C. and Smith, H. (1982) Phytochrome control of seed germination in the tropical rain-forest pioneer trees *Cecropia obtusifolia* and *Piper auritum* and its ecological significance. *New Phytologist* 92, 477–485.

Vincent, E.M. and Roberts, E.H. (1977) The interaction of light, nitrate and alternating temperature in promoting the germination of dormant seeds of common weed species. *Seed Science and Technology* 5, 659–670.

Wesson, G. and Wareing, P.F. (1969) The induction of light sensitivity in weed seeds by burial. *Journal of Experimental Botany* 20, 414–425.

Widell, K.O. and Vogelmann, T.C. (1988) Fibre optics studies of light gradients and spectral regime within *Lactuca sativa* achenes. *Physiologia Plantarum* 72, 706–712.

Willems, J.H. (1995) Soil seed bank, seedling recruitment and actual species composition in an old and isolated chalk grassland site. *Folia Geobotanica Phytotaxonomica, Praha* 30, 141–156.

Woolley, J.T. and Stoller, E.W. (1978) Light penetration and light-induced seed germination in soil. *Plant Physiology* 61, 597–600.

6 The Chemical Environment in the Soil Seed Bank

Henk W.M. Hilhorst*

Wageningen Seed Lab, Laboratory of Plant Physiology,
Wageningen University, Wageningen, the Netherlands

Introduction

Soil is the natural physical and chemical environment of most seeds. Essentially, soil is a three-phase system consisting of solids, liquids and gases in varying proportions. In most soils the solids are predominantly mineral, derived from rock materials. Minerals are defined as solid, inorganic, naturally occurring substances with a definite chemical formula and general structure. It is evident that minerals may only affect seed behaviour when they are solubilized by water that penetrates the soil. In this respect the soil pH is an important factor. The soil matrix may also contain more readily dissolvable solutes, for example salts in saline environments. Direct chemical effects of rock-derived minerals on germination of seeds in the soil seed bank are unknown. Solubilized minerals may inhibit germination non-specifically when they occur in high concentrations in soils. Also the effects of high salinity can be either osmotic or toxic. Soil may also contain organic matter. The amount of organic matter is determined by the rate at which fresh plant residues are added, and by the rate at which they are decomposed by the microflora and fauna.

Primary soil particles of different sizes are mixed in various proportions to give recognizable textural classes of soils, e.g. sand loams, sands, clay, etc. (Currie, 1973). The matrix formed by such a mixture has a fundamental pore size, which reflects the proportions of the ingredients. Thus clay soils have many small pores; sands have fewer but larger pores. Soil structure depends on the primary composition of soil, on the interaction between different soil components and, in arable land, on the cultivation methods. Therefore, structure forms a heterogeneous pattern that may show large open crevices in dry summers but may collapse under the impact of rain or pressure. Freshly ploughed fields contain clods that in turn are broken down to become the crumbs of the seedbed during subsequent cultivation. Soil structure is important to germinating seeds because it determines the distribution and availability of water, solutes and gases. The seed–soil contact and the process of water uptake have been described in detail by Hadas (1982). Moreover, soil structure also determines the depth of light penetration, which is important in the germination of light-requiring seeds in the top layer of the soil seed bank

* E-mail: henk.hilhorst@wur.nl

© CAB International 2014. *Seeds: The Ecology of Regeneration in Plant Communities,*
3rd Edition (ed. R.S. Gallagher)

(Woolley and Stoller, 1978; also see Chapter 5 of this volume).

The dissolved substances in soil that may affect germination are either inorganic or organic. Most inorganic ions do not have any specific effect on seed germination (Egley and Duke, 1985). Nitrate ions are the notable exception. Organic substances in soil that influence germination may originate from the direct vicinity of the seeds, such as neighbouring seeds and fruits, or from neighbouring plants. Mostly these organics are added to the soil solution by leakage or secretion from living underground plant organs but also by leaking from leaves or by decomposition of plant organs. The chemicals secreted by the exudation of a wide variety of compounds into the soil by plant roots influence the soil microbial community in their immediate vicinity, deter herbivores, encourage beneficial symbiosis (e.g. with mycorrhiza), locally alter the chemical and physical properties of the soil, and inhibit the growth of neighbouring plant species. The most well-known promotive action of some of these compounds is well documented for the chemical signals from roots that stimulate germination of parasitic angiosperms (Visser, 1989; Bouwmeester *et al.*, 2003). However, more recently, bioactive compounds in smoke extracts have been identified, which may play an important role in the germination of pioneer species after a forest fire.

The gas phase of soil contains the usual components of the atmosphere. The modification in its composition by the respiratory activity of soil organisms or by high water saturation may have both profound promotive and inhibitory effects on germination. Also several volatile compounds that occur in soil, such as ethylene, or products of anaerobic metabolism may influence germination. All these soil chemicals may affect seed germination directly or indirectly and, thus, play an important role in seed bank dynamics. In this chapter we will discuss the occurrence, signal translation and mechanisms of action for some of the most important chemical cues.

Inorganic Chemicals

Nitrate

Soil nitrate and dormancy

Nitrate is the major naturally occurring inorganic soil component that stimulates seed germination. This feature of the nitrate ion has been known for a long time. Lehmann (1909) was the first to report its promotive action. Since then, numerous wild (and crop) species, both monocots and dicots, have been found to be sensitive to the stimulatory action of nitrate (Roberts and Smith, 1977; Bewley and Black, 1994). In addition, the germination of fern spores may also be promoted by nitrate (Haas and Scheuerlein, 1990).

Since nitrate is central to the nitrogen cycle, most soil types contain nitrate, often at levels within the range of concentrations that are effective in laboratory germination tests. Germination of seeds is generally stimulated within a range of 0–0.05 mol l^{-1} nitrate. The nitrate concentration of the soil fluctuates within this range and therefore might play an ecological role in regulating germination of the soil seed bank. However, the ecological significance of nitrate for germination in the field cannot be treated in isolation from other biotic and abiotic factors such as water, temperature, light, responsiveness of seeds, other chemical soil constituents, etc. Indeed, interactions between some of these factors and nitrate have been described for a number of species (e.g. Vincent and Roberts, 1977). To complicate matters more, soil nitrate levels, interactions of environmental cues and responsiveness of seeds are all dynamic, showing both temporal and spatial fluctuations over shorter or longer periods and distances.

In the soil seed bank, seeds of many weedy species undergo annual cycles of dormancy (Hilhorst *et al.*, 1996; Baskin and Baskin, 1998). After shedding, seeds often possess primary dormancy (Fig. 6.1). These dormant seeds require (long-term) exposure to certain environments, either dry (afterripening) or imbibed (cold stratification) to become responsive to environmental cues that stimulate germination, including light,

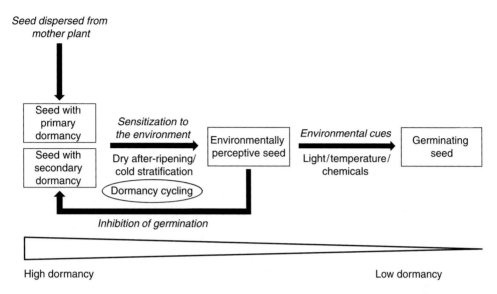

Fig. 6.1. Schematic presentation of changes in dormancy and germination.

optimal temperature and chemicals, such as nitrate. However, germination will only occur when the complete set of required cues is present. If this is not the case, germination will be inhibited and the seeds may gradually enter a new state of dormancy, called 'secondary dormancy' (see Chapter 7 of this volume). Secondary dormancy may then be broken by similar conditions as for primary dormancy. This cycle depends on seasonal fluctuations of the soil temperature and, to a lesser extent, soil moisture content (Bouwmeester and Karssen, 1993; Footitt *et al.*, 2011). Although the sensitizing conditions are generally considered to break dormancy and the environmental cues to stimulate germination, there are no obvious qualitative differences in the associated seed transcriptomes. However, these conditions and cues have a significant quantitative additive effect on the expression of gene sets that are strongly associated with dormancy (Finch-Savage *et al.*, 2007). Therefore, the successive steps that take the seed from the dormant state to germination may be considered as a gradual removal of all the blocks to germination (high to low dormancy; Fig. 6.1).

In the field situation, a seed bank consisting of seeds that undergo annual dormancy cycling will produce a (sub)population of environmentally perceptive seeds at a specific period of the year. Depending on their specific requirements and the actual environmental conditions, these seeds will either germinate or return to the dormant state. If soil nitrate levels were also to fluctuate in a seasonal pattern, it could be argued that this fluctuation interferes with seasonal patterns of seedling emergence. From the literature it appears that there is no general seasonal pattern of nitrate levels in different soil types (e.g. Runge, 1983; Rice, 1984). The rate of mineralization of nitrogen depends on temperature and moisture. Hence, a higher production of both nitrate and ammonium can be expected during the growing season. Obviously, nitrate consumption also increases during this period (Runge, 1983). Other reports show high nitrate levels during winter and early spring, followed by a steady decline, due to consumption and possibly inhibition of nitrification by root systems of established plants (Rice, 1984; Obermann, 1985). In addition, a number of site effects may strongly influence nitrate levels. These site effects may be soil type, soil pH, moisture content, burial depth, disturbance of soil, cultivation practice, etc. Therefore we can only conclude that soil nitrate is not part of the mechanism by which seeds sense the time of the year.

As mentioned earlier, temperature is the most likely candidate for this role. However, as will be discussed later, sensing of nitrate may play a role when seeds require information about the local growth conditions.

Uptake of nitrate

Seeds may take up nitrate during development on the mother plant or, after maturation and shedding, directly from the soil nitrate pool. In species such as *Sisymbrium officinale, Chenopodium album* and *Arabidopsis thaliana*, the nitrate content of the seeds on the mother plant is directly related to soil nitrate levels (Saini and Spencer, 1986; Bouwmeester *et al.,* 1994; Alboresi *et al.,* 2005) (Fig. 6.2). However, other species such as *Polygonum persicaria* and *P. lapatifolia* did not show this relationship. In those species seed nitrate levels were generally much lower. The amount of nitrate that will reach the nitrate pools in plants depends on several factors, both in the soil and the plant. One of the most important factors is the plant's capacity to assimilate nitrate via nitrate reduction. *Arabidopsis* seeds from the nitrate reductase deficient double mutant (*nia1nia2*), which retains only 0.5% nitrate reductase activity of the

wild type, contained significantly higher nitrate contents compared with the wild type grown under the same conditions (Alboresi *et al.,* 2005; Fig. 6.2b). Other known factors are the level of soil ammonium, water status of the soil, volume of the root system, competition of neighbouring plants, efficiency of the uptake system, etc. This all makes predictions of nitrate levels in seeds on the mother plant very complicated. Therefore, assessment of the ecological significance of a possible indirect sensing of the nitrogen status of the soil by seeds while on the mother plant is similarly complicated.

When a mature, dry seed is shed, it comes into direct contact with the soil (micro)environment. For uptake of nitrate from the soil several requirements have to be fulfilled. Nitrate must be available in a freely diffusible form. This implies that sufficient water must be present. The uptake of nitrate by seeds of *Sinapis arvensis* is indeed a function of the water content of the soil. It has been demonstrated that nitrate uptake is maximal at an optimal combination of nitrate and water content of the soil. Clearly, sub-optimal moisture contents will negatively influence diffusion of nitrate whereas supra-optimal water contents will dilute the available nitrate to levels below

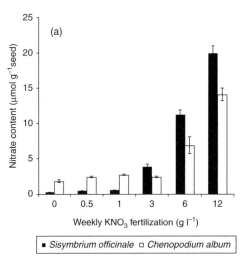
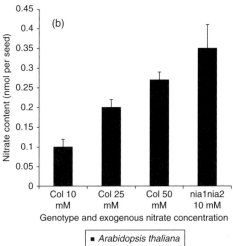

Fig. 6.2. The effect of potassium nitrate fertilization of (a) *Chenopodium album* and *Sisymbrium officinale* and (b) *Arabidopsis thaliana* plants on nitrate content of produced seeds. The plants were cultivated in plots in the open field (a) or grown on hydroculture (b). ((a) From Bouwmeester, 1990; (b) adapted from Alboresi *et al.,* 2005.)

the physiological threshold for germination (Goudey *et al.*, 1988).

Mechanism of nitrate action

A positive relationship between endogenous nitrate content and germination in water has been shown for *C. album* (Saini and Spencer, 1986), *Sisymbrium officinale* (Hilhorst and Karssen, 1990) and *A. thaliana* (Alboresi *et al.*, 2005). These species appear to be absolutely dependent on the presence of nitrate, either endogenous or exogenous. Thus, it may be concluded that the nitrate content of seeds may be a limiting factor in germination. However, most, if not all, nitrate-dependent seeds are light requiring. This suggests that a positive interaction between light and nitrate is a principal limitation to germination.

In the past various hypotheses have been proposed to account for the action of nitrate in seed germination. These include activation of the pentose phosphate pathway (Hendricks and Taylorson, 1975; Roberts and Smith, 1977), stimulation of oxygen uptake (Hilton and Thomas, 1986) and action as a co-factor of phytochrome (Hilhorst and Karssen, 1988; Hilhorst, 1990a,b). None of these have been consolidated over the past 20 years. However, recently a new model has been proposed for nitrate signalling in seed germination (Matakiadis *et al.*, 2009). Nitrate, either endogenous or exogenously applied, decreased the seed's abscisic acid (ABA) content significantly. It did this by increasing the expression of the abscisic acid catabolic gene *CYP707A2* that encodes for a cytochrome P450 that catalyses the breakdown of active ABA to 8′-hydroxy ABA. This mode of nitrate signalling fits in the proposed pathways that regulate the so-called 'GA–ABA balance' which denotes the antagonism of the germination inhibitor ABA and the germination promoter gibberellic acid (GA). It also agrees with described interactions with other environmental factors, including light, through phytochrome (Fig. 6.3).

It now seems clear that nitrate acts not only by virtue of its capacity to accept electrons. It has been shown that nitrate may act without the involvement of reduction to nitrite, nitric oxide (NO) or ammonium (Hilhorst and Karssen, 1989; Alboresi *et al.*, 2005). Thus, nitrate must be regarded as a true signalling molecule. However, there is also good evidence that nitrate may act through its reduction to nitric oxide (see section 'Nitric oxide' below).

Ecological roles of nitrate

A flush of germination of weeds in a crop field after disturbance of the soil indicates that light may be the limiting factor for germination in nitrogen-rich soils. It has been shown that during soil disturbance, very short exposure times (even less than a second) may induce germination (see information on VLFR in Chapter 5 of this volume). Germination may proceed even when seeds are returned beneath the soil surface to depths of up to 10 cm (Hartmann and Nezadal, 1990). However, it has been shown that soil disturbance may also result in considerable release of nitrate ions from the soil. This phenomenon is probably a significant factor in the promotion of weeds, such as *Avenafatua*, in North America where summer fallowing is commonly practised (Simpson, 1990). As shown in *S. officinale* and *A. thaliana* (Hilhorst and Karssen, 1988), the effect of light and nitrate on seed germination may be reciprocal. This implies that the light requirement of seeds in a soil with low nitrate is higher than that of seeds in nitrate-rich soils. Whether this has any ecological meaning remains to be shown.

In summary, seeds are equipped with a sensing mechanism for light and nitrate. This mechanism seems to be 'tuned' in such a way that (perceptive) seeds will only germinate when the combination of light quality/quantity and nitrate levels in the seed is adequate.

Assessment of the ecological significance of this sensing mechanism in this perspective can only be based on teleological considerations. Obviously, a plant that will eventually grow from a seed requires light and nitrogen for optimal development. However, a relationship between the light and nitrate levels that promote seed germination and the levels of these factors required by the growing plant has yet to be shown.

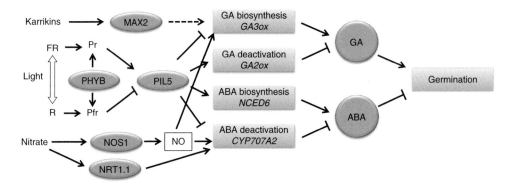

Fig. 6.3. Schematic overview of the translation of environmental cues, such as light, nitrate, karrikins and NO into internal signalling leading to the promotion or inhibition of germination of *Arabidopsis thaliana* seeds through the GA-ABA balance. MAX2, PHYB/PIL5, NOS1 and NRT1.1 are primary targets of the environmental cues, initiating signalling cascades towards promotion and inhibition of GA and ABA biosynthesis and catabolism (GA3ox, GA2ox, NCED6, CYP707A2). Arrows indicate promotive effects, whereas bars indicate inhibition. Dotted arrow indicates tentative pathway.

A more feasible explanation for the sensing mechanism is the ability of seeds to detect disturbances of light quality and quantity, and of nitrate levels, in their immediate environment. Seeds that are shaded by foliage receive light with a lower ratio of red to far-red light than direct sunlight. This is due to the leaf chlorophyll, which absorbs red light more effectively than far-red light. Consequently, the levels of P_{fr} that are established in the shaded seeds will be lower and more often below the threshold levels for germination (Frankland and Taylorson, 1983; Daws *et al.*, 2002). In this way germination beneath established plants may be prevented and competition avoided. Similarly, established plants may lower the nitrate content of the soil around their root systems. Nitrate is consumed and nitrification may be inhibited (Rice, 1984). As a result, the seeds in the immediate environment are depleted of nitrate and germination probability will be reduced (Pons, 1989).

In addition, soil temperature and diurnal temperature fluctuations are strongly influenced by the vegetation. Diurnal soil temperature fluctuations are much less pronounced beneath an established sward than in soils with no vegetation. Many weed species require diurnal temperature fluctuations for successful germination (Thompson *et al.*, 1977). Again this may be a mechanism to detect the proximity of competitors.

Organic Chemicals

Soils contain numerous organic compounds, both volatile and non-volatile. These are often the products of decaying plant and animal remains and the accompanying microorganisms. Also living plants produce a vast array of organics, usually in their root exudates. These compounds have the potential to inhibit or stimulate germination.

Allelochemicals

Burial of seeds in soil inhibits germination of a great number of species. The absence of light and diurnally fluctuating temperatures are important reasons for this inhibition. However, the germination of light-independent seeds, which are capable of germination in darkness, is often also inhibited during burial (Holm, 1972). It has often been postulated that chemical inhibition plays a major role in the regulation of germination in soil. The study of germination-affecting compounds has received much attention. We will restrict ourselves in this chapter to those organics that are released in soil and affect germination in the natural environment.

The term 'allelopathy' is derived from the Greek words allelon 'of each other' and pathos 'to suffer' and means the injurious effect of one upon the other. At present the term is generally accepted to refer to both inhibitory and stimulatory effects of one plant on another plant (Rice, 1984). It may be defined as: 'Any process involving secondary metabolites produced by plants, micro-organisms, viruses, and fungi that influence the growth and development of agricultural and biological systems (excluding animals), including positive and negative effects' (Torres *et al.*, 1996). However, allelopathy still refers mostly to detrimental effects of higher plants of one species (the donor) on the germination, growth or development of plants of another species (the recipient; Putman, 1985). In a few cases allelochemicals may be stimulatory at lower concentrations (see below). There is extensive evidence that allelopathy may contribute to interactive relationships in plants, such as those which determine species composition in community structure and species replacement in succession, etc. We will confine ourselves to effects of allelochemicals on seeds. Reviews on allelopathy have been published by Rice (1984), Putman (1985), Baskin and Baskin (1998), Kruse *et al.* (2000) and Chon and Nelson (2009).

Most allelochemicals can be classified as secondary metabolites. Many of these have unknown function but some are involved in defence against herbivores and pathogens. Allelochemicals are present in all plant parts and are released into the soil by root exudation, leaching from aboveground parts and by decomposition of plant material (Rice, 1984). The occurrence and amounts of allelopathic compounds in the soil are determined by a large number of (interacting) factors, such as the density at which leaves fall, the rate at which this material decomposes on or in the soil, the distance from other plants and rainfall. The decomposition of the plant material depends on leaf tissue composition, temperature, moisture, the presence of microorganisms, soil type and soil pH. Together with a dependency on developmental stage of the donor plants, mixture effects and conversion of allelochemicals to more or less active compounds, it is not surprising that the amount of effective concentrations of allelochemicals in the soil is highly variable.

Allelochemical inhibitors

Allelopathic inhibition of seed germination plays a major role in the regulation of plant succession. Rice (1984) describes in detail the roles of allelopathy in the development of vegetation in old fields that have been abandoned from cultivation for low fertility. Succession in such fields included four stages: (i) pioneer weed (2–3 years); (ii) annual grass (9–13 years); (iii) perennial bunch grass (for 30 years); and (iv) true prairie grasses. The weed stage that was dominated by robust plants, such as *Helianthus annuus, C. album, Sorghum halepense* and others was rapidly replaced by small annual grass species such as *Aristida oligantha*. It was found that the pioneer weeds eliminated themselves through the production of toxins that inhibited germination of their own seeds (autotoxicity) and of accompanying weeds. Germination of *A. oligantha*, a grass that often followed the weeds in succession, was not inhibited. Inhibitors were detected in extracts from decaying material from, e.g. *S. halepense* and *H. annuus*.

Several allelochemicals have been chemically identified, mostly in extracts from crop species. Very often, methanolic extracts are made from plant material and tested in the laboratory for inhibitory effects in a standard protocol, e.g. germination of lettuce seed (Kruse *et al.*, 2000). However, these tests will only indicate an allelopathic potential. Studies dealing with the natural occurrence of allelopathic substances in the soil are scarcer. For instance Numata (1982) identified *cis*-dehydroxy-matricariaester (*cis*-DME) as the active principle in the underground organs of *Solidago altissima*. This weed succeeds the pioneer *Ambrosia artemisiifolia* in old-field succession. The compound formed 2.5% of the dry weight of plant material and actively inhibited germination of the *Ambrosia* seeds. A soil block 10 cm in depth from the rhizosphere of *Solidago* contained 5 ppm *cis*-DME, a level sufficiently high to inhibit germination. The compound survived in soil for several months without decomposition by microorganisms.

Allelochemicals not only leach from leaves, stems and roots but also from seeds. *Parthenium hysterophorus*, an aggressive weed on the American continent, contains two major water soluble sesquiterpene lactones (up to 8% of its dry weight): parthenin and coronopilin (Picman and Picman, 1984). Although achenes contain lower amounts than leaves and stems, the concentrations are sufficiently high to inhibit seed germination. Autotoxicity has also been shown for seeds. Germination rates of the achenes of *P. hysterophorus* increased with decreasing achene density and increasing washing periods preceding germination. The toxins were mainly located in the coat of the achenes. The inhibitor acts as a kind of rain gauge, which determines that germination will only occur when sufficient inhibitor has been washed out. These germination inhibitors may also be effective in the formation of zones of inhibition, which deters competitors.

Seeds of *P. hysterophorus* contain the phenolic compounds caffeic, vanillic, p-coumaric, anisic, p-hydroxy-benzoic, chlorogenic and ferulic acids. These compounds are widely known for their allelopathic activities (Rice, 1984; Li *et al.*, 2010). However, removal of phenolic compounds from the extracts showed that in *P. hysterophorus* sesquiterpene lactones are the major compounds responsible for the inhibition of germination. Williams and Hoogland (1982) also expressed some doubt about the role of phenolic compounds as allelopathic agents in the control of germination. Phenolic compounds may have other functions, however. An interesting suggestion is that phenolics in seed coats help to prevent seed decay by inhibition of microbial attack (Rice, 1984). Please see Chapter 8 of this volume for a more complete review of chemical inhibitors in seeds.

Allelochemical promoters

Seeds of most angiospermous root parasites require chemical stimulants emanating from their hosts' roots to germinate. Several of these species parasitize cultivated crops heavily and cause severe damage. The most critical phase in the development of these host-specific root parasites is the early and unambiguous recognition of the correct host. The chemical communication between parasite and host is basic to this recognition. The physiology of the root parasite seed is complicated (Joel *et al.*, 1995; Matusova *et al.*, 2004). Newly shed seeds of root parasites possess primary dormancy and require a period of dry after-ripening to relieve dormancy. When brought into contact with water, a so-called conditioning period of several days is required during which the seed becomes responsive to the chemical stimulant from the host. The transition to the responsive state can be very abrupt. Contact with the host stimulant then induces germination.

The host-dependent germination of many root-parasitic weeds has been investigated extensively (reviewed by Visser, 1989; Xie *et al.*, 2010; Cardoso *et al.*, 2011). One of the first identified host-liberated chemical stimulants of germination was named *strigol*. It stimulated the germination of *Striga asiatica* seeds at extremely low concentrations (Cook *et al.*, 1972). Strigol was isolated from the false host *Gossypium hirsutum* (cotton). Chemical stimulants are not only produced by true host plants but also by false hosts, i.e. plants that cannot be parasitized. Strigol is one of the several natural strigolactones known to date (Fig. 6.4a). Most of the germination stimulants so far are strigolactones, which also function as host recognition signals for arbuscular

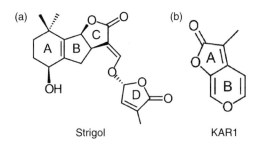

Strigol   KAR1

Fig. 6.4. Molecular structure of (a) strigol, a natural strigolactone, and (b) KAR1 (3-methyl-2H-furo[2,3-c] pyran-2-one).

mycorrhizal fungi. In addition, the strigolactones are now considered a new class of plant hormones inhibiting shoot branching (Gomez-Roldan *et al.*, 2008). The strigolactones are by far the most potent group of inhibitors, but also dihydroquinones and sesquiterpene lactones have been shown to be active. For example, a germination stimulant for *Striga* was isolated from glandular hairs in the root hair zone of a host, *Sorghum bicolor*, and identified as a labile hydroquinone, sorgoleone, that triggered the germination of *S. asiatica* seed at 10^{-7}mol l^{-1} (Chang *et al.*, 1986). Other germination stimulants include the fungal metabolites cotylenins and fusicoccins and the plant hormone jasmonic acid (Bouwmeester *et al.*, 2003).

Despite our extensive chemical knowledge of active compounds, we know relatively little about the chemical stimulation of germination in root parasites. It is difficult to understand why certain root parasites have a rather wide host range, while others exhibit an extremely narrow range. It is even more difficult to understand why false hosts also produce germination stimulants but are not infected. Perhaps this relates to other possible functions of the stimulants, e.g. in mycorrhizal symbiosis (Akiyama *et al.*, 2005).

Attempts to control parasitic weeds have largely concentrated on compounds that will stimulate germination of the parasitic weeds in the absence of a host. This phenomenon has been exploited in the case of ethylene which promotes the germination of *S. asiatica* (Eplee, 1975). The biosynthesis of ethylene appears to be part of the signal transduction to germination of the parasitic seed (Sugimoto *et al.*, 2003). Thus, ethylene, when injected into soil, will cause suicidal germination because the seedling must attach to the host root within a few days in order to survive. Therefore, ethylene-producing microorganisms may also be used to elicit suicidal germination of these root parasites, i.e. germination in the absence of host plants, which has been suggested as a potential way to reduce the seed bank of these parasitic weeds in agricultural fields (Berner *et al.*, 1999).

Smoke-derived Chemicals

Recurrent fires are an integral part of several ecosystems in the world and when such regions are protected from fire, their local ecology may become disturbed. One of the most intriguing post-fire occurrences is the very rapid germination of certain (pioneer) species that leads to a flush of new, green seedlings on a charred soil, indicating potent activation of the soil seed bank. Many potential germination stimulants are changed in the post-fire environment, including heat scarification, altered light levels (vegetation gaps) and increased nitrate in the soil. However, one of the most potent inducers of germination in post-fire environments is smoke itself. Smoke induces germination by mainly aqueous transfer from soil to seeds.

The bioactive chemical in smoke was identified as karrikinolide, previously referred to as 'butenolide', (3-methyl-2H-furo[2,3-c] pyran-2-one), a member of the class of karrikins (Flematti *et al.*, 2004). There are several active karrikins, denoted KAR1 through KAR4. These have been compared with strigolactones, which bear some structural resemblance (Fig. 6.4; Nelson *et al.*, 2009). All of the karrikins are as effective as strigolactones in the breaking of dormancy. The positive effect of smoke on seed germination is not confined to species that are native to fire-prone habitats. Smoke is now considered a generally effective chemical promoter that augments germination of some 1200 species in more than 80 genera worldwide (Dixon *et al.*, 2009). Smoke in general and its bioactive compounds, such as the karrikins, have potential to be used as management tools for land rehabilitation, plant conservation and weed control. However, crucial information for such applications is lacking, e.g. the stability of these compounds in the soil, whether they are metabolized by plants and microorganisms, and how they affect the organisms.

As to their mechanism of action, the karrikins, like light, induce indirectly an increase of expression of the two principal GA3 oxidase genes, GA3ox1 and GA3ox2 (Fig. 6.3), but do not appear to affect any of the key genes related to ABA synthesis or catabolism. Early karrikin signalling includes the MAX2 F-box protein. Interestingly, the *Arabidopsis max2* mutant has an enhanced dormancy phenotype. Despite structural similarities between karrikins and strigolactones (Fig. 6.4), recent evidence points to distinct signalling mechanisms.

Gaseous Compounds

The gaseous phase of soil occupies those pores that are not already filled with water. Movement of atmospheric gases through soil is primarily by molecular diffusion. In waterlogged soils, especially in heavy soils, the oxygen content of the gaseous phase may drop considerably below that in normal air. The gas phase is also influenced by the presence of vegetation. Roots of plants will take up oxygen and produce carbon dioxide, changing the balance between the gases. In soils having a high organic content and containing an active microflora the balance may shift in a similar way. Although oxygen levels in soils rarely drop below 19% and carbon dioxide levels rarely exceed 1%, much greater extremes may occur at microsites such as those adjacent to plant roots or decaying organic matter, and in soil of flooded areas (Roberts, 1972; Egley and Duke, 1985). In addition to oxygen, carbon dioxide and nitrogen, soils may contain several other gases and volatile compounds. Those gaseous compounds are mostly due to anaerobic conditions and the activity of microorganisms. Soils may contain methane, hydrogen sulphide, hydrogen, nitrous oxide and small amounts of carbon monoxide, ethylene and ammonia (Mayer and Poljakoff-Mayber, 1989).

Oxygen

In general, germination and early seedling growth require oxygen at atmospheric levels. Oxygen uptake usually rises rapidly during the first hours of imbibition, followed by adenosine triphosphate (ATP) formation. The rapid rise in fresh weight and respiration during imbibition is in most species

followed by a lag phase in both the uptake of water and oxygen until radicle protrusion. Oxygen diffusion can be strongly limited during the lag phase because, in a fully hydrated seed, oxygen diffusion is limited by oxygen solubility in water. Moreover, oxygen is often utilized in coats and endosperm in non-respiratory reactions. Therefore oxygen levels at the level of meristems in the embryonic axis may be rather low. For example, the glumellae adhering to the caryopses in barley (*Hordeum vulgare* L.) and oat (*Avena fatua* L.) grains impose dormancy, as germination is much improved by their removal. This effect of the glumellae on dormancy is likely the reduction of the availability of oxygen to the embryo. Fresh barley grains hardly germinated in 100% oxygen, whereas removal of the glumellae allowed complete germination in oxygen levels as low as 3% (Lenoir *et al.*, 1986). In some seeds the period of tolerance to anaerobic conditions is considerably extended. This is particularly true for seeds that germinate under water but also many terrestrial plants germinate under water. A few species of plants are known to germinate under anoxia, including four species of the genus *Echinochloa* and rice, whereas germination of others is tolerant to, or even enhanced by, reduced oxygen levels. Seeds of *E. crus-galli* germinate well under either anaerobic or aerobic conditions. This ability undoubtedly contributes to its seriousness as a water weed in rice. The seeds of *Echinochloa* species are all tolerant to the ethanol produced during anaerobic germination (Corbineau and Côme, 1995). However, the above examples are mostly exceptions to the rule that seeds require oxygen for successful germination and subsequent seedling establishment. In general, oxygen levels in soils amply fulfil the requirements.

Oxygen may also influence the effectiveness of after-ripening to relieve seed dormancy. For example, oxygen was necessary during warm temperature after-ripening of *A. fatua* and other cereal seeds (Simmonds and Simpson, 1971). Furthermore, the breaking of dormancy of rice seed by after-ripening was enhanced under high oxygen levels (Roberts, 1961). We now know that oxygen may be active in the dry dormant seed by selectively oxidizing and inactivating mRNAs and proteins that are associated with the dormant state (Bazin *et al.*, 2011). However, the role of after-ripening and the influence of oxygen and possibly other gases on seeds in the soil seed bank remain enigmatic.

Oxygen may interact with other environmental factors, including temperature and osmotic environment, as well as with seed constituents, such as the hormones ABA and GA. For example, in barley grains, ABA and GA germination response thresholds varied over several orders of magnitude in response to oxygen availability, with sensitivity to ABA increasing and sensitivity to GA decreasing with hypoxia (Bradford *et al.*, 2008).

Carbon dioxide

Similarly to oxygen, the level of carbon dioxide in the soil depends on depth, temperature, moisture, porosity and amount of biotic activity. Carbon dioxide levels increase with depth, ranging from 1 dm^3 m^{-3} at the top 10 cm of soil to 80 dm^3 m^{-3} at 50 cm (Egley, 1984). Carbon dioxide levels may increase five to tenfold when biological activity in the soil peaks in spring and summer. In general, soil carbon dioxide levels are greatly influenced by moisture content due to the restricted diffusion of gases (Yoshioka *et al.*, 1998). In addition, respiratory activity of microorganisms and actively growing plant roots, and carbon dioxide evolving from decaying plant material, increases with soil moisture content.

In general, carbon dioxide levels of 20–50 dm^3 m^{-3} stimulate germination (for references, see Baskin and Baskin, 1998). This is higher than is usually found in the soil top layer. However, it has been shown that rainfall may cause an immediate increase in carbon dioxide levels. The CO_2 concentration at a depth of 3 cm rose from 8 to 30 dm^3 m^{-3} within hours after rainfall. It was clearly demonstrated that this rise in CO_2 levels, rather than moisture content, light, ethylene or nitrate caused the intermittent flushes of germination of *Echinochloa crus-galli* seeds (Yoshioka *et al.*, 1998).

Since levels of CO_2 in the soil are generally below the levels that stimulate germination, it is not likely that the gas plays a significant role in changing the states of dormancy in the soil seed bank (Egley and Duke, 1985; Baskin and Baskin, 1998).

Ethylene

Ethylene is a common constituent of the soil atmosphere. This gas can be produced by both aerobic microorganisms and anaerobic bacteria (Pazout *et al.*, 1981) as well as by plant roots (Stumpff and Johnson, 1987). Ethylene concentrations of several parts per million have been recorded in soils (Smith and Restall, 1971). However, the concentration of ethylene may vary between different microenvironments, especially at sites where ethylene-producing microorganisms flourish.

Soil ethylene can have a strong influence on plant growth (Arshad and Frankenberger, 1990a,b). For several weed species promotive or inhibitory effects of ethylene on seed germination have been reported (Corbineau and Côme, 1995). Ethylene could enhance germination of such species as *Amaranthus caudatus, A. retroflexus, Chenopodium album* and *Rumex crispus,* but also non-weedy species, such as *Helianthus annuus* and *Lactuca sativa.* Conversely, ethylene inhibited germination of e.g. *Plantago major, P. maritime* and *Chenopodium rubrum* (Olatoye and Hall, 1973). As previously discussed, seeds of parasitic weeds, such as *Striga* spp. and *Orobanche* spp., show a positive germination response to ethylene.

Current consensus is that ethylene plays a role in the breaking of seed dormancy and induction of germination and many seeds indeed produce ethylene during germination. It is still a matter of debate whether this ethylene is a cause or a result of germination. It seems that ethylene is particularly active by virtue of its cross-talk with other signals that are relevant to dormancy and germination (Matilla and Matilla-Vázquez, 2008). In a study of the effect of ethylene on seed germination of ten grass and 33 broadleaved weed species, germination in nine species was promoted by ethylene in concentrations between 0 and 100 ppm. Germination was inhibited in two species while the other species were not affected (Taylorson, 1979). The nine species that responded positively to ethylene are also known to be responsive to nitrate. This may indicate an interaction between ethylene and nitrate. The interaction between ethylene and nitrate has been examined in more detail in *Chenopodium album* in which a clear synergism between the effects of ethylene and nitrate was found (Saini and Spencer, 1986). This was shown for both endogenous and applied nitrate. This cross-talk effect makes sense because ethylene appears to inhibit ABA signalling, presumably via the *CYP707A2* gene, which encodes for one of the enzymes that catabolize ABA (Linkies *et al.*, 2009), similar to nitrate (Fig. 6.3). This again emphasizes that environmental cues for germination of the soil seed bank, such as light, nitrate, smoke and ethylene are translated to their respective internal signals, which all seem to converge in the regulation of the ABA/GA balance of the seed (Fig. 6.3). Also, interactions between ethylene and CO_2 have been reported. It appears that elevated levels of CO_2 may enhance both the synthesis of ethylene and its effect on germination (reviewed by Corbineau and Côme, 1995).

Assessment of the ecological significance of soil ethylene for germination of seeds in the field is difficult since ethylene can both promote and inhibit germination. As to the inhibitory properties of ethylene, this again may be an adaptation to avoiding competition with established plants. Ethylene concentrations are highest in the rhizosphere because it is rich in microorganisms that are capable of synthesizing ethylene. Moreover, production of ethylene by microorganisms can be enhanced by compounds in root exudates (Arshad and Frankenberger, 1990a,b).

Nitric oxide

Nitric oxide (NO) from soil contributes up to 40% to the global budget of atmospheric NO. NO emissions from the soil are primarily caused by biological activity (nitrification and denitrification), which occurs in the topsoil, a layer often characterized by high content of organic material (Bargsten *et al.*, 2010). NO is a very reactive gaseous free radical which is a ubiquitous and potent signalling molecule in plants and animals. Among many other plant processes, NO is involved in the regulation of dormancy and germination of many species, including *Paulownia tomentosa* (Giba *et al.*, 1994), lettuce (Beligni and Lamattina, 2000), barley and *A. thaliana* (Bethke *et al.*, 2004). It is produced in plant cells from nitrite and arginine. Nitrite is the first product of the reduction of nitrate by nitrate reductase, and is then further reduced to NO via nitric oxide synthase. The NO scavenger (carboxy)2-phenyl-4,4,5,5,-tetramethylimidazoline-1-oxyl-3-oxide (PTIO) effectively counteracts the dormancy-breaking effect of nitrate in *Arabidopsis* (Bethke *et al.*, 2004) and this suggests that nitrate action may also proceed through its reduction to NO. *Arabidopsis* mutants lacking the *Nitric Oxide Synthase 1* (*AtNOS1*) gene, and which are thus deficient in endogenous NO, have increased dormancy and lower seed germination and seedling establishment rates than wild-type seeds due to enhanced ABA inhibitory action. These effects can be reversed by application of NO to the mutant seeds. There is increasing evidence that NO enhances the degradation of ABA and the biosynthesis of GA, through activation of the *CYP707A2* and *GA3ox* genes, respectively (Fig. 6.3) and thus contributes to the breaking of dormancy (Bethke *et al.*, 2007).

Conclusions

To guarantee successful emergence of seedlings in the field, germination has to occur at the proper time and place. Seeds receive information about the succession of seasons through fluctuations in temperature. In arid and semiarid zones the timing of precipitation adds important information. Seeds receive information about their depth in the soil and neighbouring vegetation through the dependency of the germination process on light and fluctuating diurnal temperatures. Therefore, germination of many seed species often only occurs at or close to the surface of the soil and in vegetation gaps.

The chemical environment provides seeds with information about the quality of their environment with respect to suitability for growth. In general, chemical factors that promote germination are also beneficial for emergence and seedling growth. The dependence of many species on nitrate for germination is a clear example of this rule. The presence of high soil nitrate levels may even stimulate the germination of the next generation of seeds, via the accumulation of nitrate during seed formation. The dependence of parasitic seeds on chemical promoters excreted by the host plant illustrates the parallelism between the stimulation of germination and seedling growth. Seedlings of the parasite also depend fully on host factors. Similarly, allelopathic substances in the soil, which inhibit germination, are generally deleterious to seedling growth.

However, most conclusions about the role of the chemical environment of seeds can only be preliminary. The list of active promoters and inhibitors is by no means complete. Our knowledge about allelopathic compounds is still very limited. Undoubtedly, many more compounds will be isolated and characterized. Also, the information about mutual interactions of the different compounds in the chemical environment is very restricted. The physiological and biochemical mechanisms underlying promotion and inhibition by these chemical factors are largely unknown. The physiological action of nitrate is so far the best explored. Structure–activity studies of stimulants of parasitic weeds, and of some organic dormancy-breaking compounds are making steady progress. However, the actual mechanisms of action at the molecular level have yet to be elucidated.

The chemical environment of seeds needs better exploration. A more detailed knowledge will undoubtedly stimulate a better understanding of the environmental control of germination and emergence. This may ultimately lead to the development of alternative methods for weed control and management.

References

Akiyama, K., Matsuzaki, K. and Hayashi, H. (2005) Plant sesquiterpenes induce hyphal branching in arbuscular mycorrhizal fungi. *Nature* 435, 824–827.

Alboresi, A., Gestin, C., Leydecker, M.T., Bedu, M., Meyer, C. and Truong, H.N. (2005) Nitrate, a signal relieving seed dormancy in Arabidopsis. *Plant, Cell & Environment* 28, 500–512.

Arshad, M. and Frankenberger, W.T., Jr. (1990a) Ethylene accumulation in soil in response to organic amendments. *Soil Science Society of America Journal* 54, 1026–1031.

Arshad, M. and Frankenberger, W.T., Jr. (1990b) Production and stability of ethylene in soil. *Biology and Fertility of Soils* 10, 29–34.

Bargsten, A., Falge, E., Pritsch, K., Huwe, B. and Meixner, F.X. (2010) Laboratory measurements of nitric oxide release from forest soil with a thick organic layer under different understory types. *Biogeosciences* 7, 1425–1441.

Baskin, C.C. and Baskin, J.M. (1998) *Seeds. Ecology, Biogeography, and Evolution of Dormancy and Germination.* Academic Press, San Diego, CA.

Bazin, J., Langlade, N., Vincourt, P., Arribat, S., Balzergue, S., El-Maarouf-Bouteau, H. and Bailly, C. (2011) Targeted mRNA oxidation regulates sunflower seed dormancy alleviation during dry after-ripening. *Plant Cell* 23, 2196–2208.

Beligni, M.V. and Lamattina, L. (2000) Nitric oxide stimulates seed germination and de-etiolation, and inhibits hypocotyl elongation, three light-inducible responses in plants. *Planta* 210, 215–221.

Berner, D.K., Schaad, N.W. and Völksch, B. (1999) Use of ethylene-producing bacteria for stimulation of *Striga* spp. seed germination. *Biological Control* 15, 274–282.

Bethke, P.C., Gubler, F., Jacobsen, J.V. and Jones, R.L. (2004) Dormancy of *Arabidopsis* seeds and barley grains can be broken by nitric oxide. *Planta* 219, 847–855.

Bethke, P.C., Libourel, I.G., Aoyama, N., Chung, Y.Y., Still, D.W. and Jones, R.L. (2007) The *Arabidopsis* aleurone layer responds to nitric oxide, gibberellin, and abscisic acid and is sufficient and necessary for seed dormancy. *Plant Physiology* 143, 1173–1188.

Bewley, J.D. and Black, M. (1994) *Seeds. Physiology of Development and Germination*, 2nd edn. Springer-Verlag, Berlin, Germany.

Bouwmeester, H.J. (1990) The effect of environmental conditions on the seasonal dormancy pattern and germination of weed seeds. PhD thesis. Wageningen University, the Netherlands.

Bouwmeester, H.J. and Karssen, C.M. (1993) Annual changes in dormancy and germination in seeds of *Sisymbrium officinale* (L.) Scop. *New Phytologist* 124, 179–191.

Bouwmeester, H.J., Derks, L., Keizer, J.J. and Karssen, C.M. (1994) Effects of endogenous nitrate content of *Sisymbrium officinale* seeds on germination and dormancy. *Acta Botanica Neerlandica* 43, 39–50.

Bouwmeester, H.J., Matusova, R., Zhongkui, S. and Beale, M.H. (2003) Secondary metabolite signalling in host–parasitic plant interactions. *Current Opinion in Plant Biology* 6, 358–364.

Bradford, K.J., Benech-Arnold, R.L., Côme, D. and Corbineau, F. (2008) Quantifying the sensitivity of barley seed germination to oxygen, abscisic acid, and gibberellin using a population-based threshold model. *Journal of Experimental Botany* 59, 335–347.

Cardoso, C., Ruyter-Spira, C. and Bouwmeester, H.J. (2011) Strigolactones and root infestation by plant-parasitic *Striga, Orobanche* and *Phelipanche* spp. *Plant Science* 180, 414–420.

Chang, M., Netzly, O.H., Butler, L.G. and Lynn, D.C. (1986) Chemical regulation of distance: characterization of the first natural host germination stimulant for *Striga asiatica. Journal of the American Chemical Society* 108, 7858–7860.

Chon, S.U. and Nelson, C.J. (2009) Allelopathy in compositae plants. A review. *Agronomy of Sustainable Development* 30, 349–358.

Cook, C.E., Whichard, L.P., Wall, M.E., Egley, G.H., Coggan, P., Luhan, B.A. and McPahil, A.T. (1972) Germination stimulants. 11. The structure of strigol – a potent seed germination stimulant for witchweed (*Strigalutea* Lour.). *Journal of the American Chemical Society* 94, 6198–6199.

Corbineau, F. and Côme, D. (1995) Control of seed germination and dormancy by the gaseous environment. In: Kigel, J. and Galili, G. (eds) *Seed Development and Germination*. Marcel Dekker, New York, pp. 397–424.

Currie, J.A. (1973) The seed-soil system. In: Heydecker, W. (ed.) *Seed Ecology*. Butterworths, London, pp. 463–480.

Daws, M.I., Burslem, D.F.R., Crabtree, L.M., Kirkman, P., Mullins, C.E. and Dalling, J.W. (2002) Differences in seed germination responses may promote coexistence of four sympatric Piper species. *Functional Ecology* 16, 258–267.

Dixon, K.W., Merritt, D.J., Flematti, G.R. and Ghisalberti, E.L. (2009) Karrikinolide: a phytoreactive compound derived from smoke with applications in horticulture, ecological restoration, and agriculture. *Acta Horticulturae* 813, 155–170.

Egley, G.H. (1984) Ethylene, nitrate and nitrite interactions in the promotion of dark germination of common purslane seeds. *Annals of Botany* 53, 833–840.

Egley, G.H. and Duke, S.O. (1985) Physiology of weed seed dormancy and germination. In: Duke, S.O. (ed.) *Weed Physiology, Vol. 1, Reproduction and Ecophysiology*. CRC Press, Boca Raton, FL, pp. 27–64.

Eplee, R.E. (1975) Ethylene, a witch weed seed germination stimulant. *Weed Science* 23, 433–436.

Finch-Savage, W.E., Cadman, C.S.C., Toorop, P.E., Lynn, J.R. and Hilhorst, H.W.M. (2007) Seed dormancy release in *Arabidopsis* Cvi by dry after-ripening, low temperature, nitrate and light shows common quantitative patterns of gene expression directed by environment specific sensing. *Plant Journal* 51, 60–78.

Flematti, G.R., Ghisalberti, E.L., Dixon, K.W. and Trengove, R.D. (2004) A compound from smoke that promotes seed germination. *Science* 305, 977.

Footitt, S., Douterelo-Soler, I., Clay, H. and Finch-Savage, W.E. (2011) Dormancy cycling in *Arabidopsis* seeds is controlled by seasonally distinct hormone-signaling pathways. *Proceedings of the National Academy of Sciences USA* 108, 20236–20241.

Frankland, B. and Taylorson, R.B. (1983) Light control of seed germination. In: Shropshire, W. and Mohr, H. (eds) *Photomorphogenesis. Encyclopedia of Plant Physiology New Series* Vol. 16A. Springer-Verlag, Berlin, Germany, pp. 428–456.

Giba, Z., Grubisic, D. and Konjevic, R. (1994) The effect of electron acceptors on the phytochrome-controlled germination of *Paulownia tomentosa* seeds. *Physiologia Plantarum* 92, 290–294.

Gomez-Roldan, V., Fermas, S., Brewer, P.B., Puech-Pagès, V., Dun, E.A., Pillot, J.P., Letisse, F., Matusova, R., Danoun, S., Portais, J.C., Bouwmeester, H., Becard, G., Beveridge, C.A., Rameau, C. and Rochange, S.F. (2008) Strigolactone inhibition of shoot branching. *Nature* 455, 189–194.

Goudey, J.S., Saini, H.S. and Spencer, M.S. (1988) Role of nitrate in regulating germination of *Sinapis arvensis* L. (wild mustard). *Plant, Cell & Environment* 11, 9–12.

Haas, C.J. and Scheuerlein, R. (1990) Phase-specific effects of nitrate on phytochrome mediated germination in spores of *Dryopteris filix-mas* L. *Photochemistry and Photobiology* 52, 67–72.

Hadas, A. (1982) Seed-soil contact and germination. In: Khan, A.A. (ed.) *The Physiology and Biochemistry of Seed Development, Dormancy and Germination*. Elsevier Biomedical Press, Amsterdam, the Netherlands, pp. 507–527.

Hartmann, K.M. and Nezadal, W. (1990) Photocontrol of weeds without herbicides. *Naturwissenschaften* 77, 158–163.

Hendricks, S.B. and Taylorson, R.B. (1975) Breaking of seed dormancy by catalase inhibition. *Proceedings of the National Academy of Sciences (USA)* 72, 306–309.

Hilhorst, H.W.M. (1990a) Dose-response analysis of factors involved in germination and secondary dormancy of seeds of *Sisymbrium officinale*. I. Phytochrome. *Plant Physiology* 94, 1090–1095.

Hilhorst, H.W.M. (1990b) Dose-response analysis of factors involved in germination and secondary dormancy of seeds of *Sisymbrium officinale*. II. Nitrate. *Plant Physiology* 94, 1096–1102.

Hilhorst, H.W.M. and Karssen, C.M. (1988) Dual effect of light on the gibberellin and nitrate-stimulated seed germination of *Sisymbrium officinale* and *Arabidopsis thaliana*. *Plant Physiology* 86, 591–597.

Hilhorst, H.W.M. and Karssen, C.M. (1989) Nitrate reductase independent stimulation of seed germination in *Sisymbrium officinale* L. (hedge mustard) by light and nitrate. *Annals of Botany* 63, 131–137.

Hilhorst, H.W.M. and Karssen, C.M. (1990) The role of light and nitrate in seed germination. In: Taylorson, R.B. (ed.) *Recent Advances in the Development and Germination of Seeds*. NATO-ASI Series A 187. Plenum Press, New York, pp. 191–206.

Hilhorst, H.W.M., Derkx, M.P.M. and Karssen, C.M. (1996) An integrating model for seed dormancy cycling: characterization of reversible sensitivity. In: Lang, G.A. (ed.) *Plant Dormancy*. CAB International, Wallingford, UK, pp. 341–360.

Hilton, J.R. and Thomas, J.A. (1986) Regulation of pregerminative rates of respiration in seeds of various seed species by potassium nitrate. *Journal of Experimental Botany* 37, 1516–1524.

Holm, R.E. (1972) Volatile metabolites controlling germination in buried weed seeds. *Plant Physiology* 50, 293–297.

Joel, D.M., Steffens, J.C. and Matthews, D.E. (1995) Germination of weedy root parasites. In: Kigel, J., Negbi, M. and Gallili, G. (eds) *Seed Development and Germination*. Marcel Dekker, New York, pp. 567–596.

Kruse, M., Strandberg, M. and Strandberg, B. (2000) *Ecological effects of allelopathic plants – a review*. National Environmental Research Institute, Silkeborg, Denmark, NERI Technical Report No. 315.

Lehmann, E. (1909) Zur Keimungsphysiologie und -biologie von *Ranunculus sceleratus* L. und einigen anderen Samen. *Berichte der Deutsche Botanische Gesellschaft* 27, 476–494.

Lenoir, C., Corbineau, F. and Côme, D. (1986) Barley (*Hordeum vulgare*) seed dormancy as related to glumella characteristics. *Physiologia Plantarum* 68, 301–307.

Li, Z.-H., Wang, Q., Ruan, X., Cun-De Pan and Jiang, D.-A. (2010) Phenolics and plant allelopathy. *Molecules* 15, 8933–8952.

Linkies, A., Müller, K., Morris, K., Turecková, V., Wenk, M., Cadman, C.S., Corbineau, F., Strnad, M., Lynn, J.R., Finch-Savage, W.E. and Leubner-Metzger, G. (2009) Ethylene interacts with abscisic acid to regulate endosperm rupture during germination: a comparative approach using *Lepidium sativum* and *Arabidopsis thaliana*. *Plant Cell* 21, 3803–3822.

Matakiadis, T., Alboresi, A., Jikumaru, Y., Tatematsu, K., Pichon, O., Renou, J.-P. *et al.* (2009) The *Arabidopsis* abscisic acid catabolic gene CYP707A2 plays a key role in nitrate control of seed dormancy. *Plant Physiology* 149, 949–960.

Matilla, A.J. and Matilla-Vázquez, M.A. (2008). Involvement of ethylene in seed physiology. *Plant Science* 175, 87–97.

Matusova, R., van Mourik, T. and Bouwmeester, H.J. (2004) Changes in the sensitivity of parasitic weed seeds to germination stimulants. *Seed Science Research* 14, 335–344.

Mayer, A.M. and Poljakoff-Mayber, A. (1989) *The Germination of Seeds*, 4th edn. Pergamon Press, Oxford, UK.

Nelson, D. C., Riseborough, J., Flematti, G.R., Stevens, J., Ghisalberti, E.L., Dixon, K.W. and Smith, S.M. (2009) Karrikins discovered in smoke trigger *Arabidopsis* seed germination by a mechanism requiring gibberellic acid synthesis and light. *Plant Physiology* 149, 863–873.

Numata, M. (1982) Weed-ecological approaches to allelopathy. In: Holzner, W. and Numata, N. (eds) *Biology and Ecology of Weeds.* Junk, The Hague, the Netherlands, pp. 169–173.

Obermann, P. (1985) Die Belastung des Grundwassers aus landwirtschaftlicher nutzung nach heutigem Kenntnisstand. In: Nieder, H. (ed.) *Nitrat im Grundwasser.* VCH Verlagsgesellschaft, Weinheim, pp. 53–64.

Olatoye, S.T. and Hall, M.A. (1973) Interaction of ethylene and light on dormant weed seeds. In: Heydecker, W. (ed.) *Seed Ecology.* Butterworths, London, UK, pp. 233–249.

Pazout, J., Wurst, M. and Vancura, V. (1981) Effect of aeration on ethylene production by soil bacteria and soil samples cultivated in a closed system. *Plant and Soil* 62, 431–437.

Picman, J. and Picman, A.K. (1984) Autotoxicity in *Parthenium hysterophorus* and its possible role in control of germination. *Biochemical Systematics and Ecology* 12, 287–292.

Pons, T.L. (1989) Breaking of seed dormancy as a gap detection mechanism. *Annals of Botany* 63, 139–143.

Putman, A.R. (1985) Weed allelopathy. In: Duke, S.O. (ed.) *Weed Physiology, Vol. 1, Reproduction and Ecophysiology.* CRC Press, Boca Raton, FL, pp. 131–155.

Rice, E.L. (1984) *Allelopathy*, 2nd edn. Academic Press, Orlando, FL.

Roberts, E.H. (1961) Dormancy in rice seed: III. The influence of temperature, moisture and gaseous environment. *Journal of Experimental Botany* 13, 75–94.

Roberts, E.H. (1972) Dormancy: a factor affecting seed survival in the soil. In: Roberts, E.H. (ed.) *Viability of Seeds.* Chapman and Hall, London, UK, pp. 321–359.

Roberts, E.H. and Smith, R.D. (1977) Dormancy and the pentose phosphate pathway. In: Khan, A.A. (ed.) *The Physiology and Biochemistry of Seed Dormancy and Germination.* Elsevier Biomedical Press, Amsterdam, the Netherlands, pp. 385–411.

Runge, M. (1983) Physiology and ecology of nitrogen nutrition. In: Lange, O.L., Nobel, P.S., Osmond, C.B. and Ziegier, H. (eds), *Physiological Plant Ecology III. Encyclopedia of Plant Physiology*, New Series, Vol. 12C. Springer-Verlag, Berlin, Germany, pp. 163–200.

Saini, H.S. and Spencer, M.S. (1986) Manipulation of seed nitrate content modulates the dormancy breaking effect of ethylene on *Chenopodium album* seed. *Canadian Journal of Botany* 65, 876–878.

Simmonds, J.A. and Simpson, C.M. (1971) Increased participation of pentose phosphate in response to after ripening and gibberellic acid treatment in caryopses of *Avenafatua. Canadian Journal of Botany* 49, 1833–1840.

Simpson, C.M. (1990) *Seed Dormancy in Grasses.* Cambridge University Press, Cambridge, UK.

Smith, K.A. and Restall, S.W.F. (1971) The occurrence of ethylene in anaerobic soil. *Journal of Soil Science* 22, 430–436.

Stumpff, N. and Johnson, J.D. (1987) Ethylene production by loblolly pine seedlings associated with water stress. *Physiologia Plantarum* 69, 167–172.

Sugimoto, Y., Ali, A.M., Yabuta, S., Kinoshita, H., Inanaga, S. and Itai, A. (2003) Germination strategy of *Striga hermonthica* involves regulation of ethylene biosynthesis. *Physiologia Plantarum* 119, 1–9.

Taylorson, R.B. (1979) Response of weed seeds to ethylene and related hydrocarbons. *Weed Science* 27, 7–10.

Thompson, K., Grime, J.P. and Mason, G. (1977) Seed germination in response to diurnal fluctuations of temperature. *Nature* 267, 147–149.

Torres, A., Oliva, R.M., Castellano, D. and Cross, P. (1996). First World Congress on Allelopathy. A Science of the Future, pp. 278. SAI (University of Cadiz), Spain, Cadiz.

Vincent, E.M. and Roberts, E.H. (1977) The interactions of light, nitrate and alternating temperatures in promoting the germination of dormant seeds of common weed species. *Seed Science and Technology* 6, 659–670.

Visser, J.H. (1989) Germination requirements of some root-parasite flowering plants. *Naturwissenschaften* 76, 253–261.

Williams, R.D. and Hoogland, R.E. (1982) The effects of naturally occurring phenolic compounds on seed germination. *Weed Science* 30, 206–210.

Woolley, J.T. and Stoller, E.W. (1978) Light penetration and light-induced seed germination in soil. *Plant Physiology* 61, 597–600.

Xie, X., Yoneyama, K. and Yoneyama, K. (2010) The strigolactone story. *Annual Review of Phytopathology* 48, 93–117.

Yoshioka, T., Satoh, S. and Yamasue, Y. (1998) Effects of increased concentration of soil CO_2 on intermittent flushes of seed germination in *Echinochloa crus-galli* var. *crus-galli. Plant, Cell & Environment* 21, 1301–1306.

7   Seed Dormancy

Alistair J. Murdoch*

School of Agriculture, Policy and Development,
University of Reading, Reading, Berkshire, UK

Introduction

The regeneration of plant communities from seed depends on seeds being in the right physiological state to germinate in the right place at the right time. In some species, this requirement is satisfied by a regeneration strategy in which seeds germinate as soon as they are shed. In others, seeds may survive for long periods in the soil seed bank with intermittent germination of a part of the population.

There are two basic physiological prerequisites for seeds to survive in soil: viability must be maintained for as long as germination is avoided by dormancy or quiescence. Moreover, for such seeds to contribute to regeneration, dormancy must be relieved and germination promoted perhaps within a limited period when there is a good chance of successful seedling establishment. In explaining how different regeneration strategies can result from varying physiology, the approach adopted in this chapter in considering dormancy is to examine how this adaptive trait influences the responses of seed populations to environmental factors, so ensuring that at least some individuals germinate in the right place and at the right time.

Defining and Measuring Dormancy

Seed dormancy is caused by one or more blocks to the process of germination within the seed. It is most easily observed, measured and defined negatively as the failure of a viable seed to germinate given moisture, air and a suitable constant temperature for radicle emergence and seedling growth (Amen, 1968; Murdoch, 2004). If these minimum requirements for germination are lacking, the seed is better described as quiescent, especially if metabolism is arrested or retarded (Bewley and Black, 1994).

One attempt to frame a positive definition of dormancy is 'a seed characteristic, the degree of which defines what conditions should be met to make the seed germinate' (Vleeshouwers *et al.*, 1995). This definition begs the question: do these 'conditions' terminate dormancy or stimulate seed germination? As pointed out by Finch-Savage and Leubner-Metzger (2006), the answer 'depends on where one chooses to draw the line between the processes of dormancy and germination'. If 'conditions', such as light or fluctuating temperatures or the nitrate ion, remove a physiological or metabolic block, they are 'the last step in the dormancy-breaking

* E-mail: a.j.murdoch@reading.ac.uk

process' (Finch-Savage and Leubner-Metzger, 2006). For example, when buried dormant, light-sensitive seeds are brought to the soil surface, the exposure to light and perhaps also greater fluctuations of temperature may relieve the final metabolic barrier(s) so that the first step of germination may occur.

'The root of the problem is that we can only measure dormancy...by testing germination' (Hilhorst, 2007), so that dormancy is measured in terms of non-germination. The simplest measure is $(V - G)$ where V is the seed lot viability and G is the proportion of seeds germinating usually at a specified constant temperature in darkness with water. This measurement incorrectly implies that dormancy is a quantal response: a seed either germinates or does not and, by inference, it might be thought it is either dormant or it is not. In reality, it is only the expression of dormancy which is recorded as a quantal response. Another problem with this measure is the imprecision of germination tests and the difficulty of distinguishing between seed lots where germination is very low due to high levels of dormancy. By analogy with the parameter K_i in the seed viability equation, which quantifies loss of seed viability in relation to temperature and moisture (Ellis and Roberts, 1980a,b, 1981), a much more accurate estimate of dormancy in a homogeneous seed lot can be obtained by probit analysis of the loss of dormancy in the seed population. For example, in *Cenchrus ciliarus* where dormancy is relieved by dry after-ripening (Fig. 7.1), the lines fitted by probit analysis are described by the equation

$$G = K_d + p/\sigma_d \qquad (1)$$

where G is the estimated germination (normal equivalent deviates, ned) following after-ripening for p (days). The slope is shown as $1/\sigma_d$, where σ_d is the standard deviation of dormancy periods (in days) within the seed population. The intercept, K_d, estimates germination at time zero (ned) more accurately than is possible in a standard germination test. The percentage dormancy, D, can then be estimated as follows:

$$D = 100 \, (1 - \Phi^{-1}(K_d)) \qquad (2)$$

where $\Phi^{-1}(K_d)$ is the inverse normal (back-) transformation of K_d (i.e. germination as a proportion). This estimate only applies to single seed lots harvested at the same place and time and without evidence of polymorphism as described later.

The intrinsic nature of dormancy in the individual seed is probably more like 'a hill that seeds approach from one side (i.e. as the requirements to break dormancy are met) and then progress down the other, gathering momentum (increasing germination rate)' (Bradford, 1995, 1996). We do not yet know how to measure the depth of dormancy in individual seeds but its variation is reflected in the seed-to-seed variation in the expression of dormancy in a seed population, which is quantified by the parameter σ_d in Equation (1). This variation leads to other measures of seed dormancy which could include the time to 50% loss of dormancy or the median dormancy period during after-ripening at constant temperature and moisture content (Fig. 7.1; Sharif-Zadeh and Murdoch, 2001), or the median base water potential of the seed lot (Bradford, 1995, 2002).

Primary Dormancy

When considering the ecology of regeneration, a simple distinction of particular value is that between primary and secondary dormancy. Primary (or innate) dormancy develops during seed maturation on the mother plant, while secondary dormancy is induced after shedding. The functions of primary dormancy as an adaptive trait appear to be twofold. First, along with the inhibition of germination of developing seeds by the mother plant, primary dormancy helps to prevent precocious germination on the mother plant (Bewley and Downie, 1996). Secondly, in many species, primary dormancy persists after maturation and shedding resulting in the temporal dispersal of seeds by preventing the immediate and approximately synchronous germination of seeds.

Many papers on primary dormancy now follow Baskin and Baskin's (1998, 2004)

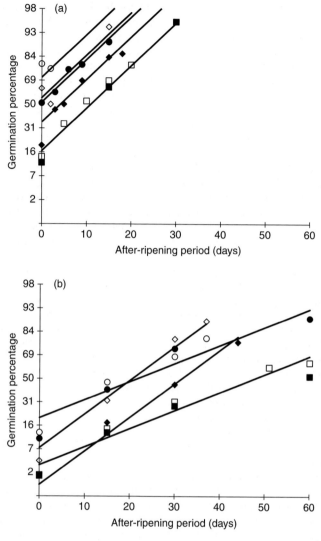

Fig. 7.1. After-ripening of (a) caryopses and (b) spikelets of *Cenchrus ciliaris*, which were matured on mother plants at temperatures of 32°C/27°C (open symbols) and 27°C/20°C (closed symbols) when they were either exposed to water stress cycles (□■), or well watered without (◇◆) or with nutrients (○●). After maturation on the mother plants, the seeds were after-ripened as spikelets at 43% r.h., 40°C. Lines were fitted by probit analysis. (From Sharif-Zadeh and Murdoch, 2000.)

classification of its putative mechanism. This classification helpfully distinguishes three basic mechanisms: 'physiological', 'morphological' (e.g. seeds with immature embryos) and 'physical' (e.g. hard, impermeable seeds), although the latter is arguably an endogenous quiescence. Clearly seeds may combine more than one of these mechanisms and so Baskin and Baskin add two

further classes for seeds which combine physiological and either morphological ('morpho-physiological') or physical ('combinational') dormancies. Less helpful in my view are some of the more detailed subdivisions of these broad categories. For example, physiological dormancy is categorized into three 'levels': non-deep, intermediate and deep with five possible 'types' of non-deep dormancy. For example, seeds with deep physiological dormancy may require 3–4 months cold stratification while intermediate ones require 2–3 months (Baskin and Baskin, 2004). While the qualitative distinctions according to mechanisms of dormancy are useful, the arbitrary timescales ignore the continuum of quantitative variation that exists in depth of dormancy in seeds harvested from the same location at the same time (e.g. Fig. 7.1).

Primary dormancy may, therefore, simply occur because the embryo is immature and in some cases largely undifferentiated such that further development is needed prior to germination. Examples of such 'morphological' dormancy include seeds with linear embryos as in the Ranunculaceae and Umbelliferae (Atwater, 1980; Ellis et al., 1985; Baskin and Baskin, 1998).

The mechanism(s) of physiological blocks to germination may reside in the seed coat (coat-imposed dormancy) and/or the embryo itself (embryo dormancy) and these mechanisms may interact strongly (Debeaujon et al., 2007).

The testa and seed covering structures may dramatically reduce germination (compare the germination of caryopses and spikelets in Figs 7.1a and b, respectively). The main mechanisms associated with seed covering structures include prevention of leakage of germination inhibitors and conversely allowing uptake of both germination inhibitors and stimulants, reducing oxygen uptake and carbon dioxide release, slowing the uptake of water during imbibition, filtering of incoming light perhaps increasing the ratio of germination-inhibiting far-red (c.730 nm) to germination-stimulating red (c.660 nm) light (see Chapter 5 of this volume), and finally, a mechanical restraint to radicle emergence (Bewley and Black, 1994). Flavonoid compounds in the seed coat and in particular, proanthocyanidins, appear to be associated with dormancy as in C. ciliaris (Fig. 7.1 and Sharif-Zadeh, 1999; reviewed by Debeaujon et al., 2007; Chapter 8 of this volume). In their oxidized form, they confer the brown colour, which often characterizes mature seeds and which is sometimes associated with dormancy polymorphism (Table 7.1; Puga-Hermida et al., 2003).

The understanding of both coat-imposed and embryo dormancy has made massive strides with the availability of mutant phenotypes especially in Arabidopsis. Such studies have confirmed the major role of abscisic acid (ABA) in induction of dormancy on the mother plant and in prevention of precocious germination and maintenance of dormancy (Nambara et al., 2010; Chapter 8 of this volume). For example, mutants deficient in ABA production or in responsiveness to ABA during seed development yield seeds which lack dormancy and germinate precociously (Karssen et al., 1983; Bewley, 1997). Clearly this dormancy must be relieved for the seeds to germinate by metabolism, decreased biosynthesis or reduced sensitivity to it, together with the positive stimulation of germination by gibberellins

Table 7.1. Dormancy polymorphism in *Chenopodium album*. Seeds were collected from various sites in England and Wales and separated according to testa colour and texture. Testa thickness and thousand-seed weights were measured and seeds were germinated at 20°C with or without prechilling of imbibed seeds for at least three weeks at 0°C. Dormancy is calculated as 100 minus the percentage germination. (From data in Williams and Harper, 1965.)

Testa colour and texture	Testa thickness, μm	Thousand-seed weight, g	Dormancy, % Unchilled	Dormancy, % Chilled
Brown	16	1.57	<10	<10
Black, reticulate	60	1.33	38	36
Black, smooth	60	1.13	68	39

(GA) (Finkelstein *et al.*, 2008; Rajjou *et al.*, 2012). Active catabolism or deactivation of ABA occurs most strongly in tissues adjacent to the radicle and coleorhiza of *Arabidopsis* and barley, respectively (Nambara *et al.*, 2010).

The overall concept is that of a balance of germination inhibition by ABA and promotion by GA. It is, however, important to note that GA and ABA are not the sole regulators of dormancy and germination. Other endogenous hormones such as auxins and cytokinins (Kucera *et al.*, 2005), and exogenous chemicals such as ethene (ethylene) have long been known to influence seed germination and dormancy and interact with ABA and GA signalling (also discussed in Chapter 6 of this volume). The role of ethene in terminating dormancy and in particular its interaction with other molecules have been reviewed by Bogatek and Gniazdowska (2012). Antagonistic responses between ABA and ethene biosynthesis are well documented (Finkelstein *et al.*, 2008). Moreover, ethene overcomes the inhibitory effects of relatively high imbibition temperatures (thermodormancy) and enhances expression of genes, the products of which induce the cell-wall loosening and cell separation that are precursors of endosperm rupture and radicle emergence (reviewed by Linkies and Leubner-Metzger, 2012). Brassinosteroids likewise antagonize the inhibitory effect of ABA and promote seed germination but by a different mechanism to GA (Kucera *et al.*, 2005; Finkelstein *et al.*, 2008). Strigolactones are another group of plant growth substances, which are exuded by roots of many plants, stimulating mycorrhizal branching (Akiyama *et al.*, 2005). Root parasitic angiosperms in the Orobanchaceae have also, however, adapted to these signals in the rhizosphere to terminate dormancy and initiate germination, thereafter attempting to parasitize the plant exuding the stimulant (Ruyter-Spira *et al.*, 2012; Bouwmeester and Yoneyama, 2013). More recent research has shown that strigolactones also stimulate germination of non-parasitic angiosperms (Toh *et al.*, 2012). Also in the soil, water soluble components of smoke may stimulate germination and hence plant regeneration (Roche *et al.*, 2008), an effect likely to occur in many ecosystems such as savannas where the vegetation is subject to burning (Bond *et al.*, 2004). The most active stimulatory component in smoke is butenolide, now known to be one of several structurally similar plant growth regulators named karrikins, whose mode of action appears to be linked to GA biosynthesis and light sensitivity (Nelson *et al.*, 2009).

Other compounds promoting germination or terminating dormancy include some nitrogen-containing compounds including nitric oxide, nitrate, nitrite, azide and cyanide. Possible mechanisms for their action(s) including stimulating the cyanide-insensitive, pentose phosphate respiratory pathway and interactions with ethene have been proposed (Roberts and Smith, 1977; Renata and Agnieszka, 2006; Finkelstein *et al.*, 2008; Bogatek and Gniazdowska, 2012; Chapter 6 of this volume).

Jasmonate and its metabolites are involved in plant responses to abiotic stress as well as plant growth regulatory functions, including during inhibition of seed germination (reviewed by Linkies and Leubner-Metzger, 2012).

Acquisition of primary dormancy on the mother plant

Heritability of dormancy is complex because parts of the seeds differ genetically. For example, a diploid embryo may receive nutrients from a triploid endosperm (with two maternal and one paternal set of chromosomes) and is surrounded by maternal tissues (testa and fruit structures). The genetics of dormancy in the Gramineae has been reviewed by Simpson (2007). Research over the past 30 years has identified many genes involved in regulation of dormancy, particularly of mutants affecting both biosynthesis and signalling of plant hormones (Graeber *et al.*, 2012), the expression and effects of the products of these genes being modified by external environmental factors, GxE interactions being expected. Thus, primary dormancy not only varies with genotype but also with maturation environment

(Fenner and Thompson, 2005, Chapter 5.5). For example, primary dormancy is acquired during grain filling of wheat and then declines progressively during pre-harvest desiccation and ripening (Fig. 7.2). Levels of dormancy are much greater in near isogenic lines of wheat containing the semi-dwarf, GA-insensitive allele, *Rht-B1c*, compared to other alleles (Fig. 7.2; Gooding *et al.*, 2012).

In *Chenopodium album*, heavier, thinner-coated non-dormant seeds were produced in short days compared to long days (Karssen,

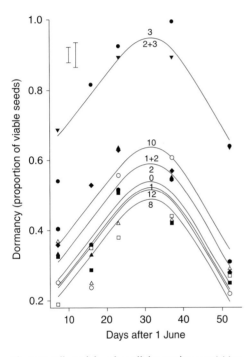

Fig. 7.2. Effect of dwarfing alleles on the acquisition and loss of dormancy during grain filling and ripening of intensively grown near isogenic lines of winter wheat in 2009. Numerals denote fitted lines for individual alleles: 0 = *Rht-B1a+D1a*, ○; 1 = *Rht-B1b*, ■; 2 = *Rht-D1b*, ▲; 1 + 2 = *Rht-B1b + D1b*, ●; 3 = *Rht-B1c*, ▼; 2 + 3 = *Rht-B1c+D1b*, ●; 10 = *Rht-D1c*, ◆; 8 = *Rht8c + Ppd-D1a*, ◆; Rht12 = △. Solid symbols are GA-insensitive alleles. Error bars are SEDs for comparing main effects of allele; left = minimum (among 0, 1, 2 and 3), right = maximum (among 8, 10 and 12). Points are means of two blocks and three (0, 1, 2 and 3), two (1 + 2 and 2 + 3) or one background (8, 10, 12). Effects of background (tall) cultivars have been removed. (From Gooding *et al.*, 2012.)

1970). Warmer temperatures during maturation usually reduce dormancy (Peters, 1982a,b) whereas the response to water stress varies with species. In *Avena fatua* and *Sorghum halepense*, dormancy is lower if seeds mature under water stress (Peters, 1982a,b; Benech-Arnold *et al.*, 1992) implying that seeds produced in warm, dry summers are thus likely to have less dormancy than those produced in cool moist ones. In *C. ciliaris*, a perennial grass which occurs in arid and semiarid environments, dormancy was also lower for seeds produced at higher temperatures and with more nitrate in the growing medium, but, by contrast, dormancy was greater with water stress, a characteristic that may be of adaptive value in plants growing in arid ecosystems (Fig. 7.1). Aspinall (1965) also observed more dormancy in barley (*Hordeum vulgare*) grains produced under water stress. Nutrition and especially nitrogenous fertilization of the mother plant during seed maturation tend to reduce seed dormancy (Fig. 7.1). Similarly, seed dormancy decreased with increase in nitrate nutrition of *Arabidopsis* mother plants from 3 to 10 to 50 mM nitrate; it being demonstrated, moreover, through the use of mutant genotypes that the effect of the nitrate ion was likely to be mediated through signalling rather than nutrition (Alboresi *et al.*, 2005). With respect to interspecific differences, seeds which mature within green tissues tend to be more light sensitive than those where chlorophyll declines at an early stage during maturation (Cresswell and Grime, 1981).

Polymorphism is shown when developmentally or morphologically different seeds differ in primary dormancy. These seeds may be produced on the same or on different plants of a given species. The classic example of *Xanthium pensylvanicum* achieves dispersal in time. Two seeds are dispersed together in each capsule, the upper seed being much more dormant than the lower one, and at least twelve months normally separate the germination of the two seeds (Esashi and Leopold, 1968; Harper, 1977). In *Chenopodium album*, between 0.2 and 5.0% of seeds collected from five localities in England and Wales were large, brown, thin coated and non-dormant in comparison with the majority, which were smaller, black, thick coated and dormant (Table 7.1). The black seeds had smooth or reticulate testae, the deeper dormancy of the former being partly relieved by chilling (Table 7.1; Williams and Harper, 1965). Two types of seeds are produced by ragwort (*Senecio jacobaea*) (McEvoy, 1984). Those from around the edge of the inflorescence (ray achenes) differ from those at the centre (disc achenes). Ray achenes are heavy, smooth, shed their long hairs and have greater dormancy. They remain on the mother plant for longer, are dispersed close to it and may therefore replace the mother plant at the same site. By contrast, the disc achenes are lighter and retain a group of long hairs which assists in wind dispersal to new sites (McEvoy, 1984). Similarly, a comparison among diverse species of *Argyranthemum* showed that ray achenes tended to show more dormancy than disc achenes, although differences were only significant in 11 of the 21 species studied, and only substantial in four species (Francisco-Ortega *et al.*, 1994). Further examples of polymorphic seeds are discussed by Fenner (1991) and Baskin and Baskin (1998).

After-ripening

Primary dormancy may decline both prior to shedding (Fig. 7.2) and subsequently. When this loss of dormancy occurs in 'air-dry' seeds, it is termed after-ripening (e.g. Fig. 7.1). The progressive loss in dormancy during after-ripening is a function of time, temperature and moisture content (Bazin *et al.*, 2011) and varies in a predictable manner. Circumstantial evidence suggests that many seeds behave like those members of the Gramineae in which the logarithm of the mean dormancy period is a negative linear function of temperature, the Q_{10} for the relation being typically in the range 2.5–3.8 (Roberts, 1965, 1988). Thus the longer and warmer the storage environment, the greater the loss of dormancy. For example, in one seed lot of *Oryza glaberrima* at 11.2% moisture

content, half the seeds lost dormancy after 65 days at 30°C while at 40°C only about 20 days were required (Ellis *et al.*, 1983); and so the Q_{10} is close to 3. A similar approach has been published for loss of primary dormancy in barley (Favier and Woods, 1993). The rate of loss in dormancy during after-ripening is only affected at certain moisture contents (Quail and Carter, 1969; Tokumasu *et al.*, 1975; Ellis *et al.*, 1983; Sharif-Zadeh and Murdoch, 2001) so that in cereals, at moisture contents below 8% the rate of dry after-ripening is minimal while it is greatest between 11 and 15% moisture content (Roberts, 1962, 1988; Ellis *et al.*, 1983). In *C. ciliaris*, evidence for an optimum moisture level for after-ripening was clear because seeds in equilibrium with 40% r.h. at 20°C after-ripened more slowly than those at 50% r.h. whereas those at 70% r.h. remained viable but largely failed to after-ripen (Sharif-Zadeh and Murdoch, 2001). Effects of temperature and moisture on the rate of after-ripening appear to be additive (Sharif-Zadeh and Murdoch, 2001). Not surprisingly, procedures used during seed collection and postharvest seed drying, together with subsequent storage prior to investigations of seed dormancy will influence dormancy. It follows that different methods of preparing seeds for burial experiments may result in differing degrees of dormancy at burial and, since dormancy is the major factor influencing survival in soil, thus influence the subsequent rate of depletion of the viable seed population in the soil seed bank. Accordingly, it is possible that reports of the longevity of seeds in soil from experiments which treated collected seeds to environments in which considerable after-ripening may have occurred could have underestimated the natural longevity under such circumstances.

Secondary Dormancy

Dormancy may be induced in both dormant and non-dormant seeds after shedding. When moist, though not necessarily fully imbibed, seeds are kept in sufficient environmental stress to prevent germination or with insufficient promotion to relieve dormancy, secondary dormancy may be induced in both dormant and non-dormant seeds. The main causes of this secondary dormancy in buried seeds were originally thought to be low oxygen and/or high carbon dioxide levels (Harper, 1957; Roberts, 1972), and the efficacy of insufficient air and anaerobiosis to induce dormancy is well known. For example, in one line of *A. fatua*, subsequent germination of caryopses decreased from 90% to 17% after imbibing the caryopses (dehulled florets) for 3 h in an anoxic atmosphere. The secondary dormancy was relieved by dry storage, nitrite, nitrate, ethanol, azide and gibberellin (GA$_3$) (Symons *et al.*, 1986).

Soil atmosphere in cultivated soils may deviate little from ambient air (Murdoch and Roberts, 1996). Nevertheless, the occurrence of anaerobic microsites (Renault and Stengel, 1994) within a largely aerobic mass has been deduced from evidence for anaerobic microbial denitrification of well aerated pasture soils (Burford and Millington, 1968; Burford and Stefanson, 1973). Very high CO$_2$

concentrations (>40%) may occur in soil with stagnating water columns (Enoch and Dasberg, 1971). Hence, anaerobiosis and hypoxic atmospheres may contribute to the induction of secondary dormancy of seeds in soil (Benvenuti and Macchia, 1995).

Secondary dormancy may also be induced in prolonged moist aerobic treatments in which germination does not occur (Totterdell and Roberts, 1979; Jones *et al.*, 1997; Kebreab and Murdoch, 1999a; Dzomeku and Murdoch, 2007a,b). During such treatments (e.g. prechilling or at higher temperatures, warm stratification), loss of primary dormancy is initially evidenced by progressive sensitization to a subsequent germination-promoting stimulus. However, prolonging the period of moist storage results in a decrease in the response to subsequent stimulation due to secondary dormancy (Fig. 7.3). Temperature has a marked effect on the induction of secondary dormancy. While an increase of temperature increases the rate of after-ripening of air-dry seeds, it also affects the rate of induction of secondary dormancy in imbibed seeds. In imbibed

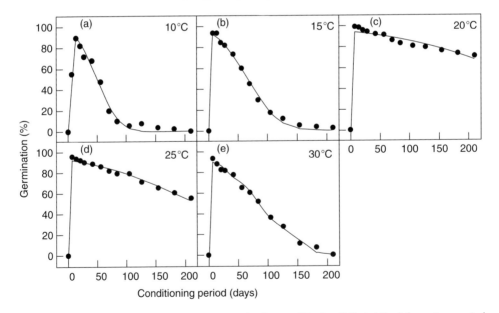

Fig. 7.3. Germination of *Phelipanche aegyptiaca* seeds after conditioning (fully imbibed) for various periods at (a) 10°C, (b) 15°C, (c) 20°C, (d) 25°C and (e) 30°C. Seeds were surface-sterilized with sodium hypochlorite solution before conditioning. Germination tests lasted 10 days at 20°C with 10^{-8} molar GR24, a synthetic strigolactone. Lines were fitted according to a multiplicative probability model. (From Kebreab and Murdoch, 1999a.)

Rumex crispus seeds for example, secondary dormancy was evident after about 21, 7–14, 7 and 3 days at 1.5, 10, 15 and 20 °C, respectively (Totterdell and Roberts, 1979; compare Totterdell and Roberts, 1981). Interestingly, for the root parasitic *Orobanche* and *Phelipanche* spp. from a warm temperate environment, rates of induction were fastest at 3°C and decreased to a minimum above about 20°C (Fig. 7.3; Kebreab and Murdoch, 1999a), whereas in *Striga hermonthica*, from the tropics, rates of induction were slowest at the relatively low temperatures of 15–20°C and increased with increase of temperature to a maximum above 25°C (Dzomeku and Murdoch, 2007a,b). These more

comprehensive studies on *S. hermonthica* also demonstrated that maximum rates of induction increased with increase in water stress (to −1.5 MPa).

In aerobic conditions, germination may also be inhibited by exposing seeds to prolonged white light especially at high radiant flux densities or to far-red light (Bartley and Frankland, 1982; Fig. 7.4). By contrast, induction of secondary dormancy may be delayed, reduced or prevented by intermittent low intensity laboratory light and nitrate in *Capsella bursa-pastoris* (Fig. 7.5) and by nitrate in *Sisymbrium officinale* (Hilhorst, 1990b).

A final and somewhat surprising circumstance in which dormancy is reimposed

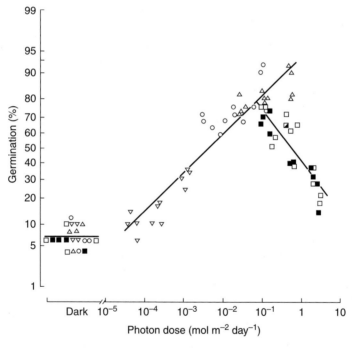

Fig. 7.4. The response of the germination (radicle emergence) of seeds of *Echinochloa turnerana* to white fluorescent light in 7-day tests at an alternating temperature of 20/30°C (16h/8h). The results from testing in the dark are shown on the left. The seeds were exposed to different photon flux densities for 24 h day^{-1} (□,■), 8 h day^{-1} (△), 1 h day^{-1} (○), or 1 min day^{-1} (▽) on filter paper moistened with water (open symbols) or a 0.02 M potassium nitrate solution (solid symbols). The lines shown are cumulative log-normal dose response curves fitted by probit analysis. The positive response, in which germination increases with increase in photon dose (the low energy reaction), shows reciprocity where daylength is ≤8 h day^{-1}. The negative response, in which germination decreased with further increase in photon dose (presumably due to the high irradiance reaction), was only detected in continuous light in this seed lot. The two slopes show the variation in the sensitivity of individual seeds to the minimum photon dose required to promote and inhibit germination, respectively. The standard deviation for promotion was 10$^{1.52}$ mol m^{-2} day^{-1}, while that for inhibition was 10$^{1.16}$ mol m^{-2} day^{-1}. (From Ellis *et al.*, 1986.)

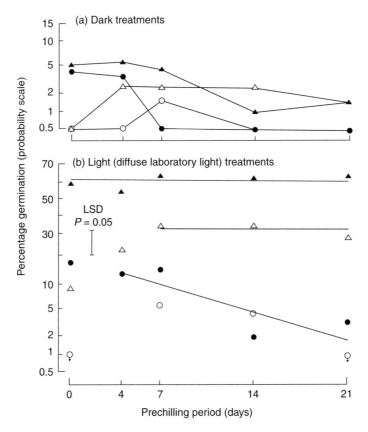

Fig. 7.5. Mean germination of seeds of *Capsella bursa-pastoris* after different periods of prechilling. Subsequent germination tests lasted 4 weeks and were carried out at alternating (10/25°C, 16h/8h; solid symbols) or at constant (25°C; open symbols) temperatures and in water (○,●) or 0.01 molar potassium nitrate solution (△,▲). Seeds were (a) kept in darkness or (b) intermittently exposed to diffuse laboratory light. In (b), lines were fitted by probit analysis with a common line for induction of secondary dormancy in water. The LSD (Least Significant Difference) value is not valid for comparisons with treatments designated (↓). A.J. Murdoch and E.H. Roberts, unpublished data, 1985.

has been shown in Sitka spruce (*Picea sitchensis*). Conditional dormancy (defined as failure to germinate at 10°C) was reimposed when prechilled seeds were then dried to 6% moisture content and stored at 10°C (Jones *et al.*, 1998). The effect was reversible and repeatable and was most rapid at lower (4–10%) compared to higher moisture contents (15–20%). This result is contrary to the conventional expectations of dry after-ripening already described.

Are the mechanisms of primary and secondary dormancies the same? Dry after-ripening and the same chemicals may relieve both primary and secondary dormancy (Symons *et al.*, 1986), implying some

similarity. It is, however, interesting that while short periods of imbibition are associated with relief of primary dormancy, longer periods lead to secondary dormancy. As shown below, empirical mathematical models imply that the dual processes of loss and induction of dormancy may be concurrent in individual seeds and so must have different physiology. In *Orobanche* and *Phelipanche* spp., the loss of primary dormancy is observed by an increased sensitivity to the exogenously supplied germination stimulants (e.g. strigolactones) while induction of secondary dormancy may be associated with accumulation of an inhibitor since in this case, the seeds must be dried to overcome secondary dormancy.

Seed-to-seed Variation in Dormancy

Unless 0% or 100% of seeds germinate, the results of germination tests divide seeds into two groups – those that germinate and those that do not. This separation may imply polymorphism but if that term is to retain its usefulness, it should only be applied where there is evidence of a discontinuous distribution of dormancy periods or in the degree of dormancy (Roberts, 1972). There is often, however, a continuous seed-to-seed variation in the dormancy of individual seeds even in apparently uniform seed lots. For example, the chilling requirements for the loss of primary dormancy of individual *Rumex* seeds varied from about 1 day to 3 weeks at 1.5 °C. After-ripening periods varied from 0 to about 130 days in rice seeds (cultivar Masalaci) when stored at 13.5% moisture content and 27°C (Roberts, 1963). In Fig. 7.4, the seed-to-seed variation in response of *Echinochloa turnerana* to white light ranges over four orders of magnitude from those requiring a photon dose of 10^{-5} mol m^{-2} day^{-1} to those requiring 10^{-1} mol m^{-2} day^{-1} (Ellis *et al.*, 1986). The variation in percentage germination as a function of the dose of dormancy-breaking treatments usually approximates to the cumulative normal frequency distribution. As with any normal distribution, the germination:dose curve can therefore be described by its mean and variance. Two crucial points are that such curves may therefore be linearized if percentage germination values are transformed to ned or probits (Figs 7.1, 7.4 and 7.5; Hewlett and Plackett, 1979), and that the slope of the probit germination:dose line therefore has the units ned per unit dose. The reciprocal of the slope quantifies the standard deviation of the dormancy response to the applied treatment in the seed population. This analysis has been found appropriate or proposed for germination responses of dormant seeds to light in the very low fluence response (Hartmann *et al.*, 1998) as well as the low energy and high irradiance reactions (Fig. 7.4; Borthwick *et al.*, 1954; Duke, 1978; Frankland and Taylorson, 1983), dry after-ripening period (Fig. 7.1; Roberts, 1961, 1965; Probert *et al.*,

1985; Favier and Woods, 1993; Sharif-Zadeh and Murdoch, 2001), imbibed pretreatments such as conditioning or chilling (Figs 7.3 and 7.5; Totterdell and Roberts, 1979; Kebreab and Murdoch, 1999a; Dzomeku and Murdoch, 2007a,b), ethylene before and after burial of seeds in soil (Schonbeck and Egley, 1980; Egley, 1989b), alternating temperatures (Probert *et al.*, 1985, 1987; Murdoch *et al.*, 1989; Kuo, 1994), gibberellin and nitrate (Hilhorst and Karssen, 1988; Hilhorst *et al.*, 1996).

Similarly, variation in percentage germination as a function of the dose of dormancy-inducing and germination-inhibiting treatments may approximate to the negative cumulative normal distribution, for example, with aerobic conditioning or chilling of imbibed seeds (Figs 7.3 and 7.5; Totterdell and Roberts, 1979; Kebreab and Murdoch, 1999a; Dzomeku and Murdoch, 2007a,b) and the high irradiance reaction in which high photon flux densities of light of any wavelength may inhibit germination (Fig. 7.4; Bartley and Frankland, 1982; see Chapter 7 of this volume). An exception with respect to induction of dormancy was an exponential decrease in light sensitivity of *Sisymbrium officinale* as a function of the duration of dark imbibition at 15°C prior to exposure to red light (Hilhorst, 1990a). Nevertheless, whichever model is used, it is clear that considerable seed-to-seed variation still exists.

Seed-to-seed variation within seed lots is therefore generally normally distributed. Sometimes, however, responses are nonlinear because more than one process may be occurring simultaneously (Fig. 7.3). The loss of primary dormancy and induction of secondary dormancy may still be normally distributed in the seed population but, because they occur concurrently, the germination after any period is given by a multiplicative probability model. The final germination is in fact the product of the probabilities that a seed: (i) has lost primary dormancy; (ii) has not entered secondary dormancy; and (iii) is still viable. For example, the seeds of the parasitic weed *P. aegyptiaca* show optimum preconditioning periods of about 3–10 days depending on temperature after which secondary dormancy is induced (Kebreab and Murdoch, 1999a; Fig. 7.3). The rate of induction of secondary dormancy appears to be minimized at 20 °C, but the rapid decline at 25°C is due to a third concurrent process of loss of viability (Kebreab & Murdoch, 1999a).

Key hypotheses of the non-linear models are: (i) seed-to-seed variation with respect to each component process is normally distributed with respect to the 'dose' variable (e.g. time); and (ii) the processes operate concurrently and independently within each individual seed. Similar processes occur in *S. hermonthica* (Dzomeku and Murdoch, 2007a,b) and further examples are also discussed by Murdoch *et al.* (2000).

Annual Cycles in Dormancy

Annual cycles in which physiologically based dormancy is relieved and induced during the course of a year occur in buried seeds of some annual plants in both temperate and tropical soil environments (Courtney, 1968; Baskin and Baskin, 1985, 1998; Bouwmeester, 1990; Murdoch, 1998; Batlla and Benech-Arnold, 2010). Germination of seeds occurs at times of low or no dormancy. Non-dormant seeds simply await the availability of moisture and suitable temperatures for germination (Roberts and Potter, 1980; Karssen, 1982; Bouwmeester, 1990; Hilhorst, 2007). Exposure to light (Taylorson, 1970, 1972), nitrate (Murdoch and Roberts, 1996), fluctuations of temperature (Benech-Arnold and Sanchez, 1995) or combinations of these treatments (Roberts *et al.*, 1987; Murdoch, 1998) may be needed to relieve residual primary and secondary dormancy at times of low dormancy. The combination of the annual dormancy cycle with the seasonal variation in temperature and moisture gives a net germinability at any given time. Sauer and Struick (1964) and Wesson and Wareing (1967, 1969a,b) produced evidence that suggested that buried seeds often acquired a positive requirement for light that prevented germination until seeds were brought to the surface. While agreeing with

the overriding importance of light, Vincent and Roberts (1977) contended that the reduced temperature fluctuations experienced by buried seeds were also likely to be important, especially because the light sensitivity of many annual weed seeds was only exhibited when they were also exposed to alternating temperatures (Roberts *et al.*, 1987). These responses help to explain the higher seedling emergence that is associated with cultivation treatments particularly at certain times of the year (Courtney, 1968; Vincent and Roberts, 1977; Roberts and Ricketts, 1979; Roberts and Potter, 1980) as well as the periodicity of seedling emergence that characterizes many species (Chepil, 1946; Roberts, 1964). An annual germinability cycle is thus superimposed as may be inferred from Allen *et al.* (2007, Fig. 4.2).

Ecologically, it is useful to identify factors in the environment that may promote germination *in situ* at any given time. An understanding of these stimuli and of how they act in seed populations with different levels of dormancy is crucial if seedling emergence from persistent soil seed banks is to be predicted (Murdoch, 1998). In the case of *Chenopodium album*, responses to light were, not surprisingly, rare in seeds retrieved from the soil surface, but their behaviour was quite interesting. Only two classes of seeds existed among the surface seeds for much of the time (Fig. 7.6): a non-dormant minority, which would presumably germinate *in situ* given moisture; and an ungerminable majority whose dormancy could seldom be broken by light or nitrate.

Various intermediate classes were found in buried seeds and proportions would germinate on retrieval given light or nitrate (Fig. 7.6). Typically, the responses to both of these factors were additive, so that positive and negative interactions were unusual (Fig. 7.6). The germinability of retrieved seeds tended to increase with depth and high proportions during the spring and summer reflect times of low dormancy. With moisture, some seeds would be expected to germinate *in situ* at 0–25 mm without nitrate or light. At their original depths of burial, however, a lower

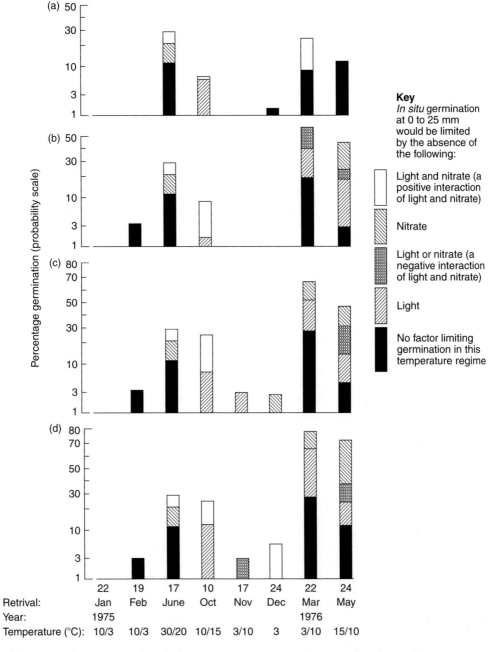

Key
In situ germination
at 0 to 25 mm
would be limited
by the absence of
the following:

Light and nitrate (a positive interaction of light and nitrate)

Nitrate

Light or nitrate (a negative interaction of light and nitrate)

Light

No factor limiting germination in this temperature regime

Percentage germination (probability scale)

Retrival:	22 Jan	19 Feb	17 June	10 Oct	17 Nov	24 Dec	22 Mar	24 May
Year:	1975						1976	
Temperature (°C):	10/3	10/3	30/20	10/15	3/10	3	3/10	15/10

Fig. 7.6. Environmental factors limiting germination of *Chenopodium album* seeds retrieved at night from (a) the soil surface or depths of (b) 25, (c) 75 and (d) 230 mm. Seeds were germinated in light or darkness and in water or 0.01 molar potassium nitrate in a 16/8h temperature regime approximating the soil temperature at 0–25 mm depth 7–10 days prior to retrieval. (Soil temperature was recorded every 20 min.) Significant effects and interactions are shown (from analysis of variance). Thus differences between depths are only shown if the effect of depth or interactions with it were significant. Germination in winter was often low and analysis was restricted to subsets of data and where applicable responses are shown. Otherwise overall mean values are given. Dark treatments were not tested on 10 October 1975. (From Murdoch, 1998.)

amplitude of temperature fluctuation might limit germination of deeply buried seeds.

The seeds that would not germinate in any treatment fell into two categories. First, there were those whose residual dormancy could not be broken in the temperature regime employed. Additional tests carried out at an approximately optimal 30/10°C (16/8h) alternation showed that this proportion was often large (data not shown in Fig. 7.6). Secondly, there were apparently viable (firm) seeds which were ungerminable in any dormancy-relieving treatment. The proportion of these decreased with depth.

Quiescence

A viable seed may therefore avoid germination by being dormant. An alternative strategy is that of quiescence, classified as physical dormancy by Baskin and Baskin (1998, 2004). Quiescence may be enforced by the environment as, for example, in air-dry seeds or in imbibed seeds below the base temperature for the rate of germination of non-dormant seed (Labouriau, 1970; Roberts, 1988; Kebreab and Murdoch, 1999a). Imbibed seeds above the ceiling temperature for the rate of germination (Roberts, 1988) are also quiescent but viability is unlikely to be maintained (Ellis *et al.*, 1987).

In the soil, exogenous quiescence – 'the inability to germinate due to an environmental restraint – shortage of water, low temperature, poor aeration' (Harper 1957, 1977) – has previously been called enforced dormancy. These restraints enforce quiescence rather than dormancy since one or more of the three minimum requirements for germination of non-dormant seeds is lacking.

The seed coats of some species may be impermeable to water and such hard seeds are intrinsically quiescent. Hardseededness, an endogenous quiescence, ensures the persistence of the seed banks of both wild (Egley and Chandler, 1983) and cultivated species (Lewis, 1973; Saunders, 1981; Egley, 1989a; Mott *et al.*, 1989; Russi *et al.*, 1992a; Standifer *et al.*, 1989). Although best known in the Leguminosae, hard impermeable seeds are found in species of several other families

including the Anacardiaceae (e.g. *Rhus* spp.), Bixaceae (e.g. *Bixa orellana*), Cannaceae (e.g. *Canna* spp.), Convolvulaceae (e.g. *Ipomoea* spp.), Ebenaceae (e.g. *Diospyros virginiana*), Geraniaceae (e.g. *Pelargonium* spp.), Liliaceae (e.g. *Asparagus densiflorus*), Malvaceae (e.g. *Gossypium* spp.), Myrtaceae (e.g. *Psidium* spp.), Rhamnaceae (e.g. *Ceanothus*), Sapindaceae (e.g. *Cardiospermum halicacabum*), Solanaceae (e.g. *Datura* spp.) and Zingiberaceae (e.g. *Elattaria cardamomum*) (Ballard, 1973; Atwater, 1980; Ellis *et al.*, 1985; Baskin *et al.*, 2000). The genetical basis of impermeability has been demonstrated in several species, but environmental conditions during seed development such as humidity, temperature, daylength and mineral nutrition modify its expression (Egley, 1989a).

Impermeability develops during maturation drying on the mother plant and it may increase after shedding in dry environments. For example, Standifer *et al.* (1989) have demonstrated that hardseededness in okra (*Abelmoschus esculentus*) increased as seeds were dried progressively from 11% to 3% moisture content. When seeds of the legumes *Trifolium pratense*, *T. repens* and *Lupinus arboreus* were dried below about 14% moisture content, intact and undamaged testae became impermeable to water (Hyde, 1954). Further drying occurs through the hilum which acts as a one-way valve permitting water vapour to escape at low ambient relative humidities, but closing at high relative humidity. Seed moisture will therefore equilibrate with the lowest relative humidity experienced. At seed moisture contents of about 10–12%, sudden increases in humidity cause the hilar fissure to close. However, a gradual increase in humidity leaves the fissure open (Hyde, 1954), and imbibition may occur. Hence, this hardseededness is reversible. Drying to even lower moisture contents (e.g. 8–9% in *Lupinus cosentini*) renders the seed absolutely and irreversibly impermeable until the hard seed coat is broken (Quinlivan, 1971). This pattern appears to be common to most if not all papilionoid legumes. A simpler mechanical closure of the hilum has been described in the Caesalpinoideae and some of the Mimosoideae (Werker, 1980/81).

Hard seeds can easily be made permeable in the laboratory by mechanical abrasion. Natural softening is often protracted, as shown later in this chapter by the longevities of hard seeds in soil. Observations of buried seeds of six papilionoid legumes (three medics and three clovers) in Syria showed that: (i) different proportions of softened seeds were found in each species each year; (ii) some species were more susceptible to softening (e.g. *Trifolium stellatum* compared to *Medicago orbicularis*); (iii) the seed coat was usually broken at the lens (or strophiole); and (iv) burial of seeds reduced the number of softened seeds present at the end of summer (Russi *et al.*, 1992a,b). The wide diurnal temperature fluctuations (15/60°C) of the dry summers of some Mediterranean climates are thought to soften legume seeds (Quinlivan, 1971; Saunders, 1981; Egley, 1989a; Russi *et al.*, 1992b;). Taylor (1981) proposed that high soil temperatures weaken the cells of the lens and that temperature fluctuations cause a gradual expansion and contraction of cells that eventually disrupts the testa at the weak points, e.g. the lens, thus relieving impermeability. Egley (1989a) also proposed that temperature fluctuations cause softening in the Malvaceae by expansion and contraction of cells in the chalazal region.

The Efficacy of Dormancy in Preservation of Seed Viability

In this section, we take the longer-term perspective of asking the question: how long can dormancy or quiescence last? because, provided a seed has an adequate potential longevity – and seeds with intermediate or recalcitrant seed storage behaviours do not – survival of seeds in the soil depends on avoiding germination by dormancy or quiescence.

The remarkable longevities of some individual seeds in the soil seed bank – the record breakers in the survival game – are legendary, implying that dormancy and quiescence may prevent germination for very long periods. In practice, this longevity is epitomized by the gardeners' adage 'One year's seeding: seven years' weeding'. In 1760, Eliot wrote in an essay on field husbandry in New England, 'The seeds of weeds are numerous and hardy, they will lye many years in the ground'. De Candolle (1832) observed that 'certain portions of soil, which, by reason of terracing work, were exposed to the air after several centuries, covered themselves the first year with a multitude of individuals belonging to certain species, sometimes uncommon in the vicinity'. He concluded that the seeds of these species had remained viable for considerable periods in the soil and there is much anecdotal evidence like this in the literature (Turner, 1933). However, long-lived seeds are not all weeds: in most temperate and tropical habitats some species form persistent seed banks (Thompson and Grime, 1979; Garwood, 1989).

Circumstantial evidence for very long dormant or quiescent periods

Qualitative circumstantial evidence for protracted periods of dormancy in soil is considerable. For example at Rothamsted Research, Hertfordshire, UK, a field called Laboratory House Meadow was established as an ungrazed hay meadow in 1859 and after 58 years without ploughing, 'a large enough number of arable weed seedlings [i.e. 30 seedlings equivalent to 320 seeds m^{-2}] appeared from the soil samples to lead one to assume that a considerable proportion must have been buried in the soil since the time of grassing down'. By contrast, four arable weed seedlings (40 m^{-2}) appeared in soil samples from Park Grass, an old pasture which had not been ploughed for at least 300 years and which had not been grazed by livestock for 40 years (Brenchley, 1918). In a similar type of study, buried seeds were examined in successional field and forest stands in the Harvard Forest, Massachusetts. The sites ranged from a field that had been abandoned for 1 year to a mature and open 80-year-old pine stand with woody seedlings and ground cover species. There were no ground cover plants in the 25 to 47-year-old plantations, but an average of 189 seedlings (2325 m^{-2}) from about 20 species appeared in soil samples taken from these sites; and many of these were characteristic field species (e.g. *Panicum capillare* and *Potentilla* spp.) rather than woodland plants. The authors proposed that 'the occurrence of viable seed in soils of forest stands of increasing age, together with the complete absence of the same species as ground cover plants is certainly indicative [that the seeds had] remained viable during long burial in the soils' (Livingston and Allessio, 1968) implying the seeds had survived for long periods, with their germination being prevented for long periods by dormancy. These studies are not unique. Where fields have been replaced by forests, disturbances such as fire, cultivation, tree felling or severe thinning are usually followed almost immediately by the appearance of earlier successional species whose reappearance sometimes relates to viable seeds (see reviews by Priestley, 1986; Garwood, 1989; Pickett and McDonnell, 1989; Rice, 1989) whose persistence depends on prevention of germination by either dormancy or quiescence.

Further circumstantial evidence for extended periods of dormancy is the presence of viable seeds in archaeologically dated soil samples. For example, Odum (1965) recorded one seed of *Chenopodium album* and three of *Spergula arvensis* germinating in a soil sample dated A.D. 200 from a depth of 120 to 146 cm at Vestervig in Denmark. Many such examples are available; and some show seeds recovered from considerable depths even though there were no seeds at shallower depths and the species was not growing on the site. For example, a soil sample from a previously undisturbed 11th century grave 1.5 m under the floor of the tower of a mediaeval church in the village of Uggelose in north Sjaelland in Denmark contained at least one viable seed of *Verbascum thapsiforme*. Odum (1974) concluded that 'the deep-lying and well-protected soil' certifies that this 'viable seed is at least 850 years old'.

Caution is needed in interpreting circumstantial evidence for long periods of dormancy. When a species is not present in the vegetation at a site, seeds in the soil at that site could have originated from nearby vegetation. For example, in a study of grassland, bracken, scrub and forest sites near Canterbury in New Zealand, an average of 35%, 60% and 72% of the seed bank species were growing 'on site', within 5 m and within 10 m, respectively (Partridge, 1989). The simplest explanation for the presence in the seed bank of about half the species not growing 'on site' is that they were dispersed to the site and so there is no conclusive evidence for long-term dormancy. Seeds may be carried into grassland or forests, or indeed any site, by birds (Piper, 1986) or on human feet or on agricultural equipment. They may be worked into the ground by animals including earthworms whose burrows may extend 210 to 240 cm below the surface (Harper, 1977).

More satisfactory evidence for viable ancient seeds is provided by radiocarbon dating of dead parts of seeds such as the testa (Priestley and Posthumus, 1982). Radiocarbon dating of two viable seeds of East Indian lotus (*Nelumbo nucifera*) from an ancient lakebed deposit at Pulantien in southern Manchuria suggested probable ages of 466 and 705 years at the time of germination (Priestley and Posthumus, 1982 and Shen-Miller *et al.*, 1983, respectively). *N. nucifera* seeds are dry fruits in which the true seed is surrounded by a highly impervious pericarp which must be scarified prior to germination demonstrating long-term quiescence.

Experiments demonstrating long dormancy periods in the soil seed bank

Experimental evidence for the long periods of dormancy or quiescence has come from the classical burial trials initiated by W.J. Beal in 1879 at East Lansing, Michigan, and by J.W.T. Duvel in 1902 at Arlington, Virginia. Beal buried 20 bottles each containing 50 seeds from 20 species. The seeds had been mixed with moist sand and the bottles (473 ml capacity) were buried at a depth of 46 cm in sandy soil with the mouth slanting downwards presumably remaining moist. After 100 years, 21 viable seeds of

Verbascum blattaria, and one each of *V. thapsus* and *Malva rotundifolia* were recovered (Kivilaan and Bandurski, 1981) while *Oenothera biennis* and *Rumex crispus* had survived 80 years (Darlington and Steinbauer, 1961). Duvel's experiment was much larger and the burial conditions more natural. Batches of 100 or 200 seeds of 107 species were mixed with sterilized soil and buried in porous flower pots covered with inverted porous saucers at three depths (20, 56 and 107 cm). Thirty-six species survived for 39 years (Toole and Brown, 1946) with over 80% survival of *Datura stramonium, Solanum nigrum* and *Phytolacca americana*. There was negligible survival of 22 crop species even for 1 year; but some seeds of tobacco (*Nicotiana tabacum*), celery (*Apium graveolens*), red clover (*Trifolium pratense*) and Kentucky blue grass (*Poa pratensis*) persisted to the end of the experiment (Toole and Brown, 1946). The short longevity of most crop species in soil seed banks, other than some forage/pasture grasses and legumes, has since been confirmed (Madsen, 1962; Rampton and Ching, 1970; Lewis, 1973) and is hardly surprising given the low levels of dormancy found in most crop seeds. These experiments were all in cool temperate environments. More recent studies are being carried out in the warm humid southern United States (Egley and Chandler, 1983) and in sub-arctic Alaska (Conn *et al.*, 2006).

Seeds may be lost from the seed bank by various means. Loss of viability is only one of several means and is likely to be greater for seeds at or near the soil surface or in semiarid or arid regions. For example, the predicted annual loss of viability for a seed lot of *Phelipanche aegyptiaca* with high initial viability ($K_i = 2.89$) was as high as 38% in Eritrea (Table 7.2; Kebreab and Murdoch, 1999b). Although predation and loss of viability can be important especially for larger seeds at the soil surface, most studies do, however, suggest that germination and hence loss of seed dormancy is the primary cause of depletion of buried seeds (Schafer and Chilcote, 1970; Taylorson, 1970, 1972; Murdoch and Roberts, 1982; Zorner *et al.*, 1984). In most investigations, large numbers of seeds disappear without emerging (Chepil, 1946; Roberts, 1964, 1979, 1986; Roberts and Chancellor, 1979; Roberts and Neilson, 1980, 1981; Roberts and Boddrell, 1983a,b, 1984a,b). Fatal germination (i.e. germination of the seed at a depth from which it fails to emerge) is the usual explanation and is supported by experimental evidence (Murdoch, 1983; Fenner, 1985; Fenner and Thompson, 2005). However, in most long-term studies of the soil seed bank, loss of viability due to disease and ageing is indistinguishable from fatal germination because dead seeds and non-emergent seedlings both decay in moist soil. For the same reason, disease and ageing cannot be separated unless losses occur over very short periods (Van Mourik *et al.*, 2005).

Seed Storage Behaviour and Dormancy

Three types of seed storage behaviour are recognized. Over 80% of seeds for which

Table 7.2. Prediction of loss of viability of *Phelipanche aegyptiaca* seeds under rain-fed farm conditions in the highlands of Eritrea in a full year. Meteorological data from FAO (1994). (From Kebreab and Murdoch, 1999b.)

Season	Mean temperature (°C)	Relative humidity (%)	Period (months)	Predicted loss of viability (%)*
Early dry season	20	53	5	3
Late dry season	26	40	5	2.5
Rainy season	25	imbibed	2	33
Total (over one year)				38.5

* Assumes $K_i = 2.89$ in the viability equation (Ellis and Roberts, 1980a).

seed storage behaviour is documented exhibit orthodox seed storage behaviour (Royal Botanic Gardens, Kew, 2008) and can be stored in a quiescent state at low moisture contents. Their longevity increases with decrease in seed storage moisture content to around −350 MPa and temperature in a quantifiable and predictable way (Roberts, 1973; Roberts and Ellis, 1982; Ellis *et al.*, 1989). Recalcitrant seeds, by contrast, do not survive desiccation (Roberts, 1973) to water potentials below about −1.5 to −5.0 MPa (Probert and Longley, 1989; Roberts and Ellis, 1989; Pritchard, 1991), i.e. roughly equivalent to the permanent wilting point of many growing tissues (Kramer, 1983). Such seeds are produced in a number of large seeded woody perennials, including some tropical plantation crops, e.g., cocoa (*Theobroma cacao*) and rubber (*Hevea bra-ziliensis*); many tropical fruits, e.g., avocado (*Persea americana*) and mango (*Mangifera indica*); a number of timber species, e.g., from the temperate latitudes, oak (*Quercus* spp.) and chestnut (*Castanea* spp.), and from the tropics, some of the Dipterocarpaceae and Araucariaceae. Such seeds are sometimes categorized as being of temperate or tropical origin, as the former can be stored for several years at near freezing temperatures whereas the latter are damaged by exposure to cool temperatures of 10–15°C or less (Bonner, 1990; Hong and Ellis, 1996).

An intermediate category of seed storage behaviour is also recognized in which dry seeds (−90 to −250 MPa) are injured by low temperatures and also by further desiccation (Ellis *et al.*, 1990). Clear experimental evidence for this category of seed storage behaviour has been shown in several species including arabica coffee (*Coffea arabica*), papaya (*Carica papaya*) and oil palm (*Elaeis guineensis*) (Ellis *et al.*, 1990, 1991a,b).

Contrasting Strategies for Plant Regeneration by Seed

The longevity of seeds in soil is probably the most useful way to classify the seed banks of individual species since the transience or persistence of seeds in soil is part of the regeneration strategy adopted by a species. However, it is possible for the same longevity of seeds in soil seed banks to be achieved by different strategies. This review has attempted to identify the key aspects of seed dormancy, viability and longevity that influence the duration of the seed–seedling phase of temporal dispersal. Individual dormant seeds of orthodox species may survive for very long periods in soil, but this is only one of four contrasting strategies for plant regeneration by seed.

Non-dormant recalcitrant seeds

Primary dormancy appears to be totally lacking in those recalcitrant seeds from the tropics that cannot be chilled below 10–15°C without damage (King and Roberts, 1979). Berjak *et al.* (1989) have proposed the hypothesis that the germination of such recalcitrant seeds (which are imbibed when shed) is initiated at or around shedding when the inhibitory effect of the maternal environment may be relieved.

Thus in extreme examples of this strategy there may be virtually no temporal dispersal, with seed development, germination and establishment representing a continuum. One of the most likely adaptive functions of primary dormancy – dispersal of germination and hence regeneration over time – may nevertheless be achieved by such species despite this strategy. For example, the large non-dormant recalcitrant seeds of the primary species of some tropical forests germinate rapidly but produce persistent seedling banks. Some of these seedlings may eventually benefit from a gap in the vegetation caused, for example, by a fallen tree (Whitmore, 1988; Garwood, 1989).

Dormant recalcitrant seeds

Some recalcitrant seeds do show dormancy, particularly those from temperate latitudes. For example, seed dormancy in sycamore (*Acer pseudoplatanus*) can be considerable when the fruits are shed (Hong and Ellis, 1990), and this is largely testa-imposed dormancy (Pinfield *et al.*, 1987, 1990). Nevertheless, the seeds' persistence is not great, as dormancy is generally lost by early spring as a result of chilling.

Non-dormant orthodox seeds

Some of the greatest longevities of seeds in the soil are achieved by seeds in which there may be no primary dormancy, but the seeds are innately quiescent because hardseededness prevents imbibition. In a sense this strategy is closest to conventional crop seed storage. Hardseededness enables the seeds to maintain their moisture content at values approaching the lower moisture content limit to the negative logarithmic relation between seed longevity and moisture, and so their longevity is considerable. Another example of this strategy is represented by non-dormant orthodox seeds in dry regions (e.g. the Sahel or in the Mediterranean summers) where persistence is associated with enforced quiescence. However, most species in this category germinate rapidly after shedding provided moisture is available, and are therefore transient in the soil seed bank.

Dormant orthodox seeds

In this strategy, the seeds may be imbibed for some or all of the time. The maintenance of viability depends upon the regular repair of deteriorative changes during periods spent at high water potentials. Longevity depends upon the primary dormancy present at shedding either not being lost or being reinforced by secondary dormancy. In this strategy, variation in the degree of dormancy among the individual seeds within each influx provides both a regular proportion of (surviving) seeds that germinate each year, and the potential for considerable persistence in extreme individuals as well as the opportunity to respond to a range of microsites (Bullied *et al.*, 2012).

Long-lived orthodox seeds (whether their longevity depends on dormancy or quiescence) can have a crucial role in regeneration of plant communities if land use changes have led to loss of those species from the vegetation.

Each of these strategies is based on differences in seed physiology and each is successful since at least some seeds are in the right physiological state to germinate and emerge in the right place at the right time. Others have quantified the relative success of different strategies (e.g. Russi *et al.*, 1992a). Using models such as those described in this chapter, it is possible to quantify and compare the variability in dormancy within and between seed populations and also to quantify responses to environmental factors. In this way the behaviour of extreme individuals can be placed in perspective and these models may also ultimately be used to predict the responses of buried seed populations in different environments.

References

Akiyama, K., Matsuzaki, K.I. and Hayashi, H. (2005) Plant sesquiterpenes induce hyphal branching in arbuscular mycorrhizal fungi. *Nature* 435, 824–827.

Alboresi, A., Gestin, C., Leydecker, M.-T., Bedu, M., Meyer, C. and Truong, H.-N. (2005) Nitrate, a signal relieving seed dormancy in *Arabidopsis. Plant, Cell & Environment* 28, 500–512.

Allen, P.S., Benech-Arnold, R.L., Batlla, D. and Bradford, K.J. (2007) Modeling of seed dormancy. In: Bradford, K.J. and Nonogaki, H. (eds) *Seed Development, Dormancy and Germination, Annual Plant Reviews* 27, 72–112, Blackwell Publishing, Oxford, UK.

Amen, R.D. (1968) A model of seed dormancy. *Botanical Review* 34, 1–31.

Aspinall, D.J. (1965) Effects of soil moisture stress on the growth of barley. III. Germination of grain from plants subject to water stress. *Journal of the Institute of Brewing* 72, 174–176.

Atwater, B.R. (1980) Germination, dormancy and morphology of the seeds of herbaceous ornamental plants. *Seed Science and Technology* 8, 523–573.

Ballard, L.A.T. (1973) Physical barriers to germination. *Seed Science and Technology* 1, 285–303.

Bartley, M.R. and Frankland, B. (1982) Analysis of the dual role of phytochrome in the photoinhibition of seed germination. *Nature* (Lond.) 300, 750–752.

Baskin, C.C. and Baskin, J.M. (1998) *Seeds: Ecology, Biogeography and Evolution of Dormancy and Germination*. Academic Press, San Diego, California, 666 pp.

Baskin, J.M. and Baskin, C.C. (1985) The annual dormancy cycle in buried weed seeds: a continuum. *Bioscience* 35, 492–498.

Baskin, J.M. and Baskin, C.C. (2004) A classification system for seed dormancy. *Seed Science Research* 14, 1–16.

Baskin, J.M., Baskin, C.C. and Li, X. (2000) Taxonomy, anatomy and evolution of physical dormancy in seeds. *Plant Species Biology* 15, 139–152.

Batlla, D. and Benech-Arnold, R.L. (2010) Predicting changes in dormancy level in natural seed soil banks. *Plant Molecular Biology* 73, 3–13.

Bazin, J., Batlla, D., Dussert, S., El-Maarouf-Bouteau, H. and Bailly, C. (2011) Role of relative humidity, temperature, and water status in dormancy alleviation of sunflower seeds during dry after-ripening. *Journal of Experimental Botany* 62, 627–640.

Benech-Arnold, R.L. and Sanchez, R.A. (1995) Modelling weed seed germination. In: Kigel, J. and Galili, G. (eds) *Seed Development and Germination*. Marcel-Dekker, New York, pp. 545–566.

Benech-Arnold, R.L., Fenner, M. and Edwards, P.J. (1992) Changes in dormancy level in *Sorghum halepense* seeds induced by water stress during seed development. *Functional Ecology* 6, 596–605.

Benvenuti, S. and Macchia, M. (1995) Effect of hypoxia on buried weed seed germination. *Weed Research* 35, 343–351.

Berjak, P., Farrant, J.M. and Pammenter, N.W. (1989). The basis of recalcitrant seed behaviour: cell biology of the homoiohydrous seed condition. In: Taylorson, R.B. (ed.) *Recent Advances in the Development and Germination of Seeds*. Plenum Press, New York, pp. 89–108.

Bewley, J.D. (1997) Seed germination and dormancy. *The Plant Cell* 9, 1055–1066.

Bewley, J.D. and Black, M. (1994) *Seeds: Physiology of Development and Germination*. Plenum Press, New York, 445 pp.

Bewley, J.D. and Downie, B. (1996) Is failure of seeds to germinate during development a dormancy-related phenomenon? In: Lang, G.A. (ed.) *Plant Dormancy: Physiology, Biochemistry and Molecular Biology*. CAB International, Wallingford, UK, pp. 17–27.

Bogatek, R. and Gniazdowska, A. (2012) Ethylene in seed development, dormancy and germination. In: McManus, M.T. (ed.) *Annual Plant Reviews, The Plant Hormone Ethylene*, 44, 189–218.

Bond, W.J., Woodward, F.I. and Midgley, G.F. (2004) The global distribution of ecosystems in a world without fire. *New Phytologist* 165, 525–538.

Bonner, F.T. (1990) Storage of seeds: Potential and limitations for germplasm conservation. *Forest Ecology Management* 35, 35–43.

Borthwick, H.A., Hendricks, S.B., Toole, E.H. and Toole, V.K. (1954) Action of light on lettuce-seed germination. *Botanical Gazetteer* 205–225.

Bouwmeester, H.J. (1990) The Effect of Environmental Conditions on the Seasonal Dormancy Pattern and Germination of Weed Seeds. Ph.D. Thesis. Agricultural University in Wageningen, the Netherlands.

Bouwmeester, H. and Yoneyama, K. (2013) Strigolactones and germination of parasitic plants. In: Joel, D.M., Gressel, J. and Musselman, L. (eds) *Biology of Root Parasitic Orobanchaceae and Control Strategies*. Springer.

Bradford, K.J. (1995) Water relations in seed germination. In: Kigel, J. and Galili, G. (eds) *Seed Development and Germination*. Marcel-Dekker, New York, pp. 351–396.

Bradford, K.J. (1996) Population-based models describing seed dormancy behaviour: Implications for experimental design and interpretation. In: Lang, G.A. (ed.) *Plant Dormancy: Physiology, Biochemistry and Molecular Biology*. CAB International, Wallingford, UK, pp. 313–339.

Bradford, K.J. (2002) Applications of hydrothermal time to quantifying and modeling seed germination and dormancy. *Weed Science* 50, 248–260.

Brenchley, W.E. (1918) Buried weed seeds. *Journal of Agricultural Science* 9, 1–31.

Bullied, W.J., Van Acker, R.C. and Bullock, P.R. (2012) Review: Microsite characteristics influencing weed seedling recruitment and implications for recruitment modeling. *Canadian Journal of Plant Science* 92, 627–650.

Burford, J.R. and Millington, R.J. (1968) Nitrous oxide in the atmosphere of a red-brown earth. *Transactions 9th International Congress of Soil Science, Adelaide* 2, 505–511.

Burford, J.R. and Stefanson, R.C. (1973) Measurement of gaseous losses of nitrogen from soils. *Soil Biology and Biochemistry* 5, 133–141.

Chepil, W.S. (1946) Germination of weed seeds. 1. Longevity, periodicity of germination and vitality of seeds in cultivated soil. *Scientific Agriculture* 26, 307–346.

Conn, J.S., Beattie, K.L. and Blanchard, A. (2006) Seed viability and dormancy of 17 weed species after 19.7 years of burial in Alaska. *Weed Science* 54, 464–470.

Courtney, A.D. (1968) Seed dormancy and field emergence in *Polygonum aviculare. Journal of Applied Ecology* 5, 675–684.

Cresswell, E.G. and Grime, J.P. (1981) Induction of a light requirement during seed development and its ecological consequences. *Nature* 291, 583–585.

Darlington, H.T. and Steinbauer, G.P. (1961) The eighty-year period for Dr. Beal's seed viability experiment. *American Journal of Botany* 48, 321–325.

Debeaujon, I., Lepiniec, L., Pourcel, L. and Routaboul, J.M. (2007) Seed coat development and dormancy. In: Bradford, K.J. and Nonogaki, H. (eds) *Seed Development, Dormancy and Germination, Annual Plant Reviews* 27, 25–49, Blackwell Publishing, Oxford, UK.

De Candolle, A.P. (1832) *Physiologie Végétale* ii, 620. (Cited by Turner, 1933.)

Duke, S.O. (1978) Significance of fluence-response data in phytochrome-initiated seed germination. *Photochemistry and Photobiology* 28, 383–388.

Dzomeku, I.K., Murdoch, A.J. (2007a) Effects of prolonged conditioning on dormancy and germination of *Striga hermonthica. Journal of Agronomy* 6, 29–36.

Dzomeku, I.K., Murdoch, A.J. (2007b) Modelling effects of prolonged conditioning on dormancy and germination of *Striga hermonthica. Journal of Agronomy* 6, 235–249.

Egley, G.H. (1989a) Water-impermeable seed coverings as barriers to germination. In: Taylorson, R.B. (ed.) *Recent Advances in the Development and Germination of Seeds*. Plenum Press, New York, pp. 207–223.

Egley, G.H. (1989b) Some effects of nitrate-treated soil upon the sensitivity of buried redroot pigweed (*Amaranthus retroflexus* L.) seeds to ethylene, temperature, light and carbon dioxide. *Plant, Cell & Environment* 12, 581–588.

Egley, G.H. and Chandler, J.M. (1983) Longevity of weed seeds after 5.5 years in the Stoneville 50-year buried-seed study. *Weed Science* 32, 264–270.

Eliot, J. (1760) *Essays upon field husbandry in New England, as it is or may be ordered. Boston, Mass. USA* (Reprinted in Rasmussen, 1975).

Ellis, R.H. and Roberts, E.H. (1980a) Improved equations for the prediction of seed longevity. *Annals of Botany* 45, 13–30.

Ellis, R.H. and Roberts, E.H. (1980b) The influence of temperature and moisture on seed viability period in barley (*Hordeum distichum* L.). *Annals of Botany* 45, 31–37.

Ellis, R.H. and Roberts, E.H. (1981) The quantification of ageing and survival in orthodox seeds. *Seed Science and Technology* 9, 373–409.

Ellis, R.H., Hong, T.D. and Roberts, E.H. (1983) Procedures for the safe removal of dormancy from rice seed. *Seed Science and Technology* 11, 72–112.

Ellis, R.H., Hong, T.D. and Roberts, E.H. (1985) *Handbook of Seed Technology for Genebanks. Volume II. Compendium of Specific Germination Information and Test Recommendations*. International Board for Plant Genetic Resources, Rome, pp. 211–667.

Ellis, R.H., Hong, T.D. and Roberts, E.H. (1986) Quantal response of seed germination in *Brachiaria humidicola, Echinochloa turnerana, Eragrostis tef* and *Panicum maximum* to photon dose for the low energy reaction and the high irradiance reaction. *Journal of Experimental Botany* 37, 742–753.

Ellis, R.H., Simon, G. and Covell, S. (1987) The influence of temperature on seed germination rate in grain legumes. III. A comparison of five faba bean genotypes using a new screening method. *Journal of Experimental Botany* 38, 1033–1043.

Ellis, R.H., Hong, T.D. and Roberts, E.H. (1989) A comparison of the low-moisture-content limit to the logarithmic relation between seed moisture and longevity in twelve species. *Annals of Botany* 63, 601–611.

Ellis, R.H., Hong, T.D. and Roberts, E.H. (1990) An intermediate category of seed storage behaviour? I. Coffee. *Journal of Experimental Botany* 41, 1167–1174.

Ellis, R.H., Hong, T.D. and Roberts, E.H. (1991a) Effect of storage temperature and moisture on the germination of papaya seeds. *Seed Science Research* 1, 69–72.

Ellis, R.H., Hong, T.D., Roberts, E.H. and Soetisna, U. (1991b) Seed storage behaviour in *Elaeis guineensis*. *Seed Science Research* 1, 99–104.

Enoch, H. and Dasberg, S. (1971) The occurrence of high CO_2 concentrations in soil air. *Geoderma* 6, 17–21.

Esashi, Y. and Leopold, C. (1968) Physical forces in dormancy and germination of *Xanthium* seeds. *Plant Physiology* 43, 871–876.

FAO (1994) *Eritrea - Agricultural Sector Review and Project Identification. Report for the Government of Eritrea*. Food and Agriculture Organization of the United Nations, Rome.

Favier, J.F. and Woods, J.L. (1993) The quantification of dormancy loss in barley (*Hordeum vulgare* L.). *Seed Science & Technology* 21, 653–674.

Fenner, M. (1985) *Seed Ecology*. Chapman and Hall, London.

Fenner, M. (1991) The effects of the parent environment on seed germinability. *Seed Science Research* 1, 75–84.

Fenner, M. and Thompson, K. (2005) *The Ecology of Seeds*. Cambridge University Press, 250 pp.

Finch-Savage, W.E. and Leubner-Metzger, G. (2006) Seed dormancy and the control of germination. *New Phytologist* 171, 501–523.

Finkelstein, R., Reeves, W., Ariizumi, T. and Steber, C. (2008) Molecular aspects of seed dormancy. *Annual Review of Plant Biology* 59, 387–415.

Francisco-Ortega, J., Ellis, R.H., González-Feria, E. and Santos-Guerra, A. (1994) Overcoming seed dormancy in *ex situ* plant germplasm conservation programmes; an example in the endemic *Argyranthemum* (Asteraceae: Anthemideae) species from the Canary Islands. *Biodiversity and Conservation* 3, 341–353.

Frankland, B. and Taylorson, R.B. (1983) Light control of seed germination. In: Shropshire, W. Jr and Mohr, H. (eds) *Encyclopaedia of Plant Physiology* 16, 428-456. Springer-Verlag, Heidelberg.

Garwood, N.C. (1989) Tropical soil seed banks: a review. In: Leck, M.A., Parker, V.T. and Simpson, R.L. (eds) *Ecology of Soil Seed Banks*. Academic Press, San Diego, California, pp. 149-209.

Gooding, M.J., Uppal, R.K., Addisu, M., Harris, K.D., Uauy, C., Simmonds, J.R. and Murdoch, A.J. (2012) Reduced height alleles (*Rht*) and Hagberg falling number of wheat. *Journal of Cereal Science* 55, 305–311.

Graeber, K., Nakabayashi, K., Miatton, E., Leubner-Metzger, G. and Soppe, W.J.J. (2012) Molecular mechanisms of seed dormancy. *Plant, Cell and Environment.* DOI: 10.1111/j.1365-3040.2012.02542.x

Harper, J.L. (1957) The ecological significance of dormancy and its importance in weed control. *International Congress of Plant Protection* 4, 415–420.

Harper, J.L. (1977) *Population Biology of Plants*. Academic Press, London.

Hartmann, K.M., Mollwo, A. and Tebbe, A. (1998) Photo control of germination by moon-and starlight. *Zeitschrift fur Pflanzenkrankheit und PflanzenSchutz* 16, 119–127.

Hewlett, P.S. and Plackett, R.L. (1979) *An Introduction to the Interpretation of Quantal Responses in Biology*. Edward Arnold, London.

Hilhorst, H.W.M. (1990a) Dose-response analysis of factors involved in germination and secondary dormancy of seeds of *Sisymbrium officinale*. I. Phytochrome. *Plant Physiology* 94, 1090–1095.

Hilhorst, H.W.M. (1990b) Dose-response analysis of factors involved in germination and secondary dormancy of seeds of *Sisymbrium officinale*. II. Nitrate. *Plant Physiology* 94, 1096–1102.

Hilhorst, H.W.M. (2007) Definitions and hypotheses of seed dormancy. In: Bradford, K.J. and H. Nonogaki, H. (eds) *Annual Plant Reviews Volume 27: Seed Development, Dormancy and Germination*. Blackwell Publishing Ltd, Oxford, UK, pp. 50–71.

Hilhorst, H.W.M. and Karssen, C.M. (1988) Dual effect of light on the gibberellin- and nitrate-stimulated seed germination of *Sisymbrium officinale* and *Arabidopsis thaliana*. *Plant Physiology* 86, 591–597.

Hilhorst, H.W.M., Derkx, M.P.M. and Karssen, C.E. (1996) An integrating model for seed dormancy cycling: Characterisation of reversible dormancy. In: Lang, G.A. (ed.) *Plant Dormancy: Physiology, Biochemistry and Molecular Biology*. CAB International, Wallingford, UK, pp. 341–360.

Hong, T.D. and Ellis, R.H. (1990) A comparison of maturation drying, germination, and desiccation tolerance between developing seeds of *Acer pseudoplatanus* L. and *Acer platanoides* L. *New Phytologist* 116, 589–596.

Hong, T.D. and Ellis, R.H. (1996) *A Protocol to Determine Seed Storage Behaviour*. Rome: International Plant Genetic Resources Institute, 64 pp.

Hyde, E.O.C. (1954) The function of the hilum in some Papilionaceae in relation to the ripening of the seed and the permeability of the testa. *Annals of Botany* 18, 241–256.

Jones, S.K., Ellis, R.H. and Gosling, P.G. (1997) Loss and induction of conditional dormancy in seeds of Sitka spruce maintained moist at different temperatures. *Seed Science Research* 7, 351–358.

Jones, S.K., Gosling, P.G. and Ellis, R.H. (1998) Reimposition of conditional dormancy during air-dry storage of prechilled Sitka spruce seeds. *Seed Science Research* 8, 113–122.

Karssen, C.M. (1970) The light promoted germination of the seeds of *Chenopodium album* L. III. Effect of the photoperiod during growth and development of the plants on the dormancy of the produced seeds. *Acta Botanica Neerlandica* 19, 81–94.

Karssen, C. M. (1982) Seasonal patterns of dormancy in weed seeds. In: Khan, A.A. (ed.) *The Physiology and Biochemistry of Seed Development, Dormancy and Germination*. Elsevier/North Holland Biomedical Press, Amsterdam, pp. 243-271.

Karssen, C.M., Brinkhorst-Van der Swan, D.L.C., Breekland, A.E. and Koorneef, M. (1983) Induction of dormancy during seed development by endogenous abscisic acid: studies on abscisic acid deficient genotypes of *Arabidopsis thaliana* (L.) Heynh. *Planta* 157, 158.

Kebreab, E. and Murdoch, A.J. (1999a) A quantitative model for loss of primary dormancy and induction of secondary in imbibed seeds of *Orobanche* spp. *Journal of Experimental Botany* 50, 211–219.

Kebreab, E. and Murdoch, A.J. (1999b) Effect of moisture and temperature on the longevity of *Orobanche* seeds. *Weed Research* 39, 199–211.

King, M.W. and Roberts, E.H. (1979) *The Storage of Recalcitrant Seeds: Achievements and Possible Approaches*. International Board for Plant Genetic Resources, Rome.

Kivilaan, A. and Bandurski, R.S. (1981) The one hundred-year period for Dr. Beal's seed viability experiment. *American Journal of Botany* 68, 1290–1292.

Kramer, P.J. (1983) *Water relations of plants*. Academic Press, New York.

Kucera, B., Cohn, M.A. and Leubner-Metzger, G. (2005) Plant hormone interactions during seed dormancy release and germination. *Seed Science Research* 15, 281–307.

Kuo, W.H.J. (1994) Seed germination of *Cyrtococcum patens* under alternating temperature regimes. *Seed Science and Technology* 22, 43–50.

Labouriau, L.G. (1970) On the physiology of seed germination in *Vicia graminea* Sm. *Anais da Academia Brasileira de Ciências* 42, 235–262.

Lewis, J. (1973) Longevity of crop and weed seeds: survival after 20 years in soil. *Weed Research* 13, 179–191.

Linkies, A. and Leubner-Metzger, G. (2012) Beyond gibberellins and abscisic acid: how ethylene and jasmonates control seed germination. *Plant Cell Reports* 31, 253–270.

Livingston, R.B. and Allessio, M.L. (1968) Buried viable seed in successional field and forest stands, Harvard Forest, Massachusetts. *Bulletin Torrey Botanical Club* 95, 58–69.

Madsen, S.B. (1962) Germination of buried and dry stored seeds, III, 1934-60. *Proceedings of the International Seed Testing Association* 27, 920–928.

McEvoy, P.B. (1984) Dormancy and dispersal in dimorphic achenes of tansy ragwort, *Senecio jacobaea* L. (Compositae). *Oecologia* 61, 160–168.

Mott, J.J., Winter, W.H. and McLean, R.W. (1989) Management options for increasing the productivity of tropical savanna pastures. IV. Population biology of introduced *Stylosanthes* spp. *Australian Journal of Agricultural Research* 40, 1227–1240.

Murdoch, A.J. (1983) Environmental control of germination and emergence in *Avena fatua*. *Aspects of Applied Biology* 4, 63–69.

Murdoch, A.J. (1998) Dormancy cycles of weed seeds in soil. *Aspects of Applied Biology* 51, 119–126.

Murdoch, A.J. (2004) Seed dormancy. In: Goodman, R.M. (ed.) *Encyclopedia of Plant & Crop Science*. Marcel Dekker, New York, pp. 1130–1133.

Murdoch, A.J. and Roberts, E.H. (1982) Biological and financial criteria of long-term control strategies for annual weeds. *Proceedings 1982 British Crop Protection Conference - Weeds*, 741–748.

Murdoch, A.J. and Roberts, E.H. (1996). Dormancy cycle of *Avena fatua* seeds in soil. *Proceedings Second International Weed Control Congress Copenhagen, Denmark* 1, 147–152.

Murdoch, A.J., Roberts, E.H. and Goedert, C.O. (1989) A model for germination responses to alternating temperatures. *Annals of Botany* 63, 97–111.

Murdoch, A.J., Sonko, L. and Kebreab, E. (2000) Population responses to temperature for loss and induction of seed dormancy and consequences for predictive empirical modelling.

In: Viémont, J.-D., Crabbé, J. (eds) *Dormancy in Plants: From Whole Plant Behaviour to Cellular Control. Proceedings 2nd International Symposium on Plant Dormancy, Angers, France.* CAB International, Wallingford, UK, pp. 57–68.

Nambara, E., Okamoto, M., Tatematsu, K., Yano, R., Seo, M. and Kamiya, Y. (2010) Abscisic acid and the control of seed dormancy and germination. *Seed Science Research* 20, 55–67.

Nelson, D.C., Riseborough, J.A., Flematti, G.R., Stevens, J., Ghisalberti, E.L., Dixon, K.W. and Smith, S.M. (2009) Karrikins discovered in smoke trigger *Arabidopsis* seed germination by a mechanism requiring gibberellic acid synthesis and light. *Plant Physiology* 149, 863–873.

Odum, S. (1965) Germination of ancient seeds. Floristical observations and experiments with archaeologically dated soil samples. *Dansk Botanisk Arkiv* 24, 1–70.

Odum, S. (1974) Seeds in ruderal soils, their longevity and contribution to the flora of disturbed ground in Denmark. *Proceedings of the 12th British Weed Control Conference*, 1131–1144.

Partridge, T.R. (1989) Soil seed banks of secondary vegetation on the Port Hills and Banks Peninsula, Canterbury, New Zealand, and their role in succession. *New Zealand Journal of Botany* 27, 421–436.

Peters, N.C.B. (1982a) Production and dormancy of wild oat (*Avena fatua*) seed from plants grown under soil waterstress. *Annals of Applied Biology* 100, 189–196.

Peters, N.C.B. (1982b) The dormancy of wild oat seed (*Avena fatua* L.) from plants grown under various temperature and soil moisture conditions. *Weed Research* 22, 205–212.

Pickett, S.T.A. and McDonnell, M.J. (1989) Seed bank dynamics in temperate deciduous forest. In: Leck, M.A., Parker, V.T. and Simpson, R.L. (eds) *Ecology of Soil Seed Banks.* Academic Press, San Diego, California, pp. 123–147.

Pinfield, N.J., Stutchbury, P.A. and Bazaid, S.M. (1987) Seed dormancy in *Acer*: is there a common mechanism for all *Acer* species and what part is played in it by abscisic acid? *Physiologia Plantarum* 71, 365–371.

Pinfield, N.J., Stutchbury, P.A., Bazaid, S.A. and Gwarazimba, V.E.E. (1990) Abscisic acid and the regulation of embryo dormancy in the genus *Acer. Tree Physiology* 6, 79–85.

Piper, J.K. (1986) Germination and growth of bird-dispersed plants: effects of seed size and light on seedling vigor and biomass allocation. *American Journal of Botany* 73, 959–965.

Priestley, D.A. (1986) *Seed Aging: Implications for Seed Storage and Persistence in the Soil.* Cornell University Press, Ithaca.

Priestley, D.A. and Posthumus, M.A. (1982) Extreme longevity of lotus seeds from Pulantien. *Nature (Lond.)* 299, 148–149.

Pritchard, H.W. (1991) Water potential and embryonic axis viability in recalcitrant seeds of *Quercus rubra. Annals of Botany* 67, 43–49.

Probert, R.J. and Longley, P.L. (1989) Recalcitrant seed storage physiology in three aquatic grasses (*Zizania palustris, Spartina anglica* and *Porteresia coarctata*). *Annals of Botany* 63, 53–63.

Probert, R.J., Smith, R.D. and Birch, P. (1985) Germination responses to light and alternating temperatures in European populations of *Dactylis glomerata* L. IV. The effects of storage. *New Phytologist* 101, 521–529.

Probert, R.J., Gajjar, K.H. and Haslam, I.K. (1987) The interactive effects of phytochrome, nitrate and thiourea on the germination response to alternating temperatures in seeds of *Ranunculus sceleratus* L. A quantal approach. *Journal of Experimental Botany* 38, 1012–1025.

Puga-Hermida, M.I., Gallardo, M., del Carmen Rodríguez-Gacio, M. and Matilla, A.J. (2003) The heterogeneity of turnip-tops (*Brassica rapa*) seeds inside the silique affects germination, the activity of the final step of the ethylene pathway, and abscisic acid and polyamine content. *Functional Plant Biology* 30, 767–775.

Quail, P.H. and Carter, O.G. (1969) Dormancy in seeds of *Avena ludoviciana* and *A. fatua. Australian Journal of Agricultural Research* 20, 1–11.

Quinlivan, B.J. (1971) Seed coat impermeability in legumes. *Journal of the Australian Institute of Agricultural Science* 37, 283–295.

Rajjou, L., Duval, M., Gallardo, K., Catusse, J., Bally, J., Job, C. and Job, D. (2012) Seed germination and vigor. *Annual Review of Plant Biology* 63, 507–533.

Rampton, H.H. and Ching, T.M. (1970) Persistence of crop seeds in soil. *Agronomy Journal* 62, 272–277.

Rasmussen, W.D. (1975) The tilling of land, 1760. In: Rasmussen, W.D. (ed.) *Agriculture in the United States: A Documentary History* 1: 188–203. Random House, Inc., New York.

Renata, B. and Agnieszka, G. (2006) Nitric oxide and HCN reduce deep dormancy of apple seeds. *Acta Physiologiae Plantarum* 28, 281–287.

Renault, P. and Stengel, P. (1994) Modeling oxygen diffusion in aggregated soils. I. Anaerobiosis inside the aggregates. *Soil Science Society of America Journal* 58, 1017–1023.

Rice, K.J. (1989) Impacts of seed banks on grassland community structure and population dynamics. In: Leck, M.A., Parker, V.T. and Simpson, R.L. (eds) *Ecology of Soil Seed Banks*. Academic Press, San Diego, California, pp. 211–230.

Roberts, E.H. (1961) Dormancy in rice seed. I. The distribution of dormancy periods. *Journal of Experimental Botany* 12, 319–329.

Roberts, E.H. (1962) Dormancy in rice seed. III. The influence of temperature, moisture, and gaseous environment. *Journal of Experimental Botany* 13, 75–94.

Roberts, E.H. (1963) An investigation of inter-varietal differences in dormancy and viability of rice seed. *Annals of Botany* 27, 365–369.

Roberts, E.H. (1965) Dormancy in rice seed. IV. Varietal responses to storage and germination temperatures. *Journal of Experimental Botany* 16, 341–349.

Roberts, E.H. (1972) Dormancy: a factor affecting seed survival in the soil. In: Roberts, E.H. (ed.) *Viability of Seeds*. Chapman and Hall, London, pp. 321–359.

Roberts, E.H. (1973) Predicting the storage life of seeds. *Seed Science and Technology* 1, 499–514.

Roberts, E.H. (1988) Temperature and seed germination. In: Long, S.P. and Woodward, F.E. (eds) *Plants and Temperature, Symposia of the Society of Experimental Biology* 42, 109–132. Company of Biologists Ltd., Cambridge, UK.

Roberts, E.H. and Ellis, R.H. (1982) Physiological, ultrastructural and metabolic aspects of seed viability. In: Khan, A.A. (ed.) *The Physiology and Biochemistry of Seed Development, Dormancy and Germination*. Elsevier/North Holland Biomedical Press, Amsterdam, pp. 465–485.

Roberts, E.H. and Ellis, R.H. (1989) Water and seed survival. *Annals of Botany* 63, 39–52.

Roberts, E.H. and Smith, R.D. (1977) Dormancy and the pentose phosphate pathway. In: Khan, A.A. (ed.) *The Physiology and Biochemistry of Seed Dormancy and Germination*. Elsevier/North Holland Biomedical Press, pp. 385–411.

Roberts, E.H., Murdoch, A.J. and Ellis, R.H. (1987) The interaction of environmental factors on seed dormancy. *Proceedings 1987 British Crop Protection Conference - Weeds* pp. 687–694.

Roberts, H.A. (1964) Emergence and longevity in cultivated soils of seeds of some annual weeds. *Weed Research* 4, 296–307.

Roberts, H.A. (1979) Periodicity of seedling emergence and seed survival in some Umbelliferae. *Journal of Applied Ecology* 16, 195–201.

Roberts, H.A. (1986) Seed persistence in soil and seasonal emergence in plant species from different habitats. *Journal of Applied Ecology* 23, 639–656.

Roberts, H.A. and Boddrell, J.E. (1983a) Seed survival and periodicity of seedling emergence in ten species of annual weeds. *Annals of Applied Biology* 102, 523–532.

Roberts, H.A. and Boddrell, J.E. (1983b) Seed survival and periodicity of seedling emergence in eight species of Cruciferae. *Annals of Applied Biology* 103, 301–304.

Roberts, H.A. and Boddrell, J.E. (1984a) Seed survival and seasonal emergence of seedlings of some ruderal plants. *Journal of Applied Ecology* 21, 617–628.

Roberts, H.A. and Boddrell, J.E. (1984b) Seed survival and periodicity of seedling emergence in four weedy species of *Papaver*. *Weed Research* 24, 195–200.

Roberts, H.A. and Chancellor, R.J. (1979) Periodicity of emergence and achene survival in some species of *Carduus, Cirsium* and *Onopordium*. *Journal of Applied Ecology* 16, 641–647.

Roberts, H.A. and Neilson, J.E. (1980) Seed survival and periodicity of seedling emergence in some species of *Atriplex, Chenopodium, Polygonum* and *Rumex*. *Annals of Applied Biology* 94, 111–120.

Roberts, H.A. and Neilson, J.E. (1981) Seed survival and periodicity of seedling emergence in twelve weedy species of Compositae. *Annals of Applied Biology* 97, 325–334.

Roberts, H.A. and Potter, M.E. (1980) Emergence patterns of weed seedlings in relation to cultivation and rainfall. *Weed Research* 20, 377–386.

Roberts, H.A. and Ricketts, M.E. (1979) Quantitative relationships between the weed flora after cultivation and the seed population in the soil. *Weed Research* 19, 269–275.

Roche, S., Koch, J.M. and Dixon, K.W. (2008) Smoke enhanced seed germination for mine rehabilitation in the southwest of Western Australia. *Restoration Ecology* 5, 191–203.

Royal Botanic Gardens Kew (2008) Seed information database. Version 7.1. Available from: http://data. kew.org/sid/ (accessed February 2012).

Russi, L., Cocks, P.S. and Roberts, E.H. (1992a) Hard-seededness and seed bank dynamics of six pasture legumes. *Seed Science Research* 2, 231–241.

Russi, L., Cocks, P.S. and Roberts, E.H. (1992b) Coat thickness and hard-seededness in some *Medicago* and *Trifolium* species. *Seed Science Research* 2, 243–249.

Ruyter-Spira, C., Al-Babili, S., van der Krol, S. and Bouwmeester, H. (2012) The biology of strigolactones. *Trends in Plant Science.* DOI: 10.1016/j.tplants.2012.10.003.

Sauer, J. and Struick, G. (1964) A possible ecological relation between soil disturbance, light-flash and seed germination. *Ecology* 45, 884–886.

Saunders, D.A. (1981) Investigations into Seed Coat Impermeability in Annual *Medicago* Species. Ph.D. Thesis, University of Reading.

Schafer, D.E. and Chilcote, D.O. (1970) Factors influencing persistence and depletion of buried seed populations. II. The effects of soil temperature and moisture. *Crop Science* 10, 342–345.

Schonbeck, M.W. and Egley, G.H. (1980) Redroot pigweed (*Amaranthus retroflexus*) seed germination responses to afterripening, temperature, ethylene, and some other environmental factors. *Weed Science* 28, 543–548.

Sharif-Zadeh, F. (1999) Mechanism and modelling of seed dormancy and germination in *Cenchrus ciliaris* as affected by different conditions of maturation and storage. PhD Thesis, The University of Reading, UK.

Sharif-Zadeh, F. and Murdoch, A.J. (2000) The effects of different maturation conditions on seed dormancy and germination of *Cenchrus ciliaris*. *Seed Science Research* 10, 447–457.

Sharif-Zadeh, F. and Murdoch, A.J. (2001) The effects of temperature and moisture on after-ripening *Cenchrus ciliaris* seeds. *Journal of Arid Environments* 49, 823–831.

Shen-Miller, J., Schopf, J.W. and Berger, R. (1983) Germination of a ca. 700 year-old lotus seed from China: evidence of exceptional longevity of seed viability. *American Journal of Botany* 70, 78.

Simpson, G.M. (2007) *Seed Dormancy in Grasses*. Cambridge University Press, 308 pp.

Standifer, L.C., Wilson, P.W. and Drummond, A. (1989) The effects of seed moisture content on hard-seededness and germination in four cultivars of okra (*Abelmoschus esculentus* (L.) Moench). *Plant Varieties and Seeds* 2, 149–154.

Symons, S.J., Naylor, J.M., Simpson, G.M. and Adkins, S.W. (1986) Secondary dormancy in *Avena fatua*: induction and characteristics in genetically pure dormant lines. *Physiologia Plantarum* 68, 27–33.

Taylor, G.B. (1981) Effect of constant temperature treatments followed by fluctuating temperatures on the softening of hard seeds of *Trifolium subterraneum* L. *Australian Journal of Plant Physiology* 35, 201–210.

Taylorson, R.B. (1970) Changes in dormancy and viability of weed seeds in soils. *Weed Science* 18, 265–269.

Taylorson, R.B. (1972) Phytochrome controlled changes in dormancy and germination of buried weed seeds. *Weed Science* 20, 417–422.

Thompson, K. and Grime, J.P. (1979). Seasonal variation in the seed banks of herbaceous species in ten contrasting habitats. *Journal of Ecology* 67, 893–921.

Toh, S., Kamiya, Y., Kawakami, N., Nambara, E., McCourt, P. and Tsuchiya, Y. (2012) Thermoinhibition uncovers a role for strigolactones in *Arabidopsis* seed germination. *Plant and Cell Physiology* 53, 107–117.

Tokumasu, S., Kata, M. and Yano, F. (1975) The dormancy of seed as affected by different humidities during storage in *Brassica*. *Japanese Journal of Breeding* 25, 197–202.

Toole, E.H. and Brown, E. (1946) Final results of the Duvel buried seed experiment. *Journal of Agricultural Research* 72, 201–210.

Totterdell, S. and Roberts, E.H. (1979) Effects of low temperatures on the loss of innate dormancy and the development of induced dormancy in seeds of *Rumex obtusifolius* L. and *Rumex crispus* L. *Plant, Cell and Environment* 2, 131–137.

Totterdell, S. and Roberts, E.H. (1981) Ontogenetic variation in response to temperature change in the control of seed dormancy of *Rumex obtusifolius* L. and *Rumex crispus* L. *Plant, Cell and Environment* 4, 75–80.

Turner, J.H. (1933) The viability of seeds. *Bulletin of Miscellaneous Information, Royal Botanic Gardens, Kew* (34), 257–269.

Van Mourik, T.A., Stomph, T.-J., Murdoch, A.J. (2005) Why high seed densities within buried mesh bags may overestimate depletion rates of soil seed banks. *Journal of Applied Ecology* 42, 299–305.

Vleeshouwers, L.M., Bouwmeester, H.J. and Karssen, C.M. (1995) Redefining seed dormancy: an attempt to integrate physiology and ecology. *Journal of Ecology* 83, 1031–1037.

Vincent, E.M. and Roberts, E.H. (1977) The interaction of light, nitrate and alternating temperature in promoting the germination of dormant seeds of common weed species. *Seed Science and Technology* 5, 659–670.

Werker, E. (1980/81) Seed dormancy as explained by the anatomy of embryo envelopes. *Israel Journal of Botany* 29, 22–44.

Wesson, G. and Wareing, P.F. (1967) Light requirements of buried seeds. *Nature* 213, 600–601.

Wesson, G. and Wareing, P.F. (1969a) The role of light in the germination of naturally occurring populations of buried weed seeds. *Journal of Experimental Botany* 20, 402–413.

Wesson, G. and Wareing, P.F. (1969b) The induction of light sensitivity in weed seeds by burial. *Journal of Experimental Botany* 20, 414–425.

Williams, J.T. and Harper, J.L. (1965) Seed polymorphism and germination. I. The influence of nitrates and low temperatures on the germination of *Chenopodium album*. *Weed Research* 5, 141–150.

Whitmore, T.C. (1988) The influence of tree population dynamics on forest species composition. In: Davy, A.J., Hutchings, M.J. and Watkinson, A.R. (eds) *Plant Population Ecology*. Blackwell Scientific Publications, Oxford, pp. 271–291.

Zorner, P.S., Zimdahl, R.L. and Schweizer, E.E. (1984) Sources of viable seed loss in buried dormant and non-dormant populations of wild oat (*Avena fatua* L.) seed in Colorado. *Weed Research* 24, 143–150.

8 The Chemical Ecology of Seed Persistence in Soil Seed Banks

Robert S. Gallagher,[1]* **Mark B. Burnham**[2] **and E. Patrick Fuerst**[3]

[1]*Independent Agricultural Consultant, Clinton, South Carolina, USA;*
[2]*Department of Biology, West Virginia University, Morgantown, West Virginia, USA;* [3]*Department of Crop and Soil Science, Washington State University, Pullman, Washington, USA*

Introduction

The endogenous chemical regulation of seed persistence in the soil seed bank is often assumed to occur by many seed and plant ecologists, but the subject receives only sparse direct coverage in the scientific literature. Priestly (1986) reviewed the experimental evidence to that date for seed persistence in the soil, but gives little mention to potential chemical regulators of that persistence. Baskin and Baskin (1998) provide a very comprehensive review of seed dormancy and germination, but their discussion of chemical regulators of seed dormancy is largely limited to plant hormones as germination inhibitors. Shirley (1998) reviewed the role of flavonoids in seeds, outlining their potential role in preventing pathogen infection, reducing lipid peroxidation (i.e. antioxidant activity) and promoting seed dormancy, but focused primarily on seeds of agronomic crops. Based on the insight from Shirley (1998), Gallagher and Fuerst (2006) provided preliminary evidence that seed chemistry could regulate seed persistence by at least three possible mechanisms: (i) preventing pathogen infection; (ii) inhibiting seed germination; and (iii) mitigation of oxidative stressors. More recently, John and Sarada (2011) and Mitrovic *et al.* (2012) reviewed the roles of phenolics in allelopathic interactions, but focused on the effects of root exudates or exogenously applied phenolic compounds on seed fate, rather than the endogenous chemistry of seeds.

In this chapter, our goal is to build on the above-mentioned reviews to provide an up to date assessment of the evidence for the chemical regulation of seed persistence in soil seed banks. In addition, we will discuss the areas where we believe more research is merited, as well as potential pitfalls associated with those research areas. Although the role of secondary metabolites (defined below) in seed persistence will be our primary focus, we will also include a discussion on the potential role of other chemicals, such as plant hormones and organic acids, as well as biochemical defence enzymes. As in Gallagher and Fuerst (2006), our review will include the discussion of chemical defence strategies against pathogens, but also include chemical defence against seed predators and pathogens, as well as a discussion of chemical germination inhibitors and antioxidants associated with maintaining seed vigour over time.

* E-mail: rsgallagher61@gmail.com

Secondary Metabolites in Seeds

The biochemistry and molecular biology of secondary metabolites has been reviewed by Croteau *et al.* (2000), which will be the source of the information in this section unless otherwise stated. Secondary metabolites are a diverse assortment of organic compounds that do not have an apparent direct role in plant growth and metabolism. They are distinguished from primary metabolites (e.g. amino acids, organic acid, phytosterols, etc.), but it is emphasized that this division is not always that clear or well defined. Key attributes to secondary metabolites are their potential role in protecting plants from herbivory and frugivory, preventing microbial infections and serving as pollination and dispersal attractants, among other roles. The three major groups of secondary metabolites are phenylpropanoids, alkaloids and terpenoids. There is an abundance of information in the literature on phenylpropanoids and alkaloids in seeds, but very limited information on terpenoids.

There are a wide variety of phenylpropanoids in plants, which are typically derived from the phenylpropanoid and phenolpropanoid acetate pathways using phenylalanine and tyrosine as precursors. Phenylpropanoids have a base phenol structure and are often configured as acids or alcohols. They may occur as simple phenolic acids and alcohols, ester-linked dimers or oligomers of phenolic acids, and the more chemically complex flavonoids, lignans and lignins. Within the flavonoid group, there are also flavonols, anthocyanins (red, blue or purple pigments) and proanthocyanidins (brown pigments also known as tannins) (Shirley, 1998). Lignans are defined as compounds where two phenylpropanoid (C_6C_3) units are linked by oxidative coupling (Begum *et al.*, 2010) and have a wide range of reported biological activities. Lignins are the most complex of the phenylpropanoids and serve primarily a structural role in plants. Hendry *et al.* (1994) introduced the term of 'orthodihydroxy phenol' into the seed ecology literature, which was also later adopted by Davis *et al.* (2008). Using standard organic chemistry nomenclature, 'orthodihydroxy phenol'

suggests the structure of pyrogallol. However, these authors are probably referring to a group of phenolic compounds that have two adjacent hydroxyl groups on the benzene ring. Such phenolics are supposedly selectively reactive with the reagents of the assay outlined in those studies, and may be the primary substrates for the polyphenol oxidase (PPO) defence systems discussed later in this chapter. The term 'polyphenols' is common in the literature, but ambiguously defined. Depending on the study, the term 'polyphenols' may refer to simple phenolic acids and alcohols, or flavonoids, or both these phenylpropanoid groups. It is not uncommon for the term to be used when data on total phenolic or flavonoid content have been determined by spectrophotometric assays, such as those reviewed by Granger *et al.* (2011).

Alkaloids are defined as nitrogen-containing secondary metabolites that are often pharmacologically active, but are not necessarily limited to plants. As indicative by this definition, the implications of alkaloids on the biology of animals have received considerably more attention than their role in plant defence. Terpenoids are comprised of an isopentane skeleton with one or more branched 5-carbon units. Terpenoids have a well demonstrated role in whole plant defence, although limited information is available on their role in seed defence.

Phenylpropanoids

Phenolic acids and alcohols

Phenolic acids and alcohols (phenolics hereafter) exist in three quantifiable forms in seeds: (i) soluble free, (ii) soluble esters and (iii) cellular bound (Krygier *et al.*, 1982). The soluble components are typically extracted initially in a polar solvent mixture (aqueous + organic). The soluble esters can be separated from the soluble free components by an acid precipitation and subsequent hydrolysis of the precipitate. The cellular-bound phenolic components are also liberated by hydrolysis with strong sodium hydroxide solution. Once all of the phenolic fractions are liberated, they can be readily identified and quantified by gas chromatography mass spectrometry (GCMS) (Gallagher *et al.*, 2010a,b; Granger *et al.*, 2011) or liquid phase techniques such as HPLC or LCMS.

It is important to scrutinize the methodologies used to extract, identify and quantify phenolics when interpreting seed phenolic studies. As outlined in Krygier *et al.* (1982), solvent mixtures have greater extraction efficiency than single solvents. However, it is common in many studies for single solvents (often methanol) to be used as the extraction media, possibly resulting in an incomplete extraction of the phenolic constituents. Secondly, total phenolic concentration can only be accurately determined when all the phenolic fractions (i.e. free simple, esters and cellular bound) have been liberated. More typically, however, the soluble free phenolic fraction is all that is quantified. As such, conclusions about the potential ecological role of phenolics may be drawn based on incomplete data. Research with wild oat (*Avena fatua* L.) helps to illustrate these points (Granger, 2012). Using the extraction protocol outlined by Krygier *et al.* (1982) and GCMS protocols outlined by Gallagher *et al.* (2010a), the free soluble phenolic fraction comprised less than 1% of total phenolics in both caryopsis and hull (palea + lemma) (Table 8.1). Liberated soluble esters comprised from 6 to 10% of total phenolics in the wild oat caryopsis, but 1% or less in the wild oat hull. The cellular-bound phenolic fraction comprised

from 89 to 99% of total phenolics as measured by GCMS. Non-hydrolysed soluble esters can be detected by liquid phase chromatography techniques (i.e. HPLC, LCMS), but need to be hydrolysed and derivatized to be detected by gas phase techniques such as GCMS due to the limited volatility of these esters. Other studies have used various spectrophotometric protocols to quantify phenolic concentrations. As previously mentioned, Hendry *et al.* (1994) and Davis *et al.* (2008) used a colorimetric method that was supposedly specific to phenolics with two adjacent hydroxyl groups to correlate the concentration of these phenolics with seed persistence. Granger *et al.* (2011), however, found that this and two other spectrophotometric methods greatly overestimated phenolic concentrations relative to those determined by the more rigorous GCMS method. Although the spectrophotometric methods were relatively simple and inexpensive to use, Granger *et al.* (2011) concluded that the methods were not suitable to make quantitative conclusions about phenolic concentrations, and do not provide any information with respect to the composition of phenolics in a sample.

Composition of phenolics in seeds

There are limited data in the literature on the composition of phenolics in non-food

Table 8.1. The concentration of phenolic acids and alcohols in two lines of wild oat seeds. The data were adapted from Granger (2012) from growth chamber-grown wild oat plants. The plants were grown under well watered conditions and ambient carbon dioxide concentrations.

Phenolic fraction	M73		SH430	
	Caryopsis	Hull	Caryopsis	Hull
	ng per seed			
Soluble free	40	19	117	322
Soluble esters	1,267	640	2,394	1,013
Cellular bound	19,220	122,570	20,205	151,168
Total	20,527	123,229	22,716	152,503
	% of total phenolic concentration			
Soluble free	<1	<1	<1	<1
Soluble esters	6	<1	10	1
Cellular bound	93	99	89	99

crop seeds. As such, we will rely on the non-food crop literature where possible, but include data from the food crops when we believe they will provide useful insights. Since seed phenolic composition data are limited, the goal here is to give the reader a preliminary assessment of phenolic compounds they are likely to encounter in seeds, rather than a complete inventory of seed phenolics.

In the seeds of three common weed species, wild oat, common lambsquarters (*Chenopodium album* L.) and broadleaf plantain (*Plantago major* L.), Granger *et al.* (2011) used GCMS to identify and quantify the phenolics (Table 8.2). Overall, 14 distinct phenolics were found, but they were not necessarily always present in each of the three species or the chemical fractions. Ferulic and vanillic acids were among the most abundant phenolics in the soluble chemical fractions (free and ester forms), whereas ferulic, *p*-coumaric and caffeic acids (lambsquarters and broadleaf plantain only) were the most abundant in the bound chemical fractions. A similar spectrum of phenolics was found in related studies with wild oat (Gallagher *et al.*, 2010a,b; Burnham, 2011; Granger, 2012), which is also similar to the spectrum of phenolics reported in cereal grain crops (Weidner *et al.*, 1999, 2000, 2001, 2002). This suite of phenolics is also similar to that reported in marama bean (*Tylosema esculentium*), although the relative phenolic compound abundances were different, with gallic and sinapic acid among the most abundant compounds in the extract fractions (Chingwaru *et al.*, 2011).

In common pea (*Pisium sativa*), Troszyñska and Ciska (2002) provide a

Table 8.2. The relative phenolic composition of the soluble and cellular-bound chemical fractions in three plant species. The data are adapted from Granger *et al.* (2011). The phenolics were extracted and quantified according to methods outlined in Gallagher *et al.* (2010a). The soluble fraction includes the soluble free phenolics and the liberated soluble esterified phenolics.

Compound	Wild oat caryopsis		Wild oat hull		Common lambsquarters		Broadleaf plantain	
	Soluble	Bound	Soluble	Bound	Soluble	Bound	Soluble	Bound
				% of total phenolic concentration				
Alcohols								
Catechol	–	<1	–	–	–	<1	–	<1
Hydroquinone	–	<1	<1	–	<1	<1	<1	<1
Benzoic acid derivatives								
Benzoic (OH)	13	<1	1	–	4	1	3	<1
Vanillic	20	4	31	<1	67	9	20	7
Protocatechuic	–	<1	–	–	–	28	1	<1
Syringic	13	2	6	–	<1	<1	2	1
Gallic	–	–	–	–	<1	–	0	<1
Cinnamic acid derivatives								
Cinnamic	1	<1	–	–	<1	2	1	–
Phenyl proprionic	7	1	1	–	5	<1	5	1
p-Coumaric	3	6	3	16	1	8	12	3
Hydrocaffeic	–	–	–	–	–	–	–	–
Ferulic	42	86	56	75	20	15	29	65
Caffeic	–	–	1	9	–	36	15	22
Sinapic	–	<1	1	–	2	–	12	–
			Total phenolic concentration (μg g tissue^{-1})					
	51	321	251	5148	9	434	348	424

comprehensive evaluation of the phenolic composition of the seed coats of two pea varieties with differing seed coat colour. In this study, protocatechuic, gentisic and vanillic acids were the predominant phenolic acids in the soluble free and esterified chemical fractions of the pea seed coats, although syringic, caffeic, ferulic and coumaric (*o* and *p*) were also detected (Fig. 8.1). In the cellular-bound fraction, only protocatechuic was detected. The pea genotype with coloured seed coats had substantially

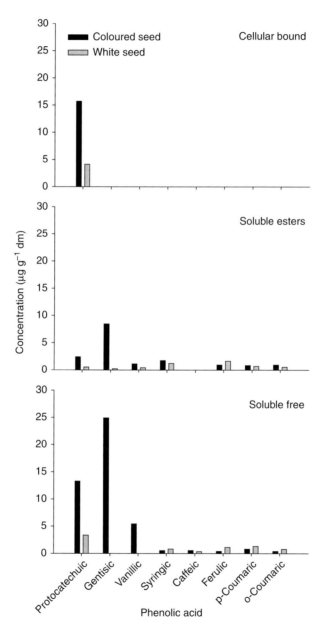

Fig. 8.1. The distribution of phenolic acids among the cellular bound, soluble ester and soluble free chemical forms in common pea seeds (*Pisium sativa*) with white and coloured seed coats. Data adapted from Troszyñska and Ciska (2002).

higher concentrations of protocatechuic, gentisic and vanillic acids than the white seed coat genotype, and gentisic and vanillic acids were absent from the soluble free chemical fraction in the white seed coat genotype. These data indicate that phenotypic colour differences in seeds may be related to the phenolic composition of the seed coat, but these authors did not speculate on the ecological implications of differential seed coat colour or seed phenolic content. In other legume species, chlorogenic, caffeic, *p*-coumaric, OH-benzoic, ferulic and vanillic acids were the predominant phenolics in groundnut (*Arachis villosulicarpa*) (Reddy *et al.*, 1977); a range of benzoic acid (gallic, protocatechuic and OH-benzoic acids) and cinnamic acid (caffeic and *p*-coumaric) derivatives were identified in three lupin species (Siger, 2010); and six distinct phenolic-sugar conjugates (i.e. glycosides) were identified in sicklepod (*Cassia tora*) (Hatano, 1999). In a survey of the data from plants from a miscellaneous assortment of families, coumarin and *o*-coumaric acid were identified in pine (*Pinus densiflora*) (Hatano, 1967); a unique phenolic ester (Fico *et al.*, 2001) and a number of quinone derivatives (Venkatachallam *et al.*, 2010) were identified in two *Nigella* spp.; ester conjugates of caffeic, *p*-coumaric and ferulic acids were identified in chicory (Jurgonski, 2011); a vanillic acid ester was identified in a Kancolla quinoa variety (Dini *et al.*, 2004); and gallic acid was identified in evening primrose (*Oenothera biennis* L.) seeds (Wettasinghe *et al.*, 2002). Recently, three unique phenolic glycosides, named itolide A, itolide B and itoside P, were identified in *Itoa orientalis* seeds (Tang *et al.*, 2012). In most of the above-mentioned studies, alcohol-based extraction protocols were used, and therefore do not provide any information about the cellular-bound phenolic fractions.

Although we have emphasized the importance of employing rigorous phenolic extraction, identification and quantification techniques, we recognize that the abundance of a particular phenolic compound may not necessarily be related to the biological/ecological importance of that compound. As will be emphasized in the following sections, pinpointing the specific role a phenolic compound plays in seed longevity is very difficult. In addition, there are certain analytical challenges associated with analysing samples with such a diverse composition of chemical constituents. For example, in the authors' work with wild oat, it was not uncommon for seed extract samples to contain over 20 compounds of interest (Gallagher *et al.*, 2010a). It is difficult to optimize the GCMS run parameters to detect and quantify compounds that ranged widely in their concentration. For example, a particular sample preparation and GC injection volume could result in near saturation of the MS detector for the most abundant compounds, such as ferulic acid, but barely permit the detection of the low abundance compounds. Although the use of selective ion monitoring or SIM (described in Gallagher *et al.*, 2010a) does permit the detection and quantification of low abundance compounds, it would be ideal to adjust the GC sample preparation and instrument methods to increase the abundance of the low concentration compounds to improve quantification accuracy. It is also worth noting that phenolic compounds with two adjacent hydroxyl groups (catechol, protocatechuic, caffeic, hydrocaffeic and gallic acids) that would be targeted by the spectrophotometric methods outlined by Hendry *et al.* (1994) and re-evaluated by Granger *et al.* (2011) represent 1% or less of total soluble phenolics in wild oat and common lambsquarters, and 16% of total soluble phenolics in broadleaf plantain (Table 8.2). This is another indication of the potential limitations of spectrophotometric extraction and quantification methods compared to GCMS, LCMS and HPLC techniques that provide information about individual compounds. Finally, it does appear that the liquid phase chromatography techniques (i.e. HPLC, LCMS) permit the detection of low volatility phenolics, phenolic esters (non-hydrolysed) and flavonoids without the need for derivatization. As such, LCMS may be a superior analytical technique compared to GCMS if one has the instrumentation available.

Composition of flavonoids and lignans in seeds

In the review of flavonoids in seeds by Shirley (1998), it is reported that predominant flavonoids were pelargonidin, cyanidin, delphinidin and aglycones in the seed coat of common bean (*Phaseolus vulgaris*); various isoflavones in the seed coat of soybean; and apigeninidin, miscellaneous flavonols and proanthocyanidins in sorghum. In other crop species, the predominant flavonoids were eapigenin in lupin species (Siger, 2010); catechin, quercetin, luteolin and apigonin in common peas (Duenas *et al.*, 2004); kaempferol and quercetin derivatives in Kancolla (Dini *et al.*, 2004); six types of flavonoids in mustard seeds (Matthaus, 1998); and kaempferol, rhamnoside and sophoroside derivatives in Ethiopian mustard seed phenotypes (Li *et al.*, 2010). In non-crop species, catechin and related derivatives were reported in mocan seeds (*Visnea mocanera* L.); catechin and epicatechin in evening primrose (Wettasinghe *et al.*, 2002); catechin, rutin, naringen, hesperidin, myricetin, quercetin, nariningenin and kaempferol in marama bean (Chingwaru *et al.*, 2011); delphinidin, cyanidin, quercetin, myricetin, catechin and epicatechin in velvetleaf (*Abutilon theiphrasti*) (Paszkowski and Kremer, 1988), and butin, vicenin and orobol in the Podalyrieæ and Lipidarieæ tribes within the legume family (De Nysschen *et al.*, 1998).

Due to the complexity of the lignan chemistry, it is more difficult to provide a meaningful summary of the predominant lignans in seeds. Likewise, research on lignan characterization appears to be more recent and less well developed than the previously discussed classes of phenylpropanoids. However, derivatives of sesemin and pinoresinol appear to be the most commonly reported lignans in the literature. Sesamin derivatives have been found in seeds of sesame (Kim *et al.*, 2006; Kumazaki *et al.*, 2009), five leaved chaste tree (*Vitex negundo*) (Zheng *et al.*, 2011), and magnolia (*Magnolia praecocissima*) (Takahashi *et al.*, 2002). Pinoresinol has been found in seeds of flax (*Linum usitatissimum*) (Herchi *et al.*, 2011) and five leaved chaste tree (Zheng *et al.*, 2011).

Alkaloids in Seeds

The alkaloid content has been characterized in the seeds of a number of plant species, but with an emphasis on the biomedical implications rather than the role of these chemicals in seed ecology of these species. Perhaps the most well known of the seed alkaloids are found in opium poppy (*Papaver somniferum*) where morphine and codeine are the predominant alkaloids, but papaverine and noscapine are also common chemical constituents (Sproll *et al.*, 2006). In the Leguminosae, the predominant seed alkaloids were lupanine, multiflorine and albine in lupin (*Lupinus* spp.) species tested (Altares *et al.*, 2005); heliotridine, axillaradine, madurensine and anacrotine in the *Crotalaria* spp. tested (Asres *et al.*, 2004); and four distinct tetrahydroisoquinoline alkaloids in velvet bean (*Mucuna pruriens*) (Misra and Wagner, 2004). Seed alkaloids in the *Erythrina* spp., also a legume, have been well characterized with 16 structurally related alkaloids initially being identified (Chawla *et al.*,1985) and later four unique alkaloids identified in seeds of *Erythrina velutina* (Ozawa *et al.*, 2011). Alkaloids found in this genus are now commonly referred to as the 'erythrina class' of alkaloids. In Jimson weed (*Datura stramonium*) and other *Datura* spp. as many as 14 structurally related alkaloids have been identified (Doncheva *et al.*, 2004). These alkaloids have a common tropane base chemical structure and therefore are commonly referred to as the 'tropane class' of alkaloids. Other plant species where the seed alkaloids have been characterized, often with traditional medicine implications, include *Colchicum cilicicum* (Sutlupinar *et al.*, 1988); *Delphinium ajacis* (Desai *et al.*, 1994); *Aspidosperma macrocarpon* (Mitaine *et al.*, 1996); *Centaurea moschata* (Sarker *et al.*, 1998); *Colchicum autumnale* (Poutaraud and Girardin, 2005); *Nelumbo nucifera* (Rai *et al.*, 2006); and *Camptotheca acuminate* (Zhang *et al.*, 2007).

However, there are few apparent commonalities of alkaloid constituents among these species or among the species where the alkaloid composition has been more elaborately outlined previously in this section, making any generalization of seed alkaloid composition difficult to make. It does appear, however, that methodologies for the extraction, purification and identification of alkaloids are well developed, particularly in the more recent publication cited here. Similar to the extraction and purification issues outlined previously in this chapter for phenolics, a complete profile of alkaloids in a sample is only achieved when multiple solvent steps are employed, as outlined in Chawla *et al.* (1985). It may very well be that the alkaloid composition of seeds is more complex than reported in studies where less rigorous extraction and purification methods were employed.

Antimicrobial Potential of Seed Chemicals

Demonstrating that endogenous phenylpropanoid constituents in seeds prevent pathogen attack in those seeds is difficult to achieve by direct measures, since there are a wide range of chemicals and chemical forms typically present. However, a few studies do exist where whole seed–pathogen interactions have been elucidated. In tame oats (*Avena sativa*) for example, Picman *et al.* (1984) showed that the hulls were more rapidly colonized by fungi after the hulls were extracted with acetone or methanol, which would remove the soluble free and esterified phenolic acids and alcohols, as well as many aliphatic organic and fatty acids (Gallagher *et al.*, 2010a). In Picman *et al.* (1984), the organic extracts showed strong inhibition of the *Bipolaris* and *Drechslera* sp. tested; moderate inhibition of *Alternaria* and *Aspergillus* sp. tested, and weak inhibition of the *Fusarium* sp. tested. These authors concluded that the oat hull probably provided both physical and chemical protection from pathogens. De Luna *et al.* (2011) demonstrated that over 800 fungal

isolates colonized field-buried wild oat seed, although the isolates varied greatly with respect to their pathogenicity. Of those isolates tested, 15% caused some level of seed decay in wild oat. Using select fungal pathogens from this study, seeds from a drought stressed and shaded maternal environment were more prone to fungal infection than the seeds grown under resource-rich conditions (A.C. Kennedy, 2010, personal communication). As outlined later in this chapter, wild oat seeds from the drought and shade environments also tended to have lower concentrations of phenolic acids (Gallagher *et al.*, 2010b). Similarly, barley mutants with testa lacking the expression of certain flavonoids were found to be substantially more susceptible to *Fusarium* sp. attack than the non-mutant barley genotypes (Skadhauge *et al.*, 1997). In velvet leaf (*Abutilon theophrasti*), seeds placed on agar cultures of a wide range of soil and seed-borne microbial species caused measureable inhibition zones in 58% of the 202 bacterial isolates tested (both Gram-positive and Gram-negative organisms) and all of the 34 fungal isolates tested (Kremer, 1986). Microbial growth inhibition was more pronounced in hard versus imbibed seeds, suggesting an interaction with other solutes leaking from the imbibed seeds. Although the specific phenolic constituents were not determined in this study, the general presence of phenolics in the inhibition zones of the culture was confirmed. In a follow-up study by Paszkowski and Kremer (1988), aqueous extracts of velvet leaf seed coats revealed six distinct flavonoids (discussed earlier in this chapter), as well as other unidentified phenolic constituents. Extraction concentrations of 50 mg seed coat material per ml water inhibited growth of *Aspergillus niger, Penicillium diversum* and *Fusarium* sp. cultures, but not *Gliocladium roseum* and *Trichoderma viride* cultures. Since *G. roseum* and *T. viride* were considered to be beneficial to the seed (i.e. presumed antibiotic activity), the authors of this study suggest that the antimicrobial activity of the phenolic constituents in the velvet leaf extractions may have some specificity against seed pathogens.

In this same study, experiments with 1 mM stock solutions of the individual flavonoids also showed inhibition of the microbial bioassay species, but this response depended on the compound being tested and the microbial bioassay. These authors did not evaluate mixtures of the stock flavonoid solutions to test for potential interactions among the flavonoids. Finally, Pons *et al.* (2008) demonstrated through *in vivo* and *in vitro* methods that the phenolic constituents in maize caryopsis may inhibit the production and accumulation of mycotoxins from *Fusarium* sp. rather than actual mycelial growth.

Most other studies demonstrating the potential antimicrobial activity of phenylpropanoids have been done solely by testing organic and aqueous seed extracts on test organisms, which may or may not be known seed pathogens. For example, a number of isolated phenolic constituents from extracts from *Itoa orientalis* showed antifungal activity at concentrations as low as 60 μM in *Sclerotium rolfsii* and 45 μM in *Rhizoctonia solani* cultures (Tang *et al.*, 2012), both known soil-borne pathogens. Most other studies, however, have had a health or food science emphasis and have been conducted with common clinical and food-borne organisms, but with mixed results. For example, extracts from grape seeds (Chedea and Braicu, 2011), curly dock (*Rumex crispus*) (Yildirim *et al.*, 2001), sea buckthorn (*Hippophaë rhamnoides*) (Michel *et al.*, 2012) and marama bean (*Tylosema esculentium*) (Chingwaru *et al.*, 2011) often showed antibacterial activity on such organisms as *Esherichia coli, Pseudomonas aeruginosa* and *Staphylococcus aureas*, among others. In curly dock, ether-based seed extracts appeared to show more inhibitory action on Gram-positive versus Gram-negative organisms (Yildirim *et al.*, 2001). In contrast, extracts from sicklepod (*Cassia tora*) seeds did not show any antibacterial activity with the clinical organisms tested (Hatano, 1999). Interpretation of seed extract studies such as these presented here is complicated by methodological issues that may or may not be under the control of the investigators. First, the extract concentrations tested are often considerably higher than the concentrations likely to be found in the spermosphere. Second, studies rarely identify and quantify all the chemical constituents of an extract. Thus, presumed effects of the target compound(s) may be due to other chemical constituents or interaction among the chemical constituents. Even with these shortcomings, these types of studies can be a valuable first step in understanding seed chemical–microorganism interactions.

Seed–Spermosphere Interactions

An important region of interaction between seeds and their surroundings is the spermosphere. Analogous to the rhizosphere around plant roots, the spermosphere is a zone around a seed that is chemically and biologically distinct from the bulk soil due to the seed's influence. Although dormant in the soil, seeds may exude a variety of compounds, which passively diffuse 5 to 10 mm into the surrounding soil (Nelson, 2004). Due to this chemically unique microenvironment, the spermosphere can be a region of elevated microbial activity and altered nutrient availability, and thus impact seed longevity in the soil seed bank.

The spermosphere may be composed of the simple phenolics and flavonoids discussed previously, as well as sugars, amino acids and fatty acids, among other compounds. The long-chain fatty acids, linoleic and palmitic acids, as well as the dicarboxylic aliphatic organic acid, azelaic acid, have been identified in the extracts of cotton seeds (Ruttledge and Nelson, 1997) and wild oat (Gallagher *et al.*, 2010a), suggesting their presence in the spermosphere of these seed species. In addition, Kovacs (1971) identified a number of phenolics and other organic acids in exudates of pea, cotton and barley seeds, including *p*-hydroxybenzoic, syringic and *p*-coumaric acids. Klejdus and Kubane (2000) also identified a number of phenolics in exudates of two species of *Festuca* seeds.

While examining the extracts of seeds may give us an idea of the potential composition of the spermosphere, soaking seeds in an aqueous solution is not necessarily a realistic model of the actual soil environment. An alternative method is to place seeds in soil or a soil-like environment, allowing them to imbibe and equilibrate with the surrounding media. Changes in the seed chemical composition can then be quantified, thus indirectly inferring spermosphere composition (Burnham, 2011). In dormant wild oat, total free (soluble) phenolic concentration in imbibed seeds was 203 ng per seed lower than unimbibed control seeds. In particular, reductions in ferulic and *p*-coumaric acids were responsible for this trend. In addition, total dicarboxylic aliphatic organic acids and fatty acid concentration was also greatly reduced in imbibed seeds compared to control seeds, largely due to the loss of succinic, malic and palmitic acids after imbibition.

The presence of seed exudates in the spermosphere has several potential implications concerning seed longevity. First, chemical changes within the seed could reduce its dormancy or viability. While the exact mechanisms of dormancy are not clearly understood and may vary among species (Baskin and Baskin, 1998), the loss of potential dormancy- or defence-related compounds (Gallagher *et al.*, 2010a) could reduce seed longevity due to earlier germination or attack by pathogenic or saprophytic microbes. In addition, seed exudation could hasten the process of seed ageing through loss of antioxidant compounds, including some phenylpropanoids, reducing the viability of the dormant seed population (Pukacka, 1991).

Second, some exuded compounds stimulate microbial growth. *Pythium ultimum*, a key oomycete pathogen that causes damping-off in numerous plant species, exhibits increased spore germination in the 2 mm surrounding bean seeds in soil and in the presence of seed extracts (Stanghellini and Hancock, 1971). More recently, Ruttledge and Nelson (1997) identified unsaturated fatty acids as important elicitors of *P. ultimum* germination, with linoleic, oleic and palmitic acids being particularly effective. In addition to *Pythium* spp., *Fusarium* spp. chlamydospores germinate to a greater extent in the presence of seed exudates (Short and

Lacy, 1974). *Rhizoctonia* spp. show increased virulence and growth around many plant roots (Weinhold *et al.*, 1972; Hietala, 1997), suggesting a similar spermosphere effect.

Microbial interactions in the spermosphere also move beyond simple stimulation of pathogenic organisms. Microbe–microbe interactions in the spermosphere of different species' seeds can also dictate the response of pathogens to seed exudates. The metabolism of fatty acids by *Enterobacter cloacae* in cucumber, cotton and wheat spermospheres reduces the germination of *P. ultimum* spores (Van Dijk and Nelson, 2000), while the higher sugar concentration in the corn spermosphere negates this effect by causing a shift in *E. cloacae* metabolism away from fatty acids (Windstam and Nelson, 2008a,b). While spermosphere interactions have received very little research, it is likely that this type of interaction is common and affects numerous plant seeds as they lie dormant in the soil.

Polyphenol Oxidase and Other Enzyme-based Defences

In proposing a 'seed defence theory', Dalling *et al.* (2011) state that seeds have four mechanisms of resistance to decay: '(i) physical barriers that render seeds impermeable to pathogens; (ii) endogenous chemical defenses of seeds; (iii) chemical defenses of beneficial seed–microbial associations; and (iv) rapid seed germination.' For seeds with physiological dormancy (such as wild oat), Dalling *et al.* (2011) predict that 'microbial and chemical defenses, if present, will be arrayed on the exterior of the seed' and there are indeed numerous reports of secondary chemical defences (e.g. phenolics, tannins) associated with seed coats as well as bacteria and fungi associated with seed surfaces (Chee-Sanford *et al.*, 2006; Chee-Sanford, 2008; Dalling *et al.*, 2011). We hypothesize that, as a fifth mechanism, biochemical defences are also arrayed on the exterior of the seed. Consistent with this hypothesis, and as described in detail below, we have shown that the defence enzyme polyphenol oxidase (PPO) is: (i) induced by

Fusarium seed-decay strains; (ii) released from the caryopsis surface following *Fusarium* challenge; and (iii) processed and activated, possibly by a protease, in the dormant wild oat line, M73. We also have preliminary data indicating that three other defence-response proteins are induced by *Fusarium* including peroxidase, oxalate oxidase and chitinase (Anderson *et al.*, 2010; Fuerst *et al.*, 2011).

Defence enzymes induced by pathogens in plants include PPO, peroxidase, oxalate oxidase and chitinase and there are multiple tissue-specific isoforms of each of these enzymes (van Loon *et al.*, 2006; Hücklehoven, 2007). Defence enzymes present near the seed surface are especially relevant to this review. PPOs oxidize phenolic compounds to the reactive semiquinone, which initiates polymerization reactions. The polymeric products of PPO may inhibit pathogens and predators by creating lignin-like physical barriers and possibly toxic products (Yoruk and Marshall, 2003). Class III peroxidases are secreted plant proteins with a remarkable number of functions including cross-linking cell wall polymers and lignification (Passardi *et al.*, 2005; Cosio and Dunand, 2009). Their role in defence is due to strengthening cell walls and massive production of reactive oxygen species. Oxalate oxidases ('germins') have two defence activities: breaking down phytotoxic oxalic acid released by fungi and production of fungicidal levels of H_2O_2 (Lane, 2002). Chitinases hydrolyse polymers containing N-acetylglucosamine such as chitin found in fungal cell walls, and are often associated with antifungal activity (Grover, 2012). Limited studies have been conducted to identify the endogenous substrates of PPO. Of the phenolics reported in Table 8.2, catechol, gallic acid, caffeic acid, and sinapic acid were considered to be substrates for wheat caryopsis PPO, but ferulic and vanillic acids were considered poor PPO substrates (Fuerst *et al.*, 2010). Other endogenous PPO substrates include chlorogenic acid, proanthocyanidins (including catechin), and tyrosine, including both free and in polypeptides. Of the phenolics considered to be substrates for wheat caryopsis PPO in this study, catechol and gallic,

caffeic and chloregenic acids would fit the previously mentioned criteria of having two adjacent hydroxyl groups on the benzene ring that would make them a so-called 'orthodihydroxy' phenolic.

The previously mentioned wild oat seed-decay fungi isolated by de Luna *et al.* (2011) are the foundation for our recent research. Since PPOs are commonly associated with plant defence (Gatehouse, 2002; Mayer, 2006) we hypothesized that pathogen induction of PPO activity is a biochemical defence mechanism in dormant wild oat seeds that may contribute to seed longevity. Consistent with this hypothesis, caryopsis PPO activity was induced by three *Fusarium* strains; however, PPO activity was inhibited by a *Pythium* strain, which may represent a mechanism of pathogenicity (Fuerst *et al.*, 2011). *Fusarium avenaceum* strain 223a (*F.a.*1) caused the most rapid decay and also caused the greatest PPO induction. We also showed that when whole wild oat seeds were incubated on *F.a.*1, PPO activity was induced in seeds, hulls (lemma and palea) and caryopses. *F.a.*1 induction of PPO in isolated caryopses exposed to *F.a.*1 was greater than in intact seeds, and occurred more rapidly as well; therefore our subsequent work focused on the *F.a.*1-wild oat caryopsis 'model system'. PPO activity of *F.a.*1-treated caryopses was readily washed off, whereas very little PPO activity could be leached from untreated caryopses. This led to a series of studies on 'caryopsis leachates', focusing on that part of the PPO that was activated by *F.a.*1.

Another interesting property of PPOs, widely reported among eukaryotic organisms, is that the mature form of the enzyme has very low activity, and the enzyme is often activated by proteolytic cleavage (van Gelder *et al.*, 1997). We hypothesized that *F.a.*1-induced activation of PPO in wild oat caryopses involved proteolytic cleavage that released active PPOs into their surrounding environment (leachate). We were able to test this hypothesis utilizing protein fractionation, immunoblots (westerns) and peptide sequencing (Anderson *et al.*, 2010). All caryopsis extracts and leachates contained an inactive ~57,000 molecular weight (MW) PPO, whereas *F.a.*1-treated leachates had an increased abundance of lower MW PPOs having greater enzymatic activity, consistent with the hypothesis that this enzyme was proteolytically activated. Peptide sequences confirmed that the inactive ~57,000 MW and active 36,000 MW wild oat proteins were homologous to known PPO sequences. Purely by chance, we also identified two other plant defence enzymes, chitinase and oxalate oxidase, by homologous peptide sequences, strongly suggesting that multiple defence enzymes were induced by *F.a.*1 in wild oat (Anderson *et al.*, 2010).

Peroxidase is another very widely studied plant defence enzyme. Preliminary data indicated that peroxidase was induced about sixfold while PPO was induced about fourfold in leachates from caryopses exposed to *F.a.*1 (E.P. Fuerst, 2013, unpublished data). Mycelia leachates also contained both PPO and peroxidase activities at lower levels. To validate the distinction between PPO and peroxidase in caryopsis leachates, we observed almost complete inhibition of PPO activity by the inhibitor tropolone and almost complete inhibition of peroxidase by catalase (which eliminates H_2O_2 co-substrate), but did not observe any inhibition of PPO by catalase or peroxidase by tropolone (E.P. Fuerst, 2013, unpublished data).

The above results strongly suggest that dormant wild oat seeds possess biochemical defences against pathogens and, more specifically, that proteolysis in the presence of fungal pathogens releases an activated form of PPO, and perhaps other defence enzymes, from the surface of wild oat caryopses and probably hulls as well. This certainly opens the door to validation of these observations among seeds of multiple plant species, and responses to multiple plant pathogens. Since these studies were all conducted in Petri dish assays, further validation of these responses is required in soil.

Chemical Defences Against Seed Predators

There are only a limited number of studies that provide some insight into possible endogenous chemical deterrents to seed predation. Under predispersal conditions, Kestring *et al.* (2009) found that fruits of *Mimosa bimucronata* infested with bruchid beetles developed seeds with higher total phenolics than the seeds from non-infested fruit. This finding suggested that there was induced *de novo* synthesis of seed phenolics in response to the bruchid attack similar to that which Burnham (2011) observed with wild oat response to *Fusarium* infection. In the infested seeds, there was a negative correlation between the total phenolic concentration of the seed and the leg length (i.e. proxy for body size) of the bruchid infesting that seed, indicating that the phenolics in the seed may negatively impact bruchid development. In *M. bimucronata*, it was concluded that the physical characteristics of the seed (i.e. size and shape) as well as the concentration of secondary metabolites determine the likelihood that a seed will be predated.

In a survey of grass and forb species of the Monte Desert in Argentina, seeds of the grass species tended to have lower total phenolic concentrations than the forb species (2.2% versus 3.0% on average, respectively) (Manuel-Rios *et al.*, 2012). Likewise, alkaloids were not detected in any of the seeds of the grasses, but commonly present in the seeds of forbs. There was not a clear relationship, however, between the concentration of secondary metabolites in seed and the feed preference of the three bird species tested (*Zonotrichia capensis, Saltatricula multicolor* and *Diuca diuca*). However, using the specific phenolic constituents found in the seeds, it was found that some phenolic compounds did deter feeding in the gramnivorous specialist bird species (*S. multicolor, D. diuca*), but not the generalist feeder (*Z. capensis*). The authors concluded that the secondary metabolite content in seed may play a role in determining seed choice in specialist seed feeders, but the overall nutritional quality of the seed is the primary factor determining seed preference in birds.

There is also some evidence of animal adaptation to seed toxins. For example, seeds of mamane (*Sophora chrysophylla*) in Hawaii are apparently protected from predation by a suite of alkaloids that can comprise up to 4% of the embryo dry weight (Banko *et al.*, 2002). Of these alkaloids, the quinolizidines are especially toxic to vertebrates, particularly when they have bioaccumulated in plant-feeding larvae. The endemic bird species, Palia (*Loxioides bailleui*), and the moth larvae (*Cydia* spp.), however, have adapted to these alkaloids, apparently through some means of metabolizing the toxins, and are among the few known species that use mamane seeds as a regular food source. Palia birds, however, will routinely discard the seed coats of mamane, which have relatively high concentrations of phenolic compounds. In contrast, Palia eat the *Cydia* larvae, which have high concentrations of both phenolics and alkaloids. The authors hypothesized that the high nutritional reward associated with the *Cydia* larvae compared to the mamane seed coat may explain this behaviour.

With rodents, Xiao *et al.* (2006) found that average time to harvest was longer in seeds of subtropical tree species with high versus low flavonoid content, but all seed types were eventually consumed. The authors concluded that seed chemistry may influence feed preferences in rodents when food sources are diverse and abundant, but not necessarily when food sources are limited. Schmidt (2000) proposed that rate of depletion of a seed patch may be a function of the type of secondary metabolites in seeds (i.e. digestibility reducers versus toxins) and the risk for the predator associated with the seed foraging. Using sunflower seeds impregnated with tannic acid to represent a digestibility reducer, and oxalic acid to represent a toxin, and fox squirrel (*Sciurus niger*) as a test species, it was demonstrated that seeds with low concentrations of digestibility inhibitors were likely to be depleted more rapidly than seeds with high concentrations of these chemicals in situations where risk to the predator was high (i.e. open habitats, long foraging times, etc.). In contrast, there was no predation rate relationship between the toxin content of seeds and predation risk. Schmidt held that digestibility reducers influence predator fitness indirectly by reducing the energy gained from consumed seed, whereas toxins reduce predator fitness directly through their biochemical action.

Chemical Regulators of Seed Dormancy

As with other chemically regulated seed quality traits, information on seed dormancy tends to come from correlative and indirect assays. For example, endogenous concentrations of the chemical inhibitor in question can be quantified within a seed population and correlated to the degree of primary dormancy among populations. Alternatives or supplements to this approach include: (i) removing the inhibitors from seeds by extraction with water or organic solvents and evaluating seed dormancy status; (ii) testing the germination inhibition capacity of seed extracts on seeds of the same species or common bioassay species such as lettuce; and (iii) testing the germination inhibition capacity of compounds identified to be in a seed on other bioassay species using dilutions of analytical grade compound or compound mixtures.

Secondary metabolites

Within the phenylpropanoid group of secondary metabolites, the potential role of flavonoids and simple phenolic acids in the regulation of seed dormancy has received the most attention. In three *Lespedeza* species, germination rate was found to be negatively correlated to the flavonoid (catechin and epicatechin) concentration in the seed (Buta and Lusby, 1986). It is not clear from this study, however, whether the inherent flavonoid concentrations and germination rates were a function of seed genetics or the environment of the mother plant. In wild soybean (*Glycine soja*), Zhou *et al.* (2010) largely corrected for this discrepancy. In this study, a gradient of hardseededness was achieved by using two soybean ecotypes grown under a range of maturation environments. Here hardseededness was positively correlated to the epicatechin concentration in the seed coat. It was also found that exogenously applied epicatechin promoted the germination of physically scarified seed at low concentrations (1 to 10 mmol l⁻¹), but inhibited germination at high concentrations (50 mmol l⁻¹ or higher). The authors of this study attributed inhibitory effects of the higher concentrations of epicatechin to the restriction of oxygen to the embryo, thus mimicking (or possibly supplementing) the function of hardseededness, but did not speculate on the hormonal type effect on seed germination of the lower epicatechin concentrations. It does seem possible, however, that as hardseededness is alleviated and epicatechin is leached from the seed, low concentrations of epicatechin serving as a germination cue could have adaptive advantages. However, the hypotheses for the functional roles of the low and high epicatechin hinge on verifying that the epicatechin concentrations per seed are consistent with concentrations used in the germination tests.

In houndstongue (*Cynoglossum officinale*), removal of the seed coat resulted in complete alleviation of primary dormancy (Qi *et al.*, 1993). Although phenolics were found to be abundant in the seed coat, the extraction of these compounds with methanol did not change the dormancy of the seed. Removal of the seed coat did, however, result in a substantial increase in the oxygen uptake of the seed. These authors concluded that the phenolics in the seed coat were not the primary regulators of dormancy in this species, but rather the restriction of oxygen uptake by the seed coat. Phenolics in the seed coat could, however, play an indirect role in regulating seed dormancy by limiting gas exchange to the embryo. For example, Briggs *et al.* (2005) determined in three *Grevillea* spp. that cellular-bound phenolics were largely located exotestal and as components of the mesotestal, and speculated that these phenolics, upon hydration-induced oxidation, would create a barrier to gas exchange. In partial contrast to the results reported by Qi *et al.* (1993), Kato *et al.* (2003) identified dihydroactinidiolide in wheat hulls, which

was shown to have strong germination inhibition of wheat caryopsis. These authors speculated that the compound was leached from the hull to the caryopsis during rainfall events, thereby preventing predispersal sprouting of the wheat seeds.

Studies testing the inhibitory effects of seed extracts on the same species, or one or more bioassay species, have produced mixed conclusions. In dormant sugar maple (*Acer saccharum*) seeds, *p*-coumaric acid concentrations dropped by nearly tenfold with seed stratification over 50 days, which also corresponded to alleviation of dormancy and induction of seed germination (Enu-Kwesi and Dumbroff, 1980). In this study, exogenous applications of *p*-coumaric and ferulic acids at concentrations of 15 to 150 µg ml⁻¹ inhibited the germination of a bioassay cress (*Lepidium sativum*) seed, although the germination inhibition at the lower concentrations tended to be short-lived. In Guayule (*Parthenium argentatum*) seed, removing the chaff from the seed alleviated primary dormancy (Naqvi and Hanson, 1982). Water extracts of the chaff were found to contain common phenolic acids, such as OH-benzoic (52%), protocatechuic (22%) and *p*-coumaric (13%), among others. Exogenous application of the extract inhibited the germination of non-dormant Guayule seeds, but had no effect on the germination of the lettuce and tomato seed bioassays. In bitter lupin (*Lupinus angustifolius*) seed, ethyl acetate extracts a wide range of phenolic acids (Stobiecki *et al.*, 1993). Exogenous additions of these extracts tended to delay germination of the lettuce bioassay seed, but did not completely inhibit germination. Exogenous additions of many of the individual phenolic compounds found in the lupin seed also tended to delay germination of lettuce, but only at 1 mM concentration or higher. These results are similar to those reported by Williams and Hoagland (1982) where exogenous applications of a wide range of phenolic acids and acid mixtures tended to delay the germination of the crop and weed seeds tested, but rarely inhibited germination.

In wild oat, Chen *et al.* (1982) demonstrated that aqueous extracts of the hulls of dormant seeds only had a weak inhibitory effect in the lettuce seed germination bioassay. As reported earlier in this chapter, the wild oat hull contained a wide variety of phenolic and short-chained fatty acids, but these chemicals were not strong regulators of germination in this species (Chen *et al.*, 1982). Similarly, sugar beet (*Beta vulgaris*) contained a number of common phenolic acids (e.g. ferulic, vanillic, benzoic, *p*-coumaric), but these compounds did not inhibit germination of this species even at the highest concentrations tested (Morris *et al.*, 1984).

Short-chained organic and fatty acids

Very limited information on the potential role of short-chained organic and fatty acids in regulating seed dormancy is available in the literature. Common short-chained organic acids found in seeds include, but are not limited to, malic, succinic, fumaric and azelaic, whereas common fatty acids include oleic, linoleic and palmitic (Gallagher *et al.*, 2010a). In wild oat, preliminary evidence suggested that short-chained fatty acids ($<C_{11}$) may play a role as germination inhibitors regulating dormancy (Berrie *et al.*, 1975), but this was not confirmed in subsequent studies with this same species (Metzger and Sebesta, 1982). In chick pea (*Cicer arietinum*), exogenous applications of short-chained fatty acids (C_5–C_{10}) at 1 mM or higher inhibited the germination of chick pea, but lower concentration had no effect on germination, *per se* (Gomez-Jimenez *et al.*, 1996). Low concentrations, however, promoted ethylene biosynthesis, a key step in the germination process, which was inhibited by 1 mM solutions. Similar to the conclusions reported for Zhou *et al.* (2010) for phenolic acids earlier in this section, Gomez-Jimenez *et al.* (1996) speculated that short-chained fatty acids play a role in retaining dormancy at high concentrations in chick pea by restricting membrane permeability. Although low concentrations of short-chained fatty acids did promote ethylene biosynthesis in the seed, and

therefore are a possible germination induction mechanism, there was no evidence that this increase in ethylene biosynthesis stimulated embryo growth in these experiments.

There are a number of methodological challenges involved with pinpointing the potential role of seed chemistry in the regulation of dormancy. Correlation-type studies are often limited by the number of biotypes or species that are being evaluated. In a common garden situation, fewer biotypes and/or species can be evaluated since maturation effects can be largely controlled. However, when common garden methods are not being employed, it is imperative that numerous populations of the species being studied are sampled so that environmental variability can be factored out. In studies where the chemical inhibitors are being extracted from the seed and subsequently tested for dormancy, there is the possibility that the extraction process itself may affect other physiological processes (e.g. enzyme activity/induction, GA biosynthesis, cuticle removal) above and beyond the removal of the inhibitory compounds from the seed. Likewise, chemicals other than those being targeted may also be unknowingly removed by the extraction process. In our experience with wild oat, we cannot be certain that the relatively chemically aggressive extraction procedure designed primarily for phenolic acids (see Gallagher *et al.*, 2010a) does not alter other non-target chemicals in the seeds. Likewise, the GC/MS separation-identification-quantification methodologies will work only for compounds that will readily volatilize (with or without derivatization). Larger molecular weight and/or low volatility constituents may require other analytical techniques, such as HPLC or LCMS. To gain a complete survey of the chemical constituents in seeds, multiple protocols are likely to be necessary. Finally, one has the challenge when applying known seed chemical constituents exogenously to bioassay seeds or calibrating an appropriate concentration and composition of those constituents to reflect the true chemical status of the seed or the seed spermosphere.

Abscisic acid

Abscisic acid (ABA) is a phytohormone that is thought to regulate many metabolic functions in plants. A thorough review of the role of ABA in the regulation of seed dormancy is provided by Kermode (2005), which will be the source of the information in this section unless otherwise stated. During seed development, ABA suppresses precocious germination until seed development is complete. ABA levels tend to be quite low in the beginning stages of seed development, peak at the time of mid-maturation, and drop off considerably in the late stages of seed development. Environmental factors, such as temperature, water availability and light quality and quantity, can influence the ABA content and sensitivity in mature seed. Based on evidence from ABA-deficient mutants in *Arabidopsis*, ABA may play a role in regulating the thickness of the mucilage layer surrounding the testa, thereby affecting oxygen and water uptake by the testa. In mature seed, ABA appears to also play a role in the inception of primary physiological dormancy. However, there is no clear relationship between the ABA content in seed and the degree of dormancy, *per se*. Rather, dormant seed will continue ABA biosynthesis during imbibition, whereas non-dormant seed will not. Since ABA is generally considered to act as an antagonist to gibberellins (GAs), ABA/GA ratios may be a regulating factor controlling the imposition or alleviation of physiological dormancy. There is evidence that other physical and chemical factors, such as light, ethylene and nitrate, may also interact with ABA and GA to regulate seed dormancy (Finch-Savage and Clay, 1994; Atia *et al.*, 2009; Subbiah and Reddy, 2010; Dong *et al.*, 2012). The role of ABA in desiccation tolerance in orthodox seed is also considered an important survival mechanism in long-lived seeds (Finkelstein *et al.*, 2008; Khandelwal *et al.*, 2010).

Chemical Maintenance of Seed Viability and Vigour

A life history characteristic of many plant species is for their seeds to persist in the soil seed bank for numerous years. Although physiological and/or physical dormancy mechanisms may prevent seeds from germinating, seeds must also remain viable and retain vigour to ensure establishment and reproduction. Seed vigour differs from seed viability in that vigour is measured as germination or emergence under (or following) stress, and is considered a better measure of potential seedling establishment in the field. Seed vigour is at its highest level at physiological maturity, and seed ageing gradually leads to the decline in seed vigour, and ultimately leads to loss of seed viability (reviewed by Priestly, 1986; Smith and Berjak, 1995; Walters, 1998; McDonald, 1999). A decline in seed vigour is typically reflected by decreased germination rates and seedling biomass accumulation, and an increase in seed and seedling susceptibility to pathogens. Reactive oxygen species (ROS) and ionizing radiation cause damage to cellular organelles, directly affecting metabolic mechanisms, whereas damage to cell membranes can result in leakage of solutes that may lead to pathogen infection. Antioxidants are chemicals that react with ROS thereby mitigating the oxidative damage they cause (reviewed in Kranner and Birtic, 2005). Seeds typically contain a wide range of chemical constituents that have antioxidant capacity. Glutathione and ascorbic acid are two of the primary water-soluble antioxidants; however, a wide range of phenylpropanoids, sugars and polyamides also have antioxidant properties.

Glutathione is thought to be particularly important in protecting seeds from oxidative stressors while they are in the dry state – a phenomenon called desiccation tolerance (Kranner and Birtic, 2005). It has been shown that glutathione concentrations will decline as seeds age if seed desiccation is maintained, but glutathione concentrations are restored with imbibition events (Bailly *et al.*, 1998, 2001; Long *et al.*, 2011). Long *et al.* (2011) proposed that seasonal wetting and drying cycles contribute to seed persistence in the soil seed bank by restoring the antioxidant mechanisms in seeds. In addition to glutathione, the raffinose series oligosaccharides have also been proposed to play an important role in desiccation tolerance in orthodox seeds (reviewed by Obendorf, 1997). For example in common bean (*Phaseolus vulgaris*), acquisition of desiccation tolerance was shown to coincide with the accumulation of raffinose and stachyose, in addition to an increase in antioxidant enzyme activity (Bailly *et al.*, 2001).

Tocopherols such as vitamin E and carotenoids such as lutein have strong antioxidant capacity and have also been implicated in maintaining seed vigour. In *Arabidopsis*, seeds from mutants deficient in vitamin E had considerably shorter longevity than the wild type seeds (Sattler *et al.*, 2004). In contrast, Garcia *et al.* (2012) demonstrated that the vitamin E concentrations in *Xyris* spp. seeds declined with burial over a two month period, but there was no discernible relationship between vitamin E concentration and seed viability. In wheat seeds, lutein concentration decreased over 36 years in dry storage and was correlated with decreased seed viability.

Considerable research has been conducted on the phenylpropanoid antioxidants in seeds, but more from nutritional and biomedical perspectives rather than a seed persistence perspective. As outlined previously in this chapter, seeds can contain a wide range of phenylpropanoids, and there is considerable documentation in the literature that seed phenylpropanoids have strong antioxidant capacity. Very little information, however, is available on the potential role of phenylpropanoids in maintaining seed vigour. Preliminary research with wild oat demonstrated that seeds from plants grown under light limitation or drought had lower concentrations of phenolic acids than seeds from the resource-rich control plants (Gallagher *et al.*, 2010b), but there was no apparent evidence of differences in seed vigour among the seed lots (Gallagher *et al.*, 2013).

There are a number of methodological considerations when evaluating changes in seed vigour over time (reviewed by Gallagher and Fuerst, 2006). Real-time field studies are rarely practical due to the long duration many seeds persist in the seed bank. Likewise, it is difficult to control environmental conditions in the field, complicating comparisons among studies. Accelerated ageing protocols under controlled laboratory conditions may have some merit. Long *et al.* (2008) proposed high humidity (60%) and high temperature (45°C) conditions as means to rapidly age seeds, showing a positive correlation between ageing dynamics and the corresponding seed persistence index among weed species. In some species, such as *Amaranthus* spp. and *Abutilon theophrastii*, temperatures above 40°C can induce secondary dormancy (R.S. Gallagher, personal experience). For wild oat, an ageing regime of 72% RH achieved with a saturated $NaNO_3$ solution (Dhingra and Sinclair, 1995) and a temperature of 35°C resulted in approximately a 15% seed water content and a gradual decline in seed vigour indicators over a 9 week period (Granger, 2012). Preliminary experiments indicated that an RH of 80% or higher generally resulted in mould growth, and temperatures lower than 35°C did not appear to appreciably age the seed, whereas temperatures 40°C or higher caused a very rapid loss of seed viability (R.S. Gallagher, 2008, unpublished data). Likewise, enriching the oxygen concentration in the ageing vessels to near 100% as outlined by Hannan and Hill (1991) did not appear to affect the ageing dynamics in our experiments as predicted. Regardless of the accelerated ageing regime being employed, we cannot be certain that the cellular degradation is similar to that which occurs naturally under field conditions.

Implications of Maternal Plant Stress on Chemical Defence in Seeds

The Resource Availability Hypothesis holds that resource-limited conditions will favour plant species with high levels of chemical defence and slow growth rates (Coley *et al.*, 1985). This hypothesis makes intuitive sense when comparing among species, but it is unclear how it applies to individuals within populations grown in physical or temporal resource gradients. It is well documented that plants within a population can have a high degree of plasticity with respect to chemical defences against herbivory (reviewed in Karban and Meyers, 1989), but this may or may not extend to defence against abiotic stress factors, such as water, nutrient and light limitations. Likewise, it is unclear whether plant defence responses to stress are similar in seeds to those that have been documented in vegetative plant tissues. Although we would expect plants growing under resource-limited conditions to produce fewer seeds, there is little information on how resource limitation affects allocation to chemical defences in seeds.

In sesame (*Sesamum indicum*), lower air temperatures (22/15°C versus 30/22°C) during seed ripening and a shorter growing season (i.e. delayed planting) tended to result in seeds with higher lignan (sesomine and sesomolin) concentration, whereas day length and soil temperature during seed ripening had no effect on seed lignan content (Kumazaki *et al.*, 2009). In wild oat, drought and light-limitation stress treatments imposed under greenhouse growing conditions resulted in seed hull fractions with 35 to 61% reductions in total phenolics compared with the non-stressed control seeds, depending on the phenolic chemical fraction (i.e. cellular bound or soluble) and the isoline tested (Gallagher *et al.*, 2010b). The most notable decrease in seed phenolics occurred with ferulic and *p*-coumaric acids, which were also the most abundant phenolic acids. The reduced allocation to phenolics in the hull coupled with the decrease in hull biomass that was also observed in the stress-grown seed suggests the chemical and physical protection provided by the hull would be compromised in these seeds. There were no discernible differences in the chemical constituents of the wild oat caryopsis among the stress treatments in this study. There was a large difference, however, in phenolic concentration of the seeds from the two growouts, with seeds from the 2006 growout having nearly five times the total phenolic concentration of the 2005 growout seeds. Circumstantially, it was noted that the 2006 growout had notable aphid infestation, although other growth environment factors may have also differed. In vegetative

tissues, aphid infestations have resulted in increased allocation to phenolics (Karban and Myers, 1998; Moran and Thompson, 2001).

If reduced allocation to defence chemicals in seeds from plants growing under abiotic stress is the norm, then this could have important implications for the persistence of seeds from these plants. For example in agriculture systems, there is some concern that threshold-based weed management systems may lead to a proliferation of the soil weed seed bank, resulting in a more aggressive weed community in the long term compared to conventional programmatic weed control approaches (Norris, 1999). However, if sub-threshold weed communities are maturing under highly competitive crop conditions, it may be that the seeds produced from these plants do not pose a serious management threat. Clearly, there is insufficient evidence at this point to make the case for or against threshold-based weed management with respect to seed bank proliferation. However, the development of herbicide resistant weed communities (Heap, 2012) and concerns over the potential non-target impact of some herbicide chemistries (Gilliom, 2007) may lead to weed management systems that are less reliant on herbicides and result in a greater number of residual weeds in crop fields. Understanding how the maturation environment of a weed influences weed seed persistence traits will be important in determining what level of weed control is needed to prevent weed seed bank proliferation. In agricultural and natural ecosystems, accounting for the potential differential in seed longevity as influenced by the seed maturation environment will be needed to make accurate predictions about plant community dynamics.

Conclusions

The endogenous chemical regulation of seed longevity in the soil seed bank is complex and not easily studied, at least not directly. Seed persistence is probably regulated by multiple chemical and biochemical mechanisms that can work together to prevent precocious germination, inhibit pathogen infection, deter seed predators and maintain seed viability and vigour over time.

The role of secondary metabolites as biochemical germination inhibitors remains unclear. There is evidence from a number of studies that phenylpropanoids common to seeds can inhibit germination within and between species when tested at high enough concentrations, usually 1 mM or higher. These concentrations appear to be in the range found in seeds. In wild oat for example, an estimated total soluble phenolic acid concentration of 2000 ng per seed (Table 8.1, caryopsis + hull), an average dry seed mass of 20 mg per seed, and a 40% seed moisture content would result in a 1.3 mM per seed phenolic concentration (based on the molecular weight of ferulic acid, 194). There is evidence that soluble chemical constituents in seeds can be leached out of imbibed seed, suggesting that any inhibitory or protective effect these chemicals serve may be short-lived. That being said, there is also evidence that *de novo* synthesis of phenylpropanoids in seeds can occur in response to predation or pathogen infection, suggesting that some of the chemical constituents leached from a seed may get replenished. It is unclear, however, how often this cycle may be repeated. In addition to the potential role of the phenylpropanoids as biochemical germination inhibitors, these chemicals, as well as other aliphatic organic acids, may physically inhibit germination by preventing gas exchange and water uptake. The cellular-bound chemical constituents are likely to be important in this role. This underscores the importance of evaluating all the chemical fractions within a seed, not just those that are readily extractable. Finally, abscisic acid (ABA) appears to be a key player in the regulation of physiological seed dormancy, although the precise physiological mechanism of ABA continues to be elucidated.

Seed pathogens are ubiquitously present in soils and we can expect that seeds in the soil seed bank will be under constant attack from these organisms. Defence mechanisms against pathogens appear to be quite effective, given the known longevity of many species in the soil seed bank. The role of passively leached secondary metabolites into the spermosphere to prevent pathogen infection is unclear. As with the inhibition of germination, there is some evidence that these chemicals can inhibit microbial growth at high enough concentrations. However, the concentration of chemical defences in the spermosphere is likely to be considerably lower than in the seed itself due to the dilution with the soil solution, and will decrease over time as chemicals diffuse into the spermosphere. *De novo* synthesis of soluble phenylpropanoids may restore the spermosphere concentrations to a degree. It would seem that the leaching of other aliphatic organic and fatty acids into the spermosphere could actually promote pathogen attack, unless the preferential growth of more benign organisms results in the competitive exclusion of seed pathogens. Based on initial results, discussed above, it seems likely that plant defence enzymes are present on seeds and their activities can be increased by pathogens; however, since these studies were conducted *in vitro*, further study in soil and *in situ* is needed. The seed–pathogen interactions in the spermosphere proposed here are very speculative, and considerably more research is needed to demonstrate such interactions.

In long-lived seeds, preventing oxidative damage to cellular membranes and organelles is likely to be an important factor in regulating persistence in the seed bank. It is unclear, however, which chemical constituents in seeds contribute to this role. There appears to be reasonable evidence for the participation of certain primary metabolites, such as glutathione, tocopherols and the raffinose series oligosaccharides in protecting some seeds from oxidative damage during the dry state and the physical damage that can occur to membranes with wetting and drying cycles. The protective role of secondary metabolites, specifically the phenylpropanoids, remains less obvious. Clearly, phenylpropanoids are abundant in seeds and typically have strong antioxidant capacity. As such, there is a good potential that they contribute to protecting seeds from oxidative stressors, but these mechanisms need further elucidation.

The analytical methodologies to study seed chemistry can be quite sophisticated, but are becoming more accessible to ecologists and physiologists who may not necessarily have experience with these techniques. The learning curve associated with mastering these techniques from the 'ground up' is steep and often expensive, but this process is feasible with a moderate familiarity with organic and biochemistry. Alternatively, researchers can search out collaborators with needed expertise and instrumentation. Although the less sophisticated analytical techniques to evaluate seed chemistry, such as those reviewed in Granger *et al.* (2011), are compelling from ease-of-use and cost perspectives, it is not clear that they always provide meaningful data. Employing the most rigorous analytical techniques possible will help reduce the risk of generating erroneous data. Finally, if there is a desire to compare chemical seed constituents among species or biotypes, it is important to minimize maternal effects, which can greatly impact the types and concentrations of chemicals in seeds. This is easily accomplished by growing the test plants under a standard common garden regime or having a sufficient sample size that cancels out maternal effects among species or biotypes.

Acknowledgements

Funding for the research by Gallagher and associates presented in this chapter was provided in part by the USDA NRI Biology of Weedy and Invasive Species Program (Award No. 2005-35320-15375), The Pennsylvania State University, and Washington State University. We greatly appreciate the GC/MS support provided by Thermo Fisher Scientific (Dr M. Bonilla, Dr G. Harkey, B. Drakontaidis and M. Hendry), the Restek Corp, and Dr E. Conklin.

References

Altares, P., Pedrosa, M.M., Burbano, C., Cuadrado, C., Goyoaga, C., Muzquiz, M., Jimenez-Martinez, C. and Davila-Ortiz, G. (2005) Alkaloid variation during germination in different lupin species. *Food Chemistry* 90, 347–355.

Anderson, J.V., Fuerst, E.P., Tedrow, T., Hulke, B., and Kennedy, A.C. (2010) Activation of polyphenol oxidase in dormant wild oat caryopses by a seed-decay isolate of *Fusarium avenaceum. Journal of Agricultural and Food Chemistry* 58, 10597–10605.

Asres, K., Sporer, F. and Wink, M. (2004) Patterns of pyrrolizidine alkaloids in 12 Ethiopian *Crotalaria* species. *Biochemical Systematics and Ecology* 32, 915–930.

Atia, A., Debez, A., Barhoumi, Z., Smaoui, A. and Abdelly, C. (2009) ABA, $GA_{(3)}$, and nitrate may control seed germination of *Crithmum maritimum* (Apiaceae) under saline conditions. *Comptes Rendus Biologies* 332, 704–710.

Bailly, C., Benamar, A., Corbineau, F. and Come, D. (1998) Free radical scavenging as affected by accelerated ageing and subsequent priming in sunflower seeds. *Physiologia Plantarum* 104, 646–652.

Bailly, C., Audigier, C., Ladonne, F., Wagner, M.H., Coste, F., Corbineau, F. and Come, D. (2001) Changes in oligosaccharide content and antioxidant enzyme activities in developing bean seeds as related to acquisition of drying tolerance and seed quality. *Journal of Experimental Botany* 52, 701–708.

Banko, P.C., Cipollini, M.L., Breton, G.W., Paulk, E., Wink, M. and Izhaki I. (2002) Seed chemistry of *Sophora chrysophylla* (mamane) in relation to diet of specialist avian seed predator *Loxioides bailleui* (palila) in Hawaii. *Journal of Chemical Ecology* 28, 1393–1410.

Baskin, C.C. and Baskin, J.M. (1998) *Seeds: Ecology, Biogeography, and Evolution of Dormancy and Germination.* Academic Press, San Diego, CA, USA.

Begum, S.A., Sahai, M. and Ray, A.B. (2010) Non-conventional lignans: coumarinolignans, flavonolignans, and stilbenolignans. In: Kinghorn, A.D., Falk, H. and Kobayashi, J. (eds.) *Progress in the Chemistry of Organic Natural Products, Vol 93.* SpringerWien, New York, New York, USA, pp. 1–70.

Berrie, A.M., Don, R., Buller, D.C., Alam, M. and Parker, W. (1975) The occurrence and function of short chain length fatty acids in plants. *Plant Science Letters* 6, 163–173.

Briggs, C.L, Morris, E.C., and Ashford, A.E. (2005) Investigations into seed dormancy in *Grevillea linearifolia, G. buxifolia* and *G. sericea*: Anatomy and histochemistry of the seed coat. *Annals of Botany* 96, 965–980.

Burnham, M.B. (2011) Seed life expectancy: The spermosphere, defense chemistry, and weed recruitment. Unpublished Master's Thesis, The Pennsylvania University, State College, Pennsylvania, USA.

Buta, J.G. and Lusby, W.R. (1986) Catechins as germination and growth inhibitors in lespedeza seeds. *Phytochemistry* 25, 93–96.

Chawla, A.S., Redha, F.M.J. and Jackson, A.H. (1985) Alkaloids in seed of 4 *Erythrina* spp. *Phytochemistry* 24, 1821–1824.

Chedea, V.S. and Braicu, C. (2011) Antibacterial action of an aqueous grape seed polyphenolic extract. *African Journal of Biotechnology* 10, 6276–6280.

Chee-Sanford, J.C. (2008) Weed seeds as nutritional resources for soil Ascomycota and characterization of specific associations between plant and fungal species. *Biology and Fertility of Soils* 44, 763–771.

Chee-Sanford, J.C., Williams, M.M., Davis, A.S., and Sims, G.K. (2006) Do microorganisms influence seed-bank dynamics? *Weed Science* 54, 575–587.

Chen, F.S., MacTaggart, J.M. and Elofson, R.M. (1982) Chemical constituents in wild oat (*Avena fatua*) hull and their effects on seed germination. *Canadian Journal of Plant Science* 62, 155–162.

Chingwaru, W.A., Duodu, G., van Zyl, Y., Schoeman, C.J., Majinda, R.T., Yeboah, S.O., Jackson, J., Kapewangolo, P.T., Kandawa-Schulz, M., Minnaar, A. and Cencic, A. (2011) Antibacterial and anticandidal activity of *Tylosema esculentum* (marama) extracts. *South African Journal of Science* 107, 79–89.

Coley, P.D., Bryant, J.P. and Chapin, F.S. (1985) Resource availability and plant antiherbivore defense. *Science* 230, 895–899.

Cosio, C., and Dunand, C. (2009) Specific functions of individual class III peroxidase genes. *Journal of Experimental Botany* 60, 391–408.

Croteau, R., Kutchan, T.M. and Lewis N.G. (2000) Natural products (secondary metabolites). In: Buchanan, B., Gruissem, W. and Jones, R. (eds) *Biochemistry & Molecular Biology of Plants*. John Wiley and Sons, Hoboken, New Jersey, USA.

Dalling, J.W., Davis, A.S., Schutte, B.J., and Arnold, A.E. (2011) Seed survival in soil: interacting effects of predation, dormancy and the soil microbial community. *Journal of Ecology* 99, 89–95.

Davis, A.S., Schutte, B.J., Iannuzzi, J., and Renner, K.A. (2008) Chemical and physical defense of weed seeds in relation to soil seedbank persistence. *Weed Science* 56, 676–684.

De Luna, L.Z, Kennedy, A.C., Hansen, J.C., Paulitz, T.C., Gallagher, R.S., and Fuerst, E.P. (2011) Mycobiota on wild oat (*Avena fatua* L.) seed and their caryopsis decay potential. *Plant Health Progress*. DOI:10.1094/PHP-2011-0210-01-RS.

De Nysschen, A., Van Wyk, B.E. and Van Heerden, F.R. (1998) Seed flavonoids of the Podalyrieae and Liparieae (Fabaceae). *Plant Systems and Evolution* 212, 1–11.

Desai, H.K., Cartwright, B.T. and Pelletier, S.W. (1994) Ajadinine: A new norditerpenoid alkaloid from the seeds of *Delphinium ajacis*: The complete NMR assignments for some lycoctonine-type alkaloids. *Journal of Natural Products* 57, 677–682.

Dini, I., Tenore, G.C. and Dini, A. (2004) Phenolic constituents of Kancolla seeds. *Food Chemistry* 84, 163–168.

Dhingra, O.D. and Sinclair, J.B. (1995) *Basic plant pathology methods*, 2nd edn. Lewis Publishers, Boca Raton, Florida, USA.

Doncheva, T., Philipov, S. and Kostova, N. (2004) Alkaloids from *Datura stramonium* L. *Dokladi na B"lgarskata Akademiya na Naukite* 57, 41–44.

Dong, T., Tong, J., Xiao, L., Cheng, H. and Song, S. (2012) Nitrate, abscisic acid and gibberellin interactions on the thermoinhibition of lettuce seed germination. *Plant Growth Regulation* 66, 191–202.

Duenas, M., Estrella, I. and Hernandez, T. (2004) Occurrence of phenolic compounds in the seed coat and the cotyledon of peas (*Pisum sativum* L.). *European Food Research and Technology* 219, 116–123.

Enu-Kwesi, L. and Dumbroff, E.B. (1980) Changes in phenolic inhibitors in seeds of *Acer saccarum* during stratification. *Journal of Experimental Botany* 31, 425–436.

Fico, G., Braca, A., Tome, F., Morelli, I. (2001) A new phenolic compound from Nigella damascena seeds. *Fitoterapia* 72, 462–463.

Finch-Savage, W.E. and Clay, H.A. (1994) Evidence that ethylene, light and abscisic acid interact to inhibit germination in the recalcitrant seeds of *Quercus robur* L. *Journal of Experimental Botany* 45, 1295–1299.

Finkelstein, R., Reeves, W., Ariizumi, T. and Steber, C. (2008) Molecular aspects of seed dormancy. *Annual Reviews of Plant Biology* 59, 387–341.

Fuerst, E.P., Anderson, J.V. and Morris, C.F. (2010) Polyphenol oxidase and darkening of Asian noodles – mechanisms, measurement and improvement. In: Hou, G., (ed.) *Asian Noodles: Science, Technology, and Processing.* John Wiley & Sons, Inc., Hoboken, New Jersey, pp. 285–312.

Fuerst, E.P., Anderson, J.V., Kennedy, A.C., and Gallagher, R.S. (2011) Induction of polyphenol oxidase activity in dormant wild oat (*Avena fatua*) seeds and caryopses: a defense response to seed decay fungi. *Weed Science* 59, 137–144.

Gallagher, R.S. and Fuerst, E.P. (2006) The ecophysiology basis of weed seed longevity in the soil. In: Basra, A.S. (ed.) *Seed Science and Technology.* Haworth Food Products Press, Binghamton, New York, USA.

Gallagher, R.S., Ananth, R., Granger, K.L., Bradley, B., Anderson, J.V. and Fuerst E.P. (2010a) Phenolic and short-chained aliphatic organic acid constituents of wild oat (*Avena fatua*) seeds. *Journal of Agricultural and Food Chemistry* 58, 218–225.

Gallagher, R.S., Granger, K.L., Keser, L.H., Rossi J., Pittmann, D., Rowland, S., Burnham, M. and Fuerst, E.P. (2010b) Shade and drought stress-induced changes in phenolic content of wild oat (*Avena fatua* L.) seeds. *Journal of Stress Physiology* 6, 90–107.

Gallagher, R.S., Granger, K.L., Snyder, A.M., Pittmann, D. and Fuerst, E.P. (2013) Implications of environmental stress during seed development on reproductive and seed bank persistence traits in wild oat (*Avena fatua* L.). *Agronomy* 3, 537–549.

Garcia, Q.S., Giorni, V.T., Mueller, M. and Munne-Bosch, S. (2012) Common and distinct responses in phytohormone and vitamin E changes during seed burial and dormancy in *Xyris bialata* and *X. peregrine. Plant Biology* 14, 347–353.

Gatehouse, J.A. (2002) Plant resistance towards insect herbivores: a dynamic interaction. *New Phytologist* 156, 145–169.

Gilliom, R.J. (2007) Pesticides in U.S. streams and groundwater. *Environmental Science & Technology* 41, 3407–3413.

Gomez-Jimenez, M., Sanchez-Calle, I., Chibi, F., and Matilla, A. (1996). Alterations in the last step of the ethylene-biosynthesis pathway provoked by short-chain saturated fatty acids in *Cicer arietinum* seeds. *Journal of Plant Physiology* 149, 626–628.

Granger, K.L. (2012) Mitigation of drought-induced reductions in wild oat seed quality by elevated carbon dioxide, Intercollege Graduate Degree Program in Ecology. Pennsylvania State University, University Park, PA, p. 248.

Granger, K.L., Gallagher, R.S, Aldredge, R., and Fuerst, E.P. (2011) Methodological comparison to quantify and identify phenolic constituents in seeds. *Methods in Ecology and Evolution.* DOI: 10.1111/j.2041-210X.2011.00120.x.

Grover, A. (2012) Plant chitinases: genetic diversity and physiological roles. *Critical Reviews in Plant Sciences* 31, 57–73.

Hannan, R.M. and Hill, H.H. (1991) Analysis of lipids in aging seed using capillary supercritical fluid chromatography. *Journal of Chromatography* 547, 393–401.

Hatano, K. (1967) Detection of coumarin and o-coumaric acid in the seed coat of *Pinus densiflora. Journal of the Japanese Forest Society* 49, 205–207.

Hatano, T. (1999) Phenolic constituents of Cassia seeds and antibacterial effect of some naphthalenes and anthraquinones on methicillin-resistant *Staphylococcus aureus. Chemical and Pharmaceutical Bulletin* 47, 1121–1127.

Heap, I.M. (2012) The international survey of herbicide resistant weeds. Available from: http://www.weedscience.org/ (verified November 26, 2012).

Hendry, G.A.F., Thompson, K., Moss, C.J., Edwards, E. and Thorpe, P.C. (1994) Seed persistence: a correlation between seed longevity in the soil and ortho-dihydroxyphenol concentration. *Functional Ecology,* 8, 658–664.

Herchi, W., Sakouhi, F., Arraez-Roman, D., Segura-Carretero, A., Boukhchina, S., Kallel, H. and Fernandez-Gutierrez, A. (2011) Changes in the content of phenolic compounds in flaxseed oil during development. *Journal of the American Oil Chemists' Society* 88,1135–1142.

Hietala, A.M. (1997). The mode of infection of a pathogenic uninucleate *Rhizoctonia* sp. in conifer seedling roots. *Canadian Journal of Forestry Research.* 27,471–480.

Hücklehoven, R. (2007) Cell wall-associated mechanisms of disease resistance and susceptibility. *Annual Review of Phytopathology* 45, 101–127.

John, J. and Sarada, S. (2011) Role of phenolics in allelopathic interactions. *Allelopathy Journal* 29, 215–230.

Jurgonski A. (2011) Composition of chicory root, peel, seed and leaf ethanol extracts and biological properties of their non-inulin fractions. *Food Technology and Biotechnology* 49, 40–47.

Karban, R. and Myers, J.H. (1989) Induced plant responses to herbivory. *Annual Reviews in Ecological Systems* 20, 331–348.

Kato, T., Imai, T., Kashimura, K., Saito, N. and Masaya, K. (2003) Germination response in wheat grains to dihydroactinidiolide, a germination inhibitor in wheat husks, and related compounds. *Journal of Agricultural and Food Chemistry* 51, 2161–2167.

Kermode, A.R. (2005) Role of abscisic acid in seed dormancy. *Journal of Plant Growth Regulation* 24, 319–344.

Kestring, D., Menezes, L., Tomaz, C.A., Lima, G. and Rossi, M. (2009) Relationship among phenolic contents, seed predation, and physical seed traits in *Mimosa bimucronata* plants. *Journal of Plant Biology* 52, 569–576.

Khandelwal, A., Cho, S.H., Marella, H., Sakata, Y., Perroud, P.F., Pan, A. and Quatrano, R.S. (2010) Role of ABA and ABI3 in desiccation tolerance. *Science* 327, 546.

Kim, K.S., Park, S H. and Choung, M G. (2006) Nondestructive determination of lignans and lignan glycosides in sesame seeds by near infrared reflectance spectroscopy. *Journal of Agricultural and Food Chemistry* 54, 4544–4550.

Klejdus, B. and Kubane, V. (2000) High performance liquid chromatographic determination of phenolic compounds in seed exudates of *Festuca arundinacea* and *F. pratense*. *Phytochemical Analysis* 11, 375–379.

Kovacs, M.F. (1971). Identification of aliphatic and aromatic acids in root and seed exudates of peas, cotton, and barley. *Plant and Soil* 34, 441–451.

Kranner, I. and Birtic, S. (2005) A modulating role for antioxidants in desiccation tolerance. *Integrative and Comparative Biology* 45, 734–740.

Kremer, R.J. (1986). Antimicrobial activity of velvetleaf (*Abutilon theophrasti*). *Weed Science* 34, 617–622.

Krygier, K, Sosulski, F, and Hogge, L. (1982). Free, esterified, and insoluble-bound phenolic acids. 1. Extraction and purification procedure. *Journal of Agriculture and Food Chemistry* 30, 330–334.

Kumazaki, T., Yamada, Y., Karaya, S., Kawamura, M., Hirano, T., Yasumoto, S., Katsuta, M. and Michiyama, H. (2009) Effects of day length and air and soil temperatures on sesamin and sesamolin contents of sesame seed. *Plant Production Science* 12, 481–491.

Lane, B.G. (2002) Oxalate, germins, and higher-plant pathogens. *IUBMB Life* 53, 67–75.

Li, W., Qiu, Y. and Patterson, C.A. (2010) The analysis of phenolic constituents in glabrous canaryseed groats. *Food Chemistry* 127, 10–20.

Long, R.L., Panetta, F.D., Steadman, K.J., Probert, R., Bekker, R.M., Brooks, S. and Adkins, S.W. (2008) Seed persistence in the field may be predicted by laboratory-controlled aging. *Weed Science* 56, 523–528.

Long, R.L., Kranner, I., Panetta, F.D., Birtic, S., Adkins, S.W. and Steadman, K.J. (2011) Wet-dry cycling extends seed persistence by re-instating antioxidant capacity. *Plant and Soil* 338, 511–519.

Manuel-Rios, J., Mangione, A. and Marone, L. (2012). Effects of nutritional and anti-nutritional properties of seeds on the feeding ecology of seed-eating birds of the Monte Desert, Argentina. *Condor* 114, 44–55.

Matthaus, B. (1998) Isolation, fractionation and HPLC analysis of neutral phenolic compounds in rapeseeds. *Nahrung* 42, 75–80.

Mayer, A.M. (2006) Polyphenol oxidases in plants and fungi: Going places? A review. *Phytochemistry* 67, 2318–2331.

McDonald, M.B. (1999). Seed deterioration: physiology, repair and assessment. *Seed Science and Technology* 27, 177–237.

Metzger, J.D. and Sebesta, D.K. (1982) Role of endogenous growth regulated in seed dormancy of *Avena fatua* 1: Short chain fatty acids. *Plant Physiology* 70, 1480–1485.

Michel, T., Destandau, E., Le Floch, G., Lucchesi, M.E. and Elfakir, C. (2012) Antimicrobial, antioxidant and phytochemical investigations of sea buckthorn (*Hippophae rhamnoides* L.) leaf, stem, root and seed. *Food Chemistry* 131, 754–760.

Misra, L. and Wagner, H. (2004) Alkaloidal constituents of *Mucuna pruriens* seeds. *Phytochemistry* 65, 2565–2567.

Mitaine, A.C., Mesbah, K., Richard, B., Petermann, C., Arrazola, S., Moretti, C., Zeches-Hanrot, M. and Le Men-Olivier, L. (1996) Alkaloids from *Aspidosperma* species from Bolivia. *Planta Medica* 62, 458–461.

Mitrovic, M., Jaric, S., Djurdjevic, L., Karadžic, B., Garjic, G., Kostic, O., Oberan, L.J., Pavlovic, M. and Pavlovic, P. (2012) Allelopathic and environmental implications of plant phenolic compounds. *Allelopathy Journal* 29, 177–198.

Moran, P.J. and Thompson, G.A. (2001) Molecular responses to aphid feeding in *Arabidopsis* in relation to plant defense pathways. *Plant Physiology* 125, 1074–1085.

Morris, P.C., Grierson, D. and Whittington, W.J. (1984) Endogenous inhibitors and germination of *Beta vulgaris*. *Journal of Experimental Botany* 35, 994–1002.

Naqvi, H.H. and Hanson, G.P. (1982) Germination and growth inhibitors in Guayule (*Parthenium argentatum*) chaff and their possible influence in seed dormancy. *American Journal of Botany* 69, 985–989.

Nelson, E.B. (2004) Microbial dynamics and interactions in the spermosphere. *Annual Reviews of Phytopathology* 42, 271–309.

Norris, R.F. (1999) Ecological implications of using thresholds for weed management. In: Buhler, D.D. (ed.) *Expanding the Context of Weed Management*. Hawthorn Press, New York, New York, USA.

Obendorf, R.L. (1997). Oligosaccharides and galactosyl cyclitols in seed desiccation tolerance. *Seed Science Research* 7, 63–74.

Ozawa, M., Kishida, A. and Ohsaki, A. (2011) Erythrinan alkaloids from seeds of *Erythrina velutina*. *Chemical and Pharmaceutical Bulletin* 59, 564–567.

Passardi, F., Cosio, C., Penel, C., and Dunand, C. (2005) Peroxidases have more functions than a Swiss army knife. *Plant Cell Reports* 24, 255–265.

Paszkowski, W.L. and Kremer, R.J. (1988) Biological activity and tentative identification of flavonoid components in velvetleaf (*Abutilon theophrasti* Medik.) seed coats. *Journal of Chemical Ecology* 14, 1573–1582.

Picman, A.K., Giaccone, R. and Ivarson, K.C. (1984) Antifungal properties of oat hulls. *Phytoprotection* 65, 9–15.

Pons, S., Pinson-Gadais, L., Atanassova, V., Boutigny, A., Roucolle, J., Carola, P. and Richard-Forget, F. (2008) Phenolic acid composition of maize kernels modulates in planta fungal growth and Trichothecenes B production by *F. graminearum*. *Biotechnology and Agronomy* 36, 459–463.

Poutaraud, A. and Girardin, P. (2005) Influence of chemical characteristics of soil on mineral and alkaloid seed contents of *Colchicum autumnale*. *Environmental and Experimental Botany* 54, 101–108.

Priestly, D.A. (1986) *Seed Aging*. Comstock Publishing Associates, Ithaca, New York, New York, USA.

Pukacka, S. (1991) Changes in membrane lipid components and antioxidant levels during natural ageing of seeds of *Acer platanoides*. *Physiologia Plantarum* 82, 306–310.

Qi, M.Q., Upadhyaya, M.K., Furness, N.H. and Ellis, B.E. (1993) Mechanism of seed dormancy in *Cynoglossum officinale* L. *Journal of Plant Physiology* 142, 325–330.

Rai, S., Wahile, A., Mukherjee, K., Saha, B.P. and Mukherjee, P.K. (2006) Antioxidant activity of *Nelumbo nucifera* (sacred lotus) seeds. *Journal of Ethnopharmacology* 104, 322–327.

Reddy, M.N., Ramagopal, G. and Rao, A.S. (1977) Phenolic acids in groundnut seed exudates. *Plant and Soil* 46, 655–658.

Ruttledge, T.R., and Nelson, E.B. (1997) Extracted fatty acids from *Gossypium hirsutum* stimulatory to the seed-rotting fungus *Pythium ultimum*. *Phytochemistry* 46, 77–82.

Sarker, S.D., Dinan, L., Sik, V., Underwood, E. and Waterman, P.G. (1998) Moschamide: An unusual alkaloid from the seeds of *Centaurea moschata*. *Tetrahedron Letters* 39, 421–1424.

Sattler, S.E., Gilliland, L.U., Magallanes-Lundback, M., Pollard, M. and DellaPenna, D. (2004) Vitamin E is essential for seed longevity, and for preventing lipid peroxidation during germination. *Plant Cell* 16, 1419–1432.

Schmidt, K.A. (2000) Interaction between food chemistry and predation risk in fox squirrels. *Ecology* 81, 2077–2085.

Shirley, B.W. (1998) Flavonoids in seeds and grains: physiological function, agronomic importance and the genetics of biosynthesis. *Seed Science Research* 8, 415–422.

Short, G.E. and Lacy, M.L. (1974) Germination of *Fusarium solani* f. sp. *pisi* chlamydospores in the spermosphere of pea. *Phytopathology* 64, 558–562.

Siger, A. (2010) Antioxidant activity and phenolic content in three lupin species. *Journal of Food Composition and Analysis* 25, 190–197.

Skadhauge,B., Thomsen, K.K. and Von Wettstein, D. (1997). The role of the barley testa layer and its flavonoid content in resistance to *Fusarium* infections. *Hereditas* 126, 147–160.

Smith, M.T. and Berjak, P. (1995) Deteriorative changes associated with the loss of viability of stored desiccation tolerant and desiccation sensitive seeds. In: Kigel, J. and Galili, G. (eds) *Seed Development and Germination.* Marcel Dekker, New York, New York, USA.

Sproll, C., Perz, R.C. and Lachenmeier, D.W. (2006) Optimized LC/MS/MS analysis of morphine and codeine in poppy seed and evaluation of their fate during food processing as a basis for risk analysis. *Journal of Agricultural and Food Chemistry* 54, 5292–5298.

Stanghellini, M.E., and Hancock, J.G. (1971) Radial extent of the bean spermosphere and its relation to the behavior of *Pythium ultimum. Phytopathology* 61,165–168.

Stobiecki, M., Ciesiotka, D., Peretiatkowicz, M. and Gulewicz, K. (1993) Phenolic compounds isolated from bitter lupine seeds and their inhibitory effects on germination and seedling growth of lettuce. *Journal of Chemical Ecology* 19, 325–338.

Subbiah, V. and Reddy, K.J. (2010) Interactions between ethylene, abscisic acid and cytokinin during germination and seedling establishment in *Arabidopsis. Journal of Biosciences* 35, 451–458.

Sutlupinar, N., Husek, A., Potesilova, H., Dvorackova, S., Hanus, V., Sedmera, P. and Simanek, V. (1988) Alkaloids and phenolics of *Colchicum cilicicum. Planta Medica* 54, 243–245.

Takahashi, H., Yoshioka, S., Kawano, S., Azuma, H., and Fukuyama, Y. (2002) Lignans and sesquiterpenes from *Magnolia praecocissima. Chemical and Pharmaceutical Bulletin* 50, 541–543.

Tang, W., Xu, H., Zeng, D. and Yu, L. (2012). The antifungal constituents from the seeds of *Itoa orientalis. Fitoterapia* 83, 513–517.

Troszyñska, A. and Ciska, E. (2002) Phenolic compounds of seed coats of white and coloured varieties of pea (*Pisum sativum* L.) and their total antioxidant activity. *Czech Journal of Food Science* 20, 15–22.

Van Dijk, K. and Nelson, E.B. (2000) Fatty acid competition as a mechanism by which *Enterobacter cloacae* suppresses *Pythium ultimum* sporangium germination and damping-off. *Applied Environmental Microbiology* 66, 5340–5347.

Van Gelder, C.W.G., Flurkey, W.H. and Wichers, H.J. (1997) Sequence and structural features of plant and fungal tyrosinases. *Phytochemistry* 45, 1309–1323.

van Loon, L.C., Rep, M. and Pieterse, C.M.J. (2006) Significance of inducible defense-related proteins in infected plants. *Annual Review of Phytopathology* 44, 135–162.

Venkatachallam, S.K.T., Hajimalang, D., Soundar, K. and Udaya, S. (2010) Chemical composition of *Nigella sativa* L. seed extracts obtained by supercritical carbon dioxide. *Journal of Food Science and Technology* 47, 598–605.

Walters, C. (1998). Understanding the mechanisms and kinetics of seed aging. *Seed Science Research* 8, 223–244.

Weidner, S., Amarowicz, R., Karamac, M. and Dąbrowski, G. (1999) Phenolic acids in caryopses of two cultivars of wheat, rye and triticale that display different resistance to pre-harvest sprouting. *European Food Research and Technology* 210, 109–113.

Weidner, S., Amarowicz, R., Karamac, M. and Frączeka, E. (2000) Changes in endogenous phenolic acids during development of *Secale cereale* caryopses and after dehydration treatment of unripe rye grains. *Plant Physiology and Biochemistry* 38, 595–602.

Weidner, S., Frączeka, E., Amarowicz, R. and Abe, S. (2001) Alternations in phenolic acids content in developing rye grains in normal environment and during enforced dehydration. *Acta Physiology Planta* 23, 475–482.

Weidner, S., Krupa, U., Amarowicz, R., Karamac, M. and Abe, S. (2002) Phenolic compounds in embryos of triticale caryopses at different stages of development and maturation in normal environment and after dehydration treatment. *Euphytica* 126, 115–122.

Weinhold, A.R., Dodman, R.L., and Bowman, T. (1972) Influence of exogenous nutrition on virulence of *Rhizoctonia solani. Phytopathology* 62, 278–281.

Wettasinghe, M., Shahidi, F. and Amarowicz, R. (2002) Identification and quantification of low molecular weight phenolic antioxidants in seeds of evening primrose (*Oenothera biennis* L.). *Journal of Agricultural and Food Chemistry* 50, 1267–1271.

Williams, R.D. and Hoagland, R.E. (1982) The effects of naturally occurring phenolic compounds on seed germination. *Weed Science* 30, 206–212.

Windstam, S., and Nelson, E.B. (2008a) Differential interference with *Pythium ultimum* sporangial activation and germination by *Enterobacter cloacae* in the corn and cucumber spermospheres. *Applied Environmental Microbiology* 74, 4285–91.

Windstam, S. and Nelson, E.B. (2008b) Temporal release of fatty acids and sugars in the spermosphere: Impacts on *Enterobacter cloacae*-induced biological control. *Applied Environmental Microbiology* 74, 4292–4299.

Xiao, Z., Wang, Y., Harris, M. and Zhang, Z. (2006) Spatial and temporal variation of seed predation and removal of sympatric large-seeded species in relation to innate seed traits in a subtropical forest, Southwest China. *Forest Ecology and Management* 222, 46–54.

Yildirim, A., Mavi, A. and Kara, A.A. (2001) Determination of antioxidant and antimicrobial activities of *Rumex crispus* L. extracts. *Journal of Agricultural and Food Chemistry* 49, 4083–4089.

Yoruk, R., and Marshall, M.R. (2003) Physicochemical properties and function of plant polyphenol oxidase: A review. *Journal of Food Biochemistry* 27, 361–422.

Zhang, H., Yu, Y., Liu, D. and Liu, Z. (2007) Extraction and composition of three naturally occurring anti-cancer alkaloids in *Camptotheca aeuminata* seed and leaf extracts. *Phytomedicine* 14, 50–56.

Zheng, C., Lan, X., Cheng, R., Huang, B., Han, T., Zhang, Q., Zhang, H., Rahman, K. and Qin, L. (2011) Furanofuran lignans from *Vitex negundo* seeds. *Phytochemistry Letters* 4, 298–300.

Zhou, S., Selizakl, H., Yang, Z., Sawa, S. and Pan, J. (2010) Phenolics in the seed coat of wild soybean (*Glycine soja*) and their significance for seed hardness and seed germination. *Journal of Agriculture and Food Chemistry* 58, 10972–10978.

9 Effects of Climate Change on Regeneration by Seeds

Rui Zhang[1]* and Kristen L. Granger[2]
[1]*Harvard Forest, Harvard University, Petersham, Massachusetts, USA;*
[2]*Department of Crop and Soil Sciences, The Pennsylvania State University, University Park, Pennsylvania, USA*

Introduction

Climate change has been shown to influence many aspects of species' life histories (Pounds *et al.*, 1999; Hughes, 2000; Walther *et al.*, 2002). Compared to the well-studied literature of how climate change affects performance of adult plants, relatively few studies have focused on the responses of seeds and seedlings, the shifts in their abundance and distributions, and changes in population dynamics and regenerations that are connected by these early life stages. As iterated in other chapters of this book, seeds play a critical role in plant regeneration. Furthermore, early life stages are expected to be more sensitive to climate change than adult stages (Lloret *et al.*, 2004; Walck *et al.*, 2011), and therefore impacts of climate change on regeneration are likely to have consequences at the population and community levels. In this chapter, we review both lab and field investigations on seed and seedling responses to climate change. There is a rich literature on how environmental factors regulate seed biology (e.g. reviewed in Baskin and Baskin, 1998; Gallagher and Fuerst, 2006), which can be also found in the other chapters and previous

editions of this book (e.g. Probert, 1992; Gutterman, 2000; Probert, 2000). Here, we focus on the linkage between climate change and plant offspring performance, as it is the key to understanding regeneration of plant communities in the future. We aim to expand the existing body of knowledge by including insights from climate change studies.

Aspects of climate change considered in this chapter include elevated atmospheric carbon dioxide (CO_2), warming, and shifted precipitation regimes, which potentially cause extreme events such as drought and/or flooding (Solomon *et al.*, 2007). Atmospheric CO_2 levels have risen since the Industrial Revolution to a current concentration of 379 ppm, and are expected to continue to increase by 1.9 ppm each year. This change in greenhouse gas concentration causes concomitant alterations in global temperature and precipitation (Solomon *et al.*, 2007). Elevated atmospheric CO_2 is generally believed to have little direct effect on soil CO_2 level, but may, however, reduce soil moisture (Manabe and Wetherald, 1987). Warming includes gradual increases of both air temperature and soil temperature. Under global warming, soil temperatures are likely to increase significantly with rising air

* E-mail: ruizhang0410@gmail.com

temperatures, especially in open and sparsely vegetated areas (Ooi, 2012). Ooi *et al.* (2012) found that an increase of 1.5°C in soil temperature corresponds to an increase of 1°C in air temperature in southeastern Australia. Other than the absolute increase in soil temperature, global warming is also likely to alter diurnal temperature range (Braganza *et al.*, 2004), and the number of warm/cold days in a season (Ha and Yun, 2012). Additionally, climate change is also expected to increase the occurrence of heatwaves in the future. Altered precipitation regimes may change the average water availability as well as the variance. More extreme events such as drought and flooding are expected. Furthermore, distribution of precipitation of different forms may also change, as winter warming is likely to increase the ratio of rain versus snow (Solomon *et al.*, 2007).

The interest in the effects of climate change on seed and seedling biology is growing. Researchers are addressing questions ranging from individual, physiological responses to community and population level dynamics, using a wide variety of approaches, including lab experiments, greenhouse and growth chamber studies, and long-term field observations and manipulations. Given the extent of the literature on climate change,

we have organized this chapter in the following ways: first, we divide effects of climate change into direct effects via offspring environment, and indirect effects via maternal environment (Fig. 9.1). Most studies have focused on the direct effects of climate change on seeds, seedlings and reproductive allocation. However, because climate change is a long-term trend that covers multiple generations, we propose that maternal environmental effects should also be addressed. Second, we further divide the discussion based on the specific climate change factor considered by the studies (e.g. warming, elevated CO_2, etc.).

Direct Effects of Climate Change on Seed and Seedling Traits

Direct effects of climate change are defined as the direct effects of changes in offspring environment, such as elevated CO_2, warming and altered precipitation and soil moisture. Compared to the maternal environmental effect discussed later, the environments directly experienced by the offspring tend to have predominant effects on plants' early life stages.

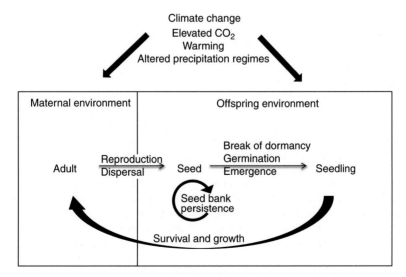

Fig. 9.1. Climate change can affect plant regeneration through directly affecting offspring environments and indirectly modifying maternal environments. Both direct and indirect effects can influence offspring performance.

Dormancy alleviation

Seed dormancy is regarded as an innate seed property that determines the environmental conditions required for germination (Finch-Savage and Leubner-Metzger, 2006; also see Chapter 7 of this volume). Alleviation of dormancy is a prerequisite for germination to happen and relies on specific environmental cues, such as light, temperature and moisture (Koller, 1972; Baskin and Baskin, 1998). Generally speaking, environmental cues serve to alleviate dormancy by signalling that environmental conditions are becoming favourable for seedling establishment and growth. Hence seed dormancy prevents seeds from germinating at inappropriate times, and is an adaptive trait that optimizes the distribution of germination time and thereby population persistence (Copete *et al.*, 2011).

Elevated carbon dioxide

It is traditionally believed that elevated atmospheric CO_2 alone may have little effect on seed dormancy, because CO_2 concentrations in the soil or at the soil surface (>1000 ppm) far exceed that in the air, and the incremental changes in atmospheric CO_2 on the scale of 350 ppm is too small to cause any effect (Edwards and Newton, 2006).

Warming

Soil temperature is the main factor controlling seed dormancy status in a majority of species (Baskin and Baskin, 1998). Temperature regulates the width of this 'germination window' and its overlaps with physical environments. Climate change can alter both the 'germination window' and the physical environment, and therefore the overlap between the two, determining the timing and the extent of germination. For after-ripening species (whose dormancy release requires warm conditions), increased temperature or prolonged warming period may enhance dormancy release, enabling seeds to germinate in a wider range of physiological environments. For example, Mira *et al.* (2011) found that increased temperature was positively

correlated with dormancy release in endangered *Silene diclinis*. Doody and O'Reilly (2011) found that prolonged warming period facilitated dormancy release of *Fraxinus excelsior*. Furthermore, Ooi *et al.* (2012) found that increased mean temperature had no effect on dormancy loss, but increased occurrence of heatwave-facilitated dormancy release of two shrub species *Acacia suaveolens* and *Dillwynia floribunda* in the seed bank. Similarly, Santana *et al.* (2010) found that summer heatwaves may break dormancy of plants in a fire-prone system. Thus, some features of climate change can cause seeds to break dormancy more readily than in the past. On the other hand, if dormancy release requires chilling (as in many spring-germinating species), increased temperature or a shortened cold period may delay dormancy break, or even completely prevent seeds from germinating. For instance, Xiao *et al.* (2010) found that shorter duration of cold stratification results in significantly decreased germinability of 11 out of 14 wetland species in the temperate region of central China. Field evidence of warming inhibiting dormancy release in seeds is rare. However, as ecophysiology of seed germination is similar to that of budburst in some species (Walck *et al.*, 2011; Vahdati *et al.*, 2012), and there is both lab and field evidence showing that budburst can be delayed by winter warming (Myking and Heide, 1995; Zhang *et al.*, 2007; Both *et al.*, 2009; Ghelardini *et al.*, 2010), we propose that warming will similarly delay seed dormancy release in some species.

Although warming can affect both germination timing and percentage, this effect may vary for plants with different life histories. Summer annuals generally emerge in the spring; seeds are dispersed in the autumn, and are generally only able to germinate at high temperatures. Autumn soil temperatures are generally lower than the seeds' required temperatures for germination, precluding emergence (Probert, 2000). Dormancy release mostly depends on cold stratification and exposure to alternating temperatures (Baskin, 2003). During cold stratification over the winter, seeds gradually lose dormancy as the minimum temperatures required for germination decrease, and the range of temperatures over which germination is possible (the 'germination window') widens. In the spring, when soil temperatures increase to a point when they overlap with the 'germination window', seed germination is initiated and life cycles start (see Fig. 11.8 in Probert, 2000). Under climate change when springs come earlier, germination may be either advanced or delayed, depending on the extent to which the chilling requirement is fulfilled (Fig. 9.2a and b).

Winter annuals are the opposite of summer annuals. Seeds are produced in the summer and their germination is typically only possible at low temperatures. Summer soil temperatures are generally higher than the seeds' required temperatures for germination, precluding emergence. Dormancy release occurs when the seeds are exposed to high environmental temperatures over the summer, during which the maximum temperatures at which the seeds can germinate increase, widening the germination window. When the germination window overlaps with autumn soil temperatures, germination is initiated (Probert, 2000). As winter approaches, temperature eventually drops below the minimum required germination temperature, and secondary dormancy is induced. Under global warming, increased autumn temperature and earlier releases of dormancy may result in little change in germination timing (Fig. 9.2c) or

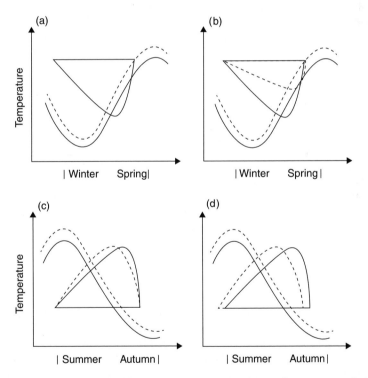

Fig. 9.2. Theoretical illustrations of possible changes in germination timing of summer annuals (a and b) and winter annuals (c and d) under warming. The figure is extrapolated based on Fig. 11.8 in Probert (2000). Solid lines represent ambient condition. Dashed lines represent warmed condition. Triangles represent temperature window for germination. Curves represent seasonal temperature fluctuation. (a) Cold stratification is fulfilled despite increased winter temperature, and seeds are expected to germinate earlier because of increased spring temperature. (b) Chilling requirement is not reached because of winter warming (narrower germination window), and germination is delayed in spite of increased spring temperature. (c) Summer warming facilitates seed dormancy loss, making seed capable of germinating at a higher temperature. But as autumn temperatures also increase, germination timing is not advanced. (d) Seeds were produced earlier because of advanced phenology of the maternal plants, leading to advanced germination timing as well.

advanced germination (Fig. 9.2d); the latter probably being attributed to advanced reproductive phenology of the maternal plants.

The germination pattern of facultative winter annuals depends on complex interactions of seasonal fluctuations of dormancy levels and germination requirements. Karlsson and Milberg (2007) used a 900-day study to examine germination patterns of four *Papaver* species under annual temperature cycles representing cold, intermediate and warm climates. They argued that *Papaver* spp. were more likely to perform as winter annuals in warmer climates, but mainly as summer annuals in colder climates, because a warmer summer can result in extensive autumn germination via accelerating dormancy loss, leaving few seeds to germinate in the following spring. Similarly, Mondoni *et al.* (2012) found that ten out of 12 alpine species whose seeds were exposed to autumn warming switched their seedling emergence from spring to autumn.

Altered precipitation regimes

Soil moisture also affects seed dormancy. Batlla and Benech-Arnold (2006) found that fluctuating soil water content promoted dormancy release in *Polygonum aviculare*. Using seeds from 158 annual species distributed along a gradient of rainfall variability and aridity in Israel, Harel *et al.* (2011) found that seed dormancy decreased with increasing aridity and rainfall variability. Cornaglia *et al.* (2009) found that flooding resulted in break of dormancy of *Paspalum dilatatum*, a dominant grass in temperate grassland in Argentina, and this effect increased with alternating temperatures. In *Lolium rigidum*, alleviation of dormancy was found to be a function of the hydration state of the seed and the temperature during after-ripening, which was collectively termed as 'hydrothermal priming time' (Gallagher *et al.*, 2004).

Germination and emergence

After dormancy alleviation, seeds have the capacity to germinate under a wide spectrum of physical environments. Basic requirements for germination include uptake of water by imbibition and exposure to appropriate oxygen and temperature levels (Finch-Savage and Leubner-Metzger, 2006). As some aspects of climate change may affect these required factors, germination percentage as well as germination rate of some species are expected to change under future climatic conditions. Not all germinated seeds are strong enough to emerge above ground. How many and how fast seedlings emerge depends on seed vigour (Baskin and Baskin, 1998), which can also be altered by climate change as discussed below.

Elevated carbon dioxide

Similar to germination, elevated atmospheric CO_2 levels are traditionally believed to have little effect on germination. For example, Edwards *et al.* (2001) detected no CO_2 effect on germination of six grassland species. In contrast, Ziska and Bunce (1993) found that doubled CO_2 level resulted in increased germination percentages of some crops and weeds. Increasing CO_2 concentration in the atmosphere is also unlikely to directly affect seedling emergence. This is confirmed by studies on an early successional plant community containing C3 and C4 plants, where elevated CO_2 did not significantly affect the timing of emergence (reviewed in Bazzaz, 1990). Instead, elevated CO_2 is more likely to have a direct effect on seedling growth after radicles emerge from the soil (see below). For freshwater wetlands, recent work demonstrated that elevated CO_2 had minimal effect on germination compared to flooding or salinity (Middleton and McKee, 2012).

Warming

Temperature, in addition to its role in regulating dormancy as discussed above, can also directly affect the process of germination. Warming is likely to affect both how many and how fast seeds germinate. But the direction and size of the effect depends on how warming relates to the minimum and maximum germination temperature as well as the optimum germination temperature (Probert, 2000). For example, germination percentage of *Ipomoea cairica* (an invasive liana plant in Southern China) in growth chambers increased as temperature increased from 22°C to 30°C (Wang *et al.*, 2011). Ramirez-Tobias *et al.* (2012) examined the germination responses of eight Agave species to a wide temperature gradient (10, 15, 20, 25, 30, 35, 40°C) and found that these congeneric species differ in their response patterns in both germination percentage and germination rate. Milbau *et al.* (2009) found that an increase of 2.5°C in summer soil temperature significantly accelerated germination of 19 out of 23 sub-arctic species, but had no effect on germination percentage. However, they also found that the positive effect of summer temperature may be diminished by reduced snow pack in the winter, which delayed germination and lowered germination percentage in about half of their study species. The interaction between warming and reduced snow pack plays a similar role in temperate alpine species, whose germination depends on the extent to which the positive effect of spring warming compensates for the negative effect of detrimental frosting associated with earlier snow melt (Inouye, 2008; Hulber *et al.*, 2011). Milbau *et al.* (2009) further showed that the effects of warming may induce divergent responses for different functional groups. Germination of trees, shrubs and grasses was improved more than that of herbs, suggesting that warmer sub-arctic climate may alter community compositions in the future. Shevtsova *et al.* (2009) found that germination and emergence of early-germinating species were more responsive to warming than late-germinating graminoids; the extents of responses also depended on the timing of warming.

Temperature can affect seedling emergence by directly influencing seedling physiology. In alpine tree and shrub communities, warmer winters have enhanced seedling and sapling establishment (Kullman, 2002). The authors note that seed viability for *Betula pubescens* and *Picea abies* has increased since the 1980s, a period which has been characterized by warmer winters.

In addition, warmer summers have enhanced seed production and emergence from the seed bank. Similarly, in sub-alpine forest hemlock populations, warmer temperatures and longer growing seasons facilitated by less snow cover enhanced hemlock seedling recruitment (Taylor, 1995). However, temperature may not have an effect on seedling emergence in all cases. In a study on a Mediterranean shrubland community, Lloret *et al.* (2009) found that a night-time warming treatment of about 0.73°C did not significantly affect the seedling establishment of woody and perennial species. The results of the study suggested that water availability rather than temperature would be more important in Mediterranean climates.

Altered precipitation regimes

Moisture is important for germination (Baskin and Baskin, 1998). Fay and Schultz (2009) evaluated how watering intervals affected the germination of two grass and six forb species commonly found in North American grasslands. They found that germination percentages peaked at intermediate watering intervals. Shifts in precipitation regimes under climate change may strongly alter germination patterns. Generally speaking, larger variations in precipitation lead to more sporadic germination patterns (Fay and Schultz, 2009). Therefore, as global precipitation patterns are becoming more variable and unpredictable (Solomon *et al.*, 2007), germination may become more sporadic both spatially and temporally. Increased occurrence of extreme events such as storms may largely alter germination patterns and the subsequent fate of plants, especially in water-limited systems, such as deserts or regions dominated by Mediterranean climates. For example, Levine *et al.* (2011) found that timing of the first storm of the growing season significantly affected the germination of an endangered annual plant, *Gilia tenuiflora*, in coastal California, with early storms enhancing its germination. Using a demographic modelling approach, they demonstrated that different timings of even a single event, in this case the first storm in the growing season, could lead to up to fourfold variation in per capita growth rate. Interestingly, they also found that the colder temperatures associated with storms largely enhanced germination of another species, *Malacothrix indecora*, due to interactions between the temperature and moisture requirements of the species.

Drought, by not providing enough water to satisfy germination requirements, can reduce seedling emergence (Lloret et al., 1999). In some cases (Sternberg et al., 1999; Lloret et al., 2009), water deficiency generally produces stronger effects than warming. For example, in a study on Mediterranean shrubland, drought caused significant reductions in seedling density in 13 woody species, while the effects of warming only affected three out of the eight study species (Lloret et al., 1999). A drought in Mediterranean shrubland caused increased emergence of woody species 2 years later, probably because the drought reduced competition from other species (del Cacho and Lloret, 2012). The same study also noted that drought did not change the emergence patterns of herbaceous species, and concluded that these species were more resilient to drought stress (del Cacho and Lloret, 2012). In contrast, another study on Mediterranean species found that habitat type (open, under shrub cover, under canopies) affected seedling emergence more than either wetter or drier conditions during the growing season (Matías et al., 2012).

Flooding has also been found to affect seedling emergence in a number of studies. A study on three Mediterranean oak species found that waterlogging reduced seed germination and root development in *Quercus suber* and *Quercus pyrenaica*. Waterlogging also increased the time to germination (Pérez-Ramos and Marañón, 2009).

Seedling performance

The direct effects of climate change are likely to change seedling responses on the physiological level, and recruitment at the community level. Therefore, seedling establishment has been a popular indication of the effects of climate change on plant communities (Lloret et al., 2009). Climate change can directly affect seedling survival, growth and competitive ability, and therefore may eventually alter community compositions.

Elevated carbon dioxide

In general, the response of seedlings to elevated CO_2 parallels the responses catalogued for adult plants. Cherry seedlings (*Prunus avium*) exposed to elevated CO_2 (350 ppm over ambient concentrations) increased their total dry mass production by 39%, exhibited a lower daily evapotranspiration rate and showed faster growth and development than their ambient counterparts (Centritto et al., 1999). In a series of studies on *Avena barbata*, Jackson et al. (1994, 1995) observed that during a drier year, more plants survived in the elevated CO_2 treatments compared to the ambient treatment. Red alder seedlings showed an increase in total plant size, basal diameter, seedling height, leaf area, bud biomass and branch number under elevated CO_2; however, the root to shoot ratio was not affected (Hibbs et al., 1995). When exposed to elevated CO_2, seedlings of invasive Kudzu exhibited higher total dry weight, more branching and taller stems (Sasek and Strain, 1988), earlier leaf expansion, larger leaf size, a faster leaf production rate, greater turgor pressure and more starch accumulation (Sasek and Strain, 1989). Such responses could make seedling establishment more successful (Sasek and Strain, 1988). Similarly, Ziska (2003) found that leaf area, leaf, stem and root biomass and net assimilation rate all increased in six weedy species (Canada thistle, field bindweed, leafy spurge, perennial sow thistle, spotted knapweed and yellow star-thistle) grown from seeds in 720 μmol/mol CO_2. Seedlings are often more responsive to elevated CO_2 when nutrients are supplied (reviewed in Krauss et al., 2008). The response to elevated CO_2 can also be affected by stressors. For example, in Australian mangrove seedlings, the net assimilation rate increased when they were exposed to elevated CO_2, but this effect was dampened when the seedlings were exposed to salinity stress (reviewed in Krauss et al.,

2008). Similarly, Middleton and McKee (2012) found that effects of flooding and salinity tended to override that of elevated CO_2 on seedling growth of dominant species in floating freshwater marshes, and species richness was insensitive to elevated CO_2. The response of seedlings to elevated CO_2 can also be tempered by the presence of competitors. A study on black mangrove (*Avicennia germinans*) seedlings showed that the seedlings accumulated biomass more rapidly under elevated than ambient CO_2 levels. However, when grown with *Spartina alterniflora* (a C4 grass), growth was reduced, even under elevated CO_2 (reviewed in Krauss *et al.*, 2008).

The effects of elevated CO_2 on seedling responses can be translated to shifts in competition outcomes at the community level. Researchers have found that elevated CO_2 can reverse normal competitive outcomes, even where the normally dominant plant has an advantage at the seedling stage of development. In studies of an annual plant community, researchers observed that larger seed and seedling size generally gave *Abutilon* (a C3 plant) a competitive advantage over *Amaranthus* (a C4 plant) at ambient conditions. However, when grown together at elevated CO_2 concentrations, *Amaranthus* exhibited a higher relative growth rate, as well as a higher nitrogen uptake rate, negating the advantage of *Abutilon*'s larger seed size (reviewed in Bazzaz, 1990). In another study on early and late successional trees, the researchers found that the seedlings of late successional trees (*Fagus, Acer saccharum* and *Tsuga*) growing in shade and elevated CO_2 accumulated more biomass than the seedlings of early successional trees (*Betula, Acer rubrum, Prunus* and *Pinus*) (reviewed in Bazzaz, 1990). An experiment on six invasive weedy species showed that Canada thistle and spotted knapweed responded more to possible future CO_2 levels (720 µmol l⁻¹) than the other five weeds, increasing their biomass to a greater extent (Ziska, 2003). In addition, Ziska (2003) found that the weedy species are more responsive to this century's increase in CO_2 (285–380 µmol l⁻¹) than plant species in general, which could explain the success of these noxious weeds (Ziska, 2003).

Warming

Depending on the optimum temperature range of the species, increased temperatures can either hinder or enhance seedling growth and development. The effect of temperature relies on the degree of the increase as well as the length of exposure. Seedlings are often more susceptible to heat than mature plants (reviewed in Krauss *et al.*, 2008). Studies on mangrove seedlings have shown that root growth can be hindered at 37°C and that 48 h in 40°C can kill seedlings (reviewed in Krauss *et al.*, 2008). In contrast, in the alpine tree and shrub community discussed above, Kullman (2002) found that warmer winters have decreased mortality due to frost injury or drought, further promoting seedling establishment and survival. Furthermore, they found that seedling growth had been enhanced, in the form of height accumulation, so that the trees no longer exhibited the truncated morphology normally associated with high elevations. Mild winters are generally beneficial for seedling survival of winter annuals. Furthermore, earlier-emerged seedlings have a longer time to grow before winter and therefore have better survival (Stanton, 1984; Rice and Dyer, 2001), which can lead to increased population growth (Zhang *et al.*, 2011).

In some cases, altered evaporation and soil moisture availability associated with warming (rather than the direct temperature effect) seem to play a large role in seedling performance. A study on calcareous grassland in the UK showed that winter warming (at an increase of 3°C) alone significantly decreased species richness, attributed to an increase in seedling mortality; however, when water was added to the warmed plots, seedling survival increased and enhanced species richness (Sternberg *et al.*, 1999). Based on their results, the authors predicted that warmer temperatures would lead to decreases in soil moisture content, cause gaps in the perennial grasses that would favour the establishment of annual species, and increase the competition between annual forbs and perennial grasses.

Increasing temperatures are likely to have an impact on plant species interactions, favouring the growth of some species but not others. Shifts in competitive ability may be especially important in agricultural systems, where weeds may respond better than crops to higher temperatures. For example, a study by Tungate *et al.* (2007) on sicklepod and prickly sida in a soybean system showed that weed growth was greater at higher temperatures than in soybean (while the maximum growth rate for soybean was achieved at 30°C, maximum growth for the weeds was achieved at daytime temperatures of 36°C). At optimal temperatures, the weeds had higher root to shoot ratios, and greater mycorrhizal colonization, which could help them outcompete soybean for water and nitrogen. This study showed that higher temperatures could change competitive dynamics by affecting resource acquisition, causing weeds to have greater impacts on crop yield in the future (Tungate *et al.*, 2007).

Altered precipitation regimes

Drought stress generally inhibits seedling growth. For example, Zavaleta (2006) observed that spring drought reduced seedling survival in *Baccharis pilularis*. In a study of the seedlings of five tropical tree species, the root to shoot ratio and net assimilation rate increased, while specific leaf area and relative growth rate declined; the decrease in plant biomass was greater in slow-growing species than in the faster-growing species (Khurana and Singh, 2004). Similarly, Sternberg *et al.* (1999) found that in a calcareous grassland, summer drought increased the growth of *Pastica sativa*, a perennial forb, while decreasing the growth of perennial grasses. They also noted that the combination of winter warming and summer drought reduced the overall plant cover in their plots. Lloret *et al.* (2004) found that warming and drought reduced seedling survival and related diversity in a Mediterranean-type community. In the study on cherry seedlings discussed above, seedlings exposed to drought stress during the growing season exhibited lower dry mass, as a result of lower root growth. Growth in elevated CO_2 environments generally did not ameliorate the negative effects of drought on seedling growth (Centritto *et al.*, 1999).

Lloret *et al.* (2009) found that seedling survival during the first summer was especially dependent on whether the seedlings were exposed to drought stress. Similarly, Matías *et al.* (2012) found that seedling survival was greatest when water was added, and decreased when the seedlings were exposed to drought conditions. Furthermore, Gómez-Aparicio *et al.* (2008) found that although water addition increased *Quercus* survival in the first year, it did not necessarily translate to survival after the second growing season. This suggested that water limitation in one summer could negate the benefits of increased water availability in previous summers. Therefore, if precipitation becomes more sporadic under future climate change scenarios, it could significantly affect *Quercus* establishment (Gómez-Aparicio *et al.*, 2008).

Water availability is also a major force in competitive interactions between plant species. Competitive outcomes can vary in dry years versus wet years. For example, a study on the Mediterranean shrub *Sarcopoterium spinosum* found that the presence of annual plants during a wet year had negative effects on *S. spinosum* establishment (as indicated by seedling density), whereas in a dry year, the presence of annual plants had a neutral or even beneficial effect on seedling establishment (Seifan *et al.*, 2010). It was hypothesized that during drought, the shrub species had less competition from the annual plants, favouring greater survival and growth (Seifan *et al.*, 2010). In this study system, the effects of competition on shrub growth were expected to be greatest in the seedling stage; at later life stages, woody shrubs are generally not as susceptible to competition from herbaceous plants (reviewed in Seifan *et al.*, 2010).

Flooding can inhibit seedling growth and development. Although the effects of flooding on seedling recruitment have not been as extensively investigated as drought, the physiological effects of waterlogging on plants in general are well known. For example, the slower diffusion of gases in water (10^4 times lower than in air) leads to decreased ATP production in affected cells, and increased levels of potentially toxic forms of iron and manganese (reviewed in Gibbs and Greenway, 2003), and reduced water and nutrient transport (reviewed in Araki *et al.*, 2012). These effects are likely to have an effect on seedling growth and survival. In fact, a study on wheat showed that seedlings grown in pots and exposed to waterlogging exhibited lower shoot and root growth, and that these effects were permanent; root mass did not recover, even after the waterlogging stress had ended (Araki *et al.*, 2012). Pérez-Ramos and Marañón (2009) found that seedlings of *Quercus* species showed lower root length and slower growth rates under waterlogging. They also suggested that the higher tolerance exhibited by *Quercus canariensis* relative to its congener could explain its position as the most abundant tree in the forest (Pérez-Ramos and Marañón, 2009). Similarly, Higa *et al.* (2012) studied the seedlings of two *Salix* and three Ulmaceae riparian tree species during and after 1, 2, 4 and 8 weeks of submergence, and found that shoot elongation and leaf production were decreased in all species during submergence. However, after submergence, the two *Salix* species were able to rapidly grow and produce leaves, while the Ulmaceae species showed lower underground biomass even after submergence. These results show that the range of riparian trees is determined by flooding via its influence on seedling growth (Higa *et al.*, 2012). Submersion in water can also affect photosynthesis by decreasing chlorophyll levels, and lowering fluorescence associated with photosystem II activity (Panda *et al.*, 2006), or increasing the quantum efficiency of photosystem II but reducing the maximum photosynthetic rate (Abbas *et al.*, 2012). Abbas *et al.* (2012) showed that emerging seedlings of the invasive cordgrass *Spartina densiflora* died after submergence in 8 cm of water. In addition, submergence in up to 4 cm of water caused lower nitrogen and pigment content in the leaves in addition to the effects on photosynthesis mentioned above. The authors concluded that submergence could be used to control cordgrass establishment in invaded salt marshes.

Like drought, flooding can influence competitive interaction between plants. Several studies in particular have focused on the use or implications of flooding in controlling or causing plant invasions. Florentine and Westbrooke (2005) found that seedling emergence of the invasive weed *Nicotiana glauca* increased in flooded plots compared to unflooded plots. They found that leachates from dry leaves and twigs of *N. glauca* significantly hindered germination of *Letuca sativa* seeds, suggesting that alleopathic characteristics of *N. glauca* could affect recruitment of native species in the field.

Seed bank longevity

Survival of seeds in the seed bank not only depends on avoiding germination, but also largely depends on preserved viability (Thompson, 2000). Once a seed enters the soil, it can be subject to any of several fates: it may leave the seed bank by germinating or dying, or it may stay viable in the seed bank (Fig. 9.1). Seed death may occur due to seed predators or pathogens, or simply due to ageing and gradual loss of viability (reviewed in Leishman *et al.*, 2000). In order to persist in the soil, seeds must be able to resist these factors. Elevated atmospheric CO_2 level is expected to have little effect on seed persistence in the soil, as explained in previous sections. Warming and altered precipitation, however, can have dramatic impact on seed longevity.

Elevated carbon dioxide

Similar to germination and seedling responses, seed bank longevity is generally believed to be insensitive to elevated CO_2 in the atmosphere.

Warming

Changes in soil temperature and moisture, as a result of ongoing climate change, may affect seed deterioration, determining seeds' potential performance (Ellis and Roberts, 1981a). The effect of temperature on seed longevity is characterized by the seed longevity equation developed by Ellis and Roberts (1980). The equation predicts that dry seeds can survive longer under cooler temperatures, which is in accordance with lab studies and field observations (Ellis and Roberts, 1981b; Dickie *et al.*, 1990; Mira *et al.*, 2011). Ooi *et al.* (2009) exposed seeds of eight ephemeral species from arid Australia to increased temperatures predicted for the future. They found dramatically declined seed viability in one species and greater germination percentages for another three species. Their results suggested that warming may hamper seed bank persistence of these species, and therefore their bet-hedging strategy to thrive in the arid habitats.

Climate change can potentially affect the reproductive pattern of masting species, which use climatic clues to initiate reproduction (reviewed in Lewis and Gripenberg, 2008). For example, elevated temperature may lead to asynchronous flowering of *Chionochloa* spp. in New England, allow populations of seed predators to build up and cause declines in plant populations (McKone *et al.*, 1998).

Altered precipitation regimes

Water availability is commonly believed to negatively affect seed longevity. The relationship between soil moisture and seed longevity is complex, because viability of dry seeds first decreases with increased seed moisture content, and then after the moisture content reaches some threshold, viability positively correlates with moisture (reviewed in Murdoch and Ellis, 2000 and Walck *et al.*, 2011). Moisture also interacts with temperature to affect seed longevity (Vertucci *et al.*, 1994). Furthermore, increased precipitation may also favour fungal pathogen activities and facilitate seed decay (reviewed in Wagner and Mitschunas, 2008). In a field study on wild oat (*Avena fatua*), Mickelson and Grey (2006) found that as soil water content increased, so did mortality of buried seeds, probably due to enhanced microbial activity. On the other hand, reduced precipitation may increase predation risk in some species. Conlisk *et al.* (2012) used a combination of species distribution models and metapopulation models to predict population dynamics of *Quercus engelmannii* under climate change. They found potential

population declines under drought and increased fire frequency, which was due to decreased frequency of masting, a strategy reducing predation risk.

Maternal Effects of Climate Change on Seed and Seedling Traits

Maternal environmental conditions can also influence offspring fitness (Roach and Wulff, 1987); therefore, climate change can not only directly affect plant performance, but also influences offspring fitness indirectly. The effects of maternal environments on offspring performance are ubiquitous in plants (reviewed in Roach and Wulff, 1987; Donohue and Schmitt, 1998). Here we define maternal environmental effects as the phenomenon in which the external ecological environment of the maternal parents influences the phenotype of the offspring (Donohue, 2009).

Because most plants are hermaphrodites, changed maternal environments often imply changed paternal environments as well, although the latter is only addressed in a few studies (Lacey, 1996; Lacey and Herr, 2000). Maternal effects take place through two processes, the maturation environment during seed development (postzygotic maternal effect), and the quality of the maternal plant itself before seed setting (prezygotic maternal effect). Environmental factors during seed development such as photoperiod, temperature and water availability can affect seeds' metabolism and gene expression, which determine seed quality and subsequently seedling performance (Donohue, 2009). In some cases, prezygotic effects alone are enough to affect seed quality (Kochanek *et al.*, 2010, 2011). For example, Kochanek *et al.* (2010) found that for a semi-arid species in Australia, a warmer prezygotic environment led to the same mean seed longevity with a smaller variance, compared with a cooler prezygotic environment. Their results suggest potential declines in seed bank persistence under global warming. However, very few studies separate the prezygotic effect from the postzygotic effect (Lacey, 1996; Lacey *et al.*, 1997; Lacey and Herr, 2000). Most experiments focus on the postzygotic maternal effect alone by manipulating the seed maturation environment (e.g. Gutterman, 2000; Steadman *et al.*, 2004; Hoyle *et al.*, 2008; Kochanek *et al.*, 2010), while others focus on the combined outcomes of prezygotic and postzygotic processes by changing the environments for the entire life of the maternal plants (e.g. Schmuths *et al.*, 2006; Hovenden *et al.*, 2008). Here we distinguish the two kinds of studies by referring to the former as 'seed matured in...' and the latter as 'seed produced by plants grown in...'. We hope future studies can explicitly address prezygotic and postzygotic maternal effects, because a thorough understanding of maternal effects caused by climate change requires knowledge on both the separate and the combined effects.

Reproductive allocation of maternal plants

Variable environmental conditions during seed maturation can affect reproductive allocation (Gallagher and Fuerst, 2006), as when resources are scarce, plants must split these resources between growth, reproduction and maintenance (Bazzaz *et al.*, 2000). Changes in resource availability have been found to affect reproductive allocation, influencing seed size, morphology and chemistry (Luzuriaga *et al.*, 2006). Much of the literature on reproductive allocation and climate change, however, has focused primarily on the number and size of produced seeds. Higher seed production can help ensure successful recruitment in the following growing season, especially in unpredictable environmental conditions (Smith and Fretwell, 1974). Furthermore, enhanced sexual reproduction may improve offspring genetic variability, which can improve species persistence under changing climates (Klady *et al.*, 2011). Seed size is also a strong indicator for seed viability within species. Large seeds could have a greater nutrient content than smaller seeds (Bazzaz, 1990), and therefore better stress

tolerance (Muller-Landau, 2010), establishment and growth (Bazzaz *et al.*, 2000), and ultimately higher fitness overall than smaller seeds from conspecifics (Aronson *et al.*, 1993).

Elevated carbon dioxide

In general, elevated CO_2 is found to increase seed production, although this effect is generally contingent on the functional groups involved and availability of other resources. In studies on wild oat, elevated CO_2 increased the seed biomass per plant (Granger, 2012) and seed number (O'Donnell and Adkins, 2001; Granger, 2012). Thürig *et al.* (2003) found that in a calcareous grassland, the number of flowering shoots and seeds increased. However, because the above-ground biomass similarly grew, the overall reproductive allocation was not changed. A meta-analysis of 79 crop and wild species showed that on average, CO_2 enrichment resulted in higher numbers of flowers, fruits and seeds, as well as large seed size; however, crops tended to allocate more mass in reproduction than wild species under elevated CO_2 (Jablonski *et al.*, 2002). HilleRisLambers *et al.* (2009) also pointed out that annuals and crop species were likely to respond to an increase in a limiting resource by increasing seed production, while perennial herbaceous species may not. In a calcareous grassland, Thürig *et al.* (2003) found that grass species increased seed production more than non-legume forbs, while seed production in legumes did not change. In contrast, Miyagi *et al.* (2007) found that leguminous crop species increased seed production more under elevated CO_2 than non-nitrogen-fixing crop and wild species in a grassland, suggesting nitrogen may be a limiting factor in plants' reproductive responses to elevated CO_2. These studies suggest potential impacts of elevated CO_2 on community structure and agriculture success.

Experiments on elevated CO_2 have revealed inconsistent effects on seed size; it can both reduce (Wulff and Alexander, 1985; Andalo *et al.*, 1998; Huxman *et al.*,

1998) and increase (Garbutt and Bazzaz, 1984; Steinger *et al.*, 2000; Thürig *et al.*, 2003) seed size.

Warming

Mild temperature increases generally stimulate plant sexual reproduction, especially in high latitudes or elevations (Price and Waser, 1998; Arft *et al.*, 1999; Klady *et al.*, 2011; Pieper *et al.*, 2011); however, responses may vary through time, across climatic zones and among functional groups. Hoyle *et al.* (2008) found that for a native Australian forb *Goodenia fascicularis*, plants grown in warmer conditions produced more seeds despite smaller sizes and less above-ground biomass, compared to those in cooler conditions. A meta-analysis on data from 13 International Tundra Experiment sites showed that warming only increased reproductive effort and success after several years' warming treatment; the lag in reproductive responses was probably due to the formation of flower buds in previous seasons (Arft *et al.*, 1999). The authors also found that reproductive responses were much stronger in High Arctic than in Low Arctic, where stronger vegetative responses were found (Arft *et al.*, 1999). Furthermore, herbs were more responsive to warming than woody plants (Arft *et al.*, 1999). On the contrary, Klady *et al.* (2011) found that shrubs and graminoids were more responsive than forbs, based on data from 12 years' experimental warming in the High Arctic. Such divergent responses of different functional groups may lead to shifts in community structures.

Although gradual warming generally positively affects plant reproduction, extreme warming events may greatly hinder plant development and reproduction. Abeli *et al.* (2012) found that summer heatwaves dramatically decreased flower production of two montane species in Northern Italy. Similarly, reduced snow packs caused by extreme winter warming events largely impacted reproduction of dominant species in sub-arctic heathland (Bokhorst *et al.*, 2008, 2011) and the reproductive output of four dwarf-shrub species in alpine tundra (Wipf *et al.*, 2009).

Maternal temperature may also affect seed size and seed viability. Gao *et al.* (2012) found that increased temperature enhanced the proportion of heavy seeds and therefore increased total dry weight of *Leymus chinensis*. On the contrary, in some species, decreasing temperature during seed development can result in larger seeds because of slow maturation, and therefore warmer temperature during seed maturation led to lighter seeds compared to colder temperature (Alexander and Wulff, 1985; Lacey, 1996). Thomas *et al.* (2009) also found that seeds produced by maternal plants under heat stress were significantly smaller. Qaderi and Reid (2008) found that *Silene noctiflora* produced fewer viable seeds under warming.

Altered precipitation regime

Drought stress generally reduces reproductive allocation. By hindering photosynthetic efficiency, drought can reduce grain fill (Aronson *et al.*, 1993; Seiler *et al.*, 2011) and reduce seed size (Sultan and Bazzaz, 1993; Luzuriaga *et al.*, 2006; Harel *et al.*, 2011; Granger, 2012). Similarly, drought and unpredictable watering regimes decrease the number of seeds per plant as well as total seed biomass per plant (Aronson *et al.*, 1993; O'Donnell and Adkins, 2001; Granger, 2012). Granger (2012) found that drought had a stronger effect on reproduction than elevated CO_2; whereas drought decreased seed biomass per plant, and seed mass in wild oat (lowering both hull and caryopsis weight), elevated CO_2 affected total seed biomass per plant but had very little influence on seed mass. Luxmoore *et al.* (1973) found that flooding during the grain filling stage reduced yield, as well as average grain weight. The severity of the effect depended on the duration of flooding. Similarly, texas-weed (*Caperonia palustris*) was found to be largely tolerant to flooding, but fruit production decreased compared with control plants.

Reproductive phenology of maternal plants

Phenology sets the clock of plant life cycles – determining the timing of shifts between growth and reproduction and also the start of subsequent generations. Ongoing climate changes can shift plant phenology, affecting both the timing and speed of regeneration.

Elevated carbon dioxide

Existing literature suggests that phenological responses to precipitation and CO_2 are highly variable and inconsistent among species (Reekie and Bazzaz, 1991; reviewed in Cleland *et al.*, 2007). For example, Cleland *et al.* (2007) found that elevated CO_2 delayed flowering in grasses, while accelerating flowering in forbs. Researchers also found that crop species were more responsive to elevated CO_2 in terms of accelerating their development than wild species (Kimball *et al.*, 2002), suggesting potential limits posed by other factors such as nutrients and light (Cleland *et al.*, 2007).

Warming

Advanced flowering phenology by warming has been widely documented, as many plants use temperature as a cue for their emergence and reproduction (Bradley *et al.*, 1999; Hughes, 2000; Fitter and Fitter, 2002; Root *et al.*, 2003; Sherry *et al.*, 2007). Advanced phenology may expose plants to adverse environments and hamper reproductive organs. For example, advanced phenology of some montane wildflowers in Colorado exposed flowering buds to frosting, causing dramatic declines in their seed production, and significantly altered population size and age structures (Inouye, 2008). Furthermore, as many insects use different cues compared to plants, potential mismatches between plants and pollinators may arise under climate change (Memmott *et al.*, 2007; Hegland *et al.*, 2009; Yang and Rudolf, 2010; Forrest and Thomson, 2011). Kudo *et al.* (2004) found that seed production of *Corydalis ambigua* dramatically decreased in warmer years as a result of phenological mismatch with bumblebees (*Bombus* spp.). Furthermore, phenological shifts may also change plant–herbivore interactions (reviewed in Donnelly *et al.*, 2011). For example, warming led to an intense overlap between flowering of *Gentiana formosa* and larva emergence of a noctuid moth (*Melanchra pisi*), causing tremendous damage to the reproduction of the plants (Liu *et al.*, 2010).

Earlier initiation of reproduction (i.e. flowering) may (but not necessarily) also lead to advanced seed setting and dispersal. Hannerz *et al.* (2002) found that earlier seed dispersal of *Pinus sylvestris* was associated with warmer climate. However, whether timings of seed maturation and dispersal are shifted, as well as the direction and extent of the shifts, also depends on the durations of developmental stages (Post *et al.*, 2008; Zhang *et al.*, 2012b). For example, Price and Waser (1998) found that earlier flowering led to advanced seed setting and seed maturation, with no change in flowering or fruiting durations. On the other hand, Hoffmann *et al.* (2010) found that a sub-alpine plant species, *Poa hiemata*, maintained the same seed maturation timing, even though flowering time was shifted by experimental warming. Similarly, Jones *et al.* (1997) found that summer warming advanced pollen dispersal and stigma receptivity of two wind-dispersed *Salix* species but not their seed dispersal timing.

Shifted seed maturation and dispersal timing may affect the subsequent fate of seeds. In winter annuals, earlier germination in the autumn may lead to a longer time for seedlings to grow and a larger size to overwinter, and therefore a better chance of survival (Zhang *et al.*, 2012a). However, earlier maturation of seeds may also suffer from the risk of mismatches with seed dispersers, especially when seed dispersers use an environmental cue different from the plants (Tylianakis *et al.*, 2008; Ruxton and Schaefer, 2012). Even when they are using the same cue, e.g. temperature, mismatch may also arise depending on the extent of their responses (Warren *et al.*, 2011).

Altered precipitation regimes

Like elevated CO_2, phenological responses to shifted precipitation are highly variable and are system specific. Precipitation tends to play an important role in arid or semiarid systems (Penuelas *et al.*, 2004; Lesica and Kittelson, 2010).

Seed chemistry

Maternal environmental conditions affect seed quality by influencing the allocation of secondary compounds and organic acids to seeds. These compounds (discussed in more detail in Chapter 8 of this volume), including phenolic, aliphatic and long-chain fatty acids, have been found to influence seed persistence. Phenolic acids have been found to play a role in seed pathogen resistance (Burnham, 2011), inhibition of germination (Gallagher *et al.*, 2010a), protection against radical oxygen species (reviewed in Shirley, 1998; Yu and Jez, 2008), and defence against insects (Abdel-Aal *et al.*, 2001). Aliphatic acids have been found to be antibacterial (Charnock *et al.*, 2004), may play a role as germination inhibitors (Hart and Berrie, 1968), or serve as a carbon source (Chia *et al.*, 2000). Long-chain fatty acids likewise may inhibit germination (Berrie, 1979; van Beckum and Wang, 1994; Bhatia *et al.*, 2005) and serve in the signalling pathway for systemic acquired resistance and programmed cell death during plant responses to pathogens (Upchurch, 2008). Clearly, changes in the production and allocation of these compounds could have profound implications for seed longevity and quality.

Elevated carbon dioxide

Many studies have established that atmospheric CO_2 levels can influence the concentrations of carbon-based secondary compounds in plant roots, shoots and leaves. However, few studies have focused on the effects of elevated CO_2 on seeds. Several experiments showed that the carbon to nitrogen (C:N) ratio increases in seeds that have matured under elevated CO_2 (Andalo *et al.*, 1998; Huxman *et al.*, 1998; He *et al.*, 2005; Steinger *et al.*, 2000; Jablonski *et al.*, 2002; Hovenden *et al.*, 2008), suggesting that seeds may act as sinks for the excess photosynthates

produced under elevated CO_2 (Steinger *et al.*, 2000). However, these studies did not examine which specific compounds were responsible for the changes in the C:N ratio. A recent study by Granger (2012) found that individual phenolic compounds, as well as aliphatic and long-chain fatty acids, were influenced by atmospheric CO_2 concentration, although these effects were not consistent across wild oat lines or replicates. (For example, the concentrations in 2009 showed more response to CO_2 and drought than in 2010. Similarly, the response of the wild oat lines M73 and SH430 often differed between 2009 and 2010, making it difficult to distinguish reliable trends.) However, the results did suggest that individual organic compounds may be differentially regulated.

The effect of maternal CO_2 concentration on seeds varies for different fatty acids, and for different plant species. Högy *et al.* (2010) found that elevated CO_2 increased the concentration of an unsaturated fatty acid (oleic acid), but reduced the levels of saturated fatty acids (linolenic, nervonic, linoleic) in oilseed rape seeds. Burkey *et al.* (2007) found concentrations of both unsaturated and saturated fatty acids were decreased under elevated CO_2 in groundnut seeds. Thomas *et al.* (2003) found that fatty acid concentrations in soybean were unaffected by atmospheric CO_2 levels.

Warming

Many studies of climate effects on seed chemistry focus on crop species. Thomas *et al.* (2009) found that soybean seeds produced from warmed mothers had high concentrations of sucrose and raffinose compared to seeds from the ambient control plants, which have been reported to be associated with decreased seed quality (Bernal-Lugo and Leopold, 1998).

Altered precipitation regimes

Drought has been proposed to affect seed chemical constituents via various mechanisms: by reducing allocation of photosynthates to seeds as a result of reduced photosynthesis and enzyme activity

(Ali *et al.*, 2010), by increasing plant consumption of storage molecules (such as fatty acids) in order to cope with drought stress (Shu *et al.*, 2011) or by increasing the expression of stress hormones that can affect the levels of compounds such as long-chain fatty acids (reviewed in Bouchereau *et al.*, 1996).

Effects of drought on seed chemical constituents depend on timing and severity of the drought stress, the species and genotype of the plant and the type of compound being studied. Drought has been found to reduce phenolic concentrations in wild oat (Gallagher *et al.*, 2010b; Granger, 2012), and in groundnut seeds, depending on the cultivar (Ross and Kvien, 1989). Increased rainfall prior to flowering, along with drought stress during seed development, caused lower levels of proanthocyanins and anthocyanins in grape seeds (Lorrain *et al.*, 2011). However, drought increased the polyphenol concentrations in cumin seeds (Rebey *et al.*, 2012). Depending on the timing of drought stress, the concentrations of simple non-cholinic phenolic compounds was reduced in *Brassica napus* seeds, although the levels of sinapine increased, decreased or showed no change. Although total phenolic concentrations decreased in maize kernels that matured under drought stress, the levels of flavonoids increased (Ali *et al.*, 2010).

Fatty acid concentrations have been found to be unaffected in maize kernels grown under drought conditions (Ali *et al.*, 2010), decrease or show no change in wild oat seeds (Granger, 2012) or increase in plant embryos (reviewed in Bouchereau *et al.*, 1996). Bouchereau *et al.* (1996) found that drought stress during flowering did not affect fatty acid content, but stress during vegetative growth caused one cultivar to reduce erucic, gadoleic and oleic acid levels. However, in sunflower seeds, drought increased the oleic acid content, accompanied by a corresponding decrease in linoleic acid; stearic acid concentrations also rose (Sezen *et al.*, 2011). Studies on wild oat seeds have found that drought has weak or little effect on aliphatic acid content (Gallagher *et al.*, 2010b; Granger, 2012).

Primary seed dormancy

Elevated carbon dioxide

Few studies have investigated the effects of maternal exposure to elevated CO_2 on seed dormancy. Granger (2012) found that elevated CO_2 during seed maturation increased the percentage of dormant (*A. fatua*) seeds after 12 weeks of after-ripening. Clearly, more experiments on the effects of elevated CO_2 on species with dormant seeds are needed.

Warming

A consistent trend is that seeds that mature in warmer conditions are often less dormant (Gutterman, 2000). For example, Hoyle *et al.* (2008) found that *Goodenia fascicularis* developed at warmer temperature were less dormant compared to those developed at cooler temperature. Zhang *et al.* (2012a) found that seeds produced by warmed mothers of *Carduus nutans* were less dormant and germinated faster than seeds produced by those grown in ambient conditions. They also found that the effect of maternal temperature was more obvious in suboptimal germination temperatures (Zhang *et al.*, 2012b). Lacey *et al.* (1997) found that cooler postzygotic temperature leads to heavier seed coats but unchanged embryo weight, which may lead to increased dormancy and lower germination percentage.

Altered precipitation regimes

Several studies have found that drought stress decreases dormancy levels in wild oat (*A. fatua*) seeds (Peters, 1982; Sawhney and Naylor, 1982; Granger, 2012; Gallagher *et al.*, 2013), *Sorghum bicolor* (Benech-Arnold *et al.*, 1992), *Goodenia fascicularis* (Hoyle *et al.*, 2008), wild radish (*Raphanus raphanistrum* L.) (Eslami *et al.*, 2010) and *Brassica tournefortii* (Gorecki *et al.*, 2012). Granger (2012) found that the effects of drought on seed dormancy could be partially mitigated by maternal exposure to elevated CO_2. Gorecki *et al.* (2012) also found that, in addition to being less dormant, seeds produced by plants under drought conditions were more responsive to a smoke-derived chemical (karrikinolide) that can promote germination.

Seed dispersal

Maternal environments may not only affect the number of seeds produced, but also how far seeds can travel. While there are a few studies examining how changed abiotic dispersing environments under climate change can affect dispersal of wind-dispersed seeds (Soons and Bullock, 2008; Kuparinen *et al.*, 2009; Nathan *et al.*, 2011), very little research has been conducted on how climate change affects seed-dispersal characteristics. Seed morphology (e.g. seed size, pappus, hooks, etc.), fruit quality and dispersal-related properties of the maternal plants can all affect the dispersal process.

Elevated carbon dioxide

McPeek and Wang (2007) found seeds produced by dandelions under elevated CO_2 have longer beaks and larger pappi, which may enhance their dispersal in windy conditions. On the other hand, as elevated CO_2 may increase seed mass in some species (Jablonski *et al.*, 2002; Qaderi and Reid, 2008), wind dispersal may be hindered.

Warming

Zhang *et al.* (2011) examined the response of another wind-dispersed species, *Carduus nutans*, to warming, and found no effects on any of the studied seed-dispersal traits, such as seed mass, pappus morphology or terminal velocity. However, they found that maternal plants grew taller under warming, which enhanced seed dispersal and therefore population spread. This positive effect of maternal warming on dispersal may be amplified by direct climatic effect on the dispersal environment. In particular, warmer climate is likely to cause more air turbulence, which may facilitate seed release and dispersal (Skarpaas *et al.*, 2006; Nathan *et al.*, 2011). On the other hand,

Qaderi and Reid (2008) found that increased temperature (28/24°C versus 22/18°C) decreased plant height and number of viable seeds of *Silene noctiflora*. For animal-dispersed species, climate change may affect dispersal efficiency by altering fruit quantity and quality (McConkey *et al.*, 2012).

Altered precipitation regimes

Case studies of altered precipitation regimes on dispersal of seeds are rare. Zhang *et al.* (2011) did not find any effect of altered precipitation on the dispersal traits of *C. nutans*. But ongoing work suggests that the proportion of released seeds in *C. nutans* can be enhanced by maternal drought (B. Teller, 2010, unpublished data).

Germination and emergence

Elevated carbon dioxide

Elevated CO_2 during seed maturation has been found to have little effect on subsequent seed viability. Stiling *et al.* (2004) found that elevated CO_2 did not affect seed viability in *Quercus* species, while Way *et al.* (2010) showed similar results in *Pinus taeda*.

The effects of elevated CO_2 on germination and emergence have been found to be dependent on the species and genotype being studied, as well as interactions with other treatments and environmental factors. *P. taeda* seeds from elevated CO_2 treatments germinated to a greater extent, and also earlier, than seeds from ambient treatments (Hussain *et al.*, 2001). This result could be because of higher lipid concentrations in the elevated CO_2 seeds, which can provide energy during germination (Hussain *et al.*, 2001). Thürig *et al.* (2003) found that maturation during exposure to elevated CO_2 had no effect on germination percentage of seeds from species in a calcareous grassland, but time to germination was shorter for two species, and longer for one species. In contrast, Stiling *et al.* (2004) found no effect of elevated CO_2 on the germination rate of four *Quercus* species. Huxman *et al.* (1998)

found that germination percentage in *Bromus tectorum* was statistically similar in seeds from ambient and elevated CO_2 treatments. A study on wheat found that the germination percentages were similar between seeds from ambient and elevated CO_2 treatments. However, when they measured germination rate, the effects of elevated CO_2 were determined by the genotype. In some genotypes, germination rates were higher in 700 and 1000 ppm relative to 350 ppm, in others, seed germination rate was reduced in elevated CO_2, and in others, the germination rate was reduced in 700 ppm but not in the higher CO_2 treatment. The germination rate was found to be negatively correlated with grain carbon content (Bai *et al.*, 2003). In wild oat, the effects of elevated CO_2 were found to interact with other environmental factors and treatments. Parental exposure to elevated CO_2 significantly affected the percentage of emerged wild oat seedlings, but these effects were inconsistent and dependent on interactions with drought and accelerated ageing treatments (Granger, 2012).

Warming

Temperature effects on seed viability vary among species (Baskin and Baskin, 1998). Hoyle *et al.* (2008) showed that *Goodenia fascicularis* grown in a cool environment produced seeds with higher viability and greater mass. Hovenden *et al.* (2008) found little effect of CO_2 or warming on the mean seed mass of *Austrodanthonia caespitosa*; however, CO_2 greatly increased the proportion of non-viable seeds.

A number of studies have found that mildly increased maternal temperature resulted in seeds with enhanced germination and emergence (reviewed in Alexander and Wulff, 1985; reviewed in Fenner, 1991; Qaderi *et al.*, 2006; Schmuths *et al.*, 2006). However, in some cases, increased maternal temperature can hinder germination and emergence (Hume, 1994; Gibson and Mullen, 1996), and in other cases maternal effect can be obscure and hard to detect (Monty *et al.*, 2009), especially compared to the overwhelming offspring environmental temperature. Zhang *et al.* (2012a) found that

maternal warming had no effect on the total percentage of emerged seedlings. However, the mean emergence time was shorter in seedlings produced by warmed mothers. In contrast, Thomas *et al.* (2009) found that maternal warming led to lower percentage of emergence and slower emergence rate in soybean.

Warming also interacts with elevated CO_2 to produce complex outcomes. For example, elevated CO_2 and warming both reduced germination of *A. caespitosa* when applied alone, whereas no effect was detected when applied together (Hovenden *et al.*, 2008). High temperature during seed maturation also interacts with moisture stress and can induce dormancy in some genotypes of *Triticum aestivum* (Biddulph *et al.*, 2007). Maternal environments not only affect the number of germinators, but can also control the timing as to when the next generation starts. This maternal effect can amplify the direct effect of warming on phenology, to produce a more dramatic outcome on the timing of population regeneration. Saarinen *et al.* (2011) found that in *Thlaspi arvense* (whose seeds can germinate in both spring and winter), an 8-day warming spell in the winter experienced by the maternal plants largely reduced the percentage of winter annuals in the offspring compared to those of summer annuals. This suggests that maternal environment may alter the population turnover of this species.

Altered precipitation regimes

Drought, especially during the reproductive stages, can significantly affect seed production and viability. Lower average daily rainfalls have been correlated to lower percentages of viable seeds in common bean (*Phaseolus vulgaris*) (Muasya *et al.*, 2008). In barley (*Hordeum vulgare*), drought conditions can stimulate dormancy release, causing pre-harvest sprouting, which can lead to a loss of seed viability (Gualano and Benech-Arnold, 2009).

Parental exposure to drought generally decreases germination and emergence. For example, germination percentages have been found to decrease with drought stress in groundnut (*Arachis hypogaea*) (Ketring, 1991), soybean (*Glycine max*) (Dornbos *et al.*, 1989), native and invasive plant species in Australia (Peréz-Fernández *et al.*, 2000) and *Centaurea hyssopifolia* (Pías *et al.*, 2010). Maternal drought has been found to decrease the germination rate in groundnut (Ketring, 1991).

Seedling performance

Elevated carbon dioxide

The results of experiments on maternal exposure to elevated CO_2 and the implications for seedling vigour have been variable, showing that CO_2 can influence seedling morphology, growth and survival, but not in all cases. Hussain *et al.* (2001) found that parental exposure to elevated CO_2 can affect seedling morphology; *P. taeda* seedlings from elevated CO_2 treatments exhibited longer roots and more needles than seedlings from ambient treatments. However, this may not translate into more vigorous seedlings. In their study on *P. taeda* seedlings, Way *et al.* (2010) found that there was no difference in survival rates for seedlings from ambient and elevated CO_2 treatments, measured 80 days after stratification. Although a smaller proportion of the seeds from the ambient treatments developed into seedlings and survived, this result was not statistically significant. Similarly, elevated CO_2 during seed maturation did not consistently affect wild oat seedling growth or resistance to accelerated ageing; the effects of CO_2 depended on an interaction with a drought treatment and the duration of ageing (Granger, 2012). Conversely, *Plantago lanceolata* seeds that matured under elevated CO_2 and were then allowed to germinate under elevated CO_2 had faster relative growth rates than those seeds grown under ambient CO_2 conditions, showing that the maternal atmospheric environment can affect seedling growth (Wulff and Alexander, 1985). Studies on *A. barbata* in elevated CO_2 showed that plant density in the second year of exposure was greater than could be accounted for by seed production alone,

indicating that either seeds survived better in the elevated CO_2 treatment, or that the parental environment had enhanced seed quality (Jackson *et al.*, 1994, 1995).

Warming

Previous studies have shown that increased maternal temperature can affect seedling size in both directions (Stearns, 1960; Thomas and Raper, 1975; Alexander and Wulff, 1985). Recently, Hovenden *et al.* (2008) found that 6-week seedlings of *Austrodanthonia caespitosa* descended from mother plants grown in warmer conditions were smaller with a higher root:shoot ratio than those from unwarmed mothers. Maternal warming also led to lower seedling mass in soybean (Thomas *et al.*, 2009).

Altered precipitation regimes

Maternal drought stress has been found to have both negative and positive effects on seedling growth and survival. Dornbos *et al.* (1989) found decreased seedling growth of soybean seeds in response to maternal drought. In the annual *Impatiens capensis*, maternal drought decreased offspring biomass, as well as carbon assimilation rates and lower stomatal conductances when seedlings were grown under well-watered conditions (Riginos *et al.*, 2007). Results on whether parental drought stress affects drought tolerance in the offspring have been mixed. Maternal drought had little effect on offspring drought tolerance in *I. capensis* (Riginos *et al.*, 2007), and *Centaurea hyssopifolia* (Pías *et al.*, 2010), but positively influenced growth and survivorship in *Polygonum persicaria* (Herman *et al.*, 2012). Interestingly, Herman *et al.* (2012) not only studied the influence of parental drought on offspring growth, but also investigated the influence of the grandparental generation. They found that seedlings from drought-stressed grandparents and parents had longer, faster-growing root systems, accumulated greater biomass and had greater survival rates under drought conditions than seedlings whose ancestors had been well watered. In addition, they found the effects of grandparental and parental environmental conditions were additive, and that the influence of the grandparental generation on seedling characteristics persisted even if the parent generation had not been exposed to stress (Herman *et al.*, 2012). The mechanisms responsible for the transmission of this phenotypic plasticity across generations is unknown, but Herman *et al.* (2012) propose several possible biochemical and epigenetic pathways, including hormones, regulatory proteins or RNA.

Seed bank longevity

Elevated carbon dioxide

Few experiments have focused on the influence of maternal exposure to elevated CO_2 on seed longevity. A study on the effects of maternal environment on wild oat found that elevated CO_2 did not significantly affect seed response to accelerated ageing; although elevated CO_2 interacted with the ageing treatment, ageing had a detrimental effect on seedling emergence and growth regardless of maternal environment (Granger, 2012). In a related experiment, the effects of maternal CO_2 on microbial decay were investigated, but elevated CO_2 had an inconsistent effect on resistance to *Fusarium culmorum*, increasing resistance in one year and decreasing resistance in another year. This effect may have been due to slightly higher growth chamber temperatures in the maternal environment in the second year, changing the allocation of phenolic compounds and fatty acids to seeds (Granger, 2012).

Warming

Kochanek *et al.* (2010) explicitly addressed prezygotic effects of temperature and soil moisture on seed longevity by growing *Wahlenbergia tumidifructa* plants in different prezygotic environments and moving them to a common environment once flowering started. They found that cold-wet plants produced seeds that were longer lived than others and also cooler temperature resulted in seeds with a wider variation

of lifespans than warmer temperature. Similar results were also found for *Plantago cunninghamii* (Kochanek *et al.*, 2011). Such results suggest that seed bank persistence of these species may be hampered under expected climate change.

Altered precipitation regimes

Maternal drought did not seem to increase the seeds' susceptibility to ageing or seeds' resistance to microbial decay (Granger, 2012). Irrigation was found to significantly affect the potential longevity of stored seeds of *Brassica campestris*, with seeds from plants that were irrigated longer achieving higher potential longevity values (Sinniah *et al.*, 1998).

Concluding Remarks

Climate change widely affects species abundances and distributions. Successful predictions of species performance under climate change require filling the gap of knowledge in plant early life-history responses. Existing literature on the ecophysiological responses of seeds and seedlings to environmental factors provides a solid basis for extrapolations into climate-change biology. At the same time, more efforts are needed to link individuals' ecophysiological responses to population and community outcomes, which had been done only in a handful of studies (Lacey and Herr, 2000; Kimball *et al.*, 2010; Levine *et al.*, 2011; Zhang *et al.*, 2011; Ooi *et al.*, 2012). Combinations of lab and field experiment, application of population and community theoretical frameworks, and long-term monitoring are useful tools to complete the missing link. For example, Lacey (1996) first tested maternal temperature effects on *Plantago lanceolata* germination and emergence in the lab, followed by field experiments to verify lab findings (Lacey and Herr, 2000). The authors then incorporated their results into matrix population models to examine changes in population growth rates and the contributions of altered early life histories to altered population growth. They found maternal warming during seed maturation led to 50% increase in population growth (Lacey and Herr, 2000). Levine *et al.* (2011) used similar approaches to examine how the population growth of three endangered species was affected by changes in germination due to shifted timing of the first storm in the growing season. Kimball *et al.* (2010) combined 25 years' demographic data with ecophysiological knowledge of individual species to examine shifts in community structure of a winter-annual community in Arizona. They found that altered germination phenology was the most important driver of community shifts under climate change. These studies all demonstrate excellent integration of knowledge and methodologies from various fields to explore the role of regeneration from seeds under climate change.

The role of seed banks in population persistence under climate change may increase as temporal climatic variations may increase. Climate change is expected to have non-negligible effects on seed bank persistence. Warming and altered precipitation may alter seed mortality due to germination (Ooi *et al.*, 2012). Soil warming and heatwaves may expedite loss of seed dormancy in the seed banks (Ooi *et al.*, 2009). Shifted precipitation may also affect seed longevity (Kochanek *et al.*, 2010) and seed quality (Qaderi and Reid, 2008). Furthermore, increased precipitation and more severe weather events may affect the range and severity of soil-borne pathogens that affect seeds (Chakraborty, 2011; Luck *et al.*, 2011), and therefore may affect the process of microbial decay of seeds (Wagner and Mitschunas, 2008). Changes in seed abundance or reproductive phenology could concomitantly affect interactions with seed predators as well (Lewis and Gripenberg, 2008). Therefore the buffering effect by seed banks against unfavourable conditions may be impacted by climate change. Studies investigating the mechanistic responses of seed bank dynamics to projected climate change are scarce, but are crucial for projecting species distributions and persistence under climate change (Ooi *et al.*, 2012).

In many cases, seed and seedling responses to climate change are species specific, and are also sensitive to the timing and intensity of climatic events (Morris *et al.*, 2008). Approaches based on functional groups (e.g. Arft *et al.*, 1999; Milbau *et al.*, 2009; Shevtsova *et al.*, 2009) can be more useful in predicting future responses of individual species as well as community consequences.

Maternal environmental effects remain a missing link in our understanding of plant regeneration via seeds (Donohue, 2009). Climate change is a long-term trend and its effects are distributed over multiple generations of plants. Interaction between offspring environment and maternal environment may drive evolutionary divergence of life-history patterns in different species, and therefore affect community composition under climate change (Walck *et al.*, 2011). Hence, multigenerational studies that examine the interactions between the direct effects climate change imposes on offspring and indirect maternal effects are worthwhile for our understanding of species adaptation to climate change.

References

Abbas, A.M., Rubio-Casal, A.E., Cires, A.D., Figueroa, M.E., Lambert, A.M. and Castillo, J.M. (2012) Effects of flooding on germination and establishment of the invasive cordgrass *Spartina densiflora*. *Weed Research* 52, 269–276.

Abdel-Aal, E.S.M., Hucl, P., Sosulski, F.W., Graf, R., Gillot, C. and Pietrzak, L. (2001) Screening spring wheat for midge resistance in relation to ferulic acid content. *Journal of Agricultural and Food Chemistry* 49, 3559–3566.

Abeli, T., Rossi, G., Gentili, R., Gandini, M., Mondoni, A. and Cristofanelli, P. (2012) Effect of the extreme summer heat waves on isolated populations of two orophitic plants in the north Apennines (Italy). *Nordic Journal of Botany* 30, 109–115.

Alexander, H.M. and Wulff, R.D. (1985) Experimental ecological genetics in *Plantago*. 10. Effects of maternal temperature on seed and seedling characters in *Plantago lanceolata*. *Journal of Ecology* 73, 271–282.

Ali, Q., Ashraf, M. and Anwar, F. (2010) Seed composition and seed oil antioxidant activity of maize under water stress. *Journal of American Oil Chemists' Society* 87, 1179–1187.

Andalo, C., Raquin, C., Machon, N., Godelle, B. and Mousseau, M. (1998) Direct and maternal effects of elevated CO_2 on early root growth and germinating *Arabidopsis thaliana* seedlings. *Annals of Botany* 81, 405–411.

Araki, H., Hossain, M.A. and Takahashi, T. (2012) Waterlogging and hypoxia have permanent effects on wheat root growth and respiration. *Journal of Agronomy and Crop Science* 198, 264–275.

Arft, A.M., Walker, M.D., Gurevitch, J., Alatalo, J.M., Bret-Harte, M.S., Dale, M., Diemer, M., Gugerli, F., Henry, G.H.R., Jones, M.H., Hollister, R.D., Jonsdottir, I.S., Laine, K., Levesque, E., Marion, G.M., Molau, U., Molgaard, P., Nordenhall, U., Raszhivin, V., Robinson, C.H., Starr, G., Stenstrom, A., Stenstrom, M., Totland, O., Turner, P.L., Walker, L.J., Webber, P.J., Welker, J.M. and Wookey, P.A. (1999) Responses of tundra plants to experimental warming: Meta-analysis of the international tundra experiment. *Ecological Monographs* 69, 491–511.

Aronson, J., Kigel, J. and Shmida, A. (1993) Reproductive allocation strategies in desert and Mediterranean populations of annual plants growth with and without water stress. *Oecologia* 93, 336–342.

Bai, Y., Tischler, C.R., Booth, D.T. and Taylor, E.M. (2003) Variations in germination and grain quality within a rust resistant common wheat germplasm as affected by parental CO2 conditions. *Environmental and Experimental Botany* 50, 159–168.

Baskin, C.C. (2003) Breaking physical dormancy in seeds – focussing on the lens. *New Phytologist* 158, 229–232.

Baskin, C.C. and Baskin, J.M. (1998) Seeds, ecology, biogeography and evolution of dormancy, and germination. Academic Press, San Diego, CA.

Batlla, D. and Benech-Arnold, R.L. (2006) The role of fluctuations in soil water content on the regulation of dormancy changes in buried seeds of *Polygonum aviculare* L. *Seed Science Research* 16, 47–59.

Bazzaz, F.A. (1990) The response of natural ecosystems to the rising global CO_2 levels. *Annual Review of Ecology and Systematics* 21, 167–196.

Bazzaz, F.A., Ackerly, D.D. and Reekie, E.G. (2000) Reproductive allocation in plants. In: Fenner, M. (ed.) *Seeds: The Ecology of Regeneration in Plant Communities*. CAB International, Wallingford, UK, pp. 1–29.

Benech-Arnold, R.L., Fenner, M. and Edwards, P.J. (1992) Changes in dormancy level in *Sorghum halepense* seeds induced by water stress during seed development. *Functional Ecology* 6, 596–605.

Bernal-Lugo, I. and Leopold, A.C. (1998) The dynamics of seed mortality. *Journal of Experimental Botany* 49, 1455–1461.

Berrie, A.M.M. (1979) Possible role of volatile fatty acids and abscisic acid in the dormancy of oats. *Plant Physiology* 63, 758–764.

Bhatia, N.P., Nkang, A.E., Walsh, K.B., Baker, A.J.M., Ashwath, N. and Midmore, D.J. (2005) Successful seed germination of the nickel hyperaccumulator *Stackhousia tryonii*. *Annals of Botany* 96, 159–163.

Biddulph, T.B., Plummer, J.A., Setter, T.L. and Mares, D.J. (2007) Influence of high temperature and terminal moisture stress on dormancy in wheat (*Triticum aestivum* L.). *Field Crops Research* 103, 139–153.

Bokhorst, S., Bjerke, J.W., Bowles, F.W., Melillo, J., Callaghan, T.V. and Phoenix, G.K. (2008) Impacts of extreme winter warming in the sub-Arctic: growing season responses of dwarf shrub heathland. *Global Change Biology* 14, 2603–2612.

Bokhorst, S., Bjerke, J.W., Street, L.E., Callaghan, T.V. and Phoenix, G.K. (2011) Impacts of multiple extreme winter warming events on sub-Arctic heathland: phenology, reproduction, growth, and CO2 flux responses. *Global Change Biology* 17, 2817–2830.

Both, C., van Asch, M., Bijlsma, R.G., van den Burg, A.B. and Visser, M.E. (2009) Climate change and unequal phenological changes across four trophic levels: constraints or adaptations? *Journal of Animal Ecology* 78, 73–83.

Bouchereau, A., Clossais-Besnard, N., Bensaoud, A., Leport, L. and Renard, M. (1996) Water stress effects on rapeseed quality. *European Journal of Agronomy* 5, 19–30.

Bradley, N.L., Leopold, A.C., Ross, J. and Huffaker, W. (1999) Phenological changes reflect climate change in Wisconsin. *Proceedings of the National Academy of Sciences of the United States of America* 96, 9701–9704.

Braganza, K., Karoly, D.J. and Arblaster, J.M. (2004) Diurnal temperature range as an index of global climate change during the twentieth century. *Geophysical Research Letters* 31.

Burkey, K.O., Booker, F.L., Pursley, W.A. and Heagle, A.S. (2007) Elevated carbon dioxide and ozone effects on peanut: II. seed yield and quality. *Crop Science* 47, 1488–1497.

Burnham, M.B. (2011) Seed life expectancy: the spermosphere, defense chemistry, and weed recruitment, Department of Crop and Soil Sciences. Pennsylvania State University, University Park, PA, p. 63.

Centritto, M., Lee, H.S.J. and Jarvis, P.G. (1999) Interactive effects of elevated CO_2 and drought on cherry (*Prunus avium*) seedlings - I. Growth, whole-plant water use efficiency and water loss. *The New Phytologist* 141, 129–140.

Chakraborty, S. (2011) Climate change and plant diseases. *Plant Pathology* 60, 1–1.

Charnock, C., Brudeli, B. and Klaveness, J. (2004) Evaluation of the antibacterial efficacy of diesters of azelaic acid. *European Journal of Pharmaceutical Sciences* 21, 589–596.

Chia, D.W., Yoder, T.J., Reiter, W.-D. and Gibson, S.I. (2000) Fumaric acid: an overlooked form of fixed carbon in *Arabidopsis* and other plant species. *Planta* 211, 743–751.

Cleland, E.E., Chuine, I., Menzel, A., Mooney, H.A. and Schwartz, M.D. (2007) Shifting plant phenology in response to global change. *Trends in Ecology & Evolution* 22, 357–365.

Conlisk, E., Lawson, D., Syphard, A.D., Franklin, J., Flint, L., Flint, A. and Regan, H.M. (2012) The roles of dispersal, fecundity, and predation in the population persistence of an oak (*Quercus engelmannii*) under global change. *PLoS ONE* 7, e36391.

Copete, E., Herranz, J.M., Ferrandis, P., Baskin, C.C. and Baskin, J.M. (2011) Physiology, morphology and phenology of seed dormancy break and germination in the endemic Iberian species *Narcissus hispanicus* (Amaryllidaceae). *Annals of Botany* 107, 1003–1016.

Cornaglia, P.S., Schrauf, G.E. and Deregibus, V.A. (2009) Flooding and grazing promote germination and seedling establishment in the perennial grass *Paspalum dilatatum*. *Austral Ecology* 34, 343–350.

del Cacho, M. and Lloret, F. (2012) Resilience of Mediterranean shrubland to a severe drought episode: the role of seed bank and seedling emergence. *Plant Biology* 14, 458–466.

Dickie, J.B., Ellis, R.H., Kraak, H.L., Ryder, K. and Tompsett, P.B. (1990) Temperature and seed storage longevity. *Annals of Botany* 65, 197–204.

Donnelly, A., Caffarra, A. and O'Neill, B.F. (2011) A review of climate-driven mismatches between interdependent phenophases in terrestrial and aquatic ecosystems. *International Journal of Biometeorology* 55, 805–817.

Donohue, K. (2009) Completing the cycle: maternal effects as the missing link in plant life histories. *Philosophical Transactions of the Royal Society B: Biological Sciences* 364, 1059–1074.

Donohue, K. and Schmitt, J. (1998) Maternal environmental effects in plants: adaptive plasticity? In: Mousseau, T., Fox, C. (eds) *Maternal effects as adaptations*. Oxford University Press, Oxford, UK, pp. 137–158.

Doody, C.N. and O'Reilly, C. (2011) Effect of long-phase stratification treatments on seed germination in ash. *Annals of Forest Science* 68, 139–147.

Dornbos, D.L., Jr., Mullen, R.E. and Shibles, R.M. (1989) Drought stress effects during seed fill on soybean seed germination and vigor. *Crop Science* 29, 476–480.

Edwards, G.R. and Newton, P.C.D. (2006) Plant performance and implications for plant population dynamics and species composition in a changing climate. *Agroecosystems in a Changing Climate*. CRC Press, pp. 189–210.

Edwards, G.R., Newton, P.C.D., Tilbrook, J.C. and Clark, H. (2001) Seedling performance of pasture species under elevated CO2. *New Phytologist* 150, 359–369.

Ellis, R.H. and Roberts, E.H. (1980) Improved equations for the prediction of seed longevity. *Annals of Botany* 45, 13–30.

Ellis, R.H. and Roberts, E.H. (1981a) The quantification of aging and survival in orthodox seeds. *Seed Science and Technology* 9, 373–409.

Ellis, R.H. and Roberts, E.H. (1981b) An investigation into the possible effects of ripeness and repeated threshing on barley seed longevity under 6 different storage environments. *Annals of Botany* 48, 93–96.

Eslami, S.V., Gill, G.S. and McDonald, G. (2010) Effect of water stress during seed development on morphometric characteristics and dormancy of wild radish (*Raphanus raphanistrum* L.) seeds. *International Journal of Plant Production* 4, 1735–8043.

Fay, P.A. and Schultz, M.J. (2009) Germination, survival, and growth of grass and forb seedlings: Effects of soil moisture variability. *Acta Oecologica–International Journal of Ecology* 35, 679–684.

Fenner, M. (1991) The effects of the parent environment on seed germinability. *Seed Science Research* 1, 75–84.

Finch-Savage, W.E. and Leubner-Metzger, G. (2006) Seed dormancy and the control of germination. *New Phytologist* 171, 501–523.

Fitter, A.H. and Fitter, R.S.R. (2002) Rapid changes in flowering time in British plants. *Science* 296, 1689–1691

Florentine, S.K. and Westbrooke, M.E. (2005) Invasion of the noxious weed *Nicotiana glauca* R. Graham after an episodic flooding event in the arid zone of Australia. *Journal of Arid Environments* 60, 531–545.

Forrest, J.R.K. and Thomson, J.D. (2011) An examination of synchrony between insect emergence and flowering in Rocky Mountain meadows. *Ecological Monographs* 81, 469–491.

Gallagher, R.S., Steadman, K.J. and Crawford, A.D. (2004) Alleviation of dormancy in annual ryegrass (*Lolium rigidum*) seeds by hydration and after-ripening. *Weed Science* 52, 968–975.

Gallagher, R.S. and Fuerst, E.P. (2006) The ecophysical basis of weed seed longevity in the soil. In: Basra, A.S. (ed.) *Handbook of Seed Science and Technology*. Haworth Food Products Press, Binghamton, New York, pp. 521–557.

Gallagher, R.S., Ananth, R., Granger, K., Bradley, B., Anderson, J.V. and Fuerst, E.P. (2010a) Phenolic and short-chained aliphatic organic acid constituents of wild oat (*Avena fatua* L.) seeds. *Journal of Agricultural and Food Chemistry* 58, 218–225.

Gallagher, R.S., Granger, K.L., Keser, L.H., Rossi, J., Pittmann, D., Rowland, S., Burnham, M. and Fuerst, E.P. (2010b) Shade and drought stress-induced changes in phenolic content of wild oat (*Avena fatua* L.) seeds. *Journal of Stress Physiology and Biochemistry* 6, 90–107.

Gallagher, R.S., Granger, K.L., Snyder, A.M., Pittmann, D. and Fuerst, E.P. (2013) Implications of environmental stress during seed development on reproductive and seed bank persistence traits in wild oat (*Avena fatua* L.). *Agronomy* 3, 537–549.

Gao, S., Wang, J.F., Zhang, Z.J., Dong, G. and Guo, J.X. (2012) Seed production, mass, germinability, and subsequent seedling growth responses to parental warming environment in *Leymus chinensis*. *Crop & Pasture Science* 63, 87–94.

Garbutt, K. and Bazzaz, F.A. (1984) The effects of elevated CO_2 on plants- III. Flower, fruit and seed production and abortion. *New Phytologist* 98, 433–446.

Ghelardini, L., Santini, A., Black-Samuelsson, S., Myking, T. and Falusi, M. (2010) Bud dormancy release in elm (*Ulmus* spp.) clones-a case study of photoperiod and temperature responses. *Tree Physiology* 30, 264–274.

Gibbs, J. and Greenway, H. (2003) Mechanisms of anoxia tolerance in plants. I. Growth, survival and anaerobic catabolism. *Functional Plant Biology* 30, 1–47.

Gibson, L.R. and Mullen, R.E. (1996) Soybean seed quality reductions by high day and night temperature. *Crop Science* 36, 1615–1619.

Gómez-Aparicio, L., Pérez-Ramos, I.M., Mendoza, L., Matías, L., Quero, J.L., Castro, J., Zamora, R. and Maranón, T. (2008) Oak seedling survival and growth along resource gradients in Mediterranean forests: implications for regeneration in current and future environmental scenarios. *Oikos* 117, 1683–1688.

Gorecki, M.J., Long, R.L., Flematti, G.R. and Stevens, J.C. (2012) Parental environment changes the dormancy state and karrikinolide response of *Brassica tournefortii* seeds. *Annals of Botany* 109, 1369–1378.

Granger, K.L. (2012) Mitigation of drought-induced reductions in wild oat seed quality by elevated carbon dioxide, Intercollege Graduate Degree Program in Ecology. Pennsylvania State University, University Park, PA, p. 248.

Gualano, N.A. and Benech-Arnold, R.L. (2009) The effect of water and nitrogen availability during grain filling on the timing of dormancy release in malting barley crops. *Euphytica* 168, 291–301.

Gutterman, Y. (2000) Maternal effects on seeds during development. In: Fenner, M. (ed.) *The Ecology of Regeneration in Plant Communities*, 2nd edn. CAB International, Wallingford, UK, pp. 63–83.

Ha, K.J. and Yun, K.S. (2012) Climate change effects on tropical night days in Seoul, Korea. *Theoretical and Applied Climatology* 109, 191–203.

Hannerz, M., Almqvist, U. and Hornfeldt, R. (2002) Timing of seed dispersal in *Pinus sylvestris* stands in central Sweden. *Silva Fennica* 36, 757–765.

Harel, D., Holzapfel, C. and Sternberg, M. (2011) Seed mass and dormancy of annual plant populations and communities decreases with aridity and rainfall predictability. *Basic and Applied Ecology* 12, 674–684.

Hart, J.M. and Berrie, A.M.M. (1968) Relationship between endogenous levels of malic acid and dormancy in grain of *Avena fatua* L. *Phytochemistry* 7, 1257–1260.

He, J.-S., Flynn, D.F.B., Wolfe-Bellin, K., Fang, J. and Bazzaz, F.A. (2005) CO_2 and nitrogen, but not population density, alter the size and C/N ratio of *Phytolacca americana* seeds. *Functional Ecology* 19, 437–444.

Hegland, S.J., Nielsen, A., Lazaro, A., Bjerknes, A.L. and Totland, O. (2009) How does climate warming affect plant-pollinator interactions? *Ecology Letters* 12, 184–195.

Herman, J.J., Sultan, S.E., Horgan-Kobelski, T. and Riggs, C. (2012) Adaptive transgenerational plasticity in an annual plant: grandparental and parental drought stress enhance performance of seedlings in dry soil. *Integrative and Comparative Biology* 52, 77–88.

Hibbs, D., Chan, S.S., Castellano, M. and Niu, C.H. (1995) Response of red alder seedlings to CO_2 enrichment and water-stress. *The New Phytologist* 129, 569–577.

Higa, M., Moriyama, T. and Ishikawa, S. (2012) Effects of complete submergence on seedling growth and survival of five riparian tree species in the warm-temperate regions of Japan. *Journal of Forest Research* 17, 129–136.

HilleRisLambers, J., Harpole, W.S., Schnitzer, S., Tilman, D. and Reich, P.B. (2009) CO_2, nitrogen, and diversity differentially affect seed production of prairie plants. *Ecology* 90, 1810–1820.

Hoffmann, A.A., Camac, J.S., Williams, R.J., Papst, W., Jarrad, F.C. and Wahren, C.H. (2010) Phenological changes in six Australian subalpine plants in response to experimental warming and year-to-year variation. *Journal of Ecology* 98, 927–937.

Högy, P., Franzaring, J., Schwadorf, K., Bruer, J., Schultze, W. and Fangmeier, A. (2010) Effects of free-air CO_2 enrichment on energy traits and seed quality of oilseed rape. *Agriculture, Ecosystems and Environment* 139, 239–244.

Hovenden, M.J., Wills, K.E., Chaplin, R.E., Schoor, J.K.V., Williams, A.L., Osanai, Y. and Newton, P.C.D. (2008) Warming and elevated CO_2 affect the relationship between seed mass, germinability and seedling growth in *Austrodanthonia caespitosa*, a dominant Australian grass. *Global Change Biology* 14, 1633–1641.

Hoyle, G.L., Steadman, K.J., Daws, M.I. and Adkins, S.W. (2008) Pre- and post-harvest influences on seed dormancy status of an Australian Goodeniaceae species, *Goodenia fascicularis*. *Annals of Botany* 102, 93–101.

Hughes, L. (2000) Biological consequences of global warming: is the signal already apparent? *Trends in Ecology & Evolution* 15, 56–61.

Hulber, K., Bardy, K. and Dulinger, S. (2011) Effects of snowmelt timing and competition on the performance of alpine snowbed plants. *Perspectives in Plant Ecology Evolution and Systematics* 13, 15–26.

Hume, L. (1994) Maternal environment effects on plant-growth and germination of 2 strains of *Thlaspo arvense* L. *International Journal of Plant Sciences* 155, 180–186.

Hussain, M., Kubiske, M.E. and Connor, K.F. (2001) Germination of CO_2-enriched *Pinus taeda* L. seeds and subsequent seedling growth responses to CO_2 enrichment. *Functional Ecology* 15, 344–350.

Huxman, T.E., Hamerlynck, E.P., Jordan, D.N., Salsman, K.J. and Smith, S.D. (1998) The effects of parental CO_2 environment on seed quality and subsequent seedling performance in *Bromus rubens*. *Oecologia* 114, 202–208.

Inouye, D.W. (2008) Effects of climate change on phenology, frost damage, and floral abundance of montane wildflowers. *Ecology* 89, 353–362.

Jablonski, J.M., Wang, X. and Curtis, P.S. (2002) Plant reproduction under elevated CO_2 conditions: a meta-analysis of reports on 79 crop and wild species. *New Phytologist* 156, 9–26.

Jackson, R.B., Sala, O.E., Field, C.B. and Mooney, H.A. (1994) CO_2 alters water use, carbon gain and yield for the dominant species in a natural grassland. *Oecologia* 98, 257–262.

Jackson, R.B., Luo, Y., Cardon, Z.G., Sala, O.E., Field, C.B. and Mooney, H.A. (1995) Photosynthesis, growth and density for the dominant species in a CO_2-enriched grassland. *Journal of Biogeography* 22, 221–225.

Jones, M.H., Bay, C. and Nordenhall, U. (1997) Effects of experimental warming on arctic willows (*Salix* spp.): A comparison of responses from the Canadian High Arctic, Alaskan Arctic, and Swedish Subarctic. *Global Change Biology* 3, 55–60.

Karlsson, L.M. and Milberg, P. (2007) A comparative study of germination ecology of four Papaver taxa. *Annals of Botany* 99, 935–946.

Ketring, D.L. (1991) Physiology of oil seeds. 9. Effects of water deficit on peanut seed quality. *Crop Science* 31, 459–463.

Khurana, E. and Singh, J.S. (2004) Germination and seedling growth of five tree species from tropical dry forest in relation to water stress: impact of seed size. *Journal of Tropical Ecology* 20, 385–396.

Kimball, B.A., Kobayashi, K. and Bindi, M. (2002) Responses of agricultural crops to free-air CO_2 enrichment. *Advances in Agronomy* 77, 293–368.

Kimball, S., Angert, A.L., Huxman, T.E. and Venable, D.L. (2010) Contemporary climate change in the Sonoran Desert favors cold-adapted species. *Global Change Biology* 16, 1555–1565.

Klady, R.A., Henry, G.H.R. and Lemay, V. (2011) Changes in high arctic tundra plant reproduction in response to long-term experimental warming. *Global Change Biology* 17, 1611–1624.

Kochanek, J., Buckley, Y.M., Probert, R.J., Adkins, S.W. and Steadman, K.J. (2010) Pre-zygotic parental environment modulates seed longevity. *Austral Ecology* 35, 837–848.

Kochanek, J., Steadman, K.J., Probert, R.J. and Adkins, S.W. (2011) Parental effects modulate seed longevity: exploring parental and offspring phenotypes to elucidate pre-zygotic environmental influences. *New Phytologist* 191, 223–233.

Koller, D. (1972) Environmental control of seed germination. In: Kozlowski, T.T. (ed.) *Seed Biology*. Academic Press, London, UK, pp. 1–101.

Krauss, K.W., Lovelock, C.E., McKee, K.L., Lopez-Hoffman, L., Ewe, S.M.L. and Sousa, W.P. (2008) Environmental drivers in mangrove establishment and early development: A review. *Aquatic Botany* 89, 105–127.

Kudo, G., Nishikawa, Y., Kasagi, T. and Kosuge, S. (2004) Does seed production of spring ephemerals decrease when spring comes early? *Ecological Research* 19, 255–259.

Kullman, L. (2002) Rapid recent range-margin rise of tree and shrub species in the Swedish Scandes. *Journal of Ecology* 90, 68–77.

Kuparinen, A., Katul, G., Nathan, R. and Schurr, F.M. (2009) Increases in air temperature can promote wind-driven dispersal and spread of plants. *Proceedings of the Royal Society B: Biological Sciences* 276, 3081–3087.

Lacey, E.P. (1996) Parental effects in *Plantago lanceolata* L .1. A growth chamber experiment to examine pre- and postzygotic temperature effects. *Evolution* 50, 865–878.

Lacey, E.P. and Herr, D. (2000) Parental effects in *Plantago lanceolata* L. III. Measuring parental temperature effects in the field. *Evolution* 54, 1207–1217.

Lacey, E.P., Smith, S. and Case, A.L. (1997) Parental effects on seed mass: seed coat but not embryo/endosperm effects. *American Journal of Botany* 84, 1617–1620.

Leishman, M.R., Masters, G.J., Clarke, I.P. and Brown, V.K. (2000) Seed bank dynamics: the role of fungal pathogens and climate change. *Functional Ecology* 14, 293–299.

Lesica, P. and Kittelson, P.M. (2010) Precipitation and temperature are associated with advanced flowering phenology in a semi-arid grassland. *Journal of Arid Environments* 74, 1013–1017.

Levine, J.M., McEachern, A.K. and Cowan, C. (2011) Seasonal timing of first rain storms affects rare plant population dynamics. *Ecology* 92, 2236–2247.

Lewis, O.T. and Gripenberg, S. (2008) Insect seed predators and environmental change. *Journal of Applied Ecology* 45, 1593–1599.

Liu, H., Feng, C.L., Luo, Y.B., Chen, B.S., Wang, Z.S. and Gu, H.Y. (2010) Potential challenges of climate change to orchid conservation in a wild orchid hotspot in southwestern China. *Botanical Review* 76, 174–192.

Lloret, F., Casanovas, C. and Penuelas, J. (1999) Seedling survival of Mediterranean shrubland species in relation to root:shoot ratio, seed size and water and nitrogen use. *Functional Ecology* 13, 210–216.

Lloret, F., Peñuelas, J. and Estiarte, M. (2004) Experimental evidence of reduced diversity of seedlings due to climate modification in a Mediterranean-type community. *Global Change Biology* 10, 248–258.

Lloret, F., Penuelas, J., Prieto, P., Llorens, L. and Estiarte, M. (2009) Plant community changes induced by experimental climate change: Seedling and adult species composition. *Perspectives in Plant Ecology Evolution and Systematics* 11, 53–63.

Lorrain, B., Chira, K. and Teissedre, P.L. (2011) Phenolic composition of Merlot and Cabernet-Sauvignon grapes from Bordeaux vineyard for the 2009-vintage: Comparison to 2006, 2007 and 2008 vintages. *Food Chemistry* 126, 1991–1999.

Luck, J., Spackman, M., Freeman, A., Trebicki, P., Griffiths, W., Finlay, K. and Chakraborty, S. (2011) Climate change and diseases of food crops. *Plant Pathology* 60, 113–121.

Luxmoore, R.J., Fischer, R.A. and Stolzy, I.H. (1973) Flooding and soil temperature efects on wheat during grain filling. *Agronomy Journal* 65.

Luzuriaga, A.L., Escudero, A. and Pérez-García, F. (2006) Environmental maternal effects on seed morphology and germination in *Sinapsis arvensis* (Cruciferae). *Weed Research* 46, 163–174.

Manabe, S. and Wetherald, R.T. (1987) Large-scale changes of soil wetness induced by an increase in atmospheric carbon dioxide. *Journal of the Atmospheric Sciences* 44, 1211–1236.

Matías, L., Zamora, R. and Castro, J. (2012) Sporadic rainy events are more critical than increasing of drought intensity for woody species recruitment in a Mediterranean community. *Oecologia* 169, 833–844.

McConkey, K.R., Prasad, S., Corlett, R.T., Campos-Arceiz, A., Brodie, J.F., Rogers, H. and Santamaria, L. (2012) Seed dispersal in changing landscapes. *Biological Conservation* 146, 1–13.

McKone, M.J., Kelly, D. and Lee, W.G. (1998) Effect of climate change on mast-seeding species: frequency of mass flowering and escape from specialist insect seed predators. *Global Change Biology* 4, 591–596.

McPeek, T.M. and Wang, X.Z. (2007) Reproduction of Dandelion (*Taraxacum officinale*) in a higher CO_2 environment. *Weed Science* 55, 334–340.

Memmott, J., Craze, P.G., Waser, N.M. and Price, M.V. (2007) Global warming and the disruption of plant-pollinator interactions. *Ecology Letters* 10, 710–717.

Mickelson, J.A. and Grey, W.E. (2006) Effect of soil water content on wild oat (*Avena fatua*) seed mortality and seedling emergence. *Weed Science* 54, 255–262.

Middleton, B.A. and McKee, K.L. (2012) Can elevated CO_2 modify regeneration from seed banks of floating freshwater marshes subjected to rising sea-level? *Hydrobiologia* 683, 123–133.

Milbau, A., Graae, B.J., Shevtsova, A. and Nijs, I. (2009) Effects of a warmer climate on seed germination in the subarctic. *Annals of Botany* 104, 287–296.

Mira, S., Gonzalez-Benito, M.E., Ibars, A.M. and Estrelles, E. (2011) Dormancy release and seed ageing in the endangered species Silene diclinis. *Biodiversity and Conservation* 20, 345–358.

Miyagi, K.-M., Kinugasa, T., Hikosaka, K. and Hirose, T. (2007) Elevated CO_2 concentration, nitrogen use, and seed production in annual plants. *Global Change Biology* 13, 2161–2170.

Mondoni, A., Rossi, G., Orsenigo, S. and Probert, R.J. (2012) Climate warming could shift the timing of seed germination in alpine plants. *Annals of Botany* 110, 155–164.

Monty, A., Lebeau, J., Meerts, P. and Mahy, G. (2009) An explicit test for the contribution of environmental maternal effects to rapid clinal differentiation in an invasive plant. *Journal of Evolutionary Biology* 22, 917–926.

Morris, W.F., Pfister, C.A., Tuljapurkar, S., Haridas, C.V., Boggs, C.L., Boyce, M.S., Bruna, E.M., Church, D.R., Coulson, T., Doak, D.F., Forsyth, S., Gaillard, J.M., Horvitz, C.C., Kalisz, S., Kendall, B.E., Knight, T.M., Lee, C.T. and Menges, E.S. (2008) Longevity can buffer plant and animal populations against changing climatic variability. *Ecology* 89, 19–25.

Muasya, R.M., Lommen, W.J.M., Muui, C.W. and Struik, P.C. (2008) How weather during development of common bean (*Phaseolus vulgaris* L.) affects the crop's maximum attainable seed quality. *Njas-Wageningen Journal of Life Sciences* 56, 85–100.

Muller-Landau, H.C. (2010) The tolerance-fecundity trade off and the maintenance of diversity in seed size. *Proceedings of the National Academy of Sciences* 107, 4242–4247.

Murdoch, A.J. and Ellis, R.H. (2000) Dormancy, viability and longevity. In: Fenner, M. (ed.) *The Ecology of Regeneration in Plant Communities*, 2nd edn. CAB International, Wallingford, UK, pp. 183–214.

Myking, T. and Heide, O.M. (1995) Dormancy release and chilling requirement of buds of latitudinal ecotypes of *Betula pendula* and *B. pubescens*. *Tree Physiology* 15, 697–704.

Nathan, R., Horvitz, N., He, Y., Kuparinen, A., Schurr, F.M. and Katul, G.G. (2011) Spread of North American wind-dispersed trees in future environments. *Ecology Letters* 14, 211–219.

O'Donnell, C.C. and Adkins, S.W. (2001) Wild oat and climate change: the effect of CO_2 concentration, temperature, and water deficit on the growth and development of wild oat in monoculture. *Weed Science* 49, 694–702.

Ooi, M.K.J. (2012) Seed bank persistence and climate change. *Seed Science Research* 22, S53-S60.

Ooi, M.K.J., Auld, T.D. and Denham, A.J. (2009) Climate change and bet-hedging: interactions between increased soil temperatures and seed bank persistence. *Global Change Biology* 15, 2375–2386.

Ooi, M.K.J., Auld, T.D. and Denham, A.J. (2012) Projected soil temperature increase and seed dormancy response along an altitudinal gradient: implications for seed bank persistence under climate change. *Plant and Soil* 353, 289–303.

Panda, D., Rao, D.N., Sharma, S.G., Strasser, R.J. and Sarkar, R.K. (2006) Submergence effects on rice genotypes during seedling stage: Probing of submergence driven changes of photosystem 2 by chlorophyll alpha fluorescence induction O-J-I-P transients. *Photosynthetica* 44, 69–75.

Penuelas, J., Filella, I., Zhang, X.Y., Llorens, L., Ogaya, R., Lloret, F., Comas, P., Estiarte, M. and Terradas, J. (2004) Complex spatiotemporal phenological shifts as a response to rainfall changes. *New Phytologist* 161, 837–846.

Pérez-Fernández, M.A., Lamont, B.B., Marwick, A.J. and Lamont, W.G. (2000). Germination of seven exotic weeds and seven native species in south-western Australia under steady and fluctuating water supply. *Acta Oecologia* 21, 323–336.

Pérez-Ramos, I.M. and Marañón, T. (2009) Effects of waterlogging on seed germination of three Mediterranean oak species: ecological implication. *Acta Oecologia* 35, 422–428.

Peters, N.C.B. (1982) Production and dormancy of wild oat (*Avena fatua*) seed from plants grown under soil water stress. *Annals of Applied Biology* 100, 189–196.

Pías, B., Matesanz, S., Herrero, A., Gimeno, T.E., Escudero, A. and Valladares, F. (2010) Transgenerational effects of three global change drivers on an endemic Mediterranean plant. *Oikos* 119, 1435–1444.

Pieper, S.J., Loewen, V., Gill, M. and Johnstone, J.F. (2011) Plant responses to natural and experimental variations in temperature in alpine tundra, Southern Yukon, Canada. *Arctic Antarctic and Alpine Research* 43, 442–456.

Post, E.S., Pedersen, C., Wilmers, C.C. and Forchhammer, M.C. (2008) Phenological sequences reveal aggregate life history response to climatic warming. *Ecology* 89, 363–370.

Pounds, J.A., Fogden, M.P.L. and Campbell, J.H. (1999) Biological response to climate change on a tropical mountain. *Nature* 398, 611–615.

Price, M.V. and Waser, N.M. (1998) Effects of experimental warming on plant reproductive phenology in a subalpine meadow. *Ecology* 79, 1261–1271.

Probert, R.J. (1992) The role of temperature in germination ecophysiology. In: Fenner, M. (ed.) *The Ecology of Regeneration in Plant Communities*, 1st edn. CAB International, Wallingford, UK, pp. 285–325.

Probert, R.J. (2000) The role of temperature in the regulation of seed dormancy and germination. In: Fenner, M. (ed.) *The Ecology of Regeneration in Plant Communities*, 2nd edn. CAB International, Wallingford, UK, pp. 261–292.

Qaderi, M.M. and Reid, D.M. (2008) Combined effects of temperature and carbon dioxide on plant growth and subsequent seed germinability of *Silene Noctiflora*. *International Journal of Plant Sciences* 169, 1200–1209.

Qaderi, M.M., Cavers, P.B., Hamill, A.S., Downs, M.P. and Bernards, M.A. (2006) Maturation temperature regulates germinability and chemical constituents of Scotch thistle (*Onopordum acanthium*) cypselas. *Canadian Journal of Botany-Revue Canadienne De Botanique* 84, 28–38.

Ramirez-Tobias, H.M., Pena-Valdivia, C.B., Aguirre, J.R., Reyes-Aguro, J.A., Sanchez-Urdaneta, A.B. and Valle, S. (2012) Seed germination temperatures of eight Mexican Agave species with economic importance. *Plant Species Biology* 27, 124–137.

Rebey, I.B., Zakhama, N., Karoui, I.J. and Marzouk, B. (2012) Polyphenol composition and antioxidant activity of cumin (*Cuminum cyminum* L.) seed extract under drought. *Journal of Food Science* 77, C734–C739.

Reekie, E.G. and Bazzaz, F.A. (1991) Phenology and growth in 4 annual species grown in ambient and elevated CO_2. *Canadian Journal of Botany-Revue Canadienne De Botanique* 69, 2475–2481.

Rice, K.J. and Dyer, A.R. (2001) Seed aging, delayed germination and reduced competitive ability in Bromus tectorum. *Plant Ecology* 155, 237–243.

Riginos, C., Heschel, M.S. and Schmitt, J. (2007) Maternal effects of drought stress and inbreeding in *Impatiens capensis* (Balsaminaceae). *American Journal of Botany* 94, 1984–1991.

Roach, D.A. and Wulff, R.D. (1987) Maternal effects in plants. *Annual Review of Ecology and Systematics* 18, 209–235.

Root, T.L., Price, J.T., Hall, K.R., Schneider, S.H., Rosenzweig, C. and Pounds, J.A. (2003) Fingerprints of global warming on wild animals and plants. *Nature* 421, 57–60.

Ross, L.F. and Kvien, C.S. (1989) The effect of drought stress on peanut seed composition. I. Soluble carbohydrates, tartaric acid, and phenolics. *Oléagineux* 44, 259–299.

Ruxton, G.D. and Schaefer, H.M. (2012) The conservation physiology of seed dispersal. *Philosophical Transactions of the Royal Society B-Biological Sciences* 367, 1708–1718.

Saarinen, T., Lundell, R., Astrom, H. and Hanninen, H. (2011) Parental overwintering history affects the responses of *Thlaspi arvense* to warming winters in the North. *Environmental and Experimental Botany* 72, 409–414.

Santana, V.M., Bradstock, R.A., Ooi, M.K.J., Denham, A.J., Auld, T.D. and Baeza, M.J. (2010) Effects of soil temperature regimes after fire on seed dormancy and germination in six Australian Fabaceae species. *Australian Journal of Botany* 58, 539–545.

Sasek, T.W. and Strain, B.R. (1988) Effects of carbon dioxide enrichment on the growth and morphology of kudzu (*Pueraria lobata*). *Weed Science* 36, 28–36.

Sasek, T.W. and Strain, B.R. (1989) Effects of carbon dioxide enrichment on the expansion and size of kudzu (*Pueraria lobata*) leaves. *Weed Science* 37, 23–28.

Sawhney, R. and Naylor, J.M. (1982) Dormancy studies in seed of *Avena fatua*. 13. Influence of drought stress during seed development on duration of seed dormancy. *Canadian Journal of Botany* 60, 1016–1020.

Schmuths, H., Bachmann, K., Weber, W.E., Horres, R. and Hoffmann, M.H. (2006) Effects of preconditioning and temperature during germination of 73 natural accessions of *Arabidopsis thaliana*. *Annals of Botany* 97, 623–634.

Seifan, M., Tielborger, K. and Kadmon, R. (2010) Direct and indirect interactions among plants explain counterintuitive positive drought effects on an eastern Mediterranean shrub species. *Oikos* 119, 1601–1609.

Seiler, C., Harshavardan, V.T., Rajesh, K., Reddy, P.S., Strickert, M., Rolletschek, H., Scholz, U., Wobus, U. and Sreenivasulu, N. (2011) ABA biosynthesis and degradation contributing to ABA homeostasis during barley seed development under control and terminal drought stress conditions. *Journal of Experimental Botany* 62, 2615–2632.

Sezen, S.M., Yazar, A. and Tekin, S. (2011) Effects of partial root zone drying and deficit irrigation on yield and oil quality of sunflower in a Mediterranean environment. *Irrigation and Drainage* 60, 499–508.

Sherry, R.A., Zhou, X.H., Gu, S.L., Arnone, J.A., Schimel, D.S., Verburg, P.S., Wallace, L.L. and Luo, Y.Q. (2007) Divergence of reproductive phenology under climate warming. *Proceedings of the National Academy of Sciences of the United States of America* 104, 198–202.

Shevtsova, A., Graae, B.J., Jochum, T., Milbau, A., Kockelbergh, F., Beyens, L. and Nijs, I. (2009) Critical periods for impact of climate warming on early seedling establishment in subarctic tundra. *Global Change Biology* 15, 2662–2680.

Shirley, B.W. (1998) Flavonoids in seeds and grains: physiological function, agronomic importance and the genetics of biosynthesis. *Seed Science Research* 8, 415–422.

Shu, L., Lou, Q., Ma, C., Ding, W., Zhou, J., Wu, J., Feng, F., Lu, X., Luo, L., Xu, G. and Mei, H. (2011) Genetic, proteomic and metabolic analysis of the regulation of energy storage in rice seedlings in response to drought. *Proteomics* 11, 4122–4138.

Sinniah, U.R., Ellis, R.H. and John, P. (1998) Irrigation and seed quality development in rapid-cycling brassica: Soluble carbohydrates and heat-stable proteins. *Annals of Botany* 82, 647–655.

Skarpaas, O., Auhl, R. and Shea, K. (2006) Environmental variability and the initiation of dispersal: turbulence strongly increases seed release. *Proceedings of the Royal Society B: Biological Sciences* 273, 751–756.

Smith, C.C. and Fretwell, S.D. (1974) The optimal balance between size and number of offspring. *The American Naturalist* 108, 499–506.

Solomon, S., Qin, D., Manning, M., Chen, Z., Marquis, M., Averyt, K.B., Tignor, M. and Miller, H.L. (eds) (2007) *Climate Change 2007: The Physical Science Basis. Contribution of Working Group I to the Fourth Assessment Report of the Intergovernmental Panel on Climate Change*. Cambridge University Press, Cambridge, United Kingdom and New York, USA.

Soons, M.B. and Bullock, J.M. (2008) Non-random seed abscission, long-distance wind dispersal and plant migration rates. *Journal of Ecology* 96, 581–590.

Stanton, M.L. (1984) Seed variation in wild radish - effect of seed size on components of seedling and adult fitness. *Ecology* 65, 1105–1112.

Steadman, K.J., Ellery, A.J., Chapman, R., Moore, A. and Turner, N.C. (2004) Maturation temperature and rainfall influence seed dormancy characteristics of annual ryegrass (*Lolium rigidum*). *Australian Journal of Agricultural Research* 55, 1047–1057.

Stearns, F. (1960) Effects of seed environment during maturation on seedling growth. *Ecology* 41, 221–222.

Steinger, T., Gall, R. and Schmid, B. (2000) Maternal and direct effects of elevated CO_2 on seed provisioning, germination and seedling growth in *Bromus erectus*. *Oecologia* 123, 475–480.

Sternberg, M., Brown, V.K., Masters, G.J. and Clarke, I.P. (1999) Plant community dynamics in a calcereous grassland under climate change manipulations. *Plant Ecology* 143, 29–37.

Stiling, P., Moon, D., Hymus, G. and Drake, B. (2004) Differential effects of elevated CO_2 on acorn density, weight, germination, and predation among three oak species in a scrub-oak forest. *Global Change Biology* 10, 228–232.

Sultan, S.E. and Bazzaz, F.A. (1993) Phenotypic plasticity in *Polygonum persicaria*. II. Norms of reaction to soil moisture and the maintenance of genetic diversity. *Evolution* 47, 1032–1049.

Taylor, A.H. (1995) Forest expansion and climate-change in the mountain hemlock (*Tsuga mertensiana*) zone, Lassen Volcanic National Park, California, USA. *Arctic and Alpine Research* 27, 207–216.

Thomas, J.F. and Raper, C.D. (1975) Differences in progeny of tobacco due to temperature treatment of the mother plant. *Tobacco Science*, 37–41.

Thomas, J.M.G., Boote, K.J., Allen, L.H., Gallo-Meagher, M. and Davis, J.M. (2003) Elevated temperature and carbon dioxide effects on soybean seed composition and transcript abundance. *Crop Science* 43, 1548–1557.

Thomas, J.M.G., Prasad, P.V.V., Boote, K.J. and Allen, L.H. (2009) Seed composition, seedling emergence and early seedling vigour of red kidney bean seed produced at elevated temperature and carbon dioxide. *Journal of Agronomy and Crop Science* 195, 148–156.

Thompson, K. (2000) The functional ecology of soil seed banks. In: Fenner, M. (ed.) *The Ecology of Regeneration in Plant Communities*, 2nd edn. CAB International, Wallingford, UK, pp. 215–235.

Thürig, B., Körner, C. and Stöcklin, J. (2003) Seed production and seed quality in a calcareous grassland in elevated CO_2. *Global Change Biology* 9, 873–884.

Tungate, K.D., Israel, D.W., Watson, D.M. and Rufty, T.W. (2007) Potential changes in weed competitiveness in an agroecological system with elevated temperatures. *Environmental and Experimental Botany* 60, 42–49.

Tylianakis, J.M., Didham, R.K., Bascompte, J. and Wardle, D.A. (2008) Global change and species interactions in terrestrial ecosystems. *Ecology Letters* 11, 1351–1363.

Upchurch, R.G. (2008) Fatty acid unsaturation, mobilization, and regulation in the response of plants to stress. *Biotechnology Letters* 30, 967–977.

Vahdati, K., Aslamarz, A.A., Rahemi, M., Hassani, D. and Leslie, C. (2012) Mechanism of seed dormancy and its relationship to bud dormancy in Persian walnut. *Environmental and Experimental Botany* 75, 74–82.

van Beckum, J.M.M. and Wang, M. (1994) Effect of short chain fatty acids on physiology of barley grains cv. Triumph with a different level of dormancy. *Plant Science* 102, 153–160.

Vertucci, C.W., Roos, E.E. and Crane, J. (1994) Theoretical basis of protocols for seed storage. 3. Optimum moisture contents for pea seeds stored at different temperatures. *Annals of Botany* 74, 531–540.

Wagner, M. and Mitschunas, N. (2008) Fungal effects on seed bank persistence and potential applications in weed biocontrol: A review. *Basic and Applied Ecology* 9, 191–203.

Walck, J.L., Hidayati, S.N., Dixon, K.W., Thompson, K.E.N. and Poschlod, P. (2011) Climate change and plant regeneration from seed. *Global Change Biology* 17, 2145–2161.

Walther, G.R., Post, E., Convey, P., Menzel, A., Parmesan, C., Beebee, T.J.C., Fromentin, J.M., Hoegh-Guldberg, O. and Bairlein, F. (2002) Ecological responses to recent climate change. *Nature* 416, 389–395.

Wang, R.L., Zeng, R.S., Peng, S.L., Chen, B.M., Liang, X.T. and Xin, X.W. (2011) Elevated temperature may accelerate invasive expansion of the liana plant Ipomoea cairica. *Weed Research* 51, 574–580.

Warren, R.J., Bahn, V. and Bradford, M.A. (2011) Temperature cues phenological synchrony in ant-mediated seed dispersal. *Global Change Biology* 17, 2444–2454.

Way, D.A., LaDeau, S.L., McCarthy, H.R., Clarke, J.S., Oren, R., Finzi, A.C. and Jackson, R.B. (2010) Greater seed production in elevated CO_2 is not accompanied by reduced seed quality in *Pinus taeda* L. *Global Change Biology* 16, 1046–1056.

Wipf, S., Stoeckli, V. and Bebi, P. (2009) Winter climate change in alpine tundra: plant responses to changes in snow depth and snowmelt timing. *Climatic Change* 94, 105–121.

Wulff, R.D. and Alexander, H.M. (1985) Intraspecific variation in the response to CO2 enrichment in seeds and seedlings of *Plantago lanceolata* L. *Oecologia* 66, 458–460.

Xiao, C., Xing, W. and Liu, G.H. (2010) Seed germination of 14 wetland species in response to duration of cold-wet stratification and outdoor burial depth. *Aquatic Biology* 11, 169–177.

Yang, L.H. and Rudolf, V.H.W. (2010) Phenology, ontogeny and the effects of climate change on the timing of species interactions. *Ecology Letters* 13, 1–10.

Yu, O. and Jez, J.M. (2008) Nature's assembly line: biosynthesis of simple phenylpropanoids and polyketides *The Plant Journal* 54, 750–762.

Zavaleta, E.S. (2006) Shrub establishment under experimental global changes in a California grassland. *Plant Ecology* 184, 53–63.

Zhang, R., Jongejans, E. and Shea, K. (2011) Warming increases the spread of an invasive thistle. *PLoS ONE* 6, e21725.

Zhang, R., Gallagher, R.S. and Shea, K. (2012a) Maternal warming affects early life stages of an invasive thistle. *Plant Biology* 14, 783–788.

Zhang, R., Post, E. and Shea, K. (2012b) Warming leads to divergent responses but similarly improved performance of two invasive thistles. *Population Ecology* 54, 583–589.

Zhang, X.Y., Tarpley, D. and Sullivan, J.T. (2007) Diverse responses of vegetation phenology to a warming climate. *Geophysical Research Letters* 34, L19405.

Ziska, L. (2003) Evaluation of the growth response of six invasive species to past, present and future atmospheric carbon dioxide. *Journal of Experimental Botany* 54, 395–404.

Ziska, L.H. and Bunce, J.A. (1993) The influence of elevated CO_2 and temperature on seed-germination and emergence from soil. *Field Crops Research* 34, 147–157.

10 The Functional Role of the Soil Seed Bank in Agricultural Ecosystems

Nathalie Colbach*

INRA, UMR1347 Agroécologie, EcolDur, Dijon, France

Introduction

In many countries with temperate climate, landscapes are highly anthropized, and arable crops constitute the major part of these landscapes. In these habitats, weeds consisting of both 'real' wild species and volunteers originating from lost crop seeds constitute the main component of wild plant biodiversity. Cropped fields differ from natural habitats by frequent disturbances (e.g. tillage, herbicides, harvest) and the presence of one (or sometimes two) dominant plant species (i.e. the crop) that usually changes every year. Though these disturbances can appear as stochastic and unpredictable from the weed's point of view, they result from the farmer's operational logic, and many of them specifically aim at controlling weeds because the latter are very harmful for agricultural production (Oerke *et al.*, 1994). In this chapter, I propose to focus on the importance of the soil seed bank for plant regeneration in this particular habitat, to identify the relevant biophysical processes interacting with agricultural practices, and, finally, to determine what kind of weed species and traits are selected in different cropping systems.

The Importance of the Soil Seed Bank for Weed Dynamics in Agricultural Ecosystems

The seed bank as a necessary adaptation to agriculture

The soil seed bank in arable fields consists of seeds produced by wild species as well as seeds lost by cropped plants before and during harvest operations. Most of these seeds survive for several years in the soil and form persistent seed banks (e.g. Roberts and Boddrell, 1983; Burnside *et al.*, 1996; Murdoch and Ellis, 2000; Conn *et al.*, 2006). Depending on their longevity in the soil, seeds can germinate many years after their production, and species can thus recolonize a field from which they had been absent for several years. Seed banks are more frequent and persisting in tilled fields because weed seeds seem to survive longer when buried deeply (e.g. Ballaré *et al.*, 1988; Mohler and Galford, 1997; Conn, 2006; Harrison *et al.*, 2007). This characteristic is probably the result of a species adaptation to cropped fields because seed-bank persistence is crucial for preserving species in this habitat where conditions vary

* E-mail: nathalie.colbach@dijon.inra.fr

considerably between years, not so much because of weather variability, but mostly because of the crop sown by the farmer and the way this crop is managed (McCloskey *et al.*, 1996; Bàrberi and Lo Cascio, 2001; Menalled *et al.*, 2001; Cardina *et al.*, 2002; Fried *et al.*, 2008). Indeed, if the newly produced seeds of a species all germinated and emerged during the year following production, the species could be completely eliminated if the sown crop and the associated management practices (e.g. tillage, herbicides) were particularly detrimental to the species.

Because of the strong interaction with the cropping system (particularly the crop), the emerged weed flora observed in a given year often only represents a tiny fraction of the species diversity present in the seed bank (e.g. Dessaint *et al.*, 1997). But not all the seeds stored in the soil will one day give rise to seedlings; they can die or be predated before encountering favourable conditions. To get an idea of the biodiversity in fields and/or the future infestation of crops, the weed flora must therefore either be monitored during several years and crops, or sufficient knowledge to predict weed emergence from the soil seed bank in cropped fields must be accumulated.

Why farmers must take account of the soil seed bank for weed management

Since the onset of agriculture, a large part of crop management aimed at eliminating weeds, both by making the seed bank germinate at a time when the resulting plants would not hinder the crop and by eliminating weed plants at those times they would compete with the crop (e.g. *Oeconomicus* by Xenophon *c.*375 BC, *De Re Rustica* by Lucius Junius Moderatus Columella *c.*AD 42). The arrival of synthetic, highly efficient herbicides in the middle of the 20th century resulted in a decrease in weed species diversity and density (e.g. Andreasen *et al.*, 1996; Andersson and Milberg, 1998; Robinson and Sutherland, 2002; Fried *et al.*, 2009), and farmers largely lost interest in other cultivation techniques and in the management of the soil seed bank. However, because these

herbicide-managed fields were highly unfavourable for weed reproduction, the soil seed bank became even more important for species persistence in fields. Indeed, a single deficiency in herbicide efficiency can be sufficient to see a field recolonized by weeds (e.g. Cardina *et al.*, 1997; Chauvel *et al.*, 2009) as the soil still comprises sufficient seeds for regenerating the flora.

Moreover, badly reasoned herbicide use (e.g. repeated applications of herbicides with the same action mode, wrongly timed and dosed applications; Powles and Yu, 2010) resulted in the selection of herbicide resistance in many weed species (http://www.weedresearch.com/in.asp) (Délye, 2005; Powles and Yu, 2010), increasing the role of the seed bank. The latter could delay herbicide resistance at the beginning by 'diluting' the newly arisen resistance alleles in a large, older pool of sensitive alleles. Indeed, deep and/or inverting tillage, which increases seed-bank creation and persistence, was reported to delay the appearance of herbicide resistance in weeds (Cavan *et al.*, 2000, 2001; Beckie *et al.*, 2008; Moss *et al.*, 2010). But with time, the seed bank can stack different herbicide-resistance alleles in populations but also in given individuals, leading to multiresistant weeds and volunteers difficult to control by the farmer and increasingly reported in literature (Messéan *et al.*, 2007; Petit *et al.*, 2010; Délye *et al.*, 2011; Délye, 2013). Gene stacking in the soil seed bank can also be a problem in the particular case of crop volunteers when managing genetic harvest quality, e.g. GM volunteers appearing in later, non-GM crops where they can cause the loss of the non-GM label for the harvest (e.g. Messéan *et al.*, 2007) or volunteers with different fatty acid profiles appearing in oilseed rape crops grown for a particular food chain (e.g. Baux *et al.*, 2011).

Today, European cropping systems must reduce their reliance on herbicides because of environmental problems (biodiversity, Marshall *et al.*, 2003; ground and surface water, IFEN, 2007; human health, Waggoner *et al.*, 2012), regulatory constraints (see recent EU directive REACH on synthetic molecules, http://ec.europa.eu/environment/chemicals/reach/reach_intro.htm)

and because weeds are considered to be the main obstacle for farmers to switch to integrated or organic agriculture (ECOPHYTO R&D, 2009). Because of the persisting seed bank, weed control must not only focus on the yield loss at an annual scale but be reasoned over several years to be efficient both at short and long terms (Clements *et al.*, 1994; Bàrberi and Lo Cascio, 2001). Knowledge of seed-bank processes, their importance for plant community establishment and their interaction with cropping systems again becomes crucial for managing weeds.

The Different Processes from the Seed to the Emerged Seedling

A succession of diverse processes

A series of common processes occurring in the soil

Figure 10.1 summarizes the major stages from the seed to the emerged seedling in annual weed species in temperate climates. Seedling emergence from buried weed seeds is the result of a succession of several processes depending on different factors. Each day, buried seeds die because of age, diseases, predation, etc. The surviving seeds can acquire or lose dormancy according to seasonal environmental fluctuations. A portion of the non-dormant seeds can germinate, depending on their depth, light stimulation and soil climate. A shoot develops from the germinated seed, depending on soil temperature. Those germinated seeds that are neither buried too deeply nor impeded by soil clods succeed in emerging in the following days, depending on soil climate, structure and seed depth. Some of the emerged seedlings succeed in reproducing, and the seed rain replenishes the seeds in the superficial layer.

These processes occur for all annual weed species but their amplitude and rate vary considerably among species, influenced by morphological, physiological and phenological characteristics (Table 10.1). In the following sections, the main life cycle processes will be detailed and related to species traits.

The particular case of parasitic weeds

The previous breakdown of processes was based on the analysis of annual non-parasitic

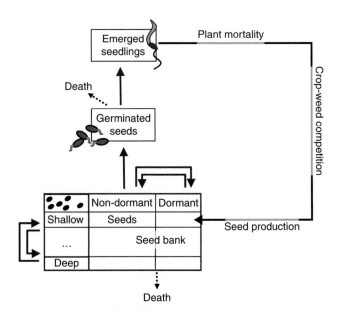

Fig. 10.1. From the seed to the emerged seedling in the weed life cycle in agricultural ecosystems. Schematic representation of the major stages (© 2012 Nathalie Colbach).

Table 10.1. Summary of functional relationships between seed-bank and emergence processes on the one hand and species seed traits on the other hand.

Process		Seed traits								
	Coat thickness	Mass	Seed shape	Lipid content	Proteins and carbohydrates	Phenolic compounds	Area/mass ratio	Taxa	Other
Seed mortality	−[a] (Mohamed-Yasseen et al., 1994; Davis et al., 2008, Gardarin et al., 2010b)	0[b] (Gardarin et al., 2010b)	0 (Gardarin et al., 2010b)	0 (Priestley, 1986; Ponquett et al., 1992; Gardarin et al., 2010b)	0 (Priestley, 1986; Ponquett et al., 1992; Gardarin et al., 2010b)	0 (Hendry et al., 1994; Davis et al., 2008)		0 (Gardarin et al., 2010b)	
Dormancy level	+[c] (Sultan, 1996; Gardarin and Colbach, in revision)	+ (young seeds) (Grime et al., 1981; Gardarin and Colbach, in revision) 0 (Saatkamp, 2009)	E<S[d] (Grime et al., 1981; Gardarin and Colbach, in revision)	0 (Gardarin and Colbach, in revision)	0 (Gardarin and Colbach, in revision)			0 (Gardarin and Colbach, in revision)	
Dormancy timing	0 (Gardarin and Colbach, in revision)	0 (Gardarin and Colbach, in revision)	0 (Gardarin and Colbach, in revision)	0 (Gardarin and Colbach, in revision)	0 (Gardarin and Colbach, in revision)			0 (Gardarin and Colbach, in revision)	
Photosensitivity	0 (Gardarin and Colbach, in revision)	− (Grime et al., 1981; Milberg et al., 2000; Saatkamp, 2009) 0 (Gardarin and Colbach, in revision)	0 (Gardarin et al., 2011)	0 (Gardarin and Colbach, in revision)	0 (Gardarin and Colbach, in revision)			0 (Gardarin and Colbach, in revision)	
Germination speed	− (Grime et al., 1981) 0 (Shipley and Parent, 1991)		S<E (Grime et al., 1981) 0 (Gardarin et al., 2011)	+ (Gardarin et al., 2011) − in anoxic conditions, (Raymond et al., 1985)	0 (Gardarin et al., 2011)		+ (Gardarin et al., 2011)	0 +(Gardarin et al., 2011)	− (base temperature, Angus et al., 1981; Gardarin et al., 2011)

Germination sensitivity to seed depth	– (via maximum shoot length, Gardarin et al., 2010a)	
Maximum shoot length (or emergence depth)	+ (Bond et al., 1999; Colbach and Dürr, 2003; Gardarin et al., 2010a)	0 (Moles and Westoby, 2006)
Maximum root length	+ (Gardarin et al., 2010a)	
Seedling sensitivity to soil clods	– (Terpstra, 1986; Gardarin et al., 2010a)	M>D[e] (Cussans et al., 1996; Gardarin et al., 2010a)

[a] –, negative correlation between process and trait. [b] 0, no correlation. [c] +, positive correlation between process and trait. [d] E, elongated and/or flattened seeds; S, spherical seeds. [e] M, monocotyledonous species; D, dicotyledonous species.

weeds. It also applies to perennials that produce seeds resulting in a soil seed bank even if these weeds mostly regenerate from rhizomes and other subterranean organs. However, the situation is quite different for parasitic weeds such as *Phelipanche, Orobanche* and *Striga* species, which are obligate parasites frequently infecting crops in the Mediterranean area (Musselman, 1980; Schneeweiss, 2007; Rubiales *et al.*, 2009) and now also spreading in arable crops more typical of temperate climates (Benharrat *et al.*, 2005; Gibot-Leclerc *et al.*, 2009). These species also constitute soil seed banks, which are even more crucial in diverse crop rotations as parasite weeds can only reproduce if fixed on a suitable host species. However, parasite seeds can only germinate in the immediate vicinity of roots of suitable host plants (including weeds, Gibot-Leclerc *et al.*, 2003; Boulet *et al.*, 2007) in response to germination stimulants exuded from host roots; germination is followed by fixation on host roots, from which parasite shoots grow toward the soil surface (Bouwmeester *et al.*, 2003; Joel *et al.*, 2007; Parker, 2009; Zwanenburg *et al.*, 2009; Gibot-Leclerc *et al.*, 2012). These complex host–parasite interactions certainly modify seed-bank processes leading to emergence and thus, the following sections will only consider annual, non-parasitic weeds.

Seed mortality

What causes seed mortality?

Seed mortality results from seed embryo ageing caused by physiological or chemical damage (Priestley, 1986), pathogen attacks caused by bacteria or fungi (Chee-Sanford *et al.*, 2006; Wagner and Mitschunas, 2008) or seed predation by vertebrates or invertebrates (Hulme, 1998). It should not be confused with seed persistence or its opposite, i.e. seed decline, which is the result of both seed mortality and germination. These processes are fundamentally different, germination seeds being potentially able to reproduce in contrast to dead seeds. Moreover, these processes are influenced differently by cropping system components (Boyd and Van Acker, 2003).

Most seed persistence studies cannot distinguish seed mortality from germination, nor do they attempt to investigate the cause of seed mortality (Telewski and Zeevaart, 2002; Conn *et al.*, 2006; Gardarin *et al.*, 2009). European studies often focus on *in situ* mortality (excluding macro-predation) of buried weed seeds while North American studies concentrate on seed predation (Gardarin *et al.*, 2009). Indeed, European cropping systems usually comprise frequent tillage operations including mouldboard ploughing (Holland, 2004; Soane *et al.*, 2012), burying seeds at a depth where they are less prone to predation and germination (e.g. Puricelli *et al.*, 2005) and thus persist longer (Sanchez del Arco *et al.*, 1995; Mohler and Galford, 1997; Omami *et al.*, 1999). Conversely, tillage is rare and shallow in North American cropping systems to limit soil erosion, leaving weed seeds close to the soil surface and thus much more exposed to predation (Holland, 2004).

Thicker seed coats are better

Various seed traits were linked to seed mortality. For instance, lipid-rich seeds have been expected to present a low survival rate because of lipid peroxidation reactions, which contribute to free radical production and thus to seed ageing (Corbineau *et al.*, 2002); phenolic compounds could protect the seed from oxidation (see Chapter 8 of this volume). However, the seed half-life was not clearly correlated to seed composition (lipids, proteins and carbohydrates; Priestley, 1986; Ponquett *et al.*, 1992; Gardarin *et al.*, 2010b). No correlation between seed mortality and seed shape or mass has been observed (Gardarin *et al.*, 2010b), though these traits influence seed persistence (Thompson *et al.*, 1993; Cerabolini *et al.*, 2003). Indeed, small and spherical seeds penetrate the soil more easily to a depth where germination is inhibited (see section 'Seed germination' below), thus resulting in a higher persistence (Bekker *et al.*, 1998).

The only consistent functional relationship identified to date shows seed mortality to decrease with increasing seed coat thickness (Mohamed-Yasseen *et al.*, 1994; Davis *et al.*, 2008; Gardarin *et al.*, 2010b). Indeed, the seed coat forms a physical and chemical barrier to preserve the seed and the embryo from parasitic attacks as well as temperature and water fluctuations (Mohamed-Yasseen *et al.*, 1994). Additional seed-coat related processes are possible. Indeed, seeds may present natural permeability sites or other structural features that enhance seed permeability (Kelly *et al.*, 1992); seed surface characteristics (e.g. smooth versus rough) may create microsites favouring seed colonization by microorganisms (Chee-Sanford *et al.*, 2006).

Seed dormancy

A crucial process for ensuring seed persistence and timing emergence

Seed dormancy is treated in detail in Chapter 8. Here, we focus on the importance of dormancy for weed dynamics. Seed dormancy is the absence of germination of a viable seed under otherwise optimal conditions (Hilhorst and Toorop, 1997). From an evolutionary point of view, dormancy can be considered as a strategy to prolong seed germination over time, and is complementary to dispersal in space (Venable and Lawlor, 1980; Venable and Brown, 1988). Therefore, dormancy contributes to species persistence in an unpredictable environment. It also prevents germination during seasons that would be unfavourable for the reproduction of the plant, i.e. germination is only possible in periods that are optimal for the plant to complete its life cycle (Baskin and Baskin, 1998). Some species such as *Aethusa cynapium* (Roberts, 1979) produce seeds with a strong primary (or innate) dormancy; other species such as *Veronica hederifolia* and *Polygonum persicaria* show a strong pattern of seasonal (or secondary) dormancy (Roberts and Lockett, 1978; Bouwmeester, 1990).

Physiological and molecular dormancy studies mostly focus on a small number of well-studied cropped species at short term (e.g. Corbineau *et al.*, 1986; Kucera *et al.*, 2005). They are of little practical use in weed science working with numerous species and focusing on multiannual dynamics. Weed dormancy has been intensively studied for the last 30 years, focusing on heritability of dormancy and/or quantifying patterns of seasonal dormancy in different species (Gardarin *et al.*, 2009). The simplest approach consists in measuring dormancy in buried seeds with time, depending on light exposure (e.g. Lonchamp *et al.*, 1984; Sester *et al.*, 2006; Gardarin and Colbach, in revision). More complex approaches take account of actual climatic and nutritional conditions; they link primary dormancy to conditions during seed production (Fawcett and Slife, 1978; Peters, 1982; Colbach and Dürr, 2003; Swain *et al.*, 2006), and subsequent dormancy to hydric time or thermal time during seed life in the soil (Allen *et al.*, 2007; Batlla and Benech-Arnold, 2007). The most mechanistic approach also integrates the existence of a seed phytochrome activated by light exposure and facilitating germination (Vleeshouwers, 1997b).

Multiple processes result in a multitude of seed trait correlations

Baskin and Baskin (1998, 2004) proposed a classification system based on dormancy mechanisms. However, as most north-west European weed species belong to the same category, i.e. species with physiological dormancy, this classification does not provide much information about weed dormancy levels and their seasonal variations. Other authors tried to relate dormancy to seed traits. A large part of dormancy is imposed by the seed coat acting as a chemical or mechanical barrier (Kelly *et al.*, 1992), resulting in a positive correlation between dormancy and seed coat thickness, both at intra (Sultan, 1996) and inter-specific levels (Gardarin and Colbach, in revision).

In addition to the traits directly implicated in dormancy mechanisms, correlations could also result from evolutionary trade-offs between dispersal in time (i.e. dormancy) and in space (generally associated with small seeds). Indeed, small and/or light seeds were generally shown to be less dormant than large and/or heavy seeds (Grime *et al.*, 1981; Gardarin and Colbach, in revision). The correlation between seed mass and photosensitivity is possibly the result of a co-selection. Indeed, germination of large-seeded species was reported to be less dependent on light than that of species with smaller seeds (Grime *et al.*, 1981; Milberg *et al.*, 2000; Saatkamp, 2009), possibly to prevent small seeds from germinating at depths (where light does not penetrate; Benvenuti, 1995) from which they could not emerge (see section 'Pre-emergent seedling mortality' below). Co-selection is possibly also the reason for the correlation of seed shape with dormancy. Elongated or flattened seeds were shown to present less dormancy than round ones (Grime *et al.*, 1981; Gardarin and Colbach, in revision). In natural habitats, elongated seeds mostly stay on the soil surface (Benvenuti, 2007) where they are more prone to predation than buried ones (Hulme and Borelli, 1999); they may thus have been selected for lack of dormancy and immediate germination to limit such mortality risks.

Seed germination

The importance of germination triggering for timing emergence

Seed germination starts with the uptake of water by the quiescent dry seed and terminates with the elongation of the embryonic axis (Bewley, 1997). The physical and chemical cues of germination are discussed in detail in Chapters 6 and 7 of this volume. In fields, these cues usually correspond to a change in environmental conditions, e.g. seed shed onto soil surface, seed displacement by tillage, moisture increase by rainfall (Froud-Williams *et al.*, 1984; Bouwmeester and Karssen, 1993; Bai *et al.*, 1995; Buhler, 1997; Colbach *et al.*, 2002; Colbach *et al.*, 2006a). Germination mainly determines when and how many non-dormant seeds succeed in emerging in the field. Weed seed germination has been analysed for many species, mostly as a function of seed dormancy at different temperatures and, to a lesser degree, moisture conditions

(Gardarin *et al.*, 2009). Germination can be characterized by two types of parameters: base temperature and soil water potential determine when soil conditions are sufficiently warm and moist for germination, whereas speed parameters determine how fast and how many seeds germinate when conditions are favourable. While base temperature has been studied for a large range of species (see references in Gardarin *et al.*, 2009), other germination parameters are only known for a small number of species, usually those studied for modelling (Wiese and Binning, 1987; Roman *et al.*, 1999; Gardarin *et al.*, 2011; Guillemin *et al.*, 2013). Germination speed has, though, been shown to increase with the proportion of non-dormant seeds in the seed populations (Courtney, 1968; Lonchamp and Gora, 1980; Baskin and Baskin, 1985; Bouwmeester, 1990; Shipley and Parent, 1991; Vleeshouwers, 1997a; Gardarin *et al.*, 2011).

Irrespective of temperature and moisture, germination of buried seeds decreases with depth (Benvenuti *et al.*, 2001; Colbach *et al.*, 2006b). The inhibition of germination with seed depth can simply be caused by the amount of soil pressing on the seed (Colbach *et al.*, 2006b; Gardarin *et al.*, 2012) in field conditions; it is, though, generally attributed to an increase in germination-inhibiting gases with increasing depth (Lonchamp, 1976; Benech-Arnold *et al.*, 2000), which could be the result of a lack of oxygen (Benvenuti and Macchia, 1995). This inhibition process reduces fatal seed germination at a depth where germinated seeds have little chance of emerging and reproducing.

Interaction with numerous and diverse seed traits

Ecologists have developed indicators of optimum environmental requirements for wild species in natural habitats (Landolt, 1977; Evans and Etherington, 1990; Ellenberg *et al.*, 1992). These indicators are, though, insufficient to discriminate weed species according to their hydrothermal preferences (Gardarin *et al.*, 2010c) as most weed species mostly belong to a limited number of categories, representing a narrow range of environmental conditions compared to the large range of ecological categories used for wild species. Indeed, weed species have been selected for the more homogeneous conditions in arable crops; for instance, both drainage and crop irrigation have reduced the range of soil moisture occurring in fields, resulting in a shift of the weed flora towards generalist species with intermediate requirements for soil moisture (Fried and Reboud, 2007).

To date, base temperature and water potential have not been correlated to any species traits, though both parameters are highly correlated, indicating that most European species needing high temperatures for germination also required high water potentials (Gardarin *et al.*, 2010c). This is surprising insofar as in European cropping systems, soil moisture is usually lowest when temperatures are highest (e.g. Gardarin *et al.*, 2010c). However, most European spring and summer annual weeds, i.e. those with high base temperatures and water potentials, originate from tropical areas (Montégut, 1984), which could explain their requirement for both high temperatures and high water potentials for germination.

Species with a high base temperature were also found to germinate faster (for crops: Angus *et al.*, 1981; for weeds: Gardarin *et al.*, 2011). This could be considered as an ecological adaptation to ensure that late-germinating species (i.e. those with a higher base temperature) germinate faster and can thus compete with earlier-germinating ones. Thus, both types of species germinate in a similar length of time despite the differences in soil temperatures at the time of their germination. In insects, this trade-off between base temperature and thermal time required for a particular process was explained by enzymes being more efficient at high temperatures (Trudgill *et al.*, 2005).

Germination speed has also been tentatively correlated to seed traits, though reports are still few and partially contradictory. The most easily explained is the earlier germination onset observed for seeds with a high area/mass ratio (Gardarin *et al.*, 2011) or the faster germination for flattened/elongated,

small and/or light seeds (Grime *et al.*, 1981). Germination timing partly depends on the time required for seed imbibition, which would be faster for seeds presenting a large area for water entrance relative to the seed water demand probably linked to the seed mass (Gardarin *et al.*, 2011). Finally, germination speed also depends on the seed reserve mobilization. A high soluble sugar content in the seed reserves is expected to hasten germination because it supplies rapidly available energy for the embryo (Kuo *et al.*, 1988; Dierking and Bilyeu, 2009) but no such effect has been found to date (Gardarin *et al.*, 2011). Conversely, in well aerated conditions, germination was shown to be faster for seeds with high lipid contents (Gardarin *et al.*, 2011), possibly because lipids generate more energy than sugar.

Pre-emergent seedling mortality

The key role of seed depth

Once a seed has germinated, it produces a shoot that must emerge fast to reach sunlight and avoid dying when the seed reserves run out. Consequently, the shorter this shoot and the deeper the seed is buried, the higher the probability of the seedling dying without emerging (e.g. Chancellor, 1964; Benvenuti *et al.*, 2001). During their pre-emergent growth, the seedlings can also be fatally blocked by soil clods because they lack the force to displace or grow around the clods, an event more frequent in compacted soil and for deeply buried seeds (Dürr and Aubertot, 2000; Vleeshouwers and Kropff, 2000a; Dorsainvil *et al.*, 2005; Colbach *et al.*, 2006b). Moreover, the shoot elongation potential can be reduced if the soil is resistant to shoot penetration (Vleeshouwers and Kropff, 2000a). Finally, seedlings close to the soil surface can also die before emergence when the superficial layers dry after germination and the root is too short to reach the underlying, moister layers (Dürr and Aubertot, 2000; Colbach *et al.*, 2006b).

Emergence has been intensively studied for all weed species (see review by

Gardarin *et al.*, 2009), though most studies focused on monitoring seedling emergence in field trials, without analysing the actual germination and pre-emergent growth leading to emergence (Debaeke, 1988; Forcella, 1998; Rasmussen *et al.*, 2002; Grundy *et al.*, 2003). The distinction between these processes was first made for crop species (Burris and Fehr, 1970; Wanjura and Buxton, 1972; Dürr *et al.*, 2001) and then inspired the first weed studies separating germination and pre-emergent growth (Oryokot *et al.*, 1997; Vleeshouwers and Kropff, 2000a; Colbach and Dürr, 2003). Pre-emergent seedling mortality as a function of soil structure has only been tackled very recently for weed species and then specifically for demographic cropping system models (Vleeshouwers and Kropff, 2000a; Sester *et al.*, 2006).

Big is beautiful!

Seedling establishment has been linked to traits, mainly seed mass, in many species (Weiher *et al.*, 1999) but only a few studies distinguished the three parameters determining pre-emergent seedling mortality, i.e. maximum shoot and root lengths as well as seedling sensitivity to soil aggregates. The general consensus, though, points to a potential shoot length increasing with seed mass, both within and between species (Table 10.1), because of a larger amount of available seed reserves and/or seed embryo for growing in the dark (Terpstra, 1986; Gardarin *et al.*, 2010a). During the dark period, seedling growth is mostly devoted to the shoot. Radicle length is thus little studied, varies little and rarely exceeds a few centimetres, though it has been shown to increase slightly with seed mass (Gardarin *et al.*, 2010a).

Correlations between seedling sensitivity to soil aggregates and species traits are even rarer. This sensitivity was, though, shown to be greater for monocotyledonous than for dicotyledonous species (Cussans *et al.*, 1996; Gardarin *et al.*, 2010a), possibly because of a different shoot morphology and a lesser emergence force exerted at the extremity of the shoot (Souty and Rode,

1994). Clod-caused mortality also decreased with increasing shoot diameter and, consequently, seed mass, both traits being positively correlated (Gardarin *et al.*, 2010a). The shoot diameter effect is again due to the larger emergence force exerted by the shoot tip (Sinha and Ghildyal, 1979).

How Are These Processes Influenced by Crop Succession and Management?

In croplands, year-to-year variations in climatic conditions, and their effects on seed-bank processes only explain a small part of weed dynamics at long term (Freckleton and Watkinson, 2002). Indeed, crop types, varieties and sowing dates usually vary between years; conditions resulting from agricultural practices such as the depth and the date of soil tillage also differ considerably between years. Table 10.2 synthesizes the known effects of cropping system components on weed seed-bank and emergence processes, either directly or via impacts on the biotic and abiotic environment of the weeds.

The choice of a crop species and its frequency in the rotation

The crop sown in a field is the major component of the cropping system influencing weed-dynamics term. Indeed, it determines the period during which weeds must emerge and reproduce: either between sowing and harvest, or between harvest and sowing. The latter is less likely, at least in tillage-based systems. Indeed, field surveys have shown the major weed species to mimic the crop insofar as they emerge with the crop and reproduce slightly earlier than the crop harvest (e.g. Fried and Reboud, 2007; Fried *et al.*, 2008; Fried *et al.*, 2009). Similarly, simulations with weed-dynamics models have shown that the crop traits most relevant for grass weed dynamics appear to be those driving crop emergence timing rather than competitive ability (Colbach *et al.*, 2013a). The successful weed species must

Table 10.2. Synthesizing effects of cropping system on seed-bank and emergence processes.

Cropping system components	Intermediate effects	Effect on weed
Crop (including undersown and temporary crops, living mulches, etc.)		Crop sowing and harvesting periods relative to dormancy cycle
	Crop–weed competition	Heavier seeds present less pre-emergent mortality
	Conditions during seed production	In some species, seeds produced in winter crops are more dormant
	Shading and water absorption	Changes the timing and amount of germination and emergence via cooler and moister conditions on soil surface, drier conditions in deeper soil layers
	Choice of cultivation techniques	See effects of techniques
Tillage (including post-sowing mechanical weeding)	Soil fragmentation	Soil compaction increases pre-emergent seedling mortality
	Soil movements = (soil structure)	Seed burial decreases germination and increases pre-emergent mortality due to insufficient seed reserve
		Seeds on soil surface germinate badly because of insufficient seed–soil contact
		Germinated seeds close to soil surface often die because the topsoil dries faster
		Exposure of imbibed seeds to light if inverting tool and tillage in moist conditions
		Triggers germination flush if the soil is tilled in moist conditions
		Uproots unemerged and emerged seedlings and/or covers them with soil
Sowing date	Date of last tillage or pre-sowing herbicide (if any)	The later the last tillage or herbicide operation, the more weed seeds have germinated and emerged already, and are killed by this operation
	Date of last tillage (if any)	If the soil is moist, triggering of a new germination flush resulting in in-crop weed emergence
	Crop emergence date	The earlier the weed seedlings emerge relative to the crop, the better they survive
Nitrogen	Conditions during seed production	Changes primary dormancy and thus timing of germination of newly produced seeds
Manure	Increases nitrogen availability	See effect of nitrogen
	Adds weed seeds to soil	Increases viable soil seed bank
Date of operations destroying mature plants (harvest, mowing, herbicide, etc.)	Seed shed onto soil surface	Triggers germination flush if the soil is moist
All	Increase soil compaction via wheel traffic	Increases mortality of germinated seeds

therefore have seeds that are little dormant when crops favourable for their reproduction are sown, and highly dormant otherwise. The seeds must also be sufficiently persistent to survive between two successive favourable crops in the rotation, and thus present a low *in situ* mortality as well as a high dormancy to avoid fatal germination.

The crop canopy also has a small effect on germination and pre-emergent growth, via its effect on environmental conditions. Indeed, shading and water absorption will change the temperature and water content in the various soil layers (Bruckler and Witono, 1989; Brisson et al., 2003). The reduction in light availability can be neglected in cultivated fields as most seeds germinate in the soil where light only penetrates a few millimetres (Benvenuti, 1995). It is, though, difficult to evaluate the impact of these effects compared to the crop interaction with the weed dormancy cycle described above, or the more direct and important effect of shading and water absorption by the crop on weed growth and development, which is not treated here. The latter may, though, influence pre-sowing processes indirectly via the mass of newly produced seeds (influencing pre-emergent seedling growth, see section 'Pre-emergent seedling mortality' above) or, in certain species, via the conditions during seed production (e.g. influencing subsequent seed dormancy, see section 'Seed dormancy' above).

Sensitivity analyses with simulation models, though, conclude that a large part of the crop effect is due to the range of possible management options (Mézière et al., 2011; Colbach et al., 2013a). For instance, there is little time to carry out false-seedbed techniques before early-sown winter crops such as oilseed rape, whereas spring crops are usually sown as early as possible to maximize their growth period.

The key role of tillage

Tillage (including post-sowing mechanical weeding) is the main variable driving seed-bank and emergence processes. It increases soil fragmentation, depending on the tool as well as soil texture and moisture (Hillel, 1971; Hughes and Baker, 1977; Roger-Estrade et al., 2000; Chatelin et al., 2005), and thus reduces pre-emergent seedling mortality (see section 'Pre-emergent seedling mortality' above). Though weed seeds are also partially buried by natural processes (e.g. Mohler et al., 2006; Westerman et al., 2009), seeds tend to remain on or close to the soil surface in untilled fields (Clements et al., 1996; Bàrberi and Lo Cascio, 2001).

Tillage is indeed the main factor for seed burial and excavation, thus determining seed depth and subsequent seed germination (see section 'Seed germination' above) as well as pre-emergent seedling mortality (see section 'Pre-emergent seedling mortality' above); burial can also protect seeds from macro-predation (e.g. Baraibar et al., 2009). The degree of seed burial greatly varies with the tool, the tillage depth and environmental conditions (Cousens and Moss, 1990; Grundy et al., 1999; Colbach et al., 2000; Roger-Estrade et al., 2001; Gruber et al., 2010) but, in contrast to natural seed burial, not with seed morphology (e.g. Moss, 1988). Therefore, these options and conditions must be carefully chosen to optimize weed management. For instance, summer tillage aiming at stimulating fatal germination (i.e. false-seedbed techniques) should be as superficial as possible to avoid inhibiting germination in buried seeds; tillage prior to crop sowing should be as deep and inverting as possible to limit germination and favour pre-emergent seedling mortality, thus reducing in-crop weed emergence. False-seedbed techniques will, though, only be successful for controlling species that present little dormancy at the time of tillage.

Tillage moreover triggers germination by improving seed–soil contact of surface seeds and by changing the environment of buried seeds, and it increases germination by exposing seeds to light (Froud-Williams et al., 1984; Bliss and Smith, 1985; Bouwmeester and Karssen, 1993; Scopel et al., 1994; Buhler, 1997; Benech-Arnold et al., 2000). This is, though, only true if seeds are imbibed, i.e. if the soil is tilled in moist conditions. To optimize weed management, the timing of tillage relative to rainfall is thus crucial. Summer tillage (e.g. stale-seedbed technique) should be delayed until after the first rainfall to maximize fatal germination whereas tillage close to sowing should be carried out in dry conditions to avoid triggering germination flushes leading to in-crop weed emergence. Last, tillage cuts off emergence flushes by destroying unemerged and emerged seedlings. Seedling mortality greatly varies with the tillage tool, tractor speed, soil moisture, weed species and stage (Rasmussen, 1992; Kurstjens and Kropff, 2001).

Tillage thus determines the timing and amplitude of weed emergence as well as its possible reproductive success. As a result, weed flora varies considerably in abundance and diversity depending on tillage tools, depths and timing (Zanin *et al.*, 1997). But tillage also has a major long-term effect, interacting with crop rotation. Indeed, those seeds buried by tillage to limit in-crop weed emergence can survive, be excavated in later years and thus emerge and potentially reproduce in subsequent crops.

Timing the crop sowing relative to weed emergence flushes

The effect of the crop sowing date is similar to the main crop effect, i.e. it influences the period during which weeds must emerge and reproduce but to a lesser extent, as the possible range of sowing dates depends on the crop choice. The later the crop is sown, the more weed seeds have germinated since the previous crop harvest and can be killed by a last tillage or pre-sowing herbicide. In tilled fields, the last tillage operation often coincides (by combining tillage and sowing tools) with the sowing, thus possibly triggering a new emergence flush resulting in weed emergence in the crop. Delaying sowing relative to the last tillage also delays the crop emergence relative to this new weed emergence flush, thus increasing the weed survival and reproduction success.

Minor effects

The other cropping system components only present minor effects on seed-bank and emergence processes. The seed bank can be increased by adding cattle manure and other organic fertilizer (Pleasant and Schlather, 1994; Miyazawa *et al.*, 2004) though it is difficult to discriminate the effect of the additional seeds from that of the additional nitrogen (Colbach *et al.*, 2013b). In some species, nitrogen addition, either via chemical or organic fertilizers, can change the dormancy and germination behaviour of the seeds produced by the weed plants growing in the fertilized field (see sections 'Seed dormancy' and 'Seed germination' above). Any operation destroying plants carrying mature seeds results in seed rain onto soil surface and can trigger a germination flush if the soil surface is sufficiently moist. Such operations probably do not affect weed dynamics via a change in timing, as the mature seeds would have shed anyway soon (though they have a large, quantitative effect by reducing seed production). Any operations can potentially increase soil compaction via the tractor wheels and thus pre-emergent seedling mortality, particularly if the soil is wet. This can reduce post-harvest emergence after late-harvested crops such as sugar beet but again does not affect weed dynamics as these seedlings would have had little chance in reproducing. Undersown crops and living mulches can have indirect effects on seed-bank processes by modifying the soil environment, e.g. by reducing light transmission (Teasdale *et al.*, 1991; Teasdale and Mohler, 1993) and water availability (Duiker and Hartwig, 2004; Teasdale *et al.*, 2007). These secondary crops can also directly disturb pre-emergent processes by liberating allelopathic molecules (Barnes and Putnam, 1983) or possibly also stimulate additional germination as in the case of parasitic weeds (see section 'A succession of diverse processes' above).

Which Species Traits are Selected in Which Cropping Systems?

Can we predict trait selection from functional relationships?

Combining our knowledge on the effects of cropping system components on these processes with the functional relationships between seed-bank and emergence processes on the one hand, and seed traits on the other hand, can already produce a few interesting conclusions. For instance, seed persistence was identified above (section 'The choice of a crop species and its frequency in the rotation') as a major prerequisite for species to be successful in diverse crop rotations. This characteristic is the result of both low seed mortality and high seed dormancy; it increases with seed coat thickness and decreases with seed shape index (Fig. 10.2). In other words, diverse crop rotations should select species with round and thick-coated seeds. Conversely, species with flattened/elongated and thin-coated seeds are more probably in monocultures of summer or early-sown autumn crops as these seeds do not persist long in the soil.

Mouldboard ploughing, which buries seeds to great depths and can excavate them years later (Clements *et al.*, 1996; Liebman *et al.*, 1996; Chauvel *et al.*, 2001; Lutman *et al.*, 2002; Cirujeda *et al.*, 2003), is another cropping system component greatly interfering with species and trait selection. Species with heavy seeds germinate well from great depths and present a low pre-emergent seedling mortality; they can therefore emerge even after burial by mouldboard ploughing (Fig. 10.3). Species with thick-coated seeds present a different strategy to persist in ploughed fields: their seeds survive for several years after burial and are still able to germinate and emerge when being excavated by a subsequent ploughing. Species with both heavy and thick-coated seeds can emerge irrespective of tillage strategy but such species are rare, indicating that this trait combination is disadvantaged during other life-cycle processes (e.g. species with heavy seeds produce fewer seeds, Storkey *et al.*, in revision). Conversely, species with light and thin-coated seeds can only emerge in fields with superficial and/or non-inverting tillage and are thus theoretically badly adapted to most European cropping systems with their frequent and deep tillage operations (Fried and Reboud, 2007).

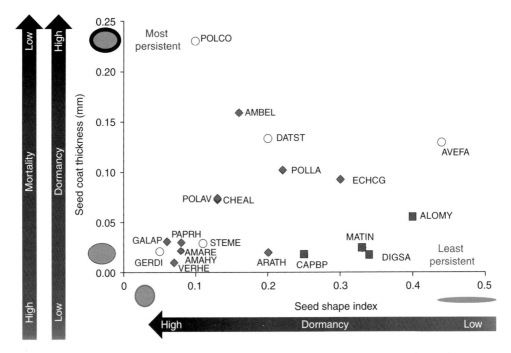

Fig. 10.2. Variation in seed persistence with two major seed traits (based on Gardarin *et al.*, 2010b; Gardarin and Colbach, in revision). Seed mortality decreases with increasing seed coat thickness; seed dormancy increases with seed coat thickness and with seed shape index (■, species with a minimum of 10% of dormant seeds when dormancy is lowest and no more than 10% of non-dormant seeds when dormancy is highest; ♦, species with less dormant seeds; ○, no data on dormancy). Species are indicated with Bayer codes (© 2012 Nathalie Colbach).

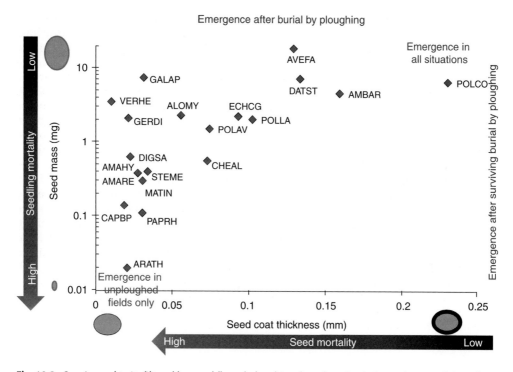

Fig. 10.3. Species and traits filtered by mouldboard ploughing (based on Gardarin *et al.*, 2010a,b). Seed mortality decreases with increasing seed coat thickness; pre-emergent seedling mortality decreases with increasing seed mass. Species are indicated with Bayer codes (© 2012 Nathalie Colbach).

The key role of models for understanding species and trait selection in cropping systems

The previous analytical approach produces interesting conclusions when looking at major factors and traits related to only one or a small number of processes. It will be much more difficult for minor factors such as delayed sowing or for traits interacting with numerous processes. This is for instance the case for seed mass. Heavy seeds should be the overall optimal strategy for colonizing all types of cropping systems because of the numerous advantages resulting from this trait. However, only a few species present heavy seeds, pointing to a trade-off between this trait and other traits as well as possible disadvantages of heavy seeds that are not visible with the present analytical approach. The prediction is further complicated because of the long-term processes behind seed-bank dynamics, and the numerous interactions with cropping system components and environmental conditions.

Models are essential to understand the effects of cropping systems and environment, and to support decision making by farmers and other stakeholders (e.g. Rossing *et al.*, 1997; Aubertot *et al.*, in revision). Mechanistic (i.e. process-based) models require the identification of major processes and factors and synthesize available data; they are thus also a valuable tool to present knowledge and will be used here to illustrate the processes leading from the seed to the seedling.

To date, only a few models use a mechanistic approach where the weed life cycle is split into sub-processes depending on direct (e.g. seedling uprooting by tillage) and indirect effects (e.g. soil fragmentation by tillage) of cropping system components, in interaction with the biological (e.g. weed stage) and physical variables (e.g. soil moisture) (Colbach and Debaeke, 1998; Colbach *et al.*, 2005; Holst *et al.*, 2007). The most detailed of these models (Vleeshouwers and Kropff, 2000b; Colbach *et al.*, 2006b; Sester *et al.*, 2007; Gardarin *et al.*, 2012) all present more or less the same structure but only FLORSYS (Gardarin *et al.*, 2012) is a truly multi-specific model, moreover integrating many of the functional relationships presented in the section 'The different processes from the seed to the emerged seedling' above, and will be used here for illustration.

Identifying species traits selected by cropping systems by simulation

To identify which species traits are selected in different cropping systems, a simulation study (Colbach *et al.*, 2012) was carried out with the FLORSYS model (Colbach *et al.*, 2010; Gardarin *et al.*, 2012; Munier-Jolain *et al.*, 2013), with a series of cropping systems typical of three French regions. The simulations showed that the weed density

and the species selected in these cropping systems depended very much on their rotation and the tillage strategy: weed density and species diversity were highest in the systems with the longest and most diverse rotations: weed density was multiplied by *c.*100 in an oilseed rape/winter wheat/sunflower/winter wheat rotation versus a maize monoculture; the number of species increased by 75% (Colbach *et al.*, 2012).

The species traits selected in these cropping systems also depended very much on their rotation and the tillage strategy (Table 10.3A). Densities increased with the seed coat thickness of the species in those systems with the longest and most diverse rotations but it decreased in monocultures. In the former, the lower seed mortality resulting from the thicker seed coat (Table 10.1) was necessary in a rotation where crops

Table 10.3. Weed species traits selected in different cropping systems via their functional relationships with life-cycle processes (based on Colbach *et al.*, 2012). Regression coefficients linking weed density to species traits in reference scenarios and after modifying agricultural practices in three regions.

			Seed traits				
			Coat thickness	Weight	Shape index[a]	Lipid content	Area
A. Control scenarios							
Burgundy	OWB[b]	1 P/3[c]	0.0517[d]	−0.219	−6.13	0.50	[e]
Poitou-Charentes	OWsW	3 P/4	0.0495	0.220	3.38	−2.84	−0.506
South-West	m	1 P/1	−0.0382	0.363	8.05		−0.284
B. Modified practices							
No plough						1.77	
Less tillage			0.0209	−0.149	−7.70		0.306
No-till			0.0313	−0.338	−6.85	1.81	0.491
Glyphosate							
Modified herbicide programme						1.33	−0.023
Early sowing			−0.0082			2.19	
Temporary crop			0.0547	−0.349	−9.11	1.45	0.209
Shorter rotation				−0.023			
Longer rotation			0.0056	−0.0612			

[a]The higher this index, the more seeds are elongated and/or flattened.
[b]O, winter oilseed rape; W, winter wheat; B, winter barley; s, sunflower; m, maize.
[c]Ploughing frequency, i.e. number of ploughing operations per number of years.
[d]In the case of a positive (negative) regression coefficient, weed density increases (decreases) with increasing trait value.
[e]Empty cases indicate effects not significant at $p = 0.01$.
These scenarios were simulated, using a weed flora consisting of seven major, mostly autumnal species (*Alopecurus myosuroides*, *Avena fatua*, *Capselle bursa-pastoris*, *Galium aparine*, *Geranium dissectum*, *Stellaria media*, *Veronica hederifolia*). Each scenario was simulated over 27 years and repeated 10 times, by randomly choosing each year annual climate series measured in the tested regions.

favourable for reproduction were rare. But a thicker seed coat also results in higher seed dormancy and thus lower germination in a given year, which appears to be a disadvantage in monocultures. Similarly, elongated/flattened seeds (which are less dormant and less persistent) were selected in monocultures (Table 10.3A). Heavier seeds were more abundant in systems with frequent mouldboard ploughing (Poitou-Charentes and South-West, Table 10.3A) because they persist better (because of increased dormancy) and they germinate and emerge better after burial (Table 10.1). When the disturbance due to ploughing was less frequent, lighter seeds were selected because they present less dormancy when young and they germinate faster (lower area/mass ratio), which was particularly important in systems with summer crops and early-sown autumn crops (Burgundy, Table 10.3A). Similarly, high-lipid seeds (which germinate faster) were also selected in these systems whereas low-lipid-content seeds with small areas (which germinate more slowly) were more abundant in systems with later sown and spring crops.

When simulating changes in these cropping systems in the three regions (Table 10.3B), the selected traits can be explained similarly. For instance, simplifying or abandoning tillage selected lighter seeds; longer rotations favoured thick-coated seeds etc. Generally, the more stable the systems (little or no tillage, continuous crop cover via rotations with summer crops or with temporary crops), the more important fast germination became (selection of light, high-lipid and large-area seeds). However, avoiding fatal germination by increased primary dormancy also seemed to be an issue as thick-coated and round seeds were also selected in these systems. Conversely, when crops were sown earlier, thin-coated and flattened/elongated seeds were preferred.

Are these conclusions confirmed in fields?

Model development and simulations present the major advantages of synthesizing knowledge and effects and account for interactions and long-term effects. Here, they were essential for demonstrating trade-offs between processes and species traits. However, despite their overwhelming advantages, models are only a partial, simplified representation of reality (e.g. seed predation was neglected in FLORSYS, the simulations of Table 10.3 only worked with seven species), and their conclusions should be confirmed with observations. The main conclusions can indeed be confirmed by field surveys and experiments. For instance, species with long-surviving seeds are reported to be more typical of diverse rotations and their management reasoned over a longer time-span (see examples in Van Acker, 2009) whereas abandoning mouldboard ploughing is reported to favour grass weeds with small and/or non-persisting seeds (e.g. Davidson, 1990; McCloskey *et al.*, 1996; Recasens *et al.*, 2001; Reddy, 2003; Douglas and Peltzer, 2004; Puricelli and Tuesca, 2005) as well as crop volunteers with similar seed characteristics (Roller *et al.*, 2002, 2003; Gruber *et al.*, 2004). However, it is difficult to progress further as field studies are usually restricted to the species level and rarely descend to the trait level. The few studies that attempt to link species dynamics and traits usually focus on post-emergence traits. For instance, the weed species increasing in sunflower since the 1970s was shown to be more nitrophilous, more heliophilous, less sensitive to sunflower herbicides and shared a rapid summer life cycle (Fried *et al.*, 2009). Thus, while many of the conclusions of the previous simulation studies are logical and consistent with existing knowledge, we have yet only fragmentary evidence from observations to confirm these conclusions.

Conclusion

A plea for modelling

Processes occurring in the soil are notoriously difficult to study because *in situ* observation in the field is often impossible.

Studies are therefore either destructive (e.g. seeds are excavated to measure viability, seedlings are uprooted to measure shoot lengths) or are carried out *in vitro* (e.g. germination studies) and thus in conditions quite different from the field. Another difficulty is the frequent confusion of processes, e.g. seed persistence studies that do not distinguish mortality and germination, though the two processes depend on different factors and their outcome is quite different, the germinated seeds being potentially able to reproduce in contrast to dead seeds. The situation is even more complicated by the numerous effects and interactions resulting from crop management operations and the long-term consequences of these choices.

Our knowledge on seed-bank and emergence processes thus either results from detailed but partial studies focusing on a single process, or from field observations with a high risk of confusing effects and factors. Consequently, mechanistic models are crucial to synthesize the results from the numerous single-factor or process studies and then to check their verisimilitude and to identify missing processes by comparing the resulting predictions with long-term field observations. This approach and its advantages are not restricted to agricultural ecosystems or to seed-bank and emergence processes; they are also valuable in natural ecosystems and for other life-cycle processes.

Extrapolation to other habitats

While most of the processes analysed here also take place in semi-natural and wild habitats, their relative importance changes, and other processes that have little effect in cultivated fields must also be accounted for (see Chapter 11 of this volume). For instance, seed loss through predation would be more important in natural habitats, because of their more abundant and diverse fauna (see Chapter 5) and the absence of tillage, leaving seeds on or close to the soil surface. The latter could also boost the inhibiting effect of existing vegetation on germination because, in contrast to buried seeds, surface seeds are

affected by the resulting decrease in light availability and red/far red ratio (see Chapter 5 of this volume). Seed dormancy should also become less important with the spread of perennials that regenerate from rhizomes and other subterranean organs. Additional processes becoming pertinent to the detriment of initially dominant processes can also occur in arable fields, for instance with the adoption of conservation tillage (any tillage leaving more than 30% of the soil surface covered by plant residues after crop sowing; Fawcett and Towery, 2004) now spreading by the advent of crops tolerant to non-selective herbicides (Fernandez-Cornejo and McBride, 2002; Trigo and Cap, 2003; Fawcett and Towery, 2004; Cerdeira and Duke, 2006; Givens *et al.*, 2009).

What, though, differs considerably in habitats outside fields is the indigenous flora, resulting from different selection processes. This not only changes the ensuing species and trait combinations, but could also modify the functional trait–process relationships. For instance, in untilled habitats where seeds are more exposed to macro-predation, *in situ* mortality might no longer be essentially correlated to seed coat thickness but rather to traits that render seeds appetizing to predators (see Chapter 5). Conversely, if co-selection was indeed responsible for round/small seeds being more dormant because of their migrating more easily to deeper soil layers (see section 'Seed dormancy' above), this relationship should be even stronger in natural habitats where no tillage could dilute the selection process by burying seeds irrespective of their size and shape. Strengthening existing trait–process correlations or, conversely, selecting different functional relationships could also happen in arable fields when agricultural practices change radically as in no-till or with living mulches. Consequently, even disregarding the perennials and parasites neglected here, the changes in species and traits resulting from modified agricultural practices will probably be much more drastic than those predicted in section 'Identifying species traits selected by cropping systems by simulation' above.

A need for a new weed management approach

The present analysis of the multiple factors and interactions governing seed-bank dynamics and seedling emergence stresses the need for basing weed management on a judicious combination of a large range of cropping system components (crop succession and the different crop management techniques). Numerous definitions have been proposed over the years for integrated crop protection or pest management (Bajwa and Kogan, 2002), one of the latest defining integrated pest management as 'a sustainable approach to managing pests by combining biological, cultural, physical, and chemical tools in a way that minimizes economic, health, and environmental risks' (Food Quality Protection Act, 1996). Combining different control measures remains necessary even if novel weed control measures such biocontrol, newly domesticated species or engineered weed-resistant crops become available (Gressel, 2011). Indeed, changes in cropping systems and climate change are expected to lessen the efficiency and predictability of individual weed management measures (Rodenburg *et al.*, 2011).

Moreover, weeds are a particular case of crop pests insofar as they are also a valuable biotic component of the agroecosystem and produce numerous agroecological services, e.g. habitats and food resources for many other guilds and species (e.g. pollinators, granivorous insectes and birds; Wilson *et al.*, 1999; Marshall *et al.*, 2003; Taylor *et al.*, 2006; Carvalheiro *et al.*, 2011; Evans *et al.*, 2011; Petit *et al.*, 2011). Consequently, managing weeds must comprise not only the control of communities, species and populations harmful for the crop but also the preservation of those producing agroecological services, requiring even more detailed knowledge of biophysical processes governing the effect of cropping system components on seed-bank dynamics and seedling emergence.

References

Allen, P.S., Benech-Arnold, R.L., Batlla, D. and Bradford, K.J. (2007) Modeling of seed dormancy. In: Bradford, K.J. and Nonogaki, H. (eds) *Seed development, dormancy and germination*. Blackwell Publishing Ltd, pp. 72–112.

Andersson, T.N. and Milberg, P. (1998) Weed flora and the relative importance of site, crop, crop rotation, and nitrogen. *Weed Science* 46, 30–38.

Andreasen, C., Stryhn, H. and Streibig, J.C. (1996) Decline of the flora in Danish arable fields. *Journal of Applied Ecology* 33, 619–626.

Angus, J.F., Cunningham, R.B., Moncur, M.W. and Mackenzie, D.H. (1981) Phasic development in field crops. I. Thermal response in the seedling phase. *Field Crops Research* 3, 365–378.

Aubertot, J.N., Lescourret, F., Bonato, O., Colbach, N., Debaeke, P., Doré, T., Fargues, J., Lô-Pelzer, E., Loyce, C. and Sauphanor, B. (in revision) How to improve pest management in cropping systems. Effects of cultural practices on pest development. A review. *Agronomy for Sustainable Development*.

Bai, Y., Romo, J.T. and Young, J.A. (1995) Influences of temperature, light and water stress on germination of fringed sage (*Artemisia frigida*). *Weed Science* 43, 219–225.

Bajwa, W.I. and Kogan, M. (2002) Compendium of IPM Definitions (CID) - What is IPM and how is it defined in the Worldwide Literature? *IPPC Publication No. 998*. Integrated Plant Protection Center (IPPC), Oregon State University, OR USA, pp. 19.

Ballaré, C.L., Scopel, A.L., Ghersa, C.M. and Sàjnchez, R.A. (1988) The fate of Datura ferox seeds in the soil as affected by cultivation, depth of burial and degree of maturity. *Annals of Applied Biology* 112, 337–345.

Baraibar, B., Westerman, P.R. and Recasens, J. (2009) Effects of tillage and irrigation in cereal fields on weed seed removal by seed predators. *Journal of Applied Ecology* 46, 380–387.

Bàrberi, P. and Lo Cascio, B. (2001) Long-term tillage and crop rotation effects on weed seedbank size and composition. *Weed Research* 41, 325–340.

Barnes, J.P. and Putnam, A.R. (1983) Rye residues contribute weed suppression in no-tillage cropping systems. *Journal of Chemical Ecology* 9, 1045–1057.

Baskin, C.C. and Baskin, J.M. (1998) *Seeds: ecology, biogeography, and evolution of dormancy and germination*. Academic Press, San Diego, USA.

Baskin, J.M. and Baskin, C.C. (1985) The annual dormancy cycle in buried weed seeds: A continuum. *Bioscience* 35, 492–498.

Baskin, J.M. and Baskin, C.C. (2004) A classification system for seed dormancy. *Seed Science Research* 14, 1–16.

Batlla, D. and Benech-Arnold, R.L. (2007) Predicting changes in dormancy level in weed seed soil banks: Implications for weed management. *Crop Protection* 26, 189–197.

Baux, A., Colbach, N. and Pellet, D. (2011) Crop management for optimal low-linolenic rapeseed oil production – field experiments and modelling. *European Journal of Agronomy* 35, 144–153.

Beckie, H.J., Leeson, J.Y., Thomas, A.G., Hall, L.M. and Brenzil, C.A. (2008) Risk assessment of weed resistance in the Canadian prairies. *Weed Technology* 22, 741–746.

Bekker, R.M., Bakker, J.P., Grandin, U., Kalamees, R., Milberg, P., Poschlod, P., Thompson, K. and Willems, J.H. (1998) Seed size, shape and vertical distribution in the soil: indicators of seed longevity. *Functional Ecology* 12, 834–842.

Benech-Arnold, R.L., Sanchez, R.A., Forcella, F., Kruk, B.C. and Ghersa, C.M. (2000) Environmental control of dormancy in weed seed banks in soil. *Field Crops Research* 67, 105–122.

Benharrat, H., Boulet, C., Theodet, C. and Thalouarn, P. (2005) Virulence diversity among branched broomrape (*Orobanche ramosa* L.) populations in France. *Agronomy for Sustainable Development* 25, 123–128.

Benvenuti, S. (1995) Soil light penetration and dormancy of Jimsonweed (*Datura stramonium*) seeds. *Weed Science* 43, 389–393.

Benvenuti, S. (2007) Natural weed seed burial: effect of soil texture, rain and seed characteristics. *Seed Science Research* 17, 211–219.

Benvenuti, S. and Macchia, M. (1995) Effect of hypoxia on buried weed seed germination. *Weed Research* 35, 343–351.

Benvenuti, S., Macchia, M. and Miele, S. (2001) Quantitative analysis of emergence of seedlings from buried weed seeds with increasing soil depth. *Weed Science* 49, 528–535.

Bewley, J.D. (1997) Seed germination and dormancy. *The Plant Cell* 9, 1055–1066.

Bliss, D. and Smith, H. (1985) Penetration of light into soil and its role in the control of seed germination. *Plant, Cell and Environment* 8, 475–483.

Bond, W.J., Honig, M. and Maze, K.E. (1999) Seed size and seedling emergence: an allometric relationship and some ecological implications. *Oecologia* 120, 132–136.

Boulet, C., Pineault, D., Benharrat, H., Simier, P. and Delavault, P. (2007) Adventices du colza et orobanche rameuse. In: *XXème Conférence du COLUMA: Journées Internationales sur la Lutte contre les Mauvaises Herbes*. AFPP, Dijon, France, pp.

Bouwmeester, H.J. (1990) The effect of environmental conditions on the seasonal dormancy pattern and germination of weed seeds. PhD thesis, Wageningen Agricultural University, Wageningen, the Netherlands.

Bouwmeester, H.J. and Karssen, C.M. (1993) Seasonal periodicity in germination of seeds of *Chenopodium album* L. *Annals of Botany* 72, 463–473.

Bouwmeester, H.J., Matusova, R., Zhongkui, S. and Beale, M.H. (2003) Secondary metabolite signalling in host-parasitic plant interactions. *Current Opinion in Plant Biology* 6, 358–364.

Boyd, N.S. and Van Acker, R.C. (2003) The effects of depth and fluctuating soil moisture on the emergence of eight annual and six perennial plant species. *Weed Science* 51, 725–730.

Brisson, N., Gary, C., Justes, E., Roche, R., Mary, B., Ripoche, D., Zimmer, D., Sierra, J., Bertuzzi, P., Burger, P., Bussiere, F., Cabidoche, Y.M., Cellier, P., Debaeke, P., Gaudillere, J.P., Henault, C., Maraux, F., Seguin, B. and Sinoquet, H. (2003) An overview of the crop model STICS. *European Journal of Agronomy* 18, 309–332.

Bruckler, L. and Witono, H. (1989) Use of remotely sensed soil moisture content as boundary conditions in soil-atmosphere water transport modeling. 2. Estimating soil water balance. *Water Resources Research* 25, 2437–2447.

Buhler, D.D. (1997) Effects of tillage and light environment on emergence of 13 annual weeds. *Weed Technology* 11, 496–501.

Burnside, O.C., Wilson, R.G., Weisberg, S. and Hubbard, K.G. (1996) Seed longevity of 41 weed species buried 17 years in eastern and western Nebraska. *Weed Science* 44, 74–86.

Burris, J.S. and Fehr, W.R. (1970) Methods for evaluation of Soybean hypocotyl length. *Transactions of the ASAE*, 116–117.

Cardina, J., Johnson, G.A. and Sparrow, D.H. (1997) The nature and consequence of weed spatial distribution. *Weed Science* 45, 364–373.

Cardina, J., Herms, C.P. and Doohan, D.J. (2002) Crop rotation and tillage system effects on weed seedbanks. *Weed Science* 50, 448–460.

Carvalheiro, L., Veldtman, R., Shenkute, G., Tesfay, G., Walter, C., Pirk, W., Donaldson, J. and Nicolson, S. (2011) Natural and within-farmland biodiversity enhances crop productivity. *Ecology Letters* 14, 251–259.

Cavan, G., Cussans, J. and Moss, S.R. (2000) Modelling different cultivation and herbicide strategies for their effect on herbicide resistance in Alopecurus myosuroides. *Weed Research* 40, 561–568.

Cavan, G., Cussans, J. and Moss, S. (2001) Managing the risks of herbicide resistance in wild oat. *Weed Science* 49, 236–240.

Cerabolini, B., Ceriani, R.M., Caccianiga, M., De Andreis, R. and Raimondi, B. (2003) Seed size, shape and persistence in soil: a test on Italian flora from Alps to Mediterranean coasts. *Seed Science Research* 13, 75–85.

Cerdeira, A.L. and Duke, S.O. (2006) The current status and environmental impacts of glyphosate-resistant crops: a review. *Journal of Environmental Quality* 35, 1633–1658.

Chancellor, R.J. (1964) The depth of weed seed germination in the field. In: *Seventh British Weed Control Conference*. Brighton, England, pp. 607–613.

Chatelin, M.H., Aubry, C., Poussin, J.C., Meynard, J.M., Masse, J., Verjux, N., Gate, P. and Le Bris, X. (2005) DéciBlé, a software package for wheat crop management simulation. *Agricultural Systems* 83, 77–99.

Chauvel, B., Guillemin, J.P., Colbach, N. and Gasquez, J. (2001) Evaluation of cropping systems for management of herbicide-resistant populations of blackgrass (*Alopecurus myosuroides* Huds.). *Crop Protection* 20, 127–137.

Chauvel, B., Guillemin, J.P. and Colbach, N. (2009) Evolution of a herbicide-resistant population of *Alopecurus myosuroides* Huds. in a long-term cropping system experiment. *Crop Protection* 28, 343–349.

Chee-Sanford, J.C., Williams, M.M., II, Davis, A.S. and Sims, G.K. (2006) Do microorganisms influence seed-bank dynamics? *Weed Science* 54, 575–587.

Cirujeda, A., Recasens, J. and Taberner, A. (2003) Effect of ploughing and harrowing on a herbicide resistant corn poppy (Papaver rhoeas) population. *Biological Agriculture & Horticulture* 21, 231–246.

Clements, D.R., Weise, S.F. and Swanton, C.J. (1994) Integrated weed management and weed species diversity. *Phytoprotection* 75, 1–18.

Clements, D.R., Benoit, D.L., Murphy, S.D. and Swanton, C.J. (1996) Tillage effects on weed seed return and seedbank composition. *Weed Science* 44, 314–322.

Colbach, N. and Debaeke, P. (1998) Integrating crop management and crop rotation effects into models of weed population dynamics: a review. *Weed Science* 46, 717–728.

Colbach, N. and Dürr, C. (2003) Effects of seed production and storage conditions on blackgrass (*Alopecurus myosuroides* Huds.) germination and shoot elongation. *Weed Science* 51, 708–717.

Colbach, N., Roger-Estrade, J., Chauvel, B. and Caneill, J. (2000) Modelling vertical and lateral seed bank movements during mouldboard ploughing. *European Journal of Agronomy* 13, 111–124.

Colbach, N., Dürr, C., Chauvel, B. and Richard, G. (2002) Effect of environmental conditions on *Alopecurus myosuroides* germination. II. Effect of moisture conditions and storage length. *Weed Research* 42, 222–230.

Colbach, N., Dürr, C., Roger-Estrade, J. and Caneill, J. (2005) How to model the effects of farming practices on weed emergence. *Weed Research* 45, 2–17.

Colbach, N., Busset, H., Yamada, O., Dürr, C. and Caneill, J. (2006a) ALOMYSYS: Modelling blackgrass (*Alopecurus myosuroides* Huds.) germination and emergence, in interaction with seed characteristics, tillage and soil climate. II. Evaluation. *European Journal of Agronomy* 24, 113–128.

Colbach, N., Dürr, C., Roger-Estrade, J., Chauvel, B. and Caneill, J. (2006b) ALOMYSYS: Modelling blackgrass (*Alopecurus myosuroides* Huds.) germination and emergence, in interaction with seed characteristics, tillage and soil climate - I. Construction. *European Journal of Agronomy* 24, 95–112.

Colbach, N., Gardarin, A. and Munier-Jolain, N.M. (2010) FLORSYS: a mechanistic model of cropping system effects on weed flora based on functional relationships with species traits. In: *15th International EWRS Symposium*. Kaposvár, Hungary, pp. 157–158.

Colbach, N., Granger, S., Guyot, S.H.M. and Mézière, D. (2012) Changing agricultural practices modifies the species and trait composition of the weed flora. A simulation study using a model of
cropping system effects on weed dynamics. In: *Proceedings of 12th Congress of the European
Society for Agronomy. Helsinki, Finland, 20–24 August 2012*, 152–153.

Colbach, N., Granger, S. and Mézière, D. (2013a) Using a sensitivity analysis of a weed dynamics model to develop sustainable cropping systems. II. Long-term effect of past crops
and management techniques on weed infestation. *Journal of Agricultural Science* 151,
247–267.

Colbach, N., Tschudy, C., Meunier, D., Houot, S. and Nicolardot, B. (2013b) Weed seeds in exogenous
organic matter and their contribution to weed dynamics in cropping systems. A simulation
approach. *European Journal of Agronomy* 45, 7–19.

Conn, J.S. (2006) Weed seed bank affected by tillage intensity for barley in Alaska. *Soil & Tillage
Research* 90, 156–161.

Conn, J.S., Beattie, K.L. and Blanchard, A. (2006) Seed viability and dormancy of 17 weed species after
19.7 years of burial in Alaska. *Weed Science* 54, 464–470.

Corbineau, F., Lecat, S. and Come, D. (1986) Dormancy of three cultivars of oat seeds (Avena sativa L.).
Seed Science and Technology 14, 725–735.

Corbineau, F., Gay-Mathieu, C., Vinel, D. and Come, D. (2002) Decrease in sunflower (*Helianthus
annuus*) seed viability caused by high temperature as related to energy metabolism, membrane
damage and lipid composition. *Physiologia Plantarum* 116, 489–496.

Courtney, A.D. (1968) Seed dormancy and field emergence in Polygonum aviculare. *Journal of Applied
Ecology* 5, 675–683.

Cousens, R. and Moss, S.R. (1990) A model of the effects of cultivation on the vertical distribution of
weed seeds within the soil. *Weed Research* 30, 61–70.

Cussans, G.W., Raudonius, S., Brain, P. and Cumberworth, S. (1996) Effects of depth of seed burial and
soil aggregate size on seedling emergence of *Alopecurus myosuroides, Galium aparine, Stellaria
media* and wheat. *Weed Research* 36, 133–141.

Davidson, R.M. (1990) Management of herbicide resistant annual ryegrass, Lolium rigidum, in crops
and pastures. In: *Proceedings of the 9th Australian Weeds Conference*, pp. 230–233.

Davis, A.S., Schutte, B.J., Iannuzzi, J. and Renner, K.A. (2008) Chemical and physical defense of weed
seeds in relation to soil seedbank persistence. *Weed Science* 56, 676–684.

Debaeke, P. (1988) Modelling the long-term evolution of the weed flora. II. Application to three annual
broad-leaved weeds on a given site. *Agronomie* 8, 767–777.

Délye, C. (2005) Weed resistance to acetyl coenzyme A carboxylase inhibitors: an update. *Weed
Science* 53, 728–746.

Délye, C. (2013) Unravelling the genetic bases of non-target-site based resistance (NTSR) to herbicides: a major challenge for weed science in the forthcoming decade. *Pest Management Science*
69(2), 176–187.

Délye, C., Gardin, J.A.C., Boucansaud, K., Chauvel, B. and Petit, C. (2011) Non-target-site-based resistance should be the centre of attention for herbicide resistance research: *Alopecurus myosuroides*
as an illustration. *Weed Research* 51, 433–437.

Dessaint, F., Chadoeuf, R. and Barralis, G. (1997) Nine years' soil seed bank and weed vegetation relationships in an arable field without weed control. *Journal of Applied Ecology* 34, 123–130.

Dierking, E.C. and Bilyeu, K.D. (2009) Raffinose and stachyose metabolism are not required for efficient soybean seed germination. *Journal of Plant Physiology* 166, 1329–1335.

Dorsainvil, F., Dürr, C., Justes, E. and Carrera, A. (2005) Characterisation and modelling of white mustard (*Sinapis alba* L.) emergence under several sowing conditions. *European Journal of Agronomy*
23, 146–158.

Douglas, A. and Peltzer, S.C. (2004) Managing herbicide resistant annual ryegrass (Lolium rigidum
Gaud.) in no-till systems in Western Australia using occasional inversion ploughing. In: *Weed
management: balancing people, planet, profit. 14th Australian Weeds Conference, Wagga Wagga,
New South Wales, Australia, 6-9 September 2004: papers and proceedings*. Weed Society of New
South Wales, pp. 300–303.

Duiker, S.W. and Hartwig, N.L. (2004) Living mulches of legumes in imidazolinone-resistant corn.
Agronomy Journal 96, 1021–1028.

Dürr, C. and Aubertot, J.N. (2000) Emergence of seedlings of sugar beet (*Beta vulgaris* L.) as affected by
the size, roughness and position of aggregates in the seedbed. *Plant and Soil* 219, 211–220.

Dürr, C., Aubertot, J.N., Richard, G., Dubrulle, P., Duval, Y. and Boiffin, J. (2001) SIMPLE: A model for SIMulation of PLant Emergence predicting the effects of soil tillage and sowing operations. *Soil Science Society of America Journal* 65, 414–423.

ECOPHYTO R&D (2009) Vers des systèmes de culture économes en produits phytosanitaires. Tome II: Analyse comparative de différents systèmes en grandes cultures. In: Guichard, L. and Savini, I. (eds) *Rapport d'expertise scientifique collective*. Paris, France, INRA, pp. 218.

Ellenberg, H., Weber, H.E., Düll, R.R., Wirth, V., Werner, W. and Paulißen, D. (1992) Zeigewerte von Pflanzen in Mitteleuropa. In: Goltze, E. (ed.) *Scripta Geobotanica*. pp. 260.

Evans, C.E. and Etherington, J.R. (1990) The effect of soil water potential on seed germination of some British plants. *New Phytologist* 115, 539–548.

Evans, D.M., Pocock, M.J.O., Brooks, J. and Memmott, J. (2011) Seeds in farmland food-webs: Resource importance, distribution and the impacts of farm management. *Biological Conservation* 144, 2941–2950.

Fawcett, R. and Towery, D. (2004) *Conservation tillage and plant biotechnology: how new technologies can improve the environment by reducing the need to plow*. Conservation Technology Information Center, West Lafayette.

Fawcett, R.S. and Slife, F.W. (1978) Effects of field applications of nitrate on weed seed germination and dormancy. *Weed Science* 26, 594–596.

Fernandez-Cornejo, J. and McBride, W.D. (2002) Genetically engineered crops: U.S. adoption & impacts. *Agricultural Outlook* 294, 24–27.

Food Quality Protection Act (1996) Food Quality Protection Act of 1996, P.L. 104- 170, Title II, Section 303, Enacted August 3, 1996. Codified in: Title 7, U.S. Code, Section 136r-1. Integrated Pest Management., pp.

Forcella, F. (1998) Real-time assessment of seed dormancy and seedling growth for weed management. *Seed Science Research* 8, 201–209.

Freckleton, R.P. and Watkinson, A.R. (2002) Are weed population dynamics chaotic? *Journal of Applied Ecology* 39, 699–707.

Fried, G. and Reboud, X. (2007) Évolution de la composition des communautés adventices des cultures de colza sous l'influence des systèmes de culture. *Oleagineux, Corps Gras, Lipides* 14, 130–138.

Fried, G., Norton, L.R. and Reboud, X. (2008) Environmental and management factors determining weed species composition and diversity in France. *Agriculture, Ecosystems & Environment* 128, 68–76.

Fried, G., Chauvel, B. and Reboud, X. (2009) A functional analysis of large-scale temporal shifts from 1970 to 2000 in weed assemblages of sunflower crops in France. *Journal of Vegetation Science* 20, 49–58.

Froud-Williams, R.J., Chancellor, R.J. and Drennan, D.S.H. (1984) The effects of seed burial and soil disturbance on emergence and survival of arable weeds in relation to minimal cultivation. *Journal of Applied Ecology* 21, 629–641.

Gardarin, A. and Colbach, N. (in revision) How much of seed dormancy in weeds can be explained by seed traits? *Weed Research*.

Gardarin, A., Dürr, C. and Colbach, N. (2009) Which model species for weed seedbank and emergence studies? A review. *Weed Research* 49, 117–130.

Gardarin, A., Dürr, C. and Colbach, N. (2010a) Effects of seed depth and soil structure on the emergence of weeds with contrasted seed traits. *Weed Research* 50, 91–101.

Gardarin, A., Dürr, C., Mannino, M.R., Busset, H. and Colbach, N. (2010b) Seed mortality in the soil is related to the seed coat thickness. *Seed Science Research* 20, 243–256.

Gardarin, A., Guillemin, J.P., Munier-Jolain, N.M. and Colbach, N. (2010c) Estimation of key parameters for weed population dynamics models: base temperature and base water potential for germination. *European Journal of Agronomy* 32, 162–168.

Gardarin, A., Dürr, C. and Colbach, N. (2011) Prediction of germination rates of weed species: relationships between germination parameters and species traits. *Ecological Modelling* 222, 626–636.

Gardarin, A., Dürr, C. and Colbach, N. (2012) Modeling the dynamics and emergence of a multispecies weed seed bank with species traits. *Ecological Modelling* 240, 123–138.

Gibot-Leclerc, S., Brault, M., Pinochet, X. and Sallé, G. (2003) Rôle potentiel des plantes adventices du colza d'hiver dans l'extension de l'orobanche rameuse en Poitou-Charentes. *Comptes Rendus de Biologie* 326, 645–658.

Gibot-Leclerc, S., Charles, J. and Dessaint, F. (2009) Sensibilité d'hôtes potentiels vis-à-vis de deux pathovars d'*Orobanche ramosa* L. In: *XIIIème Colloque International sur la Biologie des Mauvaises Herbes*. Dijon, France, pp. 446–456.

Gibot-Leclerc, S., Sallé, G., Reboud, X. and Moreau, D. (2012) What are the traits of *Phelipanche ramosa* (L.) Pomel that contribute to the success of its biological cycle on its host *Brassica napus* L.? *Flora* 207, 512–521.

Givens, W.A., Shaw, D.R., Kruger, G.R., Johnson, W.G., Weller, S.C., Young, B.G., Wilson, R.G., Owen, M.D.K. and Jordan, D. (2009) Survey of tillage trends following the adoption of glyphosate-resistant crops. *Weed Technology* 23, 150–155.

Gressel, J. (2011) Global advances in weed management. *Journal of Agricultural Science* 149, 47–53.

Grime, J.P., Mason, G., Curtis, A.V., Rodman, J., Band, S.R., Mowforth, M.A.G., Neal, A.M. and Shaw, S. (1981) A comparative study of germination characteristics in a local flora. *Journal of Ecology* 69, 1017–1059.

Gruber, S., Pekrun, C. and Claupein, W. (2004) Population dynamics of volunteer oilseed rape (Brassica napus L.) affected by tillage. *European Journal of Agronomy* 20, 351–361.

Gruber, S., Bühler, A., Möhring, J. and Claupein, W. (2010) Sleepers in the soil – vertical distribution by tillage and long-term survival of oilseed rape seeds compared with plastic pellets. *European Journal of Agronomy* 33, 81–88.

Grundy, A.C., Mead, A. and Burston, S. (1999) Modelling the effect of cultivation on seed movement with application to the prediction of weed seedling emergence. *Journal of Applied Ecology* 36, 663–678.

Grundy, A.C., Mead, A. and Burston, S. (2003) Modelling the emergence response of weed seeds to burial depth: interactions with seed density, weight and shape. *Journal of Applied Ecology* 40, 757–770.

Guillemin, J.-P., Gardarin, A., Granger, S., Reibel, C. and Colbach, N. (2013) Determination of base temperatures and base water potentials for germination of weeds. *Weed Research* 53, 76–87.

Harrison, S.K., Regnier, E.E., Schmoll, J.T. and Harrison, J.M. (2007) Seed size and burial effects on giant ragweed (*Ambrosia trifida*) emergence and seed demise. *Weed Science* 55, 16–22.

Hendry, G.A.F., Thompson, K., Moss, C.J., Edwards, E. and Thorpe, P.C. (1994) Seed persistence: a correlation between seed longevity in the soil and ortho-dihydroxyphenol concentration. *Functional Ecology* 8, 658–664.

Hilhorst, H.W.M. and Toorop, P.E. (1997) Review on dormancy, germinability, and germination in crop and weed seeds. *Advances in Agronomy* 61, 111–165.

Hillel, D. (1971) *Soil and water: Physical principles and processes*. Academic Press, New York.

Holland, J.M. (2004) The environmental consequences of adopting conservation tillage in Europe: reviewing the evidence. *Agriculture Ecosystems & Environment* 103, 1–25.

Holst, N., Rasmussen, I.A. and Bastiaans, L. (2007) Field weed population dynamics: a review of model approaches and applications. *Weed Research* 47, 1–14.

Hughes, K.A. and Baker, C.J. (1977) The effects of tillage and zero-tillage systems on soil aggregates in a silt loam. *Journal of Agricultural Engineering Research* 22, 291–301.

Hulme, P.E. (1998) Post-dispersal seed predation: consequences for plant demography and evolution. *Perspectives in Plant Ecology, Evolution and Systematics* 1, 32–46.

Hulme, P.E. and Borelli, T. (1999) Variability in post-dispersal seed predation in deciduous woodland: relative importance of location, seed species, burial and density. *Plant Ecology* 145, 149–156.

IFEN (2007) *Les pesticides dans les eaux - Données 2005*. Institut français de l'environnement, Orléans, pp. 37.

Joel, D.M., Hershenhorn, J., Eizenberg, H. and Aly, R. (2007) Biology and management of weedy root parasites. *Horticultural Reviews* 33, 267–350.

Kelly, K.M., Van Staden, J. and Bell, W.E. (1992) Seed coat structure and dormancy. *Plant Growth Regulation* 11, 201–209.

Kucera, B., Cohn, M.A. and Leubner-Metzger, G. (2005) Plant hormone interactions during seed dormancy release and germination. *Seed Science Research* 15, 281–307.

Kuo, T.M., Van Middlesworth, J.F. and Wolf, W.J. (1988) Content of raffinose oligosaccharides and sucrose in various plant seeds. *Journal of Agricultural and Food Chemistry* 36, 32–36.

Kurstjens, D.A.G. and Kropff, M.J. (2001) The impact of uprooting and soil-covering on the effectiveness of weed harrowing. *Weed Research* 41, 211–228.

Landolt, E. (1977) *Ökologischer Zeigerwerte zur Schweizer Flora*. Veröffentlichungen des Geobotanischen Institutes der Eidgenössischen Technischen Hochschule, Stiftung Rubel. pp. 208.

Liebman, M., Drummond, F.A., Corson, S. and Zhang, J.X. (1996) Tillage and rotation crop effects on weed dynamics in potato production systems. *Agronomy Journal* 88, 18–26.

Lonchamp, J.P. (1976) Effect of depth of burial on the germination of two weeds of autumn crops: *Veronica hederifolia* L. and *Viola tricolor* L. In: *Ve Colloque International sur l'Ecologie et la Biologie des Mauvaises Herbes*. Dijon, pp. 319–328.

Lonchamp, J.P. and Gora, M. (1980) Effect of burial on the germinative requirements of weeds. *Proceedings of the 6th International Colloquium on Weed Ecology, Biology and Systematics, organized by COLUMA-EWRS, Montpellier*. pp. 113–122.

Lonchamp, J.P., Chadoeuf, R. and Barralis, G. (1984) Évolution de la capacité de germination des semences de mauvaises herbes enfouies dans le sol. *Agronomie* 4, 671–682.

Lutman, P.J.W., Cussans, G.W., Wright, K.J., Wilson, B.J., Wright, G.M. and Lawson, H.M. (2002) The persistence of seeds of 16 weed species over six years in two arable fields. *Weed Research* 42, 231–241.

Marshall, E.J.P., Brown, V.K., Boatman, N.D., Lutman, P.J.W., Squire, G.R. and Ward, L.K. (2003) The role of weeds in supporting biological diversity within crop fields. *Weed Research* 43, 77–89.

McCloskey, M., Firbank, L.G., Watkinson, A.R. and Webb, D.J. (1996) The dynamics of experimental arable weed communities under different management practices. *Journal of Vegetation Science* 7, 799–808.

Menalled, F.D., Gross, K.L. and Hammond, M. (2001) Weed aboveground and seedbank community responses to agricultural management systems. *Ecological Applications* 11, 1586–1601.

Messéan, A., Sausse, C., Gasquez, J. and Darmency, H. (2007) Occurrence of genetically modified oilseed rape seeds in the harvest of subsequent conventional oilseed rape over time. *European Journal of Agronomy* 27, 115–122.

Mézière, D., Granger, S., Boissinot, F. and Colbach, N. (2011) Maîtriser les adventices graminées automnales sans herbicide: Quel est le Poids de l'histoire culturale? Evaluation avec un modèle de dynamique d'adventices. In: *AFPP – Quatrième Conférence Internationale sur les Méthodes Alternatives en Protection des Cultures*. Lille, France, pp. 774–784.

Milberg, P., Andersson, L. and Thompson, K. (2000) Large-seeded species are less dependent on light for germination than small-seeded ones. *Seed Science Research* 10, 99–104.

Miyazawa, K., Tsuji, H., Yamagata, M., Nakano, H. and Nakamoto, T. (2004) Response of weed flora to combinations of reduced tillage, biocide application and fertilization practices in a 3-year crop rotation. *Weed Biology and Management* 4, 24–34.

Mohamed-Yasseen, Y., Barringer, S.A., Splittstoesser, W.E. and Costanza, S. (1994) The role of seed coats in seed viability. *Botanical Review* 60, 426–439.

Mohler, C.L. and Galford, A.E. (1997) Weed seedling emergence and seed survival: separating the effects of seed position and soil modification by tillage. *Weed Research* 37, 147–155.

Mohler, C.L., Frisch, J.C. and McCulloch, C.E. (2006) Vertical movement of weed seed surrogates by tillage implements and natural processes. *Soil & Tillage Research* 86, 110–122.

Moles, A.T. and Westoby, M. (2006) Seed size and plant strategy across the whole life cycle. *Oikos* 113, 91–105.

Montégut, J. (1984) La levée au champ des mauvaises herbes. In: *7ème colloque international sur l'écologie, la biologie et la systématique des mauvaises herbes*. Columa/EWRS, Paris, France, pp. 121–139.

Moss, S.R. (1988) Influence of cultivations on the vertical distribution of weed seeds in the soil. *VIIIe colloque international sur la biologie, l'écologie et la systematique des mauvaises herbes*. Dijon.

Moss, S.R., Tatnell, L.V., Hull, R., Clarke, J.H., Wynn, S. and Marshall, R. (2010) Integrated management of herbicide resistance. *HGCA Project Report*, xvii + 115 pp.

Munier-Jolain, N.M., Guyot, S.H.M. and Colbach, N. (2013) A 3D model for light interception in heterogeneous crop:weed canopies. Model structure and evaluation. *Ecological Modelling*, 250, 101–110.

Murdoch, A.J. and Ellis, R.H. (2000) Dormancy, viability and longevity. In: *Seeds: The Ecology of Regeneration in Plant Communities*. CAB International, Wallingford, UK, pp. 183–214.

Musselman, L.J. (1980) The biology of *Striga, Orobanche*, and other root-parasitic weeds. *Annual Review of Phytopathology* 18, 463–489.

Oerke, E.C., Dehne, H.W., Schonbeack, F. and Weber, A. (1994) *Crop production and crop protection*. Elsevier Science, Amsterdam, Netherlands.

Omami, E.N., Haigh, A.M., Medd, R.W. and Nicol, H.I. (1999) Changes in germinability, dormancy and viability of *Amaranthus retroflexus* as affected by depth and duration of burial. *Weed Research* 39, 345–354.

Oryokot, J.O.E., Murphy, S.D., Thomas, A.G. and Swanton, C.J. (1997) Temperature- and moisture-dependent models of seed germination and shoot elongation in green and redroot pigweed (*Amaranthus powellii, A. retroflexus*). *Weed Science* 45, 488–496.

Parker, C. (2009) Observations on the current status of *Orobanche* and *Striga* problems worldwide. *Pest Management Science* 65, 453–459.

Peters, N.C.B. (1982) The dormancy of wild oat seed (*Avena fatua* L.) from plants grown under various temperature and soil moisture conditions. *Weed Research* 22, 205–212.

Petit, C., Bay, G., Pernin, F. and Délye, C. (2010) Prevalence of cross- or multiple resistance to the acetyl-coenzyme A carboxylase inhibitors fenoxaprop, clodinafop and pinoxaden in black-grass (*Alopecurus myosuroides* Huds.) in France. *Pest Management Science* 66, 168–177.

Petit, S., Boursault, A., Le Guilloux, M., Munier-Jolain, N. and Reboud, X. (2011) Weeds in agricultural landscapes. A review. *Agronomy for Sustainable Development* 31, 309–317

Pleasant, J.M. and Schlather, K.J. (1994) Incidence of weed seed in cow (*bos* sp) manure and its importance as a weed source for cropland. *Weed Technology* 8, 304–310.

Ponquett, R.T., Smith, M.T. and Ross, G. (1992) Lipid autoxidation and seed ageing: putative relationships between seed longevity and lipid stability. *Seed Science Research* 2, 51–54.

Powles, S.B. and Yu, Q. (2010) Evolution in action: Plants resistant to herbicides. In: Merchant, S., Briggs, W.R. and Ort, D. (eds) *Annual Review of Plant Biology, Vol 61*. Palo Alto, Annual Reviews, pp. 317–347.

Priestley, D.A. (1986) *Seed Aging - Implications for Seed Storage and Persistence in the Soil*. Comstock Publishing Associates, pp. 304.

Puricelli, E. and Tuesca, D. (2005) Weed density and diversity under glyphosate-resistant crop sequences. *Crop Protection* 24, 533–542.

Puricelli, E., Faccini, D., Orioli, G. and Sabbatini, M.R. (2005) Seed survival and predation of Anoda cristata in soyabean crops. *Weed Research* 45, 477–482.

Rasmussen, I.A., Holst, N., Petersen, L. and Rasmussen, K. (2002) Computer model for simulating the long term dynamics of annual weeds under different cultivation practices. In: *5th EWRS Workshop on Physical and Structural Weed Control*. Pisa, Italy, pp. 6–13.

Rasmussen, J. (1992) Testing harrows for mechanical control of annual weeds in agricultural crops. *Weed Research*. 32, 267–274.

Raymond, P., Al-Ani, A. and Pradet, A. (1985) ATP production by respiration and fermentation, and energy charge during aerobiosis and anaerobiosis in twelve fatty and starchy germinating seeds. *Plant Physiology* 79, 879–884.

Recasens, J., Planes, J., Bosque, J.L.I., Briceno, R. and Taberner, A. (2001) Management strategies for herbicide-resistant *Lolium rigidum* Gaud. populations. In: *Actas Congreso 2001 Sociedad Espanola de Malherbologia*. Sociedad Espanola de Malherbologia (Spanish Weed Science Society), Leon, Spain, pp. 117–122.

Reddy, K.N. (2003) Impact of rye cover crop and herbicides on weeds, yield, and net return in narrow-row transgenic and conventional soybean (*Glycine* max). *Weed Technology* 17, 28–35.

Roberts, H.A. (1979) Periodicity of seedling emergence and seed survival in some *Umbelliferae*. *Journal of Applied Ecology* 16, 195–201.

Roberts, H.A. and Boddrell, J.E. (1983) Seed survival and periodicity of seedling emergence in ten species of annual weeds. *Annals of Applied Biology* 102, 523–532.

Roberts, H.A. and Lockett, P.M. (1978) Seed dormancy and periodicity of seedling emergence in *Veronica hederifolia* L. *Weed Research* 18, 41–48.

Robinson, R.A. and Sutherland, W.J. (2002) Post-war changes in arable farming and biodiversity in Great Britain. *Journal of Applied Ecology* 39, 157–176.

Rodenburg, J., Meinke, H. and Johnson, D.E. (2011) Challenges for weed management in African rice systems in a changing climate. *Journal of Agricultural Science* 149, 427–435.

Roger-Estrade, J., Richard, G. and Manichon, H. (2000) A compartmental model to simulate temporal changes in soil structure under two cropping systems with annual mouldboard ploughing in a silt loam. *Soil & Tillage Research* 54, 41–53.

Roger-Estrade, J., Colbach, N., Leterme, P., Richard, G. and Caneill, J. (2001) Modelling vertical and lateral weed seed movements during moulboard ploughing with a skim-coulter. *Soil & Tillage Research* 63, 35–49.

Roller, A., Beismann, H. and Albrecht, H. (2002) Persistence of genetically modified, herbicide-tolerant oilseed rape – first observations under practically relevant conditions in South Germany. *Zeitschrift für Pflanzenkrankheiten und Pflanzenschutz*, 255–260.

Roller, A., Beismann, H. and Albrecht, H. (2003) The influence of soil cultivation on the seed bank of GM-herbicide tolerant and conventional oilseed rape. In: *Seedbanks: Determination, dynamics and management. Meeting of the Association of Applied Biologists.* Reading, UK, pp. 131–135.

Roman, E.S., Thomas, A.G., Murphy, S.D. and Swanton, C.J. (1999) Modeling germination and seedling elongation of common lambsquarters (*Chenopodium album*). *Weed Science* 47, 149–155.

Rossing, W.A.H., Meynard, J.M. and van Ittersum, M.K. (1997) Model-based explorations to support development of sustainable systems: case studies from France and the Netherlands. *European Journal of Agronomy* 7, 271–283.

Rubiales, D., Fernandez-Aparicio, M., Wegmann, K. and Joel, D.M. (2009) Revisiting strategies for reducing the seedbank of *Orobanche* and *Phelipanche* spp. Weed Research. *Weed Research* 49, 23–33.

Saatkamp, A. (2009) Population dynamics and functional traits of annual plants – a comparative study on how rare and common arable weeds persist in agroecosystems. PhD thesis, Université Paul Cézanne Aix-Marseille III, France and Universität Regensburg, Germany.

Sanchez del Arco, M.J., Torner, C. and Quintanilla, C.F. (1995) Seed dynamics in populations of Avena sterilis ssp. ludoviciana. *Weed Research* 35, 477–487.

Schneeweiss, G.M. (2007) Correlated evolution of life history and host range in the non photosynthetic parasitic flowering plants *Orobanche* and *Phelipanche* (Orobanchaceae). *Journal of Evolutionary Biology* 20, 471–478.

Scopel, A.L., Ballare, C.L. and Radosevich, S.R. (1994) Photostimulation of seed-germination during soil tillage. *New Phytologist* 126, 145–152.

Sester, M., Dürr, C., Darmency, H. and Colbach, N. (2006) Evolution of weed beet (*Beta vulgaris* L.) seed bank: quantification of seed survival, dormancy, germination and pre-emergence growth. *European Journal of Agronomy* 24, 19–25.

Sester, M., Dürr, C., Darmency, H. and Colbach, N. (2007) Modelling the effects of cropping systems on the seed bank dynamics and emergence of weed beet. *Ecological Modelling* 204, 47–58.

Shipley, B. and Parent, M. (1991) Germination responses of 64 wetland species in relation to seed size, minimum time to reproduction and seedling relative growth rate. *Functional Ecology* 5, 111–118.

Sinha, A.K. and Ghildyal, B.P. (1979) Emergence force of crop seedlings. *Plant and Soil* 51, 153–156.

Soane, B.D., Ball, B.C., Arvidsson, J., Basch, G., Moreno, F. and Roger-Estrade, J. (2012) No-till in northern, western and south-western Europe: A review of problems and opportunities for crop production and the environment. *Soil & Tillage Research* 118, 66–87.

Souty, N. and Rode, C. (1994) Seedling emergence in the field: a problem of mechanics? *Sécheresse* 5, 13–22.

Storkey, J., Holst, N., Bøjer, O., Bigongialli, F., Bocci, G., Colbach, N., Dorner, Z., Riemens, M., Sartorato, I., Sønderkøv, M. and Verschwele, A.T. (in revision) The development of a weed traits database as a resource for predicting shifts in weed communities. *Weed Research*.

Sultan, S.E. (1996) Phenotypic plasticity for offspring traits in Polygonum persicaria. *Ecology* 77, 1791–1807.

Swain, A.J., Hughes, Z.S., Cook, S.K. and Moss, S.R. (2006) Quantifying the dormancy of Alopecurus myosuroides seeds produced by plants exposed to different soil moisture and temperature regimes. *Weed Research* 46, 470–479.

Taylor, R., Maxwell, B. and Boik, R. (2006) Indirect effects of herbicides on bird food resources and beneficial arthropods. *Agriculture, Ecosystems & Environment* 116, 157–164.

Teasdale, J.R. and Mohler, C.L. (1993) Light transmittance, soil-temperature, and soil-moisture under residue of hairy vetch and rye. *Agronomy Journal* 85, 673–680.

Teasdale, J.R., Beste, C.E. and Potts, W.E. (1991) Response of weeds to tillage and cover crop residue. *Weed Science.* 39, 195–199.

Teasdale, J.R., Coffman, C.B. and Mangum, R.W. (2007) Potential long-term benefits of no-tillage and organic cropping systems for grain production and soil improvement. *Agronomy Journal* 99, 1297–1305.

Telewski, F.W. and Zeevaart, J.A.D. (2002) The 120-year period for Dr Beal's seed viability experiment. *American Journal of Botany* 89, 1285–1288.

Terpstra, R. (1986) Behavior of weed seed in soil clods. *Weed Science* 34, 889–895.

Thompson, K., Band, S.R. and Hodgson, J.G. (1993) Seed size and shape predict persistence in soil. *Functional Ecology* 7, 236–241.

Trigo, E.J. and Cap, E.J. (2003) The impact of the introduction of transgenic crops in Argentinean agriculture. *AgBioForum* 6, 87–94.

Trudgill, D.L., Honek, A., Li, D. and Van Straalen, N.M. (2005) Thermal time – concepts and utility. *Annals of Applied Biology* 146, 1–14.

Van Acker, R.C. (2009) Weed biology serves practical weed management. *Weed Research* 49, 1–5.

Venable, D.L. and Brown, J.S. (1988) The selective interactions of dispersal, dormancy, and seed size as adaptations for reducing risk in variable environments. *The American Naturalist* 131, 360–384.

Venable, D.L. and Lawlor, L. (1980) Delayed germination and dispersal in desert annuals: escape in space and time. *Oecologia* 46, 272–282.

Vleeshouwers, L.M. (1997a) Modelling the effect of temperature, soil penetration resistance, burial depth and seed weight on pre-emergence growth of weeds. *Annals of Botany* 79, 553–563.

Vleeshouwers, L.M. (1997b) Modelling weed emergence patterns. University of Wageningen, Netherlands.

Vleeshouwers, L.M. and Kropff, M.J. (2000a) Modelling field emergence patterns in arable weeds. *New Phytologist* 148, 445–457.

Vleeshouwers, L.M. and Kropff, M.J. (2000b) Modelling field emergence patterns in arable weeds. *New Phytologist* 148, 445–457.

Waggoner, J.K., Henneberger, P.K., Kullman, G.J., Umbach, D.M., Kamel, F., Beane Freeman, L.E., Alavanja, M.C.R., Sandler, D.P. and Hoppin, J.A. (2012) Pesticide use and fatal injury among farmers in the Agricultural Health Study. *Int Arch Occup Environ Health*. DOI: 10.1007/s00420-012-0752-x.

Wagner, M. and Mitschunas, N. (2008) Fungal effects on seed bank persistence and potential applications in weed biocontrol: A review. *Basic and Applied Ecology* 9, 191–203.

Wanjura, D.F. and Buxton, D.R. (1972) Hypocotyl and radicle elongation of cotton as affected by soil environment. *Agronomy Journal* 64, 431–434.

Weiher, E., van der Werf, A., Thompson, K., Roderick, M., Garnier, E. and Eriksson, O. (1999) Challenging Theophrastus: a common core list of plant traits for functional ecology. *Journal of Vegetation Science* 10, 609–620.

Westerman, P.R., Dixon, P.M. and Liebman, M. (2009) Burial rates of surrogate seeds in arable fields. *Weed Research* 49, 142–152.

Wiese, A.M. and Binning, L.K. (1987) Calculating the threshold temperature of development for weeds. *Weed Science* 35, 177–179.

Wilson, J.D., Morris, A.J., Arroyo, B.E., Clark, S.C. and Bradbury, R.B. (1999) A review of the abundance and diversity of invertebrate and plant foods of granivorous birds in northern Europe in relation to agricultural change. *Agriculture Ecosystems & Environment* 75, 13–30.

Zanin, G., Otto, S., Riello, L. and Borin, M. (1997) Ecological interpretation of weed flora dynamics under different tillage systems. *Agriculture Ecosystems & Environment* 66, 177–188.

Zwanenburg, B., Mwakaboko, A., Reizelman, A., Anilkumar, G. and Sethumadhavan, D. (2009) Structure and function of natural and synthetic signaling molecules in parasitic weed germination. *Pest Management Science* 65, 478–491.

11 The Functional Role of Soil Seed Banks in Natural Communities

Arne Saatkamp,[1]* Peter Poschlod[2] and D. Lawrence Venable[3]

[1]*Institut Méditerranéen de Biodiversité et d'Ecologie (IMBE UMR CNRS 7263), Université d'Aix-Marseille, Marseille, France;* [2]*LS Biologie VIII, Universität Regensburg, Regensburg, Germany;* [3]*Department of Ecology and Evolutionary Biology, University of Arizona, Tucson, Arizona, USA*

Introduction

When I was a child, playing in the meadows and woods, I (A.S.) was fascinated by all the seedlings coming out of seemingly lifeless soil where the ponds dried out, a new river bank was exposed or a mole built its hill. Beggarticks (*Bidens tripartita*) quickly covered the former pond; the river bank turned blue with forget-me-nots (*Myosotis pratensis*); and molehills were crowned with stitchwort (*Stellaria media*). It was a difficult experience when my parents had me weed out our overgrown vegetable garden where lambsquarters (*Chenopodium album*) from the seed bank grew faster than the radishes we had sown. I learned, however, to distinguish the few *Calendula* seedlings and to keep some flowers for my mother. Later, my fascination persisted as I asked myself why there were so many heather seedlings in the place where the pinewoods burned, but so few thistles? Why did many seedlings sometimes emerge in a footprint, but not just beside it? Why did the annual grass *Bromus rubens* show up every year, but on the same site *Glaucium corniculatum* only every other year? And why did some plants make such prominent seed banks and others none at all?

Some of us would be satisfied with answers like 'the large size of *Calendula* and *Carduus* seeds limits the number that can be produced by the plant and which will get buried' or 'decades ago heathland grew where the pinewood used to be'. But others of us, inspired by Darwin's 'three tablespoonsful of mud' from which he grew 537 plants, also want to understand the evolution of soil seed banks, pursuing the deeper sense to the 'why' question in biology that Darwin (1859) gave us. The goal of this chapter is to help to answer the questions on: (i) types and definitions of soil seed banks; (ii) how soil seed reservoirs can evolve; (iii) what functional role seed banks play in the dynamics of natural communities; and (iv) what are adaptive traits to build up soil seed banks.

By the 'functional role' of soil seed banks we mean their role in population dynamics, their adaptive role, the effect seed banks have on communities and coexistence, and the role of soil seed banks in the evolution of other plant traits through interactive selection. These aspects will

* E-mail: arne.saatkamp@imbe.fr

help us to understand the build-up and existence of soil seed banks. We use 'natural communities' in a pragmatic sense to mean any spontaneous plant assemblage. The functional role of seed banks in agroecosystems is treated in detail in Chapter 10 of this volume.

Types and Definitions of Soil Seed Banks

Soil seed banks include all living seeds in a soil profile, including those on the soil surface. Here we simply speak of seeds, although in the beginning, soil seed banks are also composed of dispersal units, which are seeds or fruits surrounded by structures serving for dispersal and sometimes contain other plant parts such as bracts or stems. Over time, the dispersal structures, as well as seed coats, can decompose, leaving only germination units. For example, *Ranunculus arvensis* has a thick seed coat and spikes which both decompose after burial in soil after a few years, leaving coatless seeds (A. Saatkamp, 2009, unpublished data). Soil seed banks resemble other biological reservoirs, such as invertebrate eggs, tubercles and bulb banks, spores of non-spermatophyte plants and fungi, or seeds retained on mother plants (serotiny). Many of these resting stages share similar evolutionary constraints and physiological functioning, in such a way that hatching of invertebrate eggs and seed germination can be modelled in the same way (Trudgill *et al.*, 2005).

Soil seed banks vary much according to seed proximity, seed persistence and physiological state. Living seeds have been found in or on the soil for different durations (Duvel, 1902; Priestley, 1986; Roberts, 1986; Poschlod *et al.*, 1998), different seasons (Roberts, 1986; Poschlod and Jackel, 1993; Milberg and Andersson, 1997), at different depths (Duvel, 1902; Grundy *et al.*, 2003; Benvenuti, 2007), in different quantities (Thompson and Grime, 1979; Thompson *et al.*, 1997) and in different states of dormancy or procession to germination (Baskin and Baskin, 1998; Walck *et al.*, 2005; Finch-Savage and

Leubner-Metzger, 2006). Seeds in the soil seed bank may occur in or on the soil, but in many situations, there is a continuity between seeds at the surface, partly buried and completely buried seeds (Thompson, 2000; Benvenuti, 2007). In practice, it is rarely possible to properly separate buried seeds from the seeds in the litter. Seeds of several plant species hardly ever enter the soil but persist at its surface or in the litter for many years, prominent examples are the large and hard fruits of *Medicago* and *Neurada*, which contain dozens of seeds and can give rise to several plants over several years.

Plants differ in the duration their seeds remain in the soil and even within a species and among seeds of the same cohort there is variability in the time they spend in the soil seed bank. Thompson and Grime (1979) proposed a system of soil seed bank types, based on the study of the seasonal dynamics and the duration of soil seed banks for the flora of Central England (Fig. 11.1). According to their data, they distinguished between transient seed banks for species that have viable seeds present for less than 1 year, and persistent seed banks for species with viable seeds that remain for more than 1 year. Persistent soil seed banks can be subdivided further into short-term persistent for seeds that are detectable for more than 1 but less than 5 years, and long-term persistent seed banks that are present for more than 5 years (Maas, 1987; Bakker, 1989; Thompson and Fenner, 1992). A classification key for the three basic types can be found in Grime (1989), which is based on the abundance and depth distribution of seeds in the soil seed bank, their seed size, their seasonality and the presence/absence of a plant in the established vegetation around the seed bank sample. More detailed classifications have been proposed but they did not gain wider usage, mostly because necessary data are rarely available (reviewed in Csontos and Tamás, 2003; e.g. Poschlod and Jackel, 1993). For temperate regions, Thompson and Grime (1979) also used seasonality to separate winter and summer seed banks for plants with autumn and spring germination (Fig 11.1). Since timing of seed dispersal and germination vary greatly among species

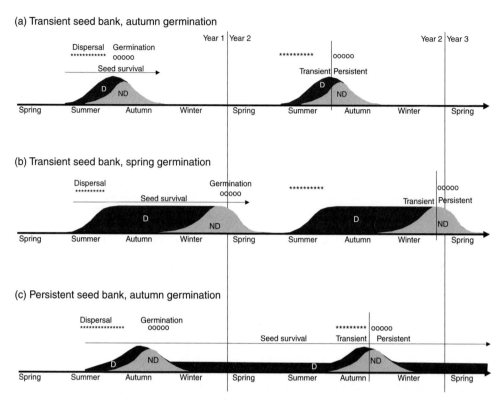

Fig. 11.1. Seed bank types according to timing of dispersal, germination and survival of seeds. (a) Transient seed bank with autumn germination; (b) transient seed bank with spring germination; (c) persistent seed bank with autumn germination. Note that the limit between transient or persistent seed banks as defined by Walck *et al.* (2005) does not coincide with a 1 year distance from the dispersal of seeds, the limit is indicated by a line (redrawn from Thompson and Grime, 1979 and Walck *et al.*, 2005). D, dormant; ND, non-dormant.

and among climates (Baskin and Baskin, 1998; Dalling *et al.*, 1998; Boedeltje *et al.*, 2004), Walck *et al.* (2005) suggested that the time between dispersal and the first germination season should be used to distinguish transient from persistent seed banks (Fig. 11.1).

Some plants produce both transient and persistent seeds, in varying ratios (Clauss and Venable, 2000; Cavieres and Arroyo, 2001; Tielbörger *et al.*, 2011) and variation in the environment leads to variable seed exit by germination from the seed bank (Meyer and Allen, 2009). Whereas simple seed bank types are useful for multispecies comparisons, we need also to consider dynamic and quantitative aspects of seed banks if we want to predict more precisely the role of seed banks. For example,

plants can build up seed banks when their seeds are buried during disturbance and stay ungerminated due to a light requirement but germinate nearly completely when they remain at the surface (see Chapters 5 and 6 of this volume). Soil seed banks are a dynamic part of plant populations with a set of factors that quantitatively influence their entry, persistence and exit, all of which vary according to plant biology, time and their environment. Such an approach will improve our ability to predict ecological outcomes in response to community disturbance and/or community invasion.

Research on soil seed banks differs in the type of data collected, sometimes consisting of (i) studying soil samples by identifying and counting seedlings, or sifting and identifying seeds, without any precise knowledge on seed ages and the size of the original seed rain; or (ii) burial experiments, which follow, in the best case, counted numbers of seeds over time under defined conditions of depth, soil type, moisture or fertility. We propose to distinguish 'persistence' of seeds in a general sense or with undefined numbers from 'survival' of individual seeds or precisely quantified seed populations. The difference between these data types needs attention, and potentially leads to contrasting conclusions with respect to the seed size–number trade-off (see below).

Evolution of Soil Seed Banks

Soil seed banks are both the outcome of environmental or plant developmental contingencies and the result of evolutionary history. Climate, herbivory and disturbances vary and lead directly to year-to-year changes in soil seed bank density and spatial heterogeneity. Some environments particularly favour the evolution of persistent soil seed banks, such as river mud flats or ephemeral ponds, forest gaps, pastures and arable fields since they are often or intensely disturbed (Ortega et al., 1997; Bekker et al., 1998c) or have very variable habitat conditions (Brock, 2011). Plants with persistent soil seed banks are some of the most characteristic species of these habitats. Many other ecosystems also contain at least a few plant species with persistent soil seed banks, either with some kind of dormancy (Keeley, 1987; Baskin and Baskin, 1998), with increased germination in presence of smoke-derived substances (Brown, 1993; Flematti et al., 2004), or with a gap detection mechanism (Thompson and Grime, 1983; Dalling et al., 1998; Pearson et al., 2003). Even if these ecosystems have low disturbance levels, they share a form of temporal and spatial unpredictability of regeneration opportunities, which may stem from disturbances including gap dynamics or climatic variability. In the following, we review theoretical works that demonstrate the adaptive

value of seed persistence, the first germination opportunity in environments with such temporal variability and also works that demonstrate how delayed germination can evolve without temporal variability. These theoretical studies will help to understand under which conditions persistent soil seed banks evolve and in which direction and relative magnitude they affect the delay of germination.

Timing of germination and fitness of individual seeds

Germination can be 'delayed' at different timescales, either from one year to later years, from one season to another season or within a given season. Also plant species differ among each other in the degree of delay at all scales. Before we discuss the evolution of persistent seed banks, let's have a look at the two shorter temporal scales. Under optimal conditions, during the appropriate germination season, early germination would seem to maximize the fitness of a seed due to longer growth and the resulting higher fecundity (Ross and Harper, 1972; Fowler, 1984; Kelly and Levin, 1997; Dyer et al., 2000; Turkington et al., 2005; Verdù and Traveset, 2005; De Luis et al., 2008), although in some cases fitness can be reduced with early germination due to high mortality of seedlings (Marks and Prince, 1981; Jones and Sharitz, 1989; Donohue, 2005). Delay in germination can delay reproduction, which could result in a longer generation time, or, for a short-lived plant, extending reproduction into an unfavourable season. Despite the manifest advantages of early germination, many plants have delayed germination due to some form of dormancy, especially in seasonal climates (Baskin and Baskin, 1998; Jurado and Flores, 2005; Merritt et al., 2007), which contributes to seed persistence in these types of ecosystems (Leck et al., 1989; Thompson et al., 1997). Within years, the optimal time for germination often differs from the season of seed production such that there is strong selection for delayed germination of fresh seeds. Therefore, germination timing must be under stabilizing selection, with fitness declining for germination that is too early or too late. Likewise, the prevalence of persistent seed banks and their association with certain habitats suggests that the proportion of germinating seeds in one season compared to those that will persist to a subsequent one also has adaptive value. It is impressive on what short timescales mixtures of genotypes of *Arabidopsis thaliana*, with or without dormancy, are sorted out according to their fitness in climates contrasting in the severity of winter conditions (Donohue *et al.*, 2005; Huang *et al.*, 2010). This rapid evolution between winter and spring germination in *Arabidopsis* is astonishing, because of the recurrent differences between warm and cold germinating species when one compares many species over larger areas and which are often related to contrasting traits (Baskin and Baskin, 1998; Merritt *et al.*, 2007).

transient seed banks (Ortega *et al.*, 1997). Predictable rainfall, e.g. in monsoon climates, and frost in arctic or alpine environments have similar effects on timing of emergence from seed banks (reviewed in Baskin and Baskin, 1998).

Sometimes disturbances are predictable at longer timescales only (10–20 years), such as fires with immediately following regeneration opportunities. This leads to seed banks that persist in the interval between fires and whose germination can be stimulated by smoke or whose dormancy is released by heat, which are highly predictive of favourable regeneration opportunities (Cowling and Lamont, 1985; Thanos *et al.*, 1992; Brown, 1993; Flematti *et al.*, 2004). In other cases, habitats with periodical flooding harbour plants that only produce transient seed banks, like the very short-lived willow seeds (*Salix*), which live only for weeks (Thompson and Grime, 1979) and for which clonal reproduction may be an important alternative to seed banks.

Seed banks and the predictability of environment

Even predictable changes in the environment can lead to formation of soil seed banks, although lasting for a shorter time. Typically more predictable environmental factors include seasonal changes in temperature, moisture (Baskin and Baskin, 1998; Jurado and Flores, 2005; Merritt *et al.*, 2007), water level in some aquatic ecosystems such as flood plains of large rivers (Leck *et al.*, 1989; Kubitzki and Ziburski, 1994), and the number of competing seeds from the same mother plant or environment (Cohen, 1967; Ellner, 1986; Tielbörger and Valleriani, 2005; Valleriani and Tielbörger, 2006). When favourable environments for germination are predictable on shorter timescales, transient rather than persistent soil seed banks tend to form with germination time determined by cues for dormancy loss and germination of non-dormant seeds (Thompson and Grime, 1979). For example, many annuals in Mediterranean-type climates that germinate with autumn and winter rains have

Unpredictable environments promote evolution of persistent seed banks

We intuitively relate the evolution of delayed germination to environmental unpredictability, without invoking competition or other density-dependent effects. A prominent example of a system where the environment (rainfall) varies unpredictably is annual plants in deserts. Desert annuals reproduce or die depending on the occurrence of unpredictable rainfall events during their one and only growing season. In response to this uncertainty, they may retain a fraction of ungerminated seeds for possible future germination opportunities in potentially more favourable years. This 'bet hedging' is understood as an insurance against reproduction failure, or more generally, as a strategy that may reduce arithmetic mean fitness, but also fitness variance and hence increase long-term fitness. For bet hedging to occur in absence of density dependence, global variation in environment quality is needed.

Even with a low frequency of total reproductive failure, populations that do not maintain a fraction of ungerminated seeds for subsequent rainfall events, would go extinct (Gutterman, 2002). Models that incorporate bet hedging and density dependence typically show the higher the variance in reproductive success, the lower the fitness-maximizing germination fraction in any given year (Cohen, 1966, 1967; Venable and Lawlor, 1980; Bulmer, 1984; Ellner, 1985a,b; Venable and Brown, 1988; Rees, 1994; Pake and Venable, 1995; Clauss and Venable, 2000; Evans and Dennehy, 2005; Venable, 2007; Tielbörger et al., 2011). The basic prediction of bet hedging, has been demonstrated empirically for different sites with differing levels of risk (Clauss and Venable, 2000; Tielbörger et al., 2011), and across species differing in risk levels at a given site (Venable, 2007). Bet hedging, in the form of risk spreading in temporally variable, unpredictable environments, is the best known evolutionary mechanism leading to delayed germination and the evolution of a persistent soil seed bank (Cohen, 1966; Venable, 2007; Tielbörger et al., 2011).

Beyond rainfall, predation in the form of herbivory can be another factor that creates temporally unpredictable risk in reproduction and thus the conditions for bet hedging, and in this way increases the adaptive value of persistent soil seed banks. This 'escape from predators' and the influence of other disturbances of biotic origin may be an important source for the evolution of soil seed banks via bet hedging, especially in desert and grassland ecosystems, which harbour a certain number of species with persistent soil seed banks.

Evolution of persistent seed banks and density dependence: competition and predation

Bet hedging explains evolution of persistent seed banks in the absence of density-dependent effects, such as competition or density-dependent seed predation. But, in many ecosystems, competition and density-dependent seed predation play an important role and this affects the evolution of soil seed banks. For example, competition can lead to deterministic fluctuations in otherwise constant environments due to high reproductive rate and deterministic growth. In this case, competition favours evolution of persistent seed banks, because variation in density creates opportunities to escape from competition (Ellner, 1987; Venable, 1989; Lalonde and Roitberg, 2006), an effect that increases evolution of persistent seed banks in absence of global temporal variation (bet hedging) or sibling competition. Competition can promote evolution of persistent seed banks also when variance in density results from other things than competition alone. Obviously, any kind of disturbance will create such variance in density. If there is environmental variation and density dependence, then escape from competition will also promote between-year delay of germination (Venable and Brown, 1988). The difference that competition makes for the evolution of persistence is that lower probability of good years will not necessarily increase the delayed germination, rather, the variability of good/bad years and the frequency of changes will increase delayed germination. In this way, theory underlines the importance of disturbances or environmental variation for the evolution of persistent soil seed banks.

Besides temporal variability, also spatial variability in habitat conditions and competition alone can trigger the evolution of delayed germination (Venable and Lawlor, 1980; Bulmer, 1984; Ellner, 1985a,b). Interestingly, a persistent soil seed bank can also evolve because a highly dormant genotype can recolonize a previously occupied safe site more easily from the seed bank in a local patch than in a distant one (Satterthwaite, 2009). Similarly, Rees (1994) showed the adaptive advantage of a persistent soil seed bank in situations with limited patches for synchronous and age-structured plants.

Furthermore, predation is influenced by density of seeds or plants. Preferential predation of first-year seeds over those in the persistent seed bank from previous years can result in the evolution of lower germination fractions and greater specialization of the growing phase plant to conditions found in favourable years, conditions that result in temporal clumping of reproduction (Brown and Venable, 1991). This mast-like clumping is especially favoured with negative density-dependent seed predation, i.e. if seed predators cannot consume the high number of seeds produced in favourable years, though it can evolve even with density-independent seed predation.

Competition among sib seedlings

During favourable years, a higher seed production potentially leads to more intense competition among sibling seedlings. Such a scenario favours differing germination percentages among seed produced in productive compared to unproductive years or for seeds from different watering conditions. One reason for this is the higher abundance of seeds from the same mother plant leading to increased competition among siblings. This suggests that seeds produced by highly fecund plants should have lower germination fractions compared to low fecundity plants (Silvertown, 1988; Venable, 1989; Nilsson *et al.*, 1994; Lundberg *et al.*, 1996; Hyatt and Evans, 1998; Tielbörger and Valleriani, 2005; Tielbörger and Petru, 2010; Eberhart and Tielbörger, 2012), an effect that promotes evolution of persistent seed banks independently from global temporal variation. This has been shown empirically in natural populations (Philippi, 1993; Zammit and Zedler, 1993). But also abiotic variation in the maternal environment, and, related to this, general levels of inter-specific competition may result in plastic increases in dormancy, as has been shown in several works of Tielbörger and co-workers (Tielbörger and Valleriani, 2005; Tielbörger and Petru, 2010). Nevertheless, seed production and levels of dormancy are not always negatively related among plants differing individually in fecundity in the field (Eberhart and Tielbörger, 2012).

Parent–offspring conflict, maternal effects and evolution of delayed germination

The genome of the seed embryo in most cases contains only half of the mother plant's genome. Therefore, delay in germination and its promoting factors do not affect the fitness of the mother plant and that of the offspring seed in the same way. For example, early germination of seeds may reduce the fitness of the mother plant because offspring plants may compete with the mother plant, but at the same time may increase the fitness of the offspring by shortening generation time. Spreading of germination (bet hedging) across time or space may increase the fitness of the mother plant, but the delay may reduce the fitness of an individual seed. Situations when individual seeds increase their fitness by delaying their germination result from predictable changes in favourability of the environment, most importantly, seasonal changes in water and temperature, and drought- and frost-free periods which can be predicted by temperature changes. Timing the germination to anticipate favourable periods for establishment maximizes fitness of both mother plant and offspring.

This discussion shows that in most situations, the maternal fitness is favoured more by delayed germination than offspring fitness is. That delayed germination evolved often in spite of this becomes plausible considering the dependence of zygotes on provisioning by the mother plant, and the many aspects of seed morphology and physiology that are controlled by the mother plant, such as the number and size of seeds and their protection and dispersal structures and depth of dormancy (Ellner, 1986; Silvertown, 1999). Seed dormancy mechanisms such as underdeveloped embryos, water impermeable seed coats formed by maternal tissues and germination inhibitors have also been interpreted in terms of maternal control of germination (Ellner, 1986; Silvertown, 1999). This is beyond what is habitually called 'maternal effects'. Maternal effects are usually defined as different seed and offspring features that stem from variation in the maternal environment, such as different levels of dormancy among seeds from genetically identical mother plants grown in different temperatures or soil moisture conditions (Guttermann, 2000; Donohue, 2009; Tielbörger and Petru, 2010). The plastic maternal effects and genetically fixed maternal influences both contribute to the control of offspring seed germination and its environment-dependent fine tuning by mother plants (Zammit and Zedler, 1993; Tielbörger and Valleriani, 2005; Tielbörger and Petru, 2010).

Evolution of persistent seed banks and relation to other traits

The evolution of delayed germination and the formation of a between-year soil seed bank are not independent from other plant traits. For example, bet hedging can also act through dispersal in space or by other alternative risk-reducing traits such as stress tolerant morphology and physiology or larger seed size (Venable and Brown, 1988). Theoretical models on the interaction during selection of alternative risk-reducing traits and of persistent seed banks show that they are often, but not always negatively related (Venable and Brown, 1988; Rees, 1994; Snyder, 2006; Vitalis et al., 2013). They do not evolve independently from each other and which trait will be more favoured depends on details of the environment. Contrastingly, when there is temporal auto-correlation in habitat quality the favoured association between dormancy and dispersal can also be positive (Snyder, 2006).

A long plant lifespan is another alternative risk-reducing trait which, similarly to persistent soil seed banks, allows survival through unfavourable periods for reproduction. Consequently, these strategies are negatively related in across-species comparisons (Rees, 1994, 1996; Tuljapurkar and Wiener, 2000). This further suggests that all plant traits that hedge against temporal or spatial habitat variability can have impacts on the evolution of persistent soil seed banks, and future work might explore how

and why plants with succulence, woodiness, clonality and underground storage organs rely comparatively less on persistent soil seed banks.

In conclusion, the models summarized here have elucidated some of the reasons for the evolution of persistent soil seed banks and define the conditions under which persistent soil seed banks contribute to the fitness of plant populations. They point to specific biotic and abiotic, and spatial and temporal environmental conditions whose effects often still need to be tested empirically. They also go a long way towards understanding the relations of persistent soil seed banks to other seed and plant traits. Moreover, evolutionary models provide us only with general predictions; they need to be empirically parameterized to show the magnitude of adaptive features in real plant populations. Some might show up in only very special situations, others only in controlled experiments, and again others might be too small to ever be detected in living plant populations. More precise comparative methods (Butler and King, 2004), and comparative investigations on closely related species (Evans et al., 2005) or populations in different environments (Donohue et al., 2005; Tielbörger and Petru, 2008; Tielbörger et al., 2011) may help us to unravel the importance of these effects.

Most evolutionary models do not explain *how* persistent soil seed banks can be realized, but they explore why a fraction of ungerminated seeds remains viable and ungerminated until subsequent germination seasons contribute to fitness. It is clear that persistent seed banks can be achieved by many different mechanisms in comparable environments (impermeable seed coats, serotiny, physiological dormancy, specific germination conditions and cues), which are discussed in the subsequent sections.

Site-to-site Variation in Soil Seed Persistence

Soil seed persistence for a given species may vary from site to site, and for several species, both persistent and transient soil seed bank types have been documented (Thompson *et al.*, 1997). Between-site variation of soil seed persistence has been attributed to variation in fungal activity, soil fertility (nitrates), oxygen supply, vegetation cover, burial depth (via different disturbance regimes or successional states), seed density and predator pressure (Wagner and Mitschunas, 2008; Koprdová *et al.*, 2010; Saatkamp *et al.*, 2011a). Moreover, since evolutionary constraints of temporal habitat variability lead to different importance of persistent seed banks, local adaptation within species is a source of site-to-site variation in soil seed persistence either directly genetically or via evolution of different levels of plasticity (Tielbörger *et al.*, 2011).

Fungi, soil fertility and moisture

Fungi, either carried by the seed itself or originating from the soil, can strongly reduce soil seed viability and modify seed germination (Wagner and Mitschunas, 2008). Both fungal sources can be additive in their detrimental effects (Kiewnick, 1964). Fungal attack on buried seed depends on soil moisture and temperature. In a series of studies it has been shown that for a given set of mesic species, seed mortality is higher in wet sites, unless fungicide is applied (Schafer and Kotanen, 2003). Also, organic matter and nitrogen content importantly influence fungal activity (Schnürer *et al.*, 1985) and together with low C:N ratio can decrease survival of seed in the soil (Pakeman *et al.*, 2012). Conversely, seedling survival is much higher in plant communities with a mycorrhiza community with affinities to the plant under consideration (reviewed by Horton and Van Der Heijden, 2008). In some plants, such as orchids or some Ericaceae, germination only occurs in the presence of symbiotic fungi in the wild (Horton and Van Der Heijden, 2008). Despite the great diversity of soil fungi and their myriad interactions with plants, studies of the role of soil fungi in soil seed bank dynamics are still scarce and more research is needed to refine this picture.

Soil fertility may also affect soil seed bank persistence. One important factor is nitrate, which promotes the germination of seeds of many species (Popay and Roberts, 1970; Hendricks and Taylorson, 1974) thereby potentially contributing to the depletion of persistent soil seed banks (Bekker *et al.*, 1998b). Stimulation of germination by nitrates may also interact with other environmental parameters such as light or fluctuating temperatures and it also depends on the dormancy state of the seed (Fenner, 1985; Benech-Arnold *et al.*, 2000).

Many plants, especially those from dry habitats, have reduced survival of seed in water-logged soils and it is argued that lack of oxygen is the proximate cause of seed mortality (Kiewnick, 1964; Wagner and Mitschunas, 2008). In contrast, some wetland species such as *Typha* specifically germinate during or after anoxic phases (Morinaga, 1926; Bonnewell *et al.*, 1983). Furthermore, some of the most long-lived seed banks are found in water-logged soils, which is sometimes related to the occurrence of physical dormancy in these habitats (Shen-Miller, 2002). Other wetland plants, such as sedges (*Carex*) show increased mortality when seeds are in a dry state for too long a time (Schütz, 2000). This suggests that wetland species have specific adaptations to survive in water-logged and anoxic conditions, and that they differ from mesic or dryland species in their pathogen defence mechanisms and in their oxygen requirements. The contrast between wetland and dryland species indicates that seeds are adapted to soil conditions of the environment they evolved in and that adaptations for long-term persistence of seeds cannot necessarily be generalized across habitats.

Vegetation cover, gap detection, depth of burial and disturbance

Dense vegetation prevents germination of some seeds. In these situations, seeds can detect vegetation cover via far-red/red light ratios at the soil surface (Kettenring *et al.*, 2006; Kruk *et al.*, 2006; Jankowska-Blaszczuk and Daws, 2007). Others sense vegetation or gaps in it from below ground via diurnal fluctuating temperatures (Thompson *et al.*, 1977). In this way, the density and height of vegetation covering the soil seed bank has impacts on the germination of seed populations from the soil. It can be hypothesized that some gap specialists or initial successional species maintain soil seed banks under dense vegetation, whereas they are depleted more rapidly in open areas. Seed banks can also accumulate under dense vegetation where it functions as a natural seed trap.

Seeds move up and down in soil profiles due to rain (Benvenuti, 2007) or soil turbation by earthworm activity (e.g. Zaller and Saxler, 2007; reviewed by Forey *et al.*, 2011). Some plants depend on light for germination and their seeds do not germinate when buried at sufficient depth (Woolley and Stoller, 1978) and others germinate only with diurnally fluctuating temperatures (Ghersa *et al.*, 1992), so that some seeds remain ungerminated in deeper soil layers (Saatkamp *et al.*, 2011a). These germination requirements may interact with disturbance types and intensities and modify the abundance of seeds in the soil.

Postdispersal seed predation and soil seed banks

Seed predation and dispersal by animals varies over time and space in relation to their abundance and activity (Hulme, 1994, 1998a; Menalled *et al.*, 2000; Westerman *et al.*, 2003; Koprdová *et al.*, 2010). Although vertebrates are thought to play the major role (Hulme, 1998a), ground dwelling arthropods such as carabid beetles, isopods and millipedes can be very effective seed predators (Tooley and Brust, 2002; Saska, 2008; Koprdová *et al.*, 2010). They can consume large numbers of seeds in a short time. Birds, rodents and probably also fish preferentially feed on large seeds (Hulme, 1998a), whereas invertebrates often show preference for smaller seeds (Koprdová *et al.*, 2010). Hulme (1998a,b) suggested that the preference

of rodents for large seeds in northern hemisphere regions decreases the evolution of soil persistence for large-seeded plants, based on the observation that rodents dig out and eat large but not small seeds and that independently, they prefer transient over persistent seeds.

Earthworms ingest and digest seeds of a range of sizes, and earthworm species have specific upper limits to seed sizes they ingest (Shumway and Koide, 1994). After ingestion, smaller seeds are also more easily digested than larger seeds (Forey *et al.*, 2011). Since earthworm abundance and activity is not equal among soil types and specifically depends on temperature, moisture and acidity (Curry, 2004), their interaction with seeds is likely to create heterogeneity among sites in seed persistence. Not only for earthworms, postdispersal seed predation varies among sites, among feeding animal species, and between seasons, and this variation has been suggested to be of sufficient importance to drive evolution of seed persistence (Hulme, 1998a,b). It would therefore be interesting to study the persistence of soil seed banks in areas with contrasting seed predator communities, or using predator exclusion, in order to explore the effects on the evolution of persistent seed banks and to test the prediction of Brown and Venable (1991) that germination fractions should decrease in response to predation on fresh seeds.

Seed density

Soil seed banks show very high spatial heterogeneity as a result of dispersal contingencies, and seed densities vary considerably over small distances, leading to dense or comparatively seed-free areas (Thompson, 1986; Benoit *et al.*, 1989; Dessaint *et al.*, 1991). Densely packed seeds experience a higher incidence of fungal attack than low-density soil seed banks (Van Mourik *et al.*, 2005), and have a higher depletion rate, hence a lower survival. Since density of seeds in the soil also determines the future competitive situation after emergence, seeds, if they sense each other, should react in two ways: either, germinate quickly to gain an advantage over slower germinating seeds, or, delay germination to another germination season in order to avoid crowding (Dyer *et al.*, 2000; Kluth and Bruelheide, 2005; Turkington *et al.*, 2005; Verdù and Traveset, 2005; Tielbörger and Prasse, 2009). It has also been suggested that delayed germination in response to high seed densities should be more readily adopted by annuals while rapid germination will be more advantageous for perennials. Working on four perennial plants in the Negev desert, Tielbörger and Prasse (2009) showed that indeed seeds sense each other below ground, leading to lower germination fractions at higher seed densities. When seedlings were not removed, their presence accelerated germination of seeds and both effects were influenced by successional position of the species in question. In this way, a late successional species, *Artemisia monosperma*, reduced germination percentages of other species and also germinated fastest, whereas germination of early successional species was suppressed. The site-to-site variation of soil seed persistence summarized here opens interesting perspectives to study the functioning of soil seed banks both in laboratory and field experiments and highlights the complex nature of soil seed-bank dynamics.

Seed Size and Number Trade-off

The soil seed bank inherits from adult plants the constraint that relates the size of a seed to the number of seeds produced per individual plant of comparable size or per canopy area (Smith and Fretwell, 1974; Jakobsson and Eriksson, 2000; Jakobsson *et al.*, 2006). As a rule of thumb, ten times smaller seeds can be produced in ten times higher number for a given canopy area (Aarssen and Jordan, 2001; Henery and Westoby, 2001; Moles and Westoby, 2002). The work of Moles and Westoby (2006) showed, in a global synthesis, that the advantage of higher numbers of small seeds is counterbalanced by

their lower survival as seedlings, and by smaller canopies and shorter reproductive lifespans. Disadvantages for small-seeded plants are detectable especially at the seedling stage and involve mortality due to drought and defoliation (Leishman *et al.*, 2000b).

How the survival of seeds in the soil is influenced by seed size is not well understood. Works using mostly seedling emergence from soil samples in temperate regions show consistently that small seeds have higher persistence in the soil in Europe and other temperate regions (Thompson and Grime, 1979; Leck *et al.*, 1989; Thompson *et al.*, 1993; Bekker *et al.*, 1998a; Moles *et al.*, 2000; Funes *et al.*, 2007). This can be explained by the fact that smaller seeds are more easily incorporated into the soil and moved to deeper soil layers (Benvenuti, 2007), which together with a higher predation pressure on large seeds prevent the evolution of persistence in large seeds (Hulme, 1998b; Thompson, 2000). In contrast, works using burial experiments with counted seed populations in arid areas showed that smaller seeds had lower survival dependent on seed size in the soil than larger seeds (Moles and Westoby, 2006; Moles *et al.*, 2003). These discrepancies among studies have been interpreted by differences in seed predators (Moles and Westoby, 2006). But also, soil factors such as moisture, organic content and seed density decrease seed survival due to enhanced fungal activity (Blaney and Kotanen, 2001; Schafer and Kotanen, 2003; Van Mourik *et al.*, 2005; Pakeman *et al.*, 2012;) and thus influence this relationship. This would probably increase mortality of small seeds more than large seeds since protection and nutrient reserves are different (Crist and Friese, 1993; Moles and Westoby, 2006). An alternative explanation is a difference in methods: seedling emergence studies do not quantify initial seed input, which is higher for small-seeded species than for large-seeded ones in many situations. Then, the sheer numbers of small seeds mean that they may be more easily detected than large seeds (Jakobsson *et al.*, 2006; Saatkamp *et al.*, 2009), leading to a higher ratio of small-seeded species being classified as having persistent seed banks. The detection of seed size–seed persistence relations is even more complicated because the ratio of small to large seeds will decrease with time due to the higher seedling mortality of small-seeded species (Leishman *et al.*, 2000b; Moles and Westoby, 2006). From current data it seems that both seed size–persistence relations occur in nature. Probably in moist ecosystems the amount of small seeds in persistent seed banks is higher, but the precise relation to soil moisture or rainfall has yet to be quantified.

This discussion shows that the soil seed bank cannot be understood disconnected from the entire plant life history, and that the size or numbers of seeds in the soil seed bank should be interpreted in the light of the size–number trade-off. Other seed traits, such as dispersal structures, seed coat thickness or phenolic content also scale importantly with seed size (Moles and Westoby, 2006; Davis *et al.*, 2008); this is

shown for seed coat thickness in Fig 11.2. This concerns also traits that have been related to the survival rates of seeds in the soil across species (Thompson *et al.*, 1993; Bekker *et al.*, 1998a; Gardarin *et al.*, 2010).

As outlined above, seed size importantly influences the survival of seedlings and this can dramatically change the effect of the soil seed bank on community composition and change the size distributions of seeds in the seed banks versus seedlings or adult plants. Data on the relative role of soil seed-bank persistence and seedling mortality in community assembly are crucial if we want to predict their utility for restoration of plant communities (Poschlod, 1995; Bakker *et al.*, 1996; Bossuyt and Honnay, 2008). Until now, studies that analyse the effect of soil seed banks on community composition and abundance *in situ* are comparatively scarce but give an important background picture to understand the role of soil seed banks in communities (e.g. Kalamees and Zobel, 2002).

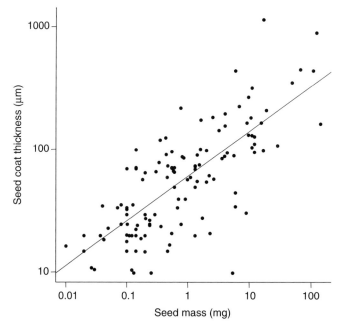

Fig. 11.2. Relation between seed coat thickness and seed weight for 123 plants of Europe and South Africa, note the logarithmic scale for both seed traits, $R^2 = 0.56$, $p < 0.001$ (A. Saatkamp, 2009, unpublished data, and data from Flynn *et al.*, 2004; Holmes and Newton, 2004; Bruun and Poschlod, 2006; Soons *et al.*, 2008; Gardarin *et al.*, 2010; Morozowska *et al.*, 2011).

Soil Seed Banks in Plant Communities

Soil seed banks and coexistence

Persistent seed banks are thought to play an important role in species coexistence through the 'storage effect' (Chesson and Warner, 1981; Facelli *et al.*, 2005; Angert *et al.*, 2009). The storage effect is a mechanism favouring coexistence of otherwise competitively excluding species due to environmental variation. Species that respond differently to environmental variation can coexist when seed banks are present to buffer them from the double disadvantage of an unfavourable environment and high competition. For example, the storage effect can promote the coexistence of dominant competitors with otherwise excluded species which differ in their reactions to disturbances, and which have a persistent soil seed bank (Fig. 11.3).

Traits that are related to different reactions of annual plants to environmental fluctuations include, among others, adaptations to cope with dry environments, which is in trade-off with their relative growth rate (Angert *et al.*, 2009). Moreover, annual plants with limited spatial dispersal and high seed mass recover more slowly from

severe disturbances than do small-seeded plants from the persistent seed bank. Seeds can play further important roles in coexistence through the storage effect since differences in germination responses to environmental variation can be the temporal niches providing the mechanism of differential species responses to the environment (Facelli *et al.*, 2005).

Most plant communities show a mix of transient and persistent soil seed banks. In dense communities of annual plants with recurrent disturbances, competition colonization trade-offs are also an important mechanism to promote coexistence. In Mediterranean cereal fields and pasture communities, for instance, this probably even plays a role within the same guild of annual plants with autumn germination and winter development. Here, low seed-longevity species such as *Agrostemma githago* and *Nigella damascena* coexist with long seed-longevity species *Adonis flammea* and *Carthamus lanatus* (Saatkamp *et al.*, 2009, 2010). Figure 11.3 shows how plants with transient and persistent seed banks can coexist through a competition–colonization trade-off. In many cases of coexisting plants with different seed bank strategies, examination of the entire plant life histories will reveal that contrasting

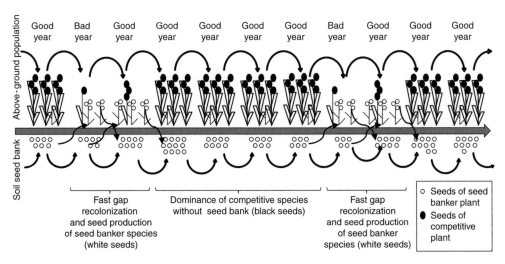

Fig. 11.3. Competition–colonization trade-off in plant communities: coexistence of a subordinate plant with a competitive plant is possible through a persistent soil seed bank of the subordinate with drought adaptation and gap detection mechanism leading to high reproduction of the subordinate in unfavourable years with less dominance of the competitive plant.

plant regeneration strategies are correlated to other (adult) plant traits, because selective interactions lead to trade-offs among risk-reducing mechanisms (Venable and Brown, 1988), and here 'storage effect' increases the possibilities of long-term coexistence (Chesson and Warner, 1981; Facelli et al., 2005; Angert et al., 2009).

Other communities with plants having apparently similar ecological niches and contrasting seed bank strategies include shorelines, with large bunches of sedges (Carex) having persistent soil seed banks (Schütz, 2000), but reed canary-grass (Phalaris arundinacea) or reed (Phragmites australis) most often having transient soil seed banks (Thompson et al. 1997). Similar contrasts exist among forest floor herbs with persistent-seeded Moehringia trinervia (Vandelook et al., 2008) but transient-seeded Oxalis acetosella (Thompson et al. 1997; Thompson, 2000). These two species have similar height, seed size and dispersal type, and one might argue that O. acetosella is a specialist of humid acidic organic soil, a perennial, and M. trinervia, an annual plant on wind-blown, bare mineral soil. The latter habitat has sufficiently unpredictable conditions to evolve persistent seed banks while in the former habitat buried seed would suffer from heavy fungi attack to prevent evolution of a persistent seed bank (Brown and Venable, 1991; Schafer and Kotanen, 2003; Wagner and Mitschunas, 2008; Pakeman et al., 2012). The cited examples show that soil seed banks contribute to coexistence either as a part of the storage effect or as an adaptation that increases niche partition between different microhabitats.

Disturbance, succession and soil seed banks

Whatever the reasons are for the coexistence of species with contrasting soil seed banks, disturbances will not equally affect the recovery of plant populations from transient compared to persistent soil seed banks (van der Valk and Pederson, 1989; Bakker et al., 1996; von Blanckenhagen and Poschlod,

2005; Bossuyt and Honnay, 2008). Plant communities also differ in the abundance of viable seeds in soil banks, and therefore the success of restoration from them varies significantly (Venable, 1989; Bekker et al., 1998c; Hopfensberger, 2007; Bossuyt and Honnay, 2008). Moreover, even plants with notoriously persistent seed banks depend crucially on time since land-use change to recover (Poschlod et al., 1998; Waldhardt et al., 2001; Mitlacher et al., 2002). The recurrent picture from dozens of works on resemblance of soil seed bank and plant communities is that frequently disturbed ecosystems or habitats with unpredictable conditions, such as arable fields, ruderal habitats, river floodplains, deserts, arid pastures and vernal pools have a high resemblance between standing vegetation and seed banks and that relatively low disturbance systems such as heathlands, mires, humid pastures, shrublands and (especially) ancient or old grown forests have comparative lower resemblance (reviewed in Hopfensberger, 2007; Thompson and Grime, 1979; Falinska, 1999; Amiaud and Touzard, 2004; Luzuriaga et al., 2005; Wellstein et al., 2007). In the very open habitats of Mediterranean matorral on gypsum soils, secondary dispersal of seeds leads to rapid local recovery of soil seed banks (Olano et al., 2012). These studies suggest a trade-off between seed persistence in the soil and adult lifespan, which was predicted by theoretical works (Rees, 1994), with short-living species relying on persistent soil seed banks in contrast to long-living species (Ehrlén and van Groenendal, 1998). Consequently, the recovery of communities after disturbances is habitat specific (Bossuyt and Honnay, 2008) and even more, it is site specific due to subtle variation in species composition and local adaptation of plants to form soil seed banks (Clauss and Venable, 2000; Tielbörger and Petru, 2008; Baldwin et al., 2010).

This picture is completed by the temporal sequence of plants in many vegetation types after disturbances, which shows a trend of early successional species having more persistent soil seed banks than late successional species (Grime, 1977, 1989; Thompson and Grime, 1979; Garwood, 1989; Butler and Chazdon, 1998; Grandin, 2001; Hopfensberger, 2007). The very difference of primary and secondary succession in plant communities lies in the relative importance of seed dispersal for primary succession (Walker *et al.*, 1986; Jumpponen *et al.*, 1999), and on persistent seed banks at least at the beginning for secondary succession (Jiménez and Armesto, 1992; Bekker *et al.*, 2000). But even for primary succession a higher importance of persistent seed banks in early compared to late stages has been shown (Marcante *et al.*, 2009; but see Grandin and Rydin, 1998; Bossuyt and Hermy, 2004). This can be seen as indirect evidence for the trade-off between spatial and temporal dispersal, which, to date, has strong theoretical (Venable and Lawlor, 1980; Venable and Brown, 1988) but still weak empirical (Ozinga *et al.*, 2007) support, and needs to be tested at the relevant temporal and spacial scales.

Persistent soil seed banks, restoration and extinction risk

Persistent seed banks have clear relevance for the restoration of plant communities. It has been shown for several communities that persistent soil seed banks are an important tool to restore local plant communities after abandonment of human use, fire, or diverse forms of direct destruction of above-ground vegetation (van der Valk and Pederson, 1989; Bakker *et al.*, 1996; Willems and Bik, 1998; von Blanckenhagen and Poschlod, 2005; Bossuyt and Honnay, 2008). As summarized above, even within communities, plants differ in their life history strategies including their dependence on persistent soil seed banks. Only plants with persistent seed banks will recover spontaneously from soil seed banks if unfavourable conditions lasted until the second subsequent germination season. Moreover, later successional species, which only regenerate when a minimum cover of vegetation already exists, will only be able to restore by later seed arrival; thus, persistent soil seed

banks can only restore a part of the community (Kiefer and Poschlod, 1996; Bekker *et al.*, 1997; Matus *et al.*, 2003; Buisson *et al.*, 2006; Valkó *et al.*, 2011; summarized by Bossuyt and Honnay, 2008; but see Bossuyt and Hermy, 2004). Many of the most endangered species do not have persistent soil seed banks. Conversely, plant populations that can be restored from persistent seed banks are often widespread or invasive species (Bossuyt and Honnay, 2008). Only in exceptional cases is restoration from seed banks effective for rare or threatened species (Poschlod, 1996; Zehm *et al.*, 2008). This seems to be the case even when local communities remain intact but are fragmented (Stöcklin and Fischer, 1999). Persistence of seeds in the soil is an important trait related to the risk of extinction of plant species (Poschlod *et al.*, 1996) since it is indicative of a spatiotemporal strategy a given species explored in its recent evolutionary history. However, the existence of a soil seed bank does not necessarily indicate its complete independence from spatial dispersal as illustrates the work of Harrison and Ray (2002) on fragmentation of vernal pool species in California.

Seedling recruitment from seed banks and species identity

Composition and abundance of species in the soil seed bank are not directly translated into adult plant communities through germination and seedling recruitment. As previously discussed, small seeds have higher mortality during seedling establishment (Moles and Westoby, 2004); this results in lower representation of small-seeded species as seedlings than could be expected from their abundance in the soil seed bank. Additionally, the importance of recruitment from seeds compared to resprouting or lateral growth from outside the gap has been shown to depend on gap size (Milberg, 1993; Dalling and Hubbell, 2002; Kalamees and Zobel, 2002). Species that regenerate in tropical forest gaps germinate in response to red/far-red light ratios, water potential and diurnal fluctuating temperatures (Pearson *et al.*, 2003; Daws *et al.*, 2008). In large tropical forest gaps, large seeds germinate faster and in drier conditions than small seeds, which are more specific to moist conditions of small gaps and near the edges, decreasing the drought risk (Daws *et al.*, 2008). However in other situations the distance to dispersing adult trees or seedling mortality/growth rates are more important for the identity of seedlings that establish in gaps (Dalling and Hubbell, 2002). Also, during the growth of crops, the changing light quality decreases germination of some weed species, leading to variable emergence in relation to crop age and density (Kruk *et al.*, 2006).

The timing of disturbances or gap creation is a second crucial factor that influences which species are recruited from the seed bank into gaps (Lavorel *et al.*, 1994; Pakeman *et al.*, 2005). This timing can be related to differences in seed availability, favouring persistent seeds when there is no seed rain (Pakeman *et al.*, 2005) or sorting species composition according to germination temperature requirements of involved species (Baskin and Baskin, 1998; Kruk *et al.*, 2006; Merritt *et al.*, 2007). Another factor that importantly impedes a direct relation between soil seed-bank composition and newly established plant communities is seed and seedling predation (Forget *et al.*, 2005).

Beyond the many filters, the recovery of species composition and abundance from soil seed banks depends in yet unpredictable fashions (Lavorel and Lebreton, 1992) on site history (Dupouey *et al.*, 2002), seed rain (Cubiña and Aide, 2001; Buisson *et al.*, 2006; Jakobsson *et al.*, 2006) and secondary dispersal (Luzuriaga *et al.*, 2005; Olano *et al.*, 2012). It has yet to be explored whether and how much stochasticity plays a role in recruitment from soil seed banks and whether aboveground communities are connected to soil seed banks as local communities are to regional species pools or metacommunities and their abundance and distance relationships (Zobel, 1997; Hubbell, 2001).

Seed banks, invasive species and climate change

Non-native, invasive species often have a large persistent soil seed bank (Newsome and Noble, 1986; Lonsdale *et al.*, 1988; D'Antonio and Meyerson, 2002). In some cases, they assemble a much larger seed bank in their new than in their native ranges (Noble, 1989). Even if they are still rare in the above-ground vegetation they already may have accumulated seeds in the soil (Drake, 1998). Therefore, restoration of native plant communities with a large number of persistent seeds of invasive plants may be impossible since the newly established vegetation would be dominated by the invasive, non-native species. This is especially the case in Mediterranean climate ecosystems such as those in South Africa (Holmes and Cowling, 1997a,b; Heelemann *et al.*, 2012) or Australia (Lunt, 1990) with major implications for restoration management (Richardson and Kluge, 2008; Heelemann *et al.*, 2012). Seed bank longevity data are critical for the management of invasive plants, because invasives with no or short-term persistent seed banks may be eliminated with only a few years of conscientious removal.

Climate change may affect soil seed bank persistence and composition in manifold ways (also reviewed in Chapter 9 of this volume). Warming may increase seed production and therefore, the input to the soil seed bank (Molau and Shaver, 1997; Totland, 1999; see also Akinola *et al.*, 1998a,b). In contrast, drought may also decrease seed production (Peñuelas *et al.*, 2004). In other cases, seed production may remain unchanged despite warmer temperatures and higher precipitation (Wookey *et al.*, 1995). Changes in precipitation will affect soil moisture and as a consequence seed persistence (Walck *et al.*, 2011), because soil moisture has important influences on fungal activity (Leishman *et al.*, 2000a; Blaney and Kotanen, 2001; Wagner and Mitschunas, 2008). Changes in temperature and soil moisture due to precipitation also change the dormancy state of buried seed populations, and in this way affect soil seed-bank composition (Walck *et al.*, 2011). Lastly, atmospheric CO_2 enrichment may affect seed traits and as a consequence soil seed longevity (Grünzweig and Dumbur, 2012). These works show that the directions of changes in soil seed banks in response to climate change depend on species, traits and factors involved and cannot be generalized at the moment.

Dynamics and Mechanisms in Soil Seed Banks

Formation of persistent soil seed banks is part of a plant's strategy in habitats with variability in rainfall, drought, flooding, vegetation gaps, disturbances or frost. Additionally, soil and climate conditions, disperser and predator communities or competitors also differ among sites and influence the survival of seed in the soil. Consequently, which traits increase seed survival in soil depends on ecosystem and species. This makes it difficult to predict features of soil seed banks from plant functional traits. Moreover, across species, only a few models for soil seed-bank dynamics exist, all to our knowledge for weeds in temperate ecosystems (Forcella, 1993, 1998; Rasmussen and Holst, 2003; Meyer and Allen, 2009; Gardarin *et al.*, 2012).

One of the mechanisms that may contribute to the persistence of seeds beyond the first possible germination season is dormancy (also reviewed in Chapter 7 of this volume). Evolutionary models often refer to 'dormancy' to speak about seeds that 'did not germinate' but are still alive and able to germinate in the future. This is not perfectly congruent with the physiological definition of dormancy which means the inability to germinate in otherwise favourable conditions in which non-dormant seeds would germinate (Baskin and Baskin, 1998; Finch-Savage and Leubner-Metzger, 2006). The delay in germination treated in these evolutionary models can be realized through different mechanisms: any dormancy mechanism, such as physical or physiological dormancy (Baskin and Baskin, 1998), underdeveloped embryos (Finch-Savage and Leubner-Metzger, 2006), delayed dispersal (Cowling and Lamont, 1985; Schwilk and Ackerly, 2001), light sensitivity cycling (Thanos and Georghiou, 1988), specific temperature and moisture

requirements (Finch-Savage and Leubner-Metzger, 2006) or sensitivity to fluctuating temperatures (Thompson and Grime, 1983; Saatkamp *et al.*, 2011a; Thompson *et al.*, 1977). Seeds with underdeveloped embryos sometimes show delayed germination and are then called morphological dormant (Baskin and Baskin, 2004). Some physiologists (Carasso *et al.*, 2011) propose to consider them non-dormant, since growth in these seeds is continuous and pre-emergence drought sensitivity appears before radicles emerge (Ali *et al.*, 2007). 'Delayed germination' and some kind of seed persistence can result from seeds being dormant, or from non-dormant seeds not getting the appropriate cues for germination, which makes it very difficult to establish an exact correspondence between dormancy and persistence of seeds in the soil (Thompson *et al.*, 2003).

Another mechanism to maintain viable soil seed banks over several years is to prevent germination in unfavourable seasons through cycling dormancy. Cycling dormancy means that seeds come out of dormancy and re-enter dormancy every year depending on levels of temperature and rainfall (e.g. Baskin *et al.*, 1993; Baskin and Baskin, 1994; reviewed in Baskin and Baskin, 1998). Thus seeds will germinate, depending on the season, either over a large range of conditions (when the following season is favourable for their development) or will germinate under a restricted range or not germinate at all (when the following season is unfavourable). Plants with different dormancy cycling coexist. Figure 11.4 shows two species with cycling dormancy, a winter annual (*Lamium purpureum*, Fig. 11.4c) and a spring annual (*Polygonum aviculare*, Fig. 4b), which are dormant in winter/spring (*L. purpureum*) or summer/autumn (*P. aviculare*). Similar seasonal cycling schemes are also known for seed coat permeability in the form of sensitivity cycling of physically dormant seeds (Jayasuriya *et al.*, 2008) and for light requirements (Thanos and Georghiou, 1988). The functional role of dormancy cycling is to maximize fitness by matching the germination to seasons with optimal seedling development. Contrastingly, in some plants like Saguaro

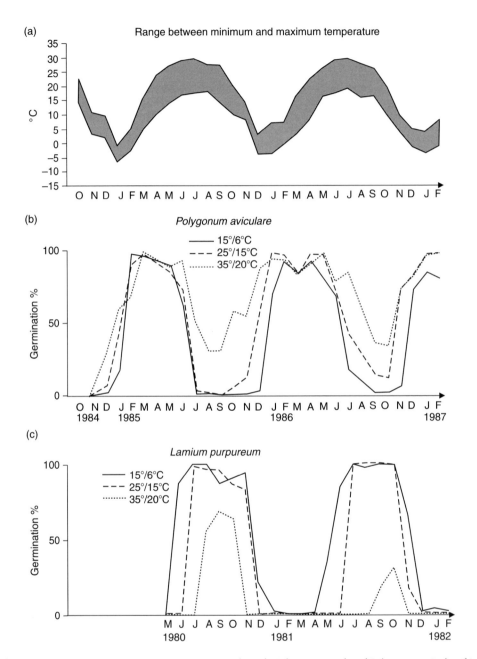

Fig. 11.4. (a) Temperature ranges in temperate regions. (b) and (c) dormancy cycles of *Polygonum aviculare* (b), a summer annual and *Lamium purpureum* (c), a winter annual; with variable germination percentages in three growth chamber conditions, seeds lots were exposed to seasonal varying temperatures (redrawn from data in Baskin and Baskin, 1984, 1990).

cactus (*Carnegia gigantean*) and Boojum (*Fouquieria columnaris*) all seeds germinate at the first opportunity or die, and they do not need dormancy cycling. Interestingly,

cycling dormancy is a necessary correlate of persistent seed banks, because all species with physiological dormancy for which dormancy cycles could be studied and which thus persisted more than one year in the experiments show dormancy cycles (Baskin and Baskin, 1998).

Mechanisms to maintain persistent soil seed banks and the traits that correlate with seed persistence may vary according to global climatic characteristics, and we will illustrate two contrasting situations in the following. Benvenuti (2007) studied how seeds with contrasting traits are buried by rain during seed-bank formation on bare soils in temperate arable land. In this case, small seeds with round shape and with smooth or alveolar surfaces are buried deeper and faster. Once buried, seed populations can be prevented from germination through a light requirement for germination (Pons, 1991; Milberg *et al.*, 2000; Saatkamp *et al.*, 2011b; Chapter 5 of this volume), detection of fluctuating temperatures (Thompson and Grime, 1983; Saatkamp *et al.*, 2011a,b), or oxygen concentrations (Benech-Arnold *et al.*, 2006). For small seeds, rapid burial also prevents predation by soil surface invertebrates and by birds, while large seeds can be dug out by rodents (Hulme, 1998a,b). Earthworms digest small seeds more easily than large ones (Forey *et al.*, 2011). In moist soils, fungi attack seeds, especially when in high density (Van Mourik *et al.*, 2005) or when organic matter content is high (Pakeman *et al.*, 2012). Seeds may differ in susceptibility to fungal attack depending on seed coat thickness (Davis *et al.*, 2008; Gardarin *et al.*, 2010) and phenolic content (Thompson, 2000; Davis *et al.*, 2008). Many seeds show cycling dormancy in response to annual temperature changes defining specific germination seasons (Baskin and Baskin, 1985, 1994, 1995, 2006; Baskin *et al.*, 1986). Cycling dormancy leads to higher depletion of soil seed reservoirs during the germination season compared to unfavourable seasons when plants die as seedlings after germination and before they could emerge at the soil surface (Saatkamp *et al.*, 2011a; Gardarin *et al.*, 2012). Desiccation sensitivity of buried seeds also changes with time after burial and can be a secondary source of mortality (Ali *et al.*, 2007). When disturbances expose non-dormant seeds from the soil bank to light and when

the progress to germination depending on temperature and moisture is sufficient (Bradford, 2002; Allen *et al.*, 2007), seeds germinate and leave the soil seed bank. This picture is drawn from temperate herbaceous communities where seeds remain in the imbibed state in the soil. Here, seed persistence in the soil can be related to smaller seed size, rounder shape, light requirements for germination, seed coat thickness and high phenol content.

In contrast to moist temperate ecosystems, in arid regions, such as Australia, fungi attack is less important and predator communities are different, in such a way that larger seeds have higher survival in the soil than small seeds (Moles *et al.*, 2003; Moles and Westoby, 2006). The difference in the relation between seed size and persistence between Australian arid areas and moist temperate areas can partly be explained by different methods that have been used to measure persistence or seed survival (Saatkamp *et al.*, 2009). In arid and semiarid climates, many species have conspicuous self-burial mechanisms such as hygroscopic appendages in *Erodium* or *Aristida*. Other plants germinate in response to chemical cues, such as smoke-derived substances from vegetation fires (Brown, 1993; Flematti *et al.*, 2004), and their absence keeps large seed reservoirs in an ungerminated state. Annual plants are comparatively rare in Australia, except in seasonally wet habitats (Brock, 2011) and longevity of seeds of woody species is lower due to the alternative risk reduction mechanism of longer lifespan (Rees, 1994, 1996; Tuljapurkar and Wiener, 2000; Campbell *et al.*, 2012). Seeds with thick impermeable seed coats with physical dormancy are common in many fire-prone arid ecosystems, and thought to have evolved in dry areas (Baskin *et al.*, 2000). Arid soil seed banks also show many seeds that germinate better in darkness than in light (Baker, 1972; Baskin and Baskin, 1998), thus germinating more easily in soil than at its surface, probably because the risk of seedling death due to drought is lower when emergence starts in deeper soil layers. The contrast between seed-bank dynamics in moist temperate and dry warm regions shows that soil seed persistence traits need to be considered in relation to a specific environment. In order to generalize this knowledge we need to study trait–environment interactions in sufficiently contrasted situations.

The understanding of soil seed banks of weeds has motivated researchers to model the dynamics of soil seed banks (Forcella, 1998; Rasmussen and Holst, 2003; Meyer and Allen, 2009; FLORSYS by Gardarin et al., 2012). They brought to light that we need to model independently the processes of germination, dormancy and 'suicide germination' (Benvenuti et al., 2001) compared to other processes such as mortality due to ageing, decay or predation (Gardarin et al., 2012). In these models, different plant traits are used to predict mortality (before germination) and germination, the first has been related to seed coat thickness (Gardarin et al., 2010), whereas the latter to base parameters of hydrothermal time models (Bradford, 2002; Allen et al., 2007). These models do not include postdispersal seed predation nor do they distinguish between seed ageing and seed decay (although FLORSYS does include mortality parameters explicitly). At least for the target species, these models predict with some accuracy abundance of seed populations in soils, their movement, dormancy state, date of germination and number of seedlings emerging (Gardarin et al., 2012). Limits of these models are the high number of input parameters – sometimes difficult to measure – and the difficulties of using them with other species and in other ecosystems.

Figure 11.5 summarizes some of the processes and traits involved in soil seed-bank dynamics in temperate ecosystems. Three main processes for the exit of seeds from the soil seed bank differ in the traits that influence persistence and adaptations: (i) germination; (ii) mortality due to ageing; or (iii) mortality due to predation including microbial or fungi attack. Traits that relate to germination do not specifically reduce mortality of seeds: for example, small embryos, high levels of abscisic acid or light requirement prevent or delay germination but do not necessarily reduce predation.

Enzymes that neutralize reactive oxygen species also do not necessarily influence predation nor germination, although when oxidated they can break dormancy (Bahin et al., 2011). Although Davis et al. (2008) concluded that ortho-dihydroxyphenols did not influence germination or ageing, but may be effective compounds for defences against microbes and fungi, Chapter 8 of this volume points out some methodological and interpretive problems associated with studies that focus on this class of phenolic compounds. It is not yet clear whether thick or impermeable seed coats influence germination as much as they influence predation, because most impermeable seed coats have specialized structures that control germination independently from coat thickness (Baskin, 2003). Moreover thick seed coats are related to larger size and hence forces of growing embryos (Mohr et al., 2010). Likewise, small seed size enhances burial speed and reduces germination (for species with a light requirement) and predation (by surface-feeding animals) but for digestion by earthworms small size is disadvantageous (Forey et al., 2011). These effects are independent from the higher number in which small seeds are produced, which independently results in a higher probability of seeds surviving. It is thus helpful to distinguish between effects of reproduction (seed number) and survival (individually) in our endeavour to understand how soil seed banks are influenced by adaptive traits in a series of environments.

Acknowledgements

We thank Robert Gallagher and Filip Vandelook for their helpful comments and corrections of an earlier version; we are grateful to Ken Thompson for stimulating discussions at the Utah Seed Ecology meeting in 2010; we thank Kristin Metzner and Marine Pouget for reading. A.S. was funded by IMBE (CNRS, Aix-Marseille University) and the region PACA (program Gévoclé).

(a) Seed input: seed production and postdispersal seed predation

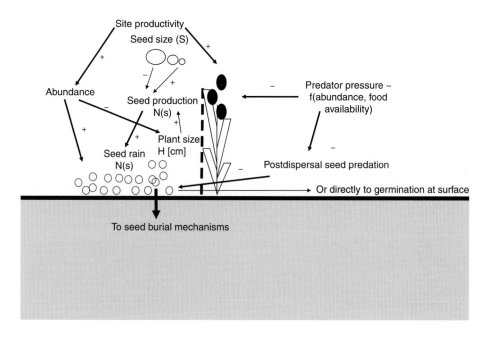

(b) Burial mechanisms and movement inside the soil profile

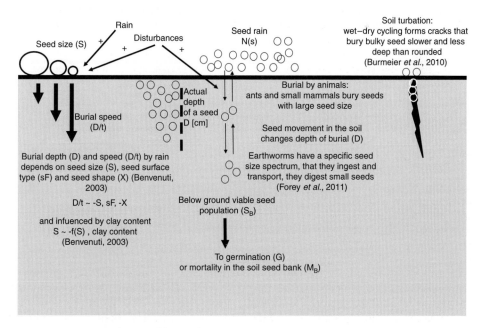

Fig. 11.5. Soil seed-bank dynamic model, with input, dormancy cycle, movement and output in three different ways, germination, death due to ageing and death due to mortality or fungi attack and the allied sets of traits and environmental influence factors (modified from Allen *et al.*, 2007; Saatkamp *et al.*, 2011b and Gardarin *et al.*, 2012).

(c) Mortality of buried seeds

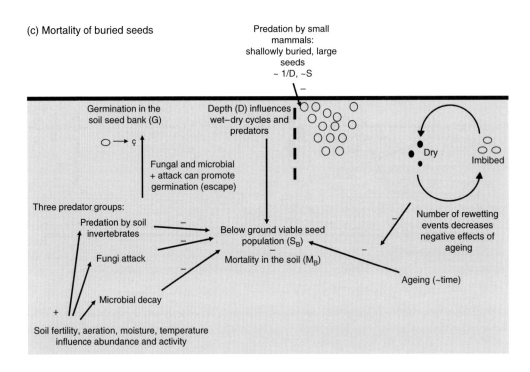

(d) Dormancy cycling, germination and gap detection

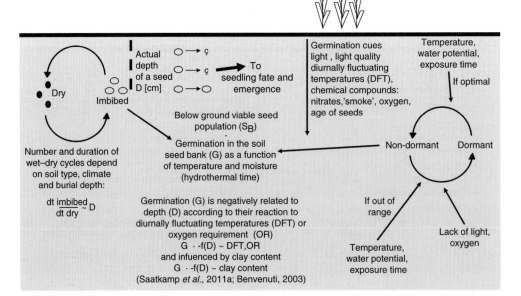

Fig. 11.5. Continued.

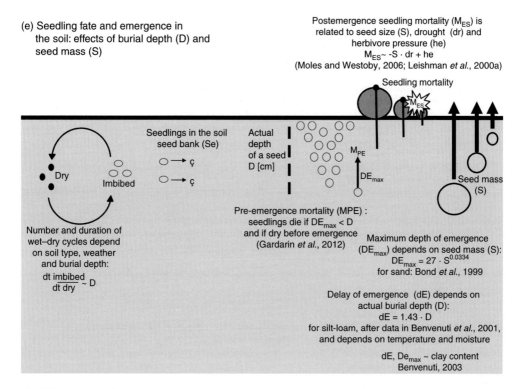

(e) Seedling fate and emergence in the soil: effects of burial depth (D) and seed mass (S)

Postemergence seedling mortality (M_{ES}) is related to seed size (S), drought (dr) and herbivore pressure (he)

$$M_{ES} \sim -S \cdot dr + he$$

(Moles and Westoby, 2006; Leishman *et al.*, 2000a)

Seedling mortality

M_{ES}

Seedlings in the soil seed bank (Se)

Dry

Imbibed

Actual depth of a seed D [cm]

M_{PE}

DE_{max}

Seed mass (S)

Number and duration of wet–dry cycles depend on soil type, weather and burial depth:

$$\frac{dt \ imbibed}{dt \ dry} \sim D$$

Pre-emergence mortality (MPE): seedlings die if $DE_{max} < D$ and if dry before emergence (Gardarin *et al.*, 2012)

Maximum depth of emergence (DE_{max}) depends on seed mass (S):

$$DE_{max} = 27 \cdot S^{0.0334}$$

for sand: Bond *et al.*, 1999

Delay of emergence (dE) depends on actual burial depth (D):

$$dE = 1.43 \cdot D$$

for silt-loam, after data in Benvenuti *et al.*, 2001, and depends on temperature and moisture

$dE, De_{max} \sim$ clay content
Benvenuti, 2003

Fig. 11.5. Continued.

References

Aarssen, L.W. and Jordan, C.Y. (2001) Between-species patterns of covariation in plant size, seed size and fecundity in monocarpic herbs. *Ecoscience* 8, 471–477.

Akinola, M.O., Thompson, K. and Buckland, S.M. (1998a) Soil seed bank of an upland calcareous grassland after 6 years of climate and management manipulations. *Journal of Applied Ecology* 35, 544–552.

Akinola, M.O., Thompson, K. and Hillier, S.H. (1998b) Development of soil seed banks beneath synthesized meadow communities after seven years of climate manipulations. *Seed Science Research* 8, 493–500.

Ali, N., Probert, R., Hay, F., Davies, H. and Stuppy, W. (2007) Post-dispersal embryo growth and acquisition of desiccation tolerance in *Anemone nemorosa* L. seeds. *Seed Science Research* 17, 155–163.

Allen, P.S., Benech-Arnold, R.L., Batlla, D. and Bradford, K.J. (2007) Modeling of seed dormancy. In: Bradford, K.J. and Nonogaki, H. (eds) *Annual Plant Reviews Volume 27: Seed Development, Dormancy and Germination.* Wiley-Blackwell, Oxford, pp. 72–112.

Amiaud, B. and Touzard, B. (2004) The relationships between soil seed bank, aboveground vegetation and disturbances in old embanked marshlands of Western France. *Flora* 199, 25–35.

Angert, A.L., Huxman, T.E., Chesson, P. and Venable, D.L. (2009) Functional tradeoffs determine species coexistence via the storage effect. *Proceedings of the National Academy of Sciences* 106, 11641–11645.

Bahin, E., Bailly, C., Sotta, B., Kranner, I., Corbineau, F. and Leymarie, J. (2011) Crosstalk between reactive oxygen species and hormonal signalling pathways regulates grain dormancy in barley. *Plant Cell and Environment* 34, 980–993.

Baker, H.G. (1972) Seed weight in relation to environmental conditions in California. *Ecology* 53, 997–1010.

Bakker, J.P. (1989) *Nature Management by Grazing and Cutting: on the Ecological Significance of Grazing and Cutting Regimes Applied to Restore Former Species-Rich Grassland Communities in the Netherlands*. Kluwer Academic Publishers, Denmark.

Bakker, J.P., Poschlod, P., Strykstra, R.J., Bekker, R.M. and Thompson, K. (1996) Seed banks and seed dispersal: important topics in restoration ecology. *Acta Botanica Neerlandica* 45, 461–490.

Baldwin, A.H., Kettenring, K.M. and Whigham, D.F. (2010) Seed banks of *Phragmites australis*-dominated brackish wetlands: Relationships to seed viability, inundation, and land cover. *Aquatic Botany* 93, 163–169.

Baskin, C.C. (2003) Breaking physical dormancy in seeds – focussing on the lens. *New Phytologist* 158, 229–232.

Baskin, C.C. and Baskin, J.M. (1994) Germination requirements of *Oenothera biennis* seeds during burial under natural seasonal temperature cycles. *Canadian Journal of Botany* 72, 779–782.

Baskin, C.C. and Baskin, J.M. (1998) *Seeds: ecology, biogeography and evolution of dormancy and germination*. Academic Press, San Diego.

Baskin, C.C. and Baskin, J.M. (2006) The natural history of soil seed banks of arable land. *Weed Science* 54, 549–557.

Baskin, C.C., Chesson, P.L. and Baskin, J.M. (1993) Annual seed dormancy cycles in two desert winter annuals. *Journal of Ecology* 81, 551–556.

Baskin, J.M. and Baskin, C.C. (1984) Role of temperature in regulating timing of germination in soil seed reserves of *Lamium purpureum* L. *Weed Science* 24, 341–349.

Baskin, J.M. and Baskin, C.C. (1985) The annual dormancy cycle in buried weed seeds: a continuum. *BioScience* 35, 492–498.

Baskin, J.M. and Baskin, C.C. (1990) The role of light and alternating temperatures on germination of *Polygonum aviculare* seeds exhumed on various dates. *Weed Research* 30, 397–402.

Baskin, J.M. and Baskin, C.C. (1995) Variation in the annual dormancy cycle in buried seeds of the weedy winter annual *Viola arvensis*. *Weed Science* 35, 353–362.

Baskin, J.M. and Baskin, C.C. (2004) A classification system for seed dormancy. *Seed Science Research* 14, 1–16.

Baskin, J.M., Baskin, C.C. and Parr, J.C. (1986) Field emergence of *Lamium amplexicaule* L. and *L. purpureum* L. in relation to the annual seed dormancy cycle. *Weed Science* 26, 185–190.

Baskin, J.M., Baskin, C.C. and Xiaojie, L. (2000) Taxonomy, anatomy and evolution of physical dormancy in seeds. *Plant Species Biology* 15, 139–152.

Bekker, R.M., Verweij, G.L., Smith, R.E.N., Reine, R., Bakker, J.P. and Schneider, S. (1997) Soil seed banks in European grasslands: does land use affect regeneration perspectives? *Journal of Applied Ecology* 34, 1293–1310.

Bekker, R.M., Bakker, J.P., Grandin, U., Kalamees, R., Milberg, P., Poschlod, P., Thompson, K. and Willems, J.H. (1998a) Seed size, shape and vertical distribution in the soil: indicators of seed longevity. *Functional Ecology* 12, 834–842.

Bekker, R.M., Knevel, I.C., Tallowin, J.B.R., Troost, E.M.L. and Bakker, J.P. (1998b) Soil nutrient input effects on seed longevity: a burial experiment with fen meadow species. *Functional Ecology* 12, 673.

Bekker, R.M., Schaminée, J.H.J., Bakker, J.P. and Thompson, K. (1998c) Seed bank characteristics of Dutch plant communities. *Acta Botanica Neerlandica* 47, 15–26.

Bekker, R.M., Verweij, G.L., Bakker, J.P. and Fresco, L.F.M. (2000) Soil seed bank dynamics in hayfield succession. *Journal of Ecology* 88, 594–607.

Benech-Arnold, R.L., Sánchez, R.A., Forcella, F., Kruk, B.C. and Ghersa, C.M. (2000) Environmental control of dormancy in weed seed banks in soil. *Field Crops Research* 67, 105–122.

Benech-Arnold, R.L., Gualano, N., Leymarie, J., Côme, D. and Corbineau, F. (2006) Hypoxia interferes with ABA metabolism and increases ABA sensitivity in embryos of dormant barley grains. *Journal of Experimental Botany* 57, 1423–1430.

Benoit, D., Kenkel, N.C. and Cavers, P.B. (1989) Factors influencing the precision of soil seed bank estimates. *Canadian Journal of Botany* 67, 2833–2840.

Benvenuti, S. (2003) Soil texture involvement in germination and emergence of buried weed seeds. *Agronomy Journal* 95, 191–198.

Benvenuti, S. (2007) Natural weed seed burial: effect of soil texture, rain and seed characteristics. *Seed Science Research* 17, 211–219.

Benvenuti, S., Macchia, M. and Miele, S. (2001) Quantitative analysis of emergence of seedlings from buried weed seeds with increasing soil depth. *Weed Science* 49, 528–535.

Blaney, C.S. and Kotanen, P.M. (2001) Effects of fungal pathogens on seeds of native and exotic plants: a test using congeneric pairs. *Journal of Applied Ecology* 38, 1104–1113.

Boedeltje, G., Bakker, J.P., Ten Brinke, A., van Groenendael, J.M. and Soesbergen, M. (2004) Dispersal phenology of hydrochorous plants in relation to discharge, seed release time and buoyancy of seeds: the flood pulse concept supported. *Journal of Ecology* 92, 786–796.

Bond, W.J., Honig, M. and Maze, K.E. (1999) Seed size and seedling emergence: an allometric relationship and some ecological implications. *Oecologia* 120, 132–136.

Bonnewell, V., Koukkari, W.L. and Pratt, D.C. (1983) Light, oxygen, and temperature requirements for *Typha latifolia* seed germination. *Canadian Journal of Botany* 61, 1330–1336.

Bossuyt, B. and Hermy, M. (2004) Seed bank assembly follows vegetation succession in dune slacks. *Journal of Vegetation Science* 15, 449–456.

Bossuyt, B. and Honnay, O. (2008) Can the seed bank be used for ecological restoration? An overview of seed bank characteristics in European communities. *Journal of Vegetation Science* 19, 875–884.

Bradford, K.J. (2002) Applications of hydrothermal time to quantifying and modeling seed germination and dormancy. *Weed Science* 50, 248–260.

Brock, M.A. (2011) Persistence of seed banks in Australian temporary wetlands. *Freshwater Biology* 56, 1312–1327.

Brown, J.S. and Venable, D.L. (1991) Life history evolution of seed-bank annuals in response to seed predation. *Evolutionary Ecology* 5, 12–29.

Brown, N.A.C. (1993) Promotion of germination of fynbos seeds by plant-derived smoke. *New Phytologist* 123, 575–583.

Bruun, H.H. and Poschlod, P. (2006) Why are small seeds dispersed through animal guts: large numbers or seed size per se? *Oikos* 113, 402–411.

Buisson, E., Dutoit, T., Torre, F., Römermann, C. and Poschlod, P. (2006) The implications of seed rain and seed bank patterns for plant succession at the edges of abandoned fields in Mediterranean landscapes. *Agriculture Ecosystems & Environment* 115, 6–14.

Bulmer, M.G. (1984) Delayed germination of seeds: Cohen's model revisited. *Theoretical Population Biology* 26, 367–377.

Burmeier, S., Eckstein, R.L., Otte, A. and Donath, T.W. (2010) Desiccation cracks act as natural seed traps in flood-meadow systems. *Plant and Soil* 333, 351–364.

Butler, B.J. and Chazdon, R.L. (1998) Species richness, spatial variation, and abundance of the soil seed bank of a secondary tropical rain forest. *Biotropica* 30, 214–222.

Butler, M.A. and King, A.A. (2004) Phylogenetic comparative analysis: A modeling approach for adaptive evolution. *American Naturalist* 164, 683–695.

Campbell, M.L., Clarke, P.J. and Keith, D.A. (2012) Seed traits and seed bank longevity of wet sclerophyll forest shrubs. *Australian Journal of Botany* 60, 96–103.

Carasso, V., Hay, F.R., Probert, R.J. and Mucciarelli, M. (2011) Temperature control of seed germination in *Fritillaria tubiformis* subsp. *moggridgei* (Liliaceae) a rare endemic of the South-west Alps. *Seed Science Research* 21, 33–38.

Cavieres, L.A. and Arroyo, M.T.K. (2001) Persistent soil seed banks in *Phacelia secunda* (Hydrophyllaceae): experimental detection of variation along an altitudinal gradient in the Andes of Central Chile. *Journal of Ecology* 89, 31–39.

Chesson, P.L. and Warner, R.R. (1981) Environmental variability promotes coexistence in lottery competitive systems. *American Naturalist* 117, 923–943.

Clauss, M.J. and Venable, D.L. (2000) Seed germination in desert annuals: an empirical test of adaptive bet hedging. *American Naturalist* 155, 168–186.

Cohen, D. (1966) Optimizing reproduction in a randomly varying environment. *Journal of Theoretical Biology* 12, 119–129.

Cohen, D. (1967) Optimizing reproduction in a randomly varying environment, when a correlation may exist between the conditions at the time a choice has to be made and the subsequent outcome. *Journal of Theoretical Biology* 16, 1–14.

Cowling, R.M. and Lamont, B.B. (1985) Variation in serotiny of three *Banksia* species along a climatic gradient. *Australian Journal of Ecology* 10, 345–350.

Crist, T.O. and Friese, C.F. (1993) The impact of fungi on soil seeds: implications for plants and granivores in a semiarid shrub-steppe. *Ecology* 74, 2231–2239.

Csontos, P. and Tamás, J. (2003) Comparisons of soil seed bank classification systems. *Seed Science Research* 13, 101–111.

Cubiña, A. and Aide, T.M. (2001) The effect of distance from forest edge on seed rain and soil seed bank in a tropical pasture. *Biotropica* 33, 260–267.

Curry, J.P. (2004) Factors affecting the abundance of earthworms in soils. In: Edwards, A.C. (ed.) *Earthworm Ecology*. CRC Press, Boca Raton, pp. 91–113.

Dalling, J.W. and Hubbell, S.P. (2002) Seed size, growth rate and gap microsite conditions as determinants of recruitment success for pioneer species. *Journal of Ecology* 90, 557–568.

Dalling, J.W., Swaine, M.D. and Garwood, N.C. (1998) Dispersal patterns and seed bank dynamics of pioneer trees in moist tropical forest. *Ecology* 79, 564–578.

D'Antonio, C. and Meyerson, L.A. (2002) Exotic plant species as problems and solutions in ecological restoration: a synthesis. *Restoration Ecology* 10, 703–713.

Darwin, C. (1859) *On the Origin of Species by Means of Natural Selection, or Preservation of Favoured Races in the Struggle for Life*, 1st edn. Murray.

Davis, A.S., Schutte, B.J., Iannuzzi, J. and Renner, K.A. (2008) Chemical and physical defense of weed seeds in relation to soil seedbank persistence. *Weed Science* 56, 676–684.

Daws, M.I., Crabtree, L.M., Dalling, J.W., Mullins, C.E. and Burslem, D.F.R.P. (2008) Germination responses to water potential in neotropical pioneers suggest large-seeded species take more risks. *Annals of Botany* 102, 945–951.

De Luis, M., Verdú, M. and Raventós, J. (2008) Early to rise makes a plant healthy, wealthy, and wise. *Ecology* 89, 3061–3071.

Dessaint, F., Chadoeuf, R. and Barralis, G. (1991) Spatial pattern analysis of weed seeds in the cultivated soil seed bank. *Journal of Applied Ecology* 28, 721.

Donohue, K. (2005) Seeds and seasons: interpreting germination timing in the field. *Seed Science Research* 15, 175–187.

Donohue, K. (2009) Completing the cycle: maternal effects as the missing link in plant life histories. *Philosophical Transactions of the Royal Society B: Biological Sciences* 364, 1059–1074.

Donohue, K., Dorn, L., Griffith, C., Kim, E., Aguilera, A., Polisetty, C.R. and Schmitt, J. (2005) Environmental and genetic influences on the germination of *Arabidopsis thaliana* in the field. *Evolution* 59, 740–757.

Drake, D. (1998) Relationships among the seed rain, seed bank and vegetation of a Hawaiian forest. *Journal of Vegetation Science* 9, 103–112.

Dupouey, J.L., Dambrine, E., Laffite, J.D. and Moares, C. (2002) Irreversible impact of past land use on forest soils and biodiversity. *Ecology* 83, 2978–2984.

Duvel, J.W.T. (1902) Seeds buried in soil. *Science* 17, 872–873.

Dyer, A.R., Fenech, A. and Rice, K.J. (2000) Accelerated seedling emergence in interspecific competitive neighbourhoods. *Ecology Letters* 3, 523–529.

Eberhart, A. and Tielbörger, K. (2012) Maternal fecundity does not affect offspring germination – an empirical test of the sibling competition hypothesis. *Journal of Arid Environments* 76, 23–29.

Ehrlén, J. and van Groenendal, J.M. (1998) The trade-off between dispersability and longevity – an important aspect of plant species diversity. *Applied Vegetation Science* 1, 29–36.

Ellner, S. (1985a) ESS germination strategies in randomly varying environments. I. Logistic-type models. *Theoretical Population Biology* 28, 50–79.

Ellner, S. (1985b) ESS germination strategies in randomly varying environments. II. Reciprocal yield-law models. *Theoretical Population Biology* 28, 80–116.

Ellner, S. (1986) Germination dimorphisms and parent offspring conflict in seed-germination. *Journal of Theoretical Biology* 123, 173–185.

Ellner, S. (1987) Competition and dormancy: a reanalysis and review. *American Naturalist* 130, 798–803.

Evans, M.E.K. and Dennehy, J.J. (2005) Germ banking: bet-hedging and variable release from egg and seed dormancy. *The Quarterly Review of Biology* 80, 431–451.

Evans, M.E.K., Hearn, D.J., Hahn, W.J., Spangle, J.M. and Venable, D.L. (2005) Climate and life-history evolution in evening primroses (*Oenothera*, Onagraceae): a phylogenetic comparative analysis. *Evolution* 59, 1914–1927.

Facelli, J.M., Chesson, P.L. and Barnes, N. (2005) Differences in seed biology of annual plants in arid lands: a key ingredient of the storage effect. *Ecology* 86, 2998–3006.

Falinska, K. (1999) Seed bank dynamics in abandoned meadows during a 20-year period in the Bialowieza National Park. *Journal of Ecology* 87, 461–475.

Fenner, M. (1985) *Seed Ecology.* Chapman and Hall, London and New York.

Finch-Savage, W.E. and Leubner-Metzger, G. (2006) Seed dormancy and the control of germination. *New Phytologist* 171, 501–523.

Flematti, G.R., Ghisalberti, E.L., Dixon, K.W. and Trengove, R.D. (2004) A compound from smoke that promotes seed germination. *Science* 305, 977.

Flynn, S., Turner, R.M. and Dickie, J.B. (2004) Seed Information Database (release 6.0, October 2004) http://www.rbgkew.org.uk/data/sid. Kew Botanical Gardens.

Forcella, F. (1993) Seedling emergence model for Velvetleaf (*Abutilon theophrasti*). *Agronomy Journal* 85, 929–933.

Forcella, F. (1998) Real-time assessment of seed dormancy and seedling growth for weed management. *Seed Science Research* 8, 201–210.

Forey, E., Barot, S., Decaëns, T., Langlois, E., Laossi, K.R., Margerie, P., Scheu, S. and Eisenhauer, N. (2011) Importance of earthworm-seed interactions for the composition and structure of plant communities: A review. *Acta Oecologica* 37, 594–603.

Forget, P.M., Lambert, J.E., Hulme, P.E. and Vander Wall, S.B. (2005) *Seed Fate: Predation, Dispersal and Seedling Establishment.* CAB International, Wallingford, UK.

Fowler, N.L. (1984) The role of germination date, spatial arrangement, and neighbourhood effects in competitive interactions in *Linum. The Journal of Ecology* 307–318.

Funes, G., Basconcelo, S., Díaz, S. and Cabido, M. (2007) Seed size and shape are good predictors of seed persistence in soil in temperate mountain grasslands of Argentina. *Seed Science Research* 9, 341–345.

Gardarin, A., Dürr, C., Mannino, M.R., Busset, H. and Colbach, N. (2010) Seed mortality in the soil is related to seed coat thickness. *Seed Science Research* 20, 243–256.

Gardarin, A., Dürr, C. and Colbach, N. (2012) Modeling the dynamics and emergence of a multispecies weed seed bank with species traits. *Ecological Modelling* 240, 123–138.

Garwood, N.C. (1989) Tropical soil seed bank: a review. In: Leck, M.A., Parker, V.T. and Simpson, R.L. (eds) *Ecology of Soil Seed Banks.* London Academic Press, London, pp. 149–209.

Ghersa, C.M., Arnold, R.L.B. and Martinez-Ghersa, M.A. (1992) The role of fluctuating temperatures in germination and establishment of *Sorghum halepense* – regulation of germination at increasing depths. *Functional Ecology* 6, 460–468.

Grandin, U. (2001) Short-term and long-term variation in seed bank/vegetation relations along an environmental and successional gradient. *Ecography* 24, 731–741.

Grandin, U. and Rydin, H. (1998) Attributes of the seed bank after a century of primary succession on islands in Lake Hjälmaren, Sweden. *Journal of Ecology* 86, 293–303.

Grime, J.P. (1977) Evidence for the existence of three primary strategies in plants and its relevance to ecological and evolutionary theory. *American Naturalist* 111, 1169–1194.

Grime, J.P. (1989) Seed banks in ecological perspective. In: Leck, M.A., Parker, V.T. and Simpson, R.L. (eds) *Ecology of Soil Seed Banks.* London Academic Press, London, pp. xv-xxii.

Grundy, A.C., Mead, A. and Burston, S. (2003) Modelling the emergence response of weed seeds to burial depth: interactions with seed density, weight and shape. *Journal of Applied Ecology* 40, 757–770.

Grünzweig, J.M. and Dumbur, R. (2012) Seed traits, seed-reserve utilization and offspring performance across pre-industrial to future CO_2 concentrations in a Mediterranean community. *Oikos* 121, 579–588.

Guttermann, Y. (2000) Maternal effects on seeds during development. In: Fenner, M. (ed.) *Seeds: The Ecology of Regeneration in Plant Communities.* CAB International, Wallingford, UK, pp. 59–84.

Gutterman, Y. (2002) *Survival Strategies of Annual Desert Plants.* Springer, New York.

Harrison, S. and Ray, C. (2002) Plant population viability and metapopulation-level processes. In: Beissinger, S. and McCullough, D.R. (eds) *Population Viability Analysis.* University of Chicago Press, Chicago, pp. 109–122.

Heelemann, S., Krug, C.B., Esler, K.J., Reisch, C. and Poschlod, P. (2012) Pioneers and perches - promising restoration methods for degraded Renosterveld habitats? *Restoration Ecology* 20, 18–23.

Hendricks, S.B. and Taylorson, R.B. (1974) Promotion of seed germination by nitrate, nitrite, hydroxylamine, and ammonium salts. *Plant Physiology* 54, 304–309.

Henery, M.L. and Westoby, M. (2001) Seed mass and seed nutrient content as predictors of seed output variation between species. *Oikos* 92, 479–490.

Holmes, P.M. and Cowling, R.M. (1997a) Diversity, composition and guild structure relationships between soil-stored seed banks and mature vegetation in alien plant-invaded South African fynbos shrubland. *Plant Ecology* 133, 107–122.

Holmes, P.M. and Cowling, R.M. (1997b) The effects of invasion by *Acacia saligna* on the guild structure and regeneration capabilities of South African fynbos shrublands. *Journal of Applied Ecology* 4, 317–332.

Holmes, P.M. and Newton, R.J. (2004) Patterns of seed persistence in South African fynbos. *Plant Ecology* 172, 143–158.

Hopfensberger, K. (2007) A review of similarity between seed bank and standing vegetation across ecosystems. *Oikos* 116, 1438–1448.

Horton, T.R. and Van Der Heijden, M.G.A. (2008) The role of symbioses in seedling establishment and survival. *Seedling Ecology & Evolution* 189–213.

Huang, X., Schmitt, J., Dorn, L., Griffith, C., Effgen, S., Takao, S., Koorneef, M. and Donohue, K. (2010) The earliest stages of adaptation in an experimental plant population: strong selection on QTLS for seed dormancy. *Molecular Ecology* 19, 1335–1351.

Hubbell, S.P. (2001) *The Unified Neutral Theory of Biodiversity and Biogeography*. Princeton University Press, Princeton.

Hulme, P.E. (1994) Post-dispersal seed predation in grassland: its magnitude and sources of variation. *Journal of Ecology* 82, 645–652.

Hulme, P.E. (1998a) Post-dispersal seed predation and seed bank persistence. *Seed Science Research* 8, 513–519.

Hulme, P.E. (1998b) Post-dispersal seed predation: consequences for plant demography and evolution. *Perspectives in Plant Ecology, Evolution and Systematics* 1, 32–46.

Hyatt, L.A. and Evans, A.S. (1998) Is decreased germination fraction associated with risk of sibling competition? *Oikos* 29–35.

Jakobsson, A. and Eriksson, O. (2000) A comparative study of seed number, seed size, seedling size and recruitment in grassland plants. *Oikos* 88, 494–502.

Jakobsson, A., Eriksson, O. and Bruun, H.H. (2006) Local seed rain and seed bank in a species-rich grassland: effects of plant abundance and seed size. *Canadian Journal of Botany* 84, 1870–1881.

Jankowska-Blaszczuk, M. and Daws, M.I. (2007) Impact of red: far red ratios on germination of temperate forest herbs in relation to shade tolerance, seed mass and persistence in the soil. *Functional Ecology* 21, 1055–1062.

Jayasuriya, K.M.G.G., Baskin, J.M. and Baskin, C.C. (2008) Cycling of sensitivity to physical dormancy-break in seeds of *Ipomoea lacunosa* (Convolvulaceae) and ecological significance. *Annals of Botany* 101, 341–352.

Jiménez, H.E. and Armesto, J.J. (1992) Importance of the soil seed bank of disturbed sites in Chilean matorral in early secondary succession. *Journal of Vegetation Science* 3, 579–586.

Jones, R.H. and Sharitz, R.R. (1989) Potential advantages and disadvantages of germinating early for trees in floodplain forests. *Oecologia* 81, 443–449.

Jumpponen, A., Võre, H., Mattson, K.G., Ohtonen, R. and Trappe, J.M. (1999) Characterization of 'safe sites' for pioneers in primary succession on recently deglaciated terrain. *Journal of Ecology* 87, 98–105.

Jurado, E. and Flores, J. (2005) Is seed dormancy under environmental control or bound to plant traits? *Journal of Vegetation Science* 16, 559–564.

Kalamees, R. and Zobel, M. (2002) The role of the seed bank in gap regeneration in a calcareous grassland community. *Ecology* 83, 1017–1025.

Keeley, J.E. (1987) Role of fire in seed-germination of woody taxa in California chaparral. *Ecology* 68, 434–443.

Kelly, M.G. and Levin, D.A. (1997) Fitness consequences and heritability aspects of emergence date in *Phlox drummondii*. *Journal of Ecology* 755–766.

Kettenring, K.M., Gardner, G.M. and Galatowitsch, S.M. (2006) Effect of light on seed germination of eight wetland *Carex* species. *Annals of Botany* 98, 869–874.

Kiefer, S. and Poschlod, P. (1996) Restoration of fallow or afforested calcareous grasslands by clear-cutting. In: *Species Survival in Fragmented Landscapes*. Kluwer, Dordrecht, pp. 209–218.

Kiewnick, L. (1964) Untersuchungen über den Einfluss der Samen-und Bodenmikroflora auf die Lebensdauer der Spelzfrüchte des Flughafers (*Avena fatua* L.). *Weed Research* 4, 31–43.

Kluth, C. and Bruelheide, H. (2005) Effects of range position, inter-annual variation and density on demographic transition rates of *Hornungia petraea* populations. *Oecologia* 145, 382–393.

Koprdová, S., Saska, P., Honek, A. and Martinková, Z. (2010) Seed consumption by millipedes. *Pedobiologia* 54, 31–36.

Kruk, B., Insausti, P., Razul, A. and Benech-Arnold, R.L. (2006) Light and thermal environments as modified by a wheat crop: effects on weed seed germination. *Journal of Applied Ecology* 43, 227–236.

Kubitzki, K. and Ziburski, A. (1994) Seed dispersal in flood plain forests of Amazonia. *Biotropica* 30–43.

Lalonde, R.G. and Roitberg, B.D. (2006) Chaotic dynamics can select for long-term dormancy. *American Naturalist* 168, 127–131.

Lavorel, S. and Lebreton, J.D. (1992) Evidence for lottery recruitment in Mediterranean old fields. *Journal of Vegetation Science* 3, 91–100.

Lavorel, S., Lepart, J., Debussche, M., Lebreton, J.D. and Beffy, J.L. (1994) Small scale disturbances and the maintenance of species diversity in Mediterranean old fields. *Oikos* 70, 455–473.

Leck, M.A., Parker, T.V. and Simpson, R.L. (1989) *Ecology of Soil Seed Banks*. London Academic Press, London.

Leishman, M.R., Masters, G.J., Clarke, I.P. and Brown, V.K. (2000a) Seed bank dynamics: the role of fungal pathogens and climate change. *Functional Ecology* 14, 293–299.

Leishman, M.R., Wright, I.J., Moles, A.T. and Westoby, M. (2000b) The evolutionary ecology of seed size. In: Fenner, M. (ed.) *Seeds: The Ecology of Regeneration in Plant Communities*. CABI, Wallingford.

Lonsdale, W.M., Harley, K.L.S. and Gillett, J.D. (1988) Seed bank dynamics in *Mimosa pigra*, an invasive tropical shrub. *Journal of Applied Ecology* 25, 963–976.

Lundberg, S., Nilsson, P. and Fagerström, T. (1996) Seed dormancy and frequency dependent selection due to sib competition: the effect of age specific gene expression. *Journal of Theoretical Biology* 183, 9–17.

Lunt, I.D. (1990) The soil seed bank of a long-grazed *Themeda triandra* grassland in Victoria (Australia). *Proceedings of the Royal Society of Victoria* 102, 53–58.

Luzuriaga, A.L., Escudero, A., Olano, J.M. and Loidi, J. (2005) Regenerative role of seed banks following an intense soil disturbance. *Acta Oecologica* 27, 57–66.

Maas, D. (1987) Keimungsansprüche von Streuwiesenpflanzen und deren Auswirkung auf das Samenpotential. PhD thesis, Technical University of Munich.

Marcante, S., Schwienbacher, E. and Erschbamer, B. (2009) Genesis of a soil seed bank on a primary succession in the Central Alps (Ötztal, Austria). *Flora* 204, 434–444.

Marks, M. and Prince, S. (1981) Influence of germination date on survival and fecundity in wild lettuce *Lactuca serriola*. *Oikos* 326–330.

Matus, G., Verhagen, R., Bekker, R.M. and Grootjans, A.P. (2003) Restoration of the Cirsio dissecti-Molinietum in The Netherlands: Can we rely on soil seed banks? *Applied Vegetation Science* 6, 73–84.

Menalled, F.D., Marino, P.C., Renner, K.A. and Landis, D.A. (2000) Post-dispersal weed seed predation in Michigan crop fields as a function of agricultural landscape structure. *Agriculture, Ecosystems & Environment* 77, 193–202.

Merritt, D.J., Turner, S.R., Clarke, S. and Dixon, K.W. (2007) Seed dormancy and germination stimulation syndromes for Australian temperate species. *Australian Journal of Botany* 55, 336–344.

Meyer, S.E. and Allen, P.S. (2009) Predicting seed dormancy loss and germination timing for *Bromus tectorum* in a semi-arid environment using hydrothermal time models. *Seed Science Research* 19, 225–239.

Milberg, P. (1993) Seed bank and seedlings emerging after soil disturbance in a wet semi-natural grassland in Sweden. *Annales Botanici Fennici* 30, 9–03.

Milberg, P. and Andersson, L. (1997) Seasonal variation in dormancy and light sensitivity in buried seeds of eight annual weed species. *Canadian Journal of Botany* 75, 1998–2004.

Milberg, P., Andersson, L. and Thompson, K. (2000) Large-seeded species are less dependent on light for germination than small-seeded ones. *Seed Science Research* 10, 99–104.

Mitlacher, K., Poschlod, P., Rosén, E. and Bakker, J.P. (2002) Restoration of wooded meadows-a comparative analysis along a chronosequence on Öland (Sweden). *Applied Vegetation Science* 5, 63–73.

Mohr, H., Schopfer, P., Lawlor, G. and Lawlor, D.W. (2010) *Plant Physiology*. Springer.

Molau, U. and Shaver, G.R. (1997) Controls on seed production and seed germinability in *Eriophorum vaginatum*. *Global Change Biology* 3, 80–88.

Moles, A.T. and Westoby, M. (2002) Seed addition experiments are more likely to increase recruitment in larger-seeded species. *Oikos* 99, 241–248.

Moles, A.T. and Westoby, M. (2004) Seedling survival and seed size: a synthesis of the literature. *Journal of Ecology* 92, 372–383.

Moles, A.T. and Westoby, M. (2006) Seed size and plant strategy across the whole life cycle. *Oikos* 113, 91–105.

Moles, A.T., Hodson, D.W. and Webb, C.J. (2000) Seed size and persistence in the soil in the New Zealand flora. *Oikos* 89, 679–685.

Moles, A.T., Warton, D.I. and Westoby, M. (2003) Seed size and survival in the soil in arid Australia. *Austral Ecology* 28, 575–585.

Morinaga, T. (1926) The favorable effect of reduced oxygen supply upon the germination of certain seeds. *American Journal of Botany* 159–166.

Morozowska, M., Czarna, A., Kujawa, M. and Jagodzinski, A.M. (2011) Seed morphology and endosperm structure of selected species of Primulaceae, Myrsinaceae, and Theophrastaceae and their systematic importance. *Plant Systematics and Evolution* 291, 159–172.

Newsome, A.E. and Noble, I.R. (1986) Ecological and physiological characters of invading species. In: Groves, R.H. and Burdon, J.J. (eds) *Ecology of Biological Invasions: an Australian Perspective*. Cambridge University Press, New York, pp. 1–20.

Nilsson, P., Fagerström, T., Tuomi, J. and Åström, M. (1994) Does seed dormancy benefit the mother plant by reducing sib competition? *Evolutionary Ecology* 8, 422–430.

Noble, I.R. (1989) Attributes of invaders and the invading process: terrestrial and vascular plants. In: Drake, J.A., Mooney, H.A., di Castri, F., Groves, R.H., Kruger, B., Rejmanek, M. and Williamson, M. (eds) *Biological Invasions: a Global Perspective*. Wiley & Sons, New York, pp. 301.

Olano, J.M., Caballero, I. and Escudero, A. (2012) Soil seed bank recovery occurs more rapidly than expected in semi-arid Mediterranean gypsum vegetation. *Annals of Botany* 109, 299–307.

Ortega, M., Levassor, C. and Peco, B. (1997) Seasonal dynamics of Mediterranean pasture seed banks along environmental gradients. *Journal of Biogeography* 24, 177–195.

Ozinga, W.A., Hennekens, S.M., Schaminée, J.H.J., Smits, N.A.C., Bekker, R.M., Romermann, C., Klimes, L., Bakker, J.P. and van Groenendael, J.M. (2007) Local above-ground persistence of vascular plants: Life-history trade-offs and environmental constraints. *Journal of Vegetation Science* 18, 489–497.

Pake, C.E. and Venable, D.L. (1995) Is coexistence of Sonoran desert annuals mediated by temporal variability in reproductive success? *Ecology* 76, 246–261.

Pakeman, R.J., Small, J.L. and Wilson, J.B. (2005) The role of the seed bank, seed rain and the timing of disturbance in gap regeneration. *Journal of Vegetation Science* 16, 121–130.

Pakeman, R.J., Small, J.L. and Torvell, L. (2012) Edaphic factors influence the longevity of seeds in the soil. *Plant Ecology* 213, 57–65.

Pearson, T.R.H., Burslem, D.F.R.P., Mullins, C.E. and Dalling, J.W. (2003) Functional significance of photoblastic germination in neotropical pioneer trees: a seed's eye view. *Functional Ecology* 17, 394–402.

Peñuelas, J., Filella, I., Zhang, X., Llorens, L., Ogaya, R., Lloret, F., Comas, P., Estiarte, M. and Terradas, J. (2004) Complex spatiotemporal phenological shifts as a response to rainfall changes. *New Phytologist* 161, 837–846.

Philippi, T. (1993) Bet-hedging germination of desert annuals: beyond the first year. *American Naturalist* 474–487.

Pons, T.L. (1991) Induction of dark dormancy in seeds – its importance for the seed bank in the soil. *Functional Ecology* 5, 669–675.

Popay, A.I. and Roberts, E.H. (1970) Factors involved in the dormancy and germination of *Capsella bursa-pastoris* (L.) Medik. and *Senecio vulgaris* L. *Journal of Ecology* 58, 103–122.

Poschlod, P. (1995) Diaspore rain and diaspore bank in raised bogs and its implication for the restoration of peat mined sites. In: Wheeler, B.D., Shaw, S.C., Fojt, W.J. and Robertson, R.A. (eds) *Restoration of Temperate Wetlands*. Wiley, Chichester, pp. 471.

Poschlod, P. (1996) Population biology and dynamics of a rare short-lived pond mud plant, *Carex bohemica* Schreber. *Verhandlungen der Gesellschaft für Ökologie* 25, 321–337.

Poschlod, P. and Jackel, A.K. (1993) Untersuchungen zur Dynamik von generativen Diasporenbanken von Samenpflanzen in Kalkmagerrasen. I. Jahreszeitliche Dynamik des Diasporenregens und der Diasporenbank auf zwei Kalkmagerrasenstandorten der Schwäbischen Alb. *Flora* 188, 49–71.

Poschlod, P., Fischer, S. and Kiefer, S. (1996) A coenotical approach of plant population viability analysis on successional and afforested calcareous grassland sites. In: Settele, J., Margules, C., Poschlod, P. and Henle, K. (eds) *Species Survival in Fragmented Landscapes*. Kluwer, Dordrecht, pp. 219–228.

Poschlod, P., Kiefer, S., Tränkle, U., Fischer, S.F. and Bonn, S. (1998) Plant species richness in calcareous grasslands as affected by dispersability in space and time. *Applied Vegetation Science* 1, 75–90.

Priestley, D.A. (1986) *Seed aging – implications for seed storage and persistence in the soil*. Comstock Publishing, New York.

Rasmussen, I.A. and Holst, N. (2003) Computer model for simulating the long-term dynamics of annual weeds: from seedlings to seeds. *Aspects of Applied Biology* 277–284.

Rees, M. (1994) Delayed germination of seeds: a look at the effects of adult longevity, the timing of reproduction, and population age/stage structure. *American Naturalist* 43–64.

Rees, M. (1996) Evolutionary ecology of seed dormancy and seed size. *Philosophical Transactions of the Royal Society of London, Series B* 351, 1299–1308.

Richardson, D.M. and Kluge, R.L. (2008) Seedbanks of invasive Australian Acacia species in South Africa: Role in invasiveness and options for management. *Perspectives in Plant Ecology, Evolution and Systematics* 10, 161–177.

Roberts, H.A. (1986) Seed persistence in soil and seasonal emergence in plant species from different habitats. *Journal of Applied Ecology* 23, 639–656.

Ross, M.A. and Harper, J.L. (1972) Occupation of biological space during seedling establishment. *The Journal of Ecology* 77–88.

Saatkamp, A., Affre, L., Dutoit, T. and Poschlod, P. (2009) The seed bank longevity index revisited: limited reliability evident from a burial experiment and database analyses. *Annals of Botany* 104, 715–724.

Saatkamp, A., Römermann, C. and Dutoit, T. (2010) Plant functional traits show non-linear response to grazing. *Folia Geobotanica* 45, 239–252.

Saatkamp, A., Affre, L., Baumberger, T., Dumas, P.J., Gasmi, A., Gachet, S. and Arène, F. (2011a) Soil depth detection by seeds and diurnally fluctuating temperatures: different dynamics in 10 annual plants. *Plant and Soil* 349, 331–340.

Saatkamp, A., Affre, L., Dutoit, T. and Poschlod, P. (2011b) Germination traits explain soil seed persistence across species: the case of Mediterranean annual plants in cereal fields. *Annals of Botany* 107, 415.

Saska, P. (2008) Granivory in terrestrial isopods. *Ecological Entomology* 33, 742–747.

Satterthwaite, W.H. (2009) Competition for space can drive the evolution of dormancy in a temporally invariant environment. *Plant Ecology* 208, 167–185.

Schafer, M. and Kotanen, P.M. (2003) The influence of soil moisture on losses of buried seeds to fungi. *Acta Oecologica* 24, 255–263.

Schnürer, J., Clarholm, M. and Rosswall, T. (1985) Microbial biomass and activity in an agricultural soil with different organic matter contents. *Soil Biology and Biochemistry* 17, 611–618.

Schütz, W. (2000) Ecology of seed dormancy and germination in sedges (*Carex*). *Perspectives in Plant Ecology, Evolution and Systematics* 3, 67–89.

Schwilk, D.W. and Ackerly, D.D. (2001) Flammability and serotiny as strategies: correlated evolution in pines. *Oikos* 94, 326–336.

Shen-Miller, J. (2002) Sacred lotus, the long-living fruits of China Antique. *Seed Science Research* 12, 131–144.

Shumway, D.L. and Koide, R.T. (1994) Preferences of *Lumbricus terrestris* L. *Applied Soil Ecology* 1, 11–15.

Silvertown, J. (1988) The demographic and evolutionary consequences of seed dormancy. In: Davy, A.J., Hutchings, M.J. and Watkinson, A.R. (eds) *Plant population ecology*. Blackwell, London.

Silvertown, J. (1999) Seed ecology, dormancy and germination: a modern synthesis from Baskin and Baskin. *American Journal of Botany* 86, 903–905.

Smith, C.C. and Fretwell, S.D. (1974) The optimal balance between size and number of offspring. *American Naturalist* 499–506.

Snyder, R.E. (2006) Multiple risk reduction mechanisms: can dormancy substitute for dispersal? *Ecology Letters* 9, 1106–1114.

Soons, M.B., van der Vlugt, C., van Lith, B., Heil, G.W. and Klaassen, M. (2008) Small seed size
    increases the potential for dispersal of wetland plants by ducks. *Journal of Ecology* 96, 619–627.
Stöcklin, J. and Fischer, M. (1999) Plants with longer-lived seeds have lower local extinction rates in
    grassland remnants 1950-1985. *Oecologia* 120, 539–543.
Thanos, C.A. and Georghiou, K. (1988) On the mechanism of skotodormancy induction in grand rapids
    lettuce (*Lactuca sativa* L) Seeds. *Journal of Plant Physiology* 133, 580–584.
Thanos, C.A., Georghiou, K., Kadis, C. and Pantazi, C. (1992) Cistaceae – a plant family with hard
    seeds. *Israel Journal of Botany* 41, 251–263.
Thompson, K. (1986) Small-scale heterogeneity in the seed bank of an acidic grassland. *Journal of
    Ecology* 74, 733–738.
Thompson, K. (2000) The functional ecology of seed banks. In: Fenner, M. (ed.) *Seeds: The Ecology of
    Regeneration in Plant Communities.* CAB International, Wallingford, UK, pp. 215–235.
Thompson, K. and Fenner, M. (1992) The functional ecology of seed banks. In: Fenner, M. (ed.) *Seeds:
    The Ecology of Regeneration in Plant Communities.* CAB International, Wallingford, UK,
    pp. 231–258.
Thompson, K. and Grime, J.P. (1979) Seasonal variation in the seed banks of herbaceous species in ten
    contrasting habitats. *Journal of Ecology* 67, 893–921.
Thompson, K. and Grime, J.P. (1983) A comparative study of germination responses to diurnally fluc-
    tuating temperatures. *Journal of Applied Ecology* 20, 141–156.
Thompson, K., Grime, J.P. and Mason, G. (1977) Seed germination in response to diurnal fluctuations
    of temperature. *Nature* 267, 147–149.
Thompson, K., Band, S.R. and Hodgson, J.G. (1993) Seed size and shape predict persistence in soil.
    Functional Ecology 7, 236–241.
Thompson, K., Bakker, J.P. and Bekker, R.M. (1997) *The Soil Seed Banks of North West Europe:
    Methodology, Density and Longevity.* Cambridge University Press, Cambridge.
Thompson, K., Ceriani, R.M., Bakker, J.P. and Bekker, R.M. (2003) Are seed dormancy and persistence
    in soil related? *Seed Science Research* 13, 97–100.
Tielbörger, K. and Petru, M. (2008) Germination behaviour of annual plants under changing climatic
    conditions: separating local and regional environmental effects. *Oecologia* 155, 717–728.
Tielbörger, K. and Petru, M. (2010) An experimental test for effects of the maternal environment on
    delayed germination. *Journal of Ecology* 98, 1216–1223.
Tielbörger, K. and Prasse, R. (2009) Do seeds sense each other? Testing for density-dependent germina-
    tion in desert perennial plants. *Oikos* 118, 792–800.
Tielbörger, K. and Valleriani, A. (2005) Can seeds predict their future? Germination strategies of
    density-regulated desert annuals. *Oikos* 111, 235–244.
Tielbörger, K., Petru, M. and Lampei, C. (2011) Bet-hedging germination in annual plants: a sound
    empirical test of the theoretical foundations. *Oikos* 121, 1860–1868.
Tooley, J. and Brust, G.E. (2002) Weed seed predation by carabid beetles. In: *The Agroecology of
    Carabid Beetles.* pp. 215–229.
Totland, Ø. (1999) Effects of temperature on performance and phenotypic selection on plant traits in
    alpine *Ranunculus acris. Oecologia* 120, 242–251.
Trudgill, D.L., Honek, A., Li, D. and Straalen, N.M. (2005) Thermal time-concepts and utility. *Annals
    of Applied Biology* 146, 1–14.
Tuljapurkar, S. and Wiener, P. (2000) Escape in time: stay young or age gracefully? *Ecological Modelling*
    133, 143–159.
Turkington, R., Goldberg, D.E., Olsvig-Whittaker, L. and Dyer, A.R. (2005) Effects of density on timing
    of emergence and its consequences for survival and growth in two communities of annual plants.
    Journal of Arid Environments 61, 377–396.
Valkó, O., Török, P., Tóthmérész, B. and Matus, G. (2011) Restoration potential in seed banks of acidic
    fen and dry-mesophilous meadows: can restoration be based on local seed banks? *Restoration
    Ecology* 19, 9–15.
Valleriani, A. and Tielbörger, K. (2006) Effect of age on germination of dormant seeds. *Theoretical
    Population Biology* 70, 1–9.
van der Valk, A.G. and Pederson, R.L. (1989) Seed banks and the management and restoration of natu-
    ral vegetation. In: Leck, M.A., Parker, V.T. and Simpson, R.L. (eds) *Ecology of Soil Seed Banks.*
    London Academic Press, London, pp. 329–346.

Van Mourik, T.A., Stomph, T.J. and Murdoch, A.J. (2005) Why high seed densities within buried mesh bags may overestimate depletion rates of soil seed banks. *Journal of Applied Ecology* 42, 299–305.

Vandelook, F., Van de Moer, D. and Van Assche, J.A. (2008) Environmental signals for seed germination reflect habitat adaptations in four temperate Caryophyllaceae. *Functional Ecology* 22, 470–478.

Venable, D.L. (1989) Modelling the evolutionary ecology of soil seed banks. In: Leck, M.A., Parker, V.T. and Simpson, R.L. (eds) *Ecology of Soil Seed Banks*. London Academic Press, London, pp. 67–87.

Venable, D.L. (2007) Bet hedging in a guild of desert annuals. *Ecology* 88, 1086–1090.

Venable, D.L. and Brown, J.S. (1988) The selective interaction of dispersal, dormancy and seed size as adaptations for reducing risks in variable environments. *American Naturalist* 131, 360–384.

Venable, D.L. and Lawlor, L. (1980) Delayed germination and dispersal in desert annuals: escape in space and time. *Oecologia* 46, 272–282.

Verdù, M. and Traveset, A. (2005) Early emergence enhances plant fitness: a phylogenetically controlled meta-analysis. *Ecology* 86, 1385–1394.

Vitalis, R., Rousset, F., Kobayashi, Y., Olivieri, I. and Gandon, S. (2013) The joint evolution of dispersal and dormancy in a metapopulation with local extinctions and kin competition. *Evolution* 67, 1676–1691.

von Blanckenhagen, B. and Poschlod, P. (2005) Restoration of calcareous grasslands: the role of the soil seed bank and seed dispersal for recolonisation processes. *Biotechnology, Agronomy, Society and Environment* 9, 143–149.

Wagner, M. and Mitschunas, N. (2008) Fungal effects on seed bank persistence and potential applications in weed biocontrol a review. *Basic and Applied Ecology* 9, 191–203.

Walck, J.L., Baskin, J.M., Baskin, C.C. and Hidayati, S.N. (2005) Defining transient and persistent seed banks in species with pronounced seasonal dormancy and germination patterns. *Seed Science Research* 15, 189–196.

Walck, J.L., Hidayati, S.N., Dixon, K.W., Thompson, K. and Poschlod, P. (2011) Climate change and plant regeneration from seed. *Global Change Biology* 17, 2145–2161.

Waldhardt, R., Fuhr-Bossdorf, K. and Otte, A. (2001) The significance of the seed bank as a potential for the reestablishment of arable-land vegetation in a marginal cultivated landscape. *Web Ecology* 2, 83–87.

Walker, L.R., Zasada, J.C. and Chapin III, F.S. (1986) The role of life history processes in primary succession on an Alaskan floodplain. *Ecology* 67, 1243–1253.

Wellstein, C., Otte, A. and Waldhardt, R. (2007) Seed bank diversity in mesic grasslands in relation to vegetation type, management and site conditions. *Journal of Vegetation Science* 18, 153–162.

Westerman, P.R., Hofman, A., Vet, L. and van der Werf, W. (2003) Relative importance of vertebrates and invertebrates in epigeaic weed seed predation in organic cereal fields. *Agriculture, Ecosystems & Environment* 95, 417–425.

Willems, J.H. and Bik, L.P.M. (1998) Restoration of high species density in calcareous grassland: the role of seed rain and soil seed bank. *Applied Vegetation Science* 1, 91–100.

Wookey, P.A., Robinson, C.H., Parsons, A.N., Welker, J.M., Press, M.C., Callaghan, T.V. and Lee, J.A. (1995) Environmental constraints on the growth, photosynthesis and reproductive development of *Dryas octopetala* at a high Arctic polar semi-desert, Svalbard. *Oecologia* 102, 478–489.

Woolley, J.T. and Stoller, E.W. (1978) Light penetration and light-induced seed germination in soil. *Plant Physiology* 61, 597–600.

Zaller, J.G. and Saxler, N. (2007) Selective vertical seed transport by earthworms: Implications for the diversity of grassland ecosystems. *European Journal of Soil Biology* 43, Supplement 1, S86-S91.

Zammit, C.A. and Zedler, P.H. (1993) Size structure and seed production in even-aged populations of *Ceanothus greggii* in mixed chaparral. *Journal of Ecology* 81, 499–511.

Zehm, A., von Brackel, W. and Mitlacher, K. (2008) Hochgradig bedrohte Strandrasenarten - Artenhilfsprogramm am bayerischen Bodenseeufer unter besonderer Berücksichtung der Diasporenbank. *Naturschutz und Landschaftspflege* 40, 73–80.

Zobel, M. (1997) The relative role of species pools in determining plant species richness. An alternative explanation of species coexistence? *Trends in Ecology & Evolution* 12, 266–269.

Index

Page numbers in **bold** refer to illustrations and tables.